T0298552

Differential Geometry and Its Visualization

Differential Geometry and Its Visualization is suitable for graduate level courses in differential geometry, serving both students and teachers. It can also be used as a supplementary reference for research in mathematics and the natural and engineering sciences.

Differential geometry is the study of geometric objects and their properties using the methods of mathematical analysis. The classical theory of curves and surfaces in three-dimensional Euclidean space is presented in the first three chapters. The abstract and modern topics of tensor algebra, Riemannian spaces and tensor analysis are studied in the last two chapters. A great number of illustrating examples, visualizations and genuine figures created by the authors' own software are included to support the understanding of the presented concepts and results, and to develop an adequate perception of the shapes of geometric objects, their properties and the relations between them.

Features
- Extensive, full colour visualisations
- Numerous exercises
- Self-contained and comprehensive treatment of the topic

Differential Geometry and Its Visualization

Eberhard Malkowsky
State University of Novi Pazar, Serbia
Ćemal Doličanin
State University of Novi Pazar, Serbia
Vesna Veličković
University of Niš, Serbia

CRC Press
Taylor & Francis Group
Boca Raton London New York

CRC Press is an imprint of the
Taylor & Francis Group, an **informa** business

A CHAPMAN & HALL BOOK

First edition published 2024
by CRC Press
6000 Broken Sound Parkway NW, Suite 300, Boca Raton, FL 33487-2742

and by CRC Press
4 Park Square, Milton Park, Abingdon, Oxon, OX14 4RN

CRC Press is an imprint of Taylor & Francis Group, LLC

© 2024 Eberhard Malkowsky, Ćemal Dolićanin, and Vesna Veličković

ISBN: 978-1-032-43666-1 (hbk)
ISBN: 978-1-032-44119-1 (pbk)
ISBN: 978-1-00337056-7 (ebk)

DOI: 10.1201/9781003370567

Typeset in Latin Modern font
by KnowledgeWorks Global Ltd.

Publisher's note: This book has been prepared from camera-ready copy provided by the authors

Contents

List of Figures

Preface

This book is based on the authors' work, courses and workshops on differential geometry, its visualizations and software development at various universities in Germany, Serbia and South Africa over the period of more than thirty years. It contains the fundamental topics from the local theory of curves and surfaces in three-dimensional Euclidean space, tensor algebra and Riemannian geometry, and tensor analysis. Since visualization strongly supports the understanding of concepts and results in differential geometry, a great number of graphical representations and enlightening examples were included in the book. All the graphics in the book were generated by the authors' own software package.

This book is intended for students and teachers of courses in differential geometry at graduate level as well as in doctoral studies. It is also useful for mathematicians, physicists and engineers who need or are interested in its topics. The material of the book is presented in a comprehensive and self-contained way and strongly supported by a great number of examples and visualizations.

The prerequisites are a solid background from the undergraduate courses in analytical geometry and analysis. As a reminder, the most important results from these areas are listed in Section 1.2.

The book contains five chapters with a total of 54 sections including 215 figures. Major parts of the first three chapters are translations from Serbian of the authors' book *Diferencijalna geometrija i njena vizuelizacija* [17].

Chapter 1 contains the study of the basic local geometric properties of curves in three-dimensional Euclidean $\mathbb{E}^3$. The topics of special interest include the parametric representation of curves, the arc length of a curve as its natural parameter, many examples of planar curves and the geometrical principles of their constructions, the frame or the vectors of the trihedra of curves, Frenet's formulae, the curvature and torsion of curves and their geometric significance, characteristic shape of a curve in a neighbourhood of any of its points, osculating circles and spheres, involutes and evolutes of curves, the fundamental theorem of curves and the natural or intrinsic equations of curves, lines of constant slope and finally spherical images of curves.

Chapter 2 deals with the local differential geometry of surfaces in three-dimensional Euclidean space $\mathbb{E}^3$. This means the study of the geometric shape of surfaces in the neighbourhood of an arbitrary one of their points. The most important concepts arising in this task are those of the normal, principal, Gaussian and mean curvature. Fundamental important topics are curves on surfaces, tangent planes and normal vectors of surfaces, first and second fundamental coefficients, Meusnier's theorem, principal directions, Euler's formula, the local shape of surfaces and Dupin's indicatrix, lines of curvature and asymptotic lines, triple orthogonal systems and finally the Gauss and Weingarten equations.

Chapter 3 deals with the intrinsic geometry of surfaces, that is, with the geometric properties of surfaces that only depend on measurements on the surface itself, namely on its first fundamental coefficients and their derivatives. The most important topics and subjects are the Christoffel symbols of first and second kind and the geodesic curvature, Liouville's theorem, geodesic lines on surfaces with orthogonal parameters and on surfaces of revolution, the minimum property of geodesic lines, orthogonal and geodesic parameters, geodesic parallel and geodesic polar coordinates, the Levi-Cività parallelism, the derivation formulae by Gauss, and Mainardi and Codazzi, the Theorema egregium, conformal, isometric and area preserving maps of surfaces, the Gauss-Bonnet theorem and finally minimal surfaces and the Weierstrass formulae.

Chapter 4 deals with the redevelopment of the intrinsic geometry of surfaces independently of their embedding in three-dimensional space, based only on the definition of measurements of lengths in a point set. This concept was originally introduced by Riemann in 1854. For this purpose, the concept of a manifold in n–dimensional space is needed. Surfaces and n–dimensional Euclidean spaces are obtained as special cases of the manifolds. The main topics and subjects of this chapter are the definition of n–dimensional manifolds of class C^k, the translation formulae between the bases of a finite dimensional vector space, and between the components of vectors with respect to bases, spaces V^* of linear functionals on finite dimensional vector spaces V, the isometry of V and V^*, the transformation formulae of the elements of V^* and their components, the concepts of contravariant and covariant vectors. Additional topics are the introduction of tensors of the second order and the transformation formulae for their components with respect to different bases, symmetric bilinear forms and inner products, the introduction of tensors of arbitrary order, the transformation formulae for their components, the characterization of tensors by the transformation properties of their components, and the identification of tensors with their components, sums, outer products, contractions and inner products of tensors, symmetric and anti-symmetric tensors of arbitrary order and the independence of their properties from the choice of the coordinate system, Riemann spaces and the metric tensor, the Christoffel symbols in a Riemann space and their transformation formulae.

Chapter 5 deals with the study of the fundamentals of tensor analysis. Tensors first appeared in the theory of surfaces. Typical examples were the first and second fundamental coefficients. A tensor of a given order was assigned to every point of a manifold. It also turned out that certain derivatives of tensors were needed in various problems of differential geometry. It seems natural to try and define derivatives of a tensor in a way such that the result again is a tensor. This guarantees the independence of a derivative of the choice of a coordinate system. The most important topics and subjects are the study of covariant derivatives of contravariant, covariant vectors and $(1,1)$–tensors, covariant derivatives of (r,s)–tensors, their basic properties and the Ricci identity for the covariant derivatives of the metric tensor, the mixed Riemann tensor of curvature and the interchange of the order of its covariant differentiation in Ricci's identity, the Bianchi identities for the derivatives of the mixed Riemann tensor of curvature and the covariant Riemann tensor of curvature, the

Beltrami differentiator of first order in a Riemann space, the divergence of contravariant and covariant vectors, a geometric meaning of the covariant differentiation, the Levi-Cività parallelism, the fundamental theorem of the theory of surfaces, a geometric interpretation of the Riemann tensor of curvature, spaces with vanishing tensor of curvature, the existence and uniqueness of autoparallel curves in Riemann spaces, an extension of Frenet's formulae for curves in a Riemann spaces, Riemann normal coordinates and the curvature of spaces and the Bertrand-Puiseux theorem.

Eberhard Malkowsky   Novi Pazar, Serbia
Ćemal Dolićanin   Novi Pazar, Serbia
Vesna Veličković   Niš, Serbia

January, 2023

Symbol Description

$\mathbb{N}$ $=\{1,2,\dots\}$, the set of natural numbers

$\mathbb{Z}$ $=\{-\mathbb{N} \cup \{0\} \cup \mathbb{N}$, the set of integers

$\mathbb{Q}$ $=\{p/q : p \in \mathbb{Z}, q \in \mathbb{N}\}$, the set of rational numbers

$\mathbb{R}$ the set of real numbers

$\mathbb{C}$ the set of complex numbers

$\mathbb{E}^3$ three-dimensional (real) Euclidean space

$\mathbb{R}^n$ n–dimensional (real) point space

x^k k^{th} coordinate of a point in $\mathbb{R}^n$

d Euclidean metric in $\mathbb{R}^n$

$d(X,Y)$ Euclidean distance of the points X and Y in $\mathbb{R}^n$

$\mathbb{V}^n$ n–dimensional (real) vector space

x^k k^{th} component of a vector in $\mathbb{V}^n$

$\vec{x} + \vec{y}$ sum of the vectors $\vec{x}, \vec{y} \in \mathbb{V}^n$

$\lambda\vec{y}$ scalar multiple of $\lambda \in \mathbb{R}$ and $\vec{x} \in \mathbb{V}^n$

$\|\cdot\|$ Euclidean norm in $\mathbb{V}^n$

$\|\vec{x}\|$ Euclidean norm (length) of a vector $\vec{x}$ in $\mathbb{V}^n$

$\mathbb{E}^n$ n–dimensional (real) Euclidean space

$\overrightarrow{PQ}$ vector that maps the point P to the point Q

$\overrightarrow{OX}$ position vector of the point X

$\vec{x} \bullet \vec{y}$ inner product of the vectors $\vec{x}, \vec{y} \in \mathbb{V}^n$

$\vec{x} \perp \vec{y}$ the vectors $\vec{x}$ and $\vec{y}$ are orthogonal

$\vec{y}_{\vec{x}}$ projection of the vector $\vec{x}$ to $\vec{y}$

$\vec{x} \times \vec{y}$ vector product (or outer product) of the vectors $\vec{x}, \vec{y} \in \mathbb{V}^3$

$\vec{N} \perp \gamma$ $\vec{N}$ is orthogonal to the straight line γ in $\mathbb{E}^2$

$\vec{N} \perp Pl$ $\vec{N}$ is orthogonal to the plane Pl in $\mathbb{E}^3$

$\vec{f}$ $=\{f^1, \dots, f^n\}$, vector-valued function

f^k k^{th} component function of the vector-valued function $\vec{f}$

$D_{\vec{f}}$ domain of the function $\vec{f}$

$\vec{f}(D_{\vec{f}})$ $=\{f(t) : t \in D_{\vec{f}}\}$, range of the function $\vec{f}$

$N_\varepsilon(X_0)$ $=\{X \in \mathbb{R}^n : d(X,X_0) < \varepsilon\}$, the ε–open ball of the point X_0

$N_\varepsilon(\vec{x}_0)$ $=\{\vec{x} \in \mathbb{E}^n : \|\vec{x} - \vec{x}_0\| < \varepsilon\}$, the ε–open ball of the vector $\vec{x}_0$

$\dot{N}_\varepsilon(X_0)$ $=\{X \in \mathbb{R}^n : 0 < d(X,X_0) < \varepsilon\}$, the ε–deleted open ball of the point X_0

$\dot{N}_\varepsilon(\vec{x}_0)$ $=\{\vec{x} \in \mathbb{E}^n : 0 < \|\vec{x} - \vec{x}_0\| < \varepsilon\}$, the punctured ε–deleted ball of the vector $\vec{x}_0$

$\bar{S}$ closure of the set S

$\lim_{t \to t_0} \vec{f}(t)$ limit of the function $\vec{f}$ as t approaches t_0

$\vec{f}'$ derivative of the vector function $\vec{f}$

$\vec{f}'(t_0)$ — derivative of the vector function $\vec{f}$ at t_0

$\dfrac{d\vec{f}}{dt}(t_0)$ — $=\vec{f}'(t_0)$

$\vec{f}''$ — second order derivative of the function $\vec{f}$

$\vec{f}'''$ — third order derivative of the function $\vec{f}$

$\vec{f}^{(m)}$ — m^{th} order derivative of the function $\vec{f}$

$\vec{f}^{(0)}$ — $=\vec{f}$

$C^0(I)$ — class of all continuous functions on the interval I

$C^m(I)$ — class of all functions that have continuous m^{th} order partial derivatives on the interval I

$C^\infty(I)$ — class of all functions that have continuous partial derivatives of any order on the interval I

$(\vec{x}(t), I)$ — parametric representation of a curve in $\mathbb{V}^n$

s — arc length along a curve; natural parameter of a curve

$\dot{\vec{x}}(s)$ — $=\dfrac{d\vec{x}}{ds}(s)$, derivative of the parametric representation of a curve with respect to s, the arc length of the curve

$\overline{PQ}$ — straight line through the points $P, Q \in \mathbb{R}^n$

$\vec{t}(t_0)$ — (unit) tangent vector to a curve at t_0

$\vec{x}_{t_0}$ — parametric representation of the tangent to a curve at t_0

$\vec{v}_1$ — tangent unit vector of a curve

$\vec{v}_2$ — principal normal vector of a curve

$\vec{v}_3$ — $=\vec{v}_1 \times \vec{v}_2$, binormal vector of a curve

$\vec{v}_1(s)$ — $=\dot{\vec{x}}(s)$ tangent unit vector at s of a curve with a parametric representation $\vec{x}(s)$

$\vec{v}_2(s)$ — principal normal vector of a curve at s

$\vec{v}_3(s)$ — $=\vec{v}_1(s) \times \vec{v}_2(s)$, binormal vector of a curve at s

$\ddot{\vec{x}}$ — vector of curvature of a curve

$\ddot{\vec{x}}(s)$ — vector of curvature of a curve at s

κ — $=\|\ddot{\vec{x}}\|$, curvature of a curve

$\kappa(s)$ — $=\|\ddot{\vec{x}}(s)\|$, curvature of a curve at s

ρ — $=1/\kappa$, radius of curvature of a curve

$\rho(s)$ — $=1/\kappa(s)$, radius of curvature of a curve at s

δ_{ik} — Kronecker symbol; $\delta_{ik} = 1$ for $i = k$ and $\delta_{ik} = 0$ for $i \neq k$

τ — torsion of a curve

$\tau(s)$ — torsion of a curve at s

$C^r(D)$ — class of all functions that have continuous r^{th} order partial derivatives on the domain D

(f, D) — parametric representation of a surface given by a function $f: D \subset \mathbb{R}^2 \to \mathbb{R}^3$

$\vec{x}(u^i)$ — parametric representation of a surface in terms of the position vectors of its points with respect to the parameters u^i $(i = 1, 2)$

$\dfrac{\partial(u^i)}{\partial(u^{*k})}$ — Jacobian of the functions $u^i = u^i(u^{*k})$

$\vec{x}_1(u^i)$ — $=\partial\vec{x}(u^i)/\partial u^1$, vector in the direction of the u^1–line of the surface with the parametric representation $\vec{x}(u^i)$

$\vec{x}_2(u^i)$ — $=\partial\vec{x}(u^i)/\partial u^2$, vector in the direction of the u^2–line of the surface with the parametric representation $\vec{x}(u^i)$

$\vec{N}(u^i)$ — $=(\vec{x}_1(u^i) \times \vec{x}_1(u^i))/\|\vec{x}_1(u^i) \times \vec{x}_1(u^i)\|$, surface normal vector of the surface with

	the parametric representation $\vec{x}(u^i)$	R^m_{ikj}	Riemannian tensor of curvature
g_{ik}	$=\vec{x}_1 \bullet \vec{x}_2$, first fundamental coefficients of the surface with the parametric representation $\vec{x}$	$(X,\mathcal{T})$	topological space
		$\mathcal{T}_S$	relative topology for a subset S of a topological space $(X,\mathcal{T})$
$(ds)^2$	$=g_{ik}(u^j)du^i du^k$, first fundamental form of the surface with the parametric representation $\vec{x}(u^i)$	$SC(P)$	set of all smooth curves through a point P
		$\mathcal{R}$	relation on $SC(P)$ defined by $\gamma\mathcal{R}\gamma^*$ if and only if γ and γ^* have the same tangents at P
(g_{ik})	matrix of the first fundamental coefficients of a surface		
g	$=\det(g_{ik})$, determinant of the matrix of the first fundamental coefficients of a surface	$T(P)$	set of all equivalent classes with respect to $\mathcal{R}$ of tangent vectors through P
(g^{ik})	$=(g_{ik})^{-1}$, the inverse of the matrix (g_{ik})	$\vec{b}_1,\ldots,\vec{b}_n$	basis vectors of an n–dimensional vector space
κ_g	geodesic curvature of a curve on a surface	$\vec{x}=x^k\vec{b}_k$	representation of the vector $\vec{x}$ with respect to the basis vectors $\vec{b}_k$ $(k=1,\ldots,n)$; x^k $(k=1,\ldots,n)$ the components of $\vec{x}$ with respect to this representation
κ_n	normal curvature of a curve on a surface		
L_{ik}	$=\vec{N}\bullet\vec{x}_{ik}=-\vec{N}_k\bullet\vec{x}_i$, the second fundamental coefficients of a surface		
L	$=\det(L_{ik})$, determinant of the matrix of the second fundamental coefficients of a surface	V^*	dual space of the vector space V, that is, the space of all continuous linear functionals on V
κ_1,κ_2	principal curvature of a surface	$\vec{b}^1,\ldots,\vec{b}^n$	basis vectors of the dual space $V*$ of an n–dimensional vector space V
H	$=(\kappa_1+\kappa_2)/2$, the mean curvature of a surface	$l=l_k\vec{b}^k$	representation of the vector $l \in V^*$ with respect to the basis vectors $\vec{b}^k$ $(k=1,\ldots,n)$; l_k $(k=1,\ldots,n)$ the components of l with respect to this representation
K	$=\kappa_1\cdot\kappa_2$, the Gaussian curvature of a surface		
$L_{ik}du^i du^k$	second fundamental form		
L^i_k	$=g^{ij}L_{jk}$		
c_{ik}	$= \vec{N}_i \bullet \vec{N}_k$, the third fundamental coefficients of a surace	$\vec{x} \in V$	contravariant vector with components x^k with respect to a basis
$[ikl]$	$=\vec{x}_{ik}\bullet\vec{x}_l$, the first Christoffel symbols of a surface	$\vec{x} \in V^*$	covariant vector with components x_k with respect to a basis
$\{^r_{ik}\}$	$=g^{lr}[ikl]$, the second Christoffel symbols of a surface	b_{ik}	components of a covariant tensor B of second order

b^{ik} components of a contravariant tensor B of second order

b_i^k components of a mixed tensor B of second order

$T_{k_1...k_r}^{i_1...i_s}$ components of an (r,s) – tensor T, that is, T is contravariant of order r and covariant of order s

$T + U$ sum of (r,s)–tensors

$T * U$ outer product of (r,s)– tensors

$T_{k_1...k_{n-1},k_{n+1}...k_r}^{i_1...i_{m-1},i_{m+1}...i_s} = T_{k_1...k_{n-1}jk_{n+1}...k_r}^{i_1...i_{m-1}ji_{m+1}...i_s}$, contraction of the tensor T

G metric tensor with the components g_{ik}

g^{ik} components of the conjugate tensor of G

$[ijk]$ $= (1/2) \cdot (g_{ik,j} - g_{ij,k} + g_{jk,i})$ first Christoffel symbols

$\left\{ \begin{smallmatrix} k \\ ij \end{smallmatrix} \right\}$ $= g^{kl}[ijl]$ second Christoffel symbols

$T_{;j}^i$ $= \dfrac{\partial T^i}{\partial x^j} + \left\{ \begin{smallmatrix} i \\ jk \end{smallmatrix} \right\} T^k$ covariant derivative of a contravariant vector with the components T^i

$T_{i;j}$ $= \dfrac{\partial T_i}{\partial x^j} - \left\{ \begin{smallmatrix} k \\ ij \end{smallmatrix} \right\} T_k$ covariant derivative of a covariant vector T with the components T_i

$T_{m;r}^n$ $= \dfrac{\partial T_m^n}{\partial x^r} - T_l^n \left\{ \begin{smallmatrix} l \\ mr \end{smallmatrix} \right\} + T_m^l \left\{ \begin{smallmatrix} n \\ rl \end{smallmatrix} \right\}$ covariant derivative of a $(1,1)$–tensor with the components T_m^n

$T_{i_1...i_r;k}^{j_1...j_s}$ $= \dfrac{\partial T_{i_1...i_r}^{j_1...j_s}}{\partial x^k}$
$- \sum_{l=1}^r T_{i_1...i_{l-1}mi_{l+1}...i_r}^{j_1...j_s} \left\{ \begin{smallmatrix} m \\ i_l k \end{smallmatrix} \right\}$
$+ \sum_{n=1}^s T_{i_1...i_r}^{j_1...j_{n-1}pj_{n+1}...j_s} \left\{ \begin{smallmatrix} j_n \\ pk \end{smallmatrix} \right\}$
covariant derivative of an (r,s)-tensor T with the components $T_{i_1...i_r}^{j_1...j_s}$

R_{ilk}^j $= \dfrac{\partial}{\partial x^k} \left(\left\{ \begin{smallmatrix} j \\ il \end{smallmatrix} \right\} \right) - \dfrac{\partial}{\partial x^l} \left(\left\{ \begin{smallmatrix} j \\ ik \end{smallmatrix} \right\} \right)$
$+ \left\{ \begin{smallmatrix} m \\ il \end{smallmatrix} \right\} \left\{ \begin{smallmatrix} j \\ mk \end{smallmatrix} \right\} - \left\{ \begin{smallmatrix} m \\ ik \end{smallmatrix} \right\} \left\{ \begin{smallmatrix} j \\ ml \end{smallmatrix} \right\}$
mixed Riemann tensor of curvature in a Riemann space with the metric tensor g_{ik}

R_{ijkl} $= g_{ir} R_{jkl}^r$

$\nabla(\Phi, \Psi)$ $= g^{ik} \Phi_{;i} \Psi_{;k}$ Beltrami differentiator of first order of two real–valued functions of class $r \geq 2$ in a Riemann space

$\nabla \Phi$ $= \nabla(\Phi, \Phi)$

div T^k $= T_{;k}^k$ divergence of a contravariant vector T with the components T^k

div T_k $= g^{ki} T_{k;i}$ divergence of a covariant vector T with the components T_k

$\triangle F$ $=$ div F_i, where F is a real-valued function and $F_i = \dfrac{\partial F}{\partial x^i}$; Laplacian operator

$\dfrac{\delta \xi^\alpha}{\delta s}$ $= \dfrac{d\xi^\alpha}{ds} + \xi^\beta \left\{ \begin{smallmatrix} \alpha \\ \beta\gamma \end{smallmatrix} \right\} \dot{u}^\gamma = 0$ $(\alpha = 1,2)$; differential equations for the components of a surface vector under parallel movement along a geodesic line

$\dfrac{\delta \xi^\alpha}{\delta s}$ $= \xi_{;\sigma}^\alpha \dot{u}^\sigma = \dfrac{d\xi^\alpha}{ds} + \xi^\delta \left\{ \begin{smallmatrix} \alpha \\ \delta\gamma \end{smallmatrix} \right\} \dot{u}^\gamma$ $= 0$ $(\alpha = 1,2)$, parallel movement by Levi–Cività of a surface vector with the components ξ^α along an arbitrary curve given by $(u^1(s), u^2(s))$

Authors' Biographies

Eberhard Malkowsky is a Full Professor of Mathematics in retirement at the State University of Novi Pazar in Serbia. He earned his Ph.D. degree and habilitation at the Department of Mathematics of the Justus-Liebig Universität Giessen in Germany in 1982 and 1988, respectively. He was a professor of mathematics at universities in Germany, South Africa, Jordan, Turkey and Serbia, and a visiting professor in the USA, India, Hungary and France. Furthermore, he participated as an invited or keynote speaker with more than 100 lectures in international scientific conferences and congresses. He is a member of the editorial boards of twelve journals of international repute. His list of publications contains 175 research papers in international journals. He is the author or co-author of nine books, and the editor or co-editor of six proceedings of international conferences. He supervised 6 Ph.D. theses and a great number of B.Sc. and M.Sc. theses in mathematics. His research and work areas include functional analysis, differential geometry and software development for the visualization of mathematics.

Ćemal Dolićanin is a Professor Emeritus at the Department of Sciences and Mathematics at the State University of Novi Pazar in Serbia and a member of the Serbian Academy of Non-linear Sciences. He was dean of the Electro-Technical Faculty and the Faculty of Technical Sciences in Priština, vice-rector of the University of Priština and founder and rector of the State University of Novi Pazar, Serbia. He obtained his M.Sc. degree at the Faculty of Mathematics of the University of Belgrade in 1974, and his Ph.D. degree at the Faculty of Sciences and Mathematics of the University of Priština in 1980. He published more than 20 books, 23 papers in national scientific journals, 41 papers in international scientific journals, and gave more than 50 lectures at international scientific conferences.

He was a visiting professor in Germany, Belorussia and Russia. He supervised 14 PhD theses and a great number of B.Sc. and M.Sc. theses in mathematics. His research and work areas include Euclidean and non-Euclidean geometry, differential geometry and applied mathematics. He is very active in promoting mathematics, and has established the Center for the Advancement and Popularization of Mathematics at the State University of Novi Pazar. He participated in the implementation of several national scientific projects TEMPUS projects and was the coordinator of Master's study programs with the World University Service Austria (WUS).

Vesna Veličković is a Professor at the Department of Computer Science at the Faculty of Sciences and Mathematics of the University of Niš, Serbia. She obtained her magister degree at the Faculty of Mathematics of the University of Belgrade in 1996, and her Ph.D. degree at the Faculty of Sciences and Mathematics of the University of Niš in 2012. She published 3 books, 30 papers in international scientific journals, 12 papers in national scientific journals, and gave more than 50 lectures at international scientific conferences. She participated in 3 international scientific and 4 software projects, and 7 study visits in Serbia, Germany, Bulgaria, Romania and Turkey. Together with Professor Malkowsky, she is developing the software *MV-Graphics* for visualization of mathematics. With large-format graphics, they participated in three exhibitions of mathematical art. For ten years she worked with pupils talented in mathematics and programming. Her students won a number of medals at International Olympiads in Informatics. She is still very active in organizing contests in programming and promoting of mathematics. Her research areas are software development, computer graphics and visualization of mathematics.

CHAPTER 1

Curves in Three-Dimensional Euclidean Space

Elementary differential geometry is the study of geometric objects using the methods of analysis. In particular, the introductory theory studies curves and surfaces embedded in three-dimensional Euclidean space $\mathbb{E}^3$.

The first chapter deals with the basic geometric properties of curves in three-dimensional Euclidean $\mathbb{E}^3$. The studies related to topics of special interest include the

- *parametric representation of curves* in Definition 1.3.1 of Section 1.3

- *geometric principles and construction of important planar curves* in Section 1.4

- *arc length of a curve as its natural parameter* in Section 1.5

- *the vectors of the trihedra of curves* in Definition 1.6.3 of Section 1.6

- *Frenet's formulae* in Theorem 1.7.1 of Section 1.7

- *the curvature and torsion of curves* in Definition 1.6.3 of Section 1.6 and in Definition 1.7.2 of Section 1.7 and *their geometric significance* in Remark 1.6.4 of Section 1.6 and in Theorem 1.8.1 of Section 1.8

- *the characteristic shape of a curve in a neighbourhood of any of its points* in Section 1.8

- *osculating circles and spheres* in Section 1.9

- *involutes and evolutes of curves* in Section 1.10

- *fundamental theorem of curves*, Theorem 1.11.1 and the *natural or intrinsic equations of curves* in Definition 1.11.4 of Section 1.11

DOI: 10.1201/9781003370567-1

- *lines of constant slope* in Section 1.12

- *spherical images of curves* in Section 1.13.

Throughout, we use the standard notations $\mathbb{N} = \{1, 2, \dots\}$, $\mathbb{Z}$, $\mathbb{Q}$, $\mathbb{R}$ and $\mathbb{C}$ for the sets of all *nonnegative integers (natural numbers)*, *integers, rational, real* and *complex numbers*, respectively.

1.1 POINTS AND VECTORS

Points and vectors are the simplest geometric objects. Therefore, we recall the definitions of *n–dimensional point* and *vector spaces*.

For any nonnegative integer n, the set

$$\mathbb{R}^n = \left\{ X = \left(x^1, x^2, \dots, x^n \right) : x^k \in \mathbb{R} \ (k = 1, 2, \dots, n) \right\}$$

of all n–tuples of real numbers, x^k is called *n–dimensional (real) point space*; its elements X are called *points*, and the reals x^k are referred to as the *coordinates of X*.

The map $d : \mathbb{R}^n \times \mathbb{R}^n \to [0, \infty)$ defined by

$$d(X, Y) = \sqrt{\sum_{k=1}^{n} (x^k - y^k)^2} \text{ for all } X, Y \in \mathbb{R}^n \text{ (Figure 1.1)}$$

is called the *Euclidean metric* and the real number $d(X, Y)$ is called the *Euclidean distance* of the points X and Y (Figure 1.1).

The set

$$\mathbb{V}^n = \left\{ \vec{x} = \left\{ x^1, x^2, \dots, x^n \right\} : x^k \in \mathbb{R} \ (k = 1, 2, \dots, n) \right\}$$

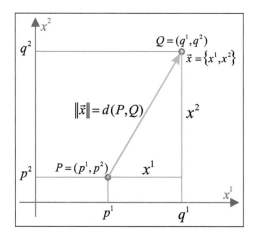

Figure 1.1 The Euclidean distance of two points and the length of a vector

of all n–tuples of real numbers x^k together with the operations

$$\vec{x} + \vec{y} = \left\{x^1 + y^1, x^2 + y^2, \ldots, x^n + y^n\right\} \text{ for all } \vec{x}, \vec{y} \in \mathbb{V}^n$$

and

$$\lambda\vec{x} = \left\{\lambda x^1, \lambda x^2, \ldots, \lambda x^n\right\} \text{ for all } \lambda \in \mathbb{R} \text{ and all } \vec{x} \in \mathbb{V}^n$$

is called the *n–dimensional (real) vector space*; its elements $\vec{x}$ are called *vectors*; the real numbers x^k are referred to as the *components of $\vec{x}$*.

In contrast to the familiar notation in analysis to write the indices of coordinates or components as lower indices or subscripts, it is usual in differential geometry to place indices of coordinates or components as upper indices or superscripts. This notation distinguishes indices of coordinates and components from the lower indices of partial derivatives. It is helpful in the use of *Einstein's convention of summation* in the theory of surfaces in Chapter 2.

The *Euclidean norm* or *length* $\|\vec{x}\|$ of a vector $\vec{x} \in \mathbb{V}^n$ is given by

$$\|\vec{x}\| = \sqrt{\sum_{k=1}^{n} (x^k)^2} \quad \text{(Figure 1.1)}.$$

The space $\mathbb{V}^n$ together with this norm is called the *n–dimensional Euclidean vector space* and denoted by $\mathbb{E}^n$.

Every vector $\vec{x} = \{x^1, x^2, \ldots, x^n\} \in \mathbb{V}^n$ may be interpreted as a map $\vec{x} : \mathbb{R}^n \to \mathbb{R}^n$ that assigns to every point $P = (p^1, p^2, \ldots, p^n)$ the point

$$Q = \vec{x}(P) = \left(q^1, q^2, \ldots, q^n\right), \text{ where } q^k = p^k + x^k \ (k = 1, 2, \ldots, n).$$

The vector that maps P to Q will be denoted by $\overrightarrow{PQ}$ (left in Figure 1.2). Thus the Euclidean distance of the points P and Q is equal to the Euclidean norm of the vector $\overrightarrow{PQ}$, that is, $d(P,Q) = \|\overrightarrow{PQ}\|$. If the origin of $\mathbb{R}^n$ is denoted by O, then the *position vector* $\vec{x}$ of a point $X \in \mathbb{R}^n$ is given by $\vec{x} = \overrightarrow{OX}$. If a point X has the coordinates x^k $(k = 1, 2, \ldots, n)$, then its position vector $\vec{x}$ has the components x^k $(k = 1, 2, \ldots, n)$; in this sense, we may – and indeed shall frequently do so – identify points with their position vectors (right in Figure 1.2).

A vector $\vec{x}$ with Euclidean norm $\|\vec{x}\| = 1$ is called a *unit vector*. If $\vec{x} \neq \vec{0}$, then the length of the vector $\vec{u} = \vec{x}/\|\vec{x}\|$ is

$$\|\vec{u}\| = \left\|\frac{\vec{x}}{\|\vec{x}\|}\right\| = \frac{1}{\|\vec{x}\|}\|\vec{x}\| = 1,$$

and $\vec{u}$ is a unit vector in the direction of $\vec{x}$.

Let $\vec{x} = \{x^1, x^2, \ldots, x^n\}$ and $\vec{y} = \{y^1, y^2, \ldots, y^n\}$ be vectors in $\mathbb{V}^n$, then the real number

$$\vec{x} \bullet \vec{y} = \sum_{k=1}^{n} x^k y^k$$

is called *scalar* or *inner* or *dot product of $\vec{x}$ and $\vec{y}$*. It is easy to see that $\bullet : \mathbb{V}^n \times \mathbb{V}^n \to \mathbb{R}$ is a *symmetric bilinear map*. Obviously $\vec{x} \bullet \vec{x} = \|\vec{x}\|^2$ for all $\vec{x} \in \mathbb{E}^n$.

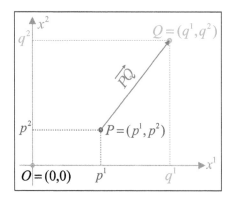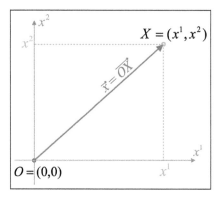

Figure 1.2 Vector and position vector

Proposition 1.1.1 (Cauchy–Schwarz inequality).

We have

$$|\vec{x} \bullet \vec{y}| \le \|\vec{x}\|\|\vec{y}\| \ \text{for all} \ \vec{x}, \vec{y} \in \mathbb{E}^n. \tag{1.1}$$

Proof. The inequality obviously holds for $\vec{y} = \vec{0}$.
If $\vec{y} = \vec{0}$, then we put $\lambda = (\vec{x} \bullet \vec{y})\|\vec{y}\|^{-2}$, and obtain

$$0 \le (\vec{x} - \lambda\vec{y}) \bullet (\vec{x} - \lambda\vec{y}) = \|\vec{x}\|^2 - 2\lambda(\vec{x} \bullet \vec{y}) + \lambda^2\|\vec{y}\|^2$$

$$= \|\vec{x}\|^2 - 2\frac{(\vec{x} \bullet \vec{y})^2}{\|\vec{y}\|^2} + \frac{(\vec{x} \bullet \vec{y})^2}{\|\vec{y}\|^2} = \|\vec{x}\|^2 - \frac{(\vec{x} \bullet \vec{y})^2}{\|\vec{y}\|^2}.$$

This implies $(\vec{x} \bullet \vec{y})^2 \le (\|\vec{x}\| \|\vec{y}\|)^2$, and the inequality in (1.1) is an immediate consequence. □

The *angle* $\alpha \in [0, \pi)$ *between two vectors* $\vec{x}, \vec{y} \in \mathbb{V}^n \setminus \{\vec{0}\}$ is defined by

$$\cos \alpha = \frac{\vec{x} \bullet \vec{y}}{\|\vec{x}\|\|\vec{y}\|}.$$

The vectors $\vec{x}, \vec{y} \in \mathbb{V}^n$ are called *orthogonal* if $\vec{x} \bullet \vec{y} = 0$; we denote this by $\vec{x} \perp \vec{y}$. Orthogonal unit vectors are said to be *orthonormal*.
We use the scalar product and orthogonality to show two well–known elementary results.

Example 1.1.2. *(a) A parallelogram is a rectangle if and only if its diagonals have equal lengths (Figure 1.3).*
(b) The heights in a triangle intersect in one point (Figure 1.4).

Proof. (a) Let $\vec{x}$ and $\vec{y}$ be the vectors that span the parallelogram. Then its diagonals are given by the vectors $\vec{x} + \vec{y}$ and $\vec{y} - \vec{x}$ and we have

$$\|\vec{x} + \vec{y}\|^2 - \|\vec{y} - \vec{x}\|^2 = (\vec{x} + \vec{y}) \bullet (\vec{x} + \vec{y}) - (\vec{y} - \vec{x}) \bullet (\vec{y} - \vec{x}) = 4\vec{x} \bullet \vec{y} = 0$$

if and only if $\vec{x} \perp \vec{y}$. This shows Part (a).

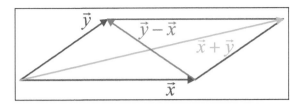 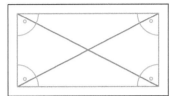

Figure 1.3 The diagonals in a parallelogram

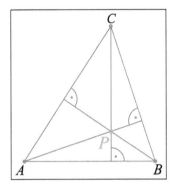 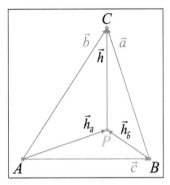

Figure 1.4 The heights in a triangle

(b) Let A, B and C denote the vertices of a triangle and P be the point of inter-section of the heights on $\overline{AB}$ and $\overline{AC}$. We put $\vec{a} = \overrightarrow{BC}$, $\vec{b} = \overrightarrow{AC}$, $\vec{c} = \overrightarrow{AB}$, $\vec{h}_{\vec{a}} = \overrightarrow{AP}$, $\vec{h}_{\vec{b}} = \overrightarrow{BP}$ and $\vec{h} = \overrightarrow{PC}$ (Figure 1.4). Then we obtain $\vec{c} = \vec{b} - \vec{a}$, $\vec{h} = \vec{a} - \vec{h}_{\vec{b}} = \vec{b} - \vec{h}_{\vec{a}}$, and $\vec{h}_{\vec{a}} \perp \vec{a}$ and $\vec{h}_{\vec{b}} \perp \vec{b}$, hence

$$\vec{c} \bullet \vec{h} = (\vec{b} - \vec{a}) \bullet \vec{h} = \vec{b} \bullet (\vec{a} - \vec{h}_{\vec{b}}) - \vec{a} \bullet (\vec{b} - \vec{h}_{\vec{a}})$$
$$= \vec{b} \bullet \vec{a} - \vec{b} \bullet \vec{h}_{\vec{b}} - \vec{a} \bullet \vec{b} + \vec{a} \bullet \vec{h}_{\vec{a}} = \vec{a} \bullet \vec{b} - \vec{b} \bullet \vec{a} = 0.$$

Thus we have $\vec{h} \perp \vec{c}$. This shows Part (b).

□

Let $\vec{x}, \vec{y} \in \mathbb{V}^n$, $\vec{x} \neq \vec{0}$ and $\vec{u}$ be the unit vector in the direction of $\vec{x}$. Then the vector $\vec{y}_{\vec{x}} = (\vec{y} \bullet \vec{u})\vec{u}$ is called the *projection of $\vec{y}$ to $\vec{x}$*. If α denotes the angle between the vectors $\vec{x}, \vec{y} \neq \vec{0}$, then the projection $\vec{y}_{\vec{x}}$ of $\vec{y}$ to $\vec{x}$ is given by

$$\vec{y}_{\vec{x}} = (\vec{y} \bullet \vec{x})\frac{\vec{u}}{\|\vec{x}\|} = \|\vec{y}\| \cos \alpha \, \vec{u} \ \text{(Figure 1.5)}.$$

For any two vectors $\vec{x} = \{x^1, x^2, x^3\}, \vec{y} = \{y^1, y^2, y^3\} \in \mathbb{V}^3$, the vector

$$\vec{x} \times \vec{y} = \{x^2 y^3 - x^3 y^2, x^3 y^1 - x^1 y^3, x^1 y^2 - x^2 y^1\}$$

is called the *vector* or *outer product of $\vec{x}$ and $\vec{y}$*.

The following results are well known from analytical geometry.

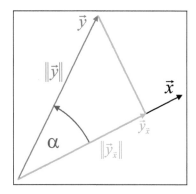

Figure 1.5 The angle between two vectors $\vec{x}$ and $\vec{y}$ and the projection of $\vec{y}$ on $\vec{x}$

Proposition 1.1.3. *We have $\vec{x} \times \vec{y} = \vec{0}$ if and only if $\vec{x}$ and $\vec{y}$ are linearly dependent. Furthermore, the following identities hold for all $\vec{x}, \vec{y}, \vec{z} \in \mathbb{V}^n$ and for all $\lambda \in \mathbb{R}$*

$$\vec{x} \times \vec{y} = -(\vec{y} \times \vec{x}) \qquad \text{(anticommutative law)}, \qquad (1.2)$$

$$\vec{x} \times (\vec{y} + \vec{z}) = \vec{x} \times \vec{y} + \vec{x} \times \vec{z} \qquad \text{(distributive law)}, \qquad (1.3)$$

$$(\lambda \vec{x}) \times \vec{y} = \lambda(\vec{x} \times \vec{y}), \qquad (1.4)$$

$$\vec{x} \times (\vec{y} \times \vec{z}) = (\vec{x} \bullet \vec{z})\vec{y} - (\vec{x} \bullet \vec{y})\vec{z} \qquad \text{(Grassmann identity)}, \qquad (1.5)$$

$$(\vec{x} \times \vec{y}) \bullet \vec{z} = \det \begin{pmatrix} x^1 & y^1 & z^1 \\ x^2 & y^2 & z^2 \\ x^3 & y^3 & z^3 \end{pmatrix} \qquad (1.6)$$

and

$$(\vec{x} \times \vec{y}) \bullet (\vec{z} \times \vec{w}) = (\vec{x} \bullet \vec{z})(\vec{y} \bullet \vec{w}) - (\vec{x} \bullet \vec{w})(\vec{y} \bullet \vec{z}). \qquad (1.7)$$

Proof. (i) The identities in (1.2), (1.3) and (1.4) are immediate consequences of the definition of the vector product.

(ii) To prove (1.5), we put $\vec{w} = \vec{y} \times \vec{z}$ and obtain

$$\vec{x} \times (\vec{y} \times \vec{z}) = \{x^2 w^3 - x^3 w^2, x^3 w^1 - x^1 w^3, x^1 w^2 - x^2 w^1\} =$$
$$= \{x^2(y^1 z^2 - y^2 z^1) - x^3(y^3 z^1 - y^1 z^3), x^3(y^2 z^3 - y^3 z^2) - x^1(y^1 z^2 - y^2 z^1),$$
$$x^1(y^3 z^1 - y^1 z^3) - x^2(y^2 z^3 - y^3 z^2)\} =$$
$$= \{y^1(x^1 z^1 + x^2 z^2 + x^3 z^3) - z^1(x^1 y^1 + x^2 y^2 + x^3 z^2),$$
$$y^2(x^1 z^1 + x^2 z^2 + x^3 z^3) - z^2(x^1 y^1 + x^2 y^2 + x^3 y^3),$$
$$y^3(x^1 z^1 + x^2 z^2 + x^3 z^3) - z^3(x^1 y^1 + x^2 y^2 + x^3 y^3)\} =$$
$$= \vec{y}(\vec{x} \bullet \vec{z}) - \vec{z}(\vec{x} \bullet \vec{y}).$$

(iii) The identity (1.6) follows from

$$\det \begin{pmatrix} x^1 & y^1 & z^1 \\ x^2 & y^2 & z^2 \\ x^3 & y^3 & z^3 \end{pmatrix} = x^1(y^2 z^3 - y^3 z^2) - x^2(y^1 z^3 - y^3 z^1) + x^3(y^1 z^2 - y^2 z^1)$$

$$= z^1(x^2 y^3 - x^3 y^2) + z^2(x^3 y^1 - x^1 y^3) + z^3(x^1 y^2 - x^2 y^1)$$

$$= \vec{z} \bullet (\vec{x} \times \vec{y}).$$

(iv) Finally, it follows from (1.6) that $(\vec{x} \times \vec{y}) \bullet \vec{v} = -(\vec{x} \times \vec{v}) \bullet \vec{y}$ for all $\vec{x}, \vec{y}, \vec{v} \in \mathbb{V}^3$. Applying this with $\vec{v} = \vec{z} \times \vec{w}$ and using (1.5), we obtain

$$(\vec{x} \times \vec{y}) \bullet (\vec{z} \times \vec{w}) = -(\vec{x} \times (\vec{z} \times \vec{w})) \bullet \vec{y}$$

$$= -((\vec{x} \bullet \vec{w})\vec{z} + (\vec{x} \bullet \vec{z})\vec{w}) \bullet \vec{y}$$

$$= (\vec{x} \bullet \vec{z})(\vec{y} \bullet \vec{w}) - (\vec{x} \bullet \vec{w})(\vec{y} \bullet \vec{z}).$$

$\square$

We state the following useful and frequently used results which are immediate consequences of Proposition 1.1.3.

Remark 1.1.4. *Using (1.6) in Proposition 1.1.3 with $\vec{z} = \vec{x}$ and with $\vec{z} = \vec{y}$, we see that $(\vec{x} \times \vec{y}) \perp \vec{x}, \vec{y}$ for all $\vec{x}, \vec{y} \in \mathbb{V}^3$. Furthermore, it follows from (1.6) in Proposition 1.1.3 that*

$$(\vec{x} \times \vec{y}) \bullet \vec{z} = (\vec{z} \times \vec{x}) \bullet \vec{y} = (\vec{y} \times \vec{z}) \bullet \vec{x} = -(\vec{y} \times \vec{x}) \bullet \vec{z} = -(\vec{x} \times \vec{z}) \bullet \vec{y}$$

$$= -(\vec{z} \times \vec{y}) \bullet \vec{x} \ \text{for all } \vec{x}, \vec{y}, \vec{z} \in V^3.$$

Example 1.1.5. *(a) Let γ be a straight line in $\mathbb{E}^2$, given by an equation $ax^1 + bx^2 = d$, and $P = (p^1, p^2)$ and $X = (x^1, x^2)$ be a given and an arbitrary point of γ. We write $\vec{n} = \{a, b\}$. Since $P, X \in \gamma$, it follows that*

$$\vec{n} \bullet \left(\overrightarrow{OX} - \overrightarrow{OP} \right) = \{a, b\} \bullet \left(\{x^1, x^2\} - \{p^1, p^2\} \right)$$

$$= ax^1 + bx^2 - (ap^1 + bp^2) = d - d = 0.$$

Thus the vector $\vec{n}$ is orthogonal to every vector of the straight line γ; we say that $\vec{n}$ is orthogonal to γ and denote this by $\vec{n} \perp \gamma$. The unit vector $\vec{N}$ in the direction of $\vec{n}$ is called normal vector *of γ, and given by*

$$\vec{N} = \frac{\{a, b\}}{\sqrt{a^2 + b^2}}.$$

Any straight line in $\mathbb{E}^2$ has exactly two unit normal vectors $\vec{N}_1$ and $\vec{N}_2$; they have opposite directions, that is, $\vec{N}_1 = -\vec{N}_2$.

(b) The distance *ρ of a point $P = (p^1, p^2) \in \mathbb{E}^2$ from a straight line $\gamma \in \mathbb{E}^2$ is the length of the projection of any vector $\overrightarrow{PX}$ ($X \in \gamma$) to a normal vector $\vec{N}$ of γ. If*

$\vec{p}$ and $\vec{x}$ denote the position vectors of the points P and X, and γ is defined by an equation as in Part (a), then ρ is given by

$$\rho = \|\overrightarrow{PX}_{\vec{N}}\| = \|(\overrightarrow{PX} \bullet \vec{N})\vec{N}\| = \frac{|\overrightarrow{PX} \bullet \vec{n}|}{\|\vec{n}\|}$$
$$= \frac{|\vec{p} \bullet \vec{n} - \vec{x} \bullet \vec{n}|}{\|\vec{n}\|} = \frac{|p^1 a + p^2 b - d|}{\sqrt{a^2 + b^2}};$$

if we replace $X \in \gamma$ by $X' \in \gamma$ in our last identities, then we see that our definition of the distance of a point from a straight line is independent of the choice of the point X on γ. If we choose $P = 0$, then the distance of γ from the origin is

$$\rho = \frac{|d|}{\sqrt{a^2 + b^2}}.$$

Now we extend Example 1.1.5 to planes in $\mathbb{E}^3$.

Visualization 1.1.6. *(a) Let Pl be a plane in $\mathbb{E}^3$, given by an equation $ax^1 + bx^2 + cx^3 + d = 0$, and let $P = (p^1, p^2, p^3)$ and $X = (x^1, x^2, x^3)$ be a given and an arbitrary point of Pl. We write $\vec{n} = \{a, b, c\}$.*
Then it follows as in Example 1.1.5 (a) that the vector $\vec{n} = \{a, b, c\}$ is orthogonal to every vector of the plane Pl; we say that $\vec{n}$ is orthogonal to Pl and denote this by $\vec{n} \perp Pl$. The unit vector $\vec{N}$ in the direction of $\vec{n}$ is called normal vector *of Pl and given by*

$$\vec{N} = \frac{\{a, b, c\}}{\sqrt{a^2 + b^2 + c^2}}.$$

Any plane in $\mathbb{E}^3$ has exactly two unit normal vectors $\vec{N}_1$ and $\vec{N}_2$; they have opposite directions, that is, $\vec{N}_1 = -\vec{N}_2$.
(b) The distance *ρ of a plane $Pl \in \mathbb{E}^3$ from a point $P = (p^1, p^2, p^3) \in \mathbb{E}^3$ is the length of the projection of any vector $\overrightarrow{PX}$ $(X \in Pl)$ to a normal vector $\vec{N}$ of Pl. As in Example 1.1.5 (b), we obtain*

$$\rho = \frac{|p^1 a + p^2 b + p^3 c - d|}{\sqrt{a^2 + b^2 + c^2}};$$

and we also see that our definition of the distance of a point from a plane is independent of the choice of the point X in the plane. If we choose $P = 0$, then the distance of Pl from the origin is

$$\rho = \frac{|d|}{\sqrt{a^2 + b^2 + c^2}} \quad (Figure \ 1.6).$$

(c) Let $P_k = (p_k^1, p_k^2, p_k^3)$ be non–collinear points with position vectors $\vec{p}_k$ $(k = 1, 2, 3)$ that span a plane Pl, and $X = (x^1, x^2, x^3)$ be an arbitrary point of Pl. Then $(\vec{x} - \vec{p_1}) \times (\vec{x} - \vec{p_2})$ is a vector orthogonal to Pl, and so

$$((\vec{x} - \vec{p_1}) \times (\vec{x} - \vec{p_2})) \bullet (\vec{x} - \vec{p_3}) = 0$$

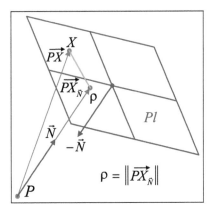

Figure 1.6 Distance of a point from a plane

is an equation of Pl. We may rewrite this by identity (1.6) in Proposition 1.1.3 as

$$\det \begin{pmatrix} x^1 - p_1^1 & x^1 - p_2^1 & x^1 - p_3^1 \\ x^2 - p_1^2 & x^2 - p_2^1 & x^2 - p_3^2 \\ x^3 - p_1^3 & x^3 - p_2^3 & x^3 - p_3^3 \end{pmatrix} = 0.$$

Visualization 1.1.7. *Let Pg be the parallelogram spanned by the vectors $\vec{x} = \{x^1, x^2\}$ and $\vec{y} = \{y^1, y^2\}$. Then the vector $\vec{x}^\perp = \{-x^2, x^1\}$ is orthogonal to $\vec{x}$, and the surface area of Pg is*

$$A(Pg) = \|\vec{x}\| \left| \vec{y} \bullet \frac{\vec{x}^\perp}{\|\vec{x}^\perp\|} \right| = |\vec{y} \bullet \vec{x}^\perp|,$$

since $\|\vec{x}\| = \|\vec{x}^\perp\|$. If ϕ denotes the angle between $\vec{x}$ and $\vec{y}$, then

$$|\vec{y} \bullet \vec{x}^\perp| = \|\vec{x}\| \cdot \|\vec{y}\| \left| \cos\left(\frac{\pi}{2} - \phi\right) \right| = \|\vec{x}\| \cdot \|\vec{y}\| \, |\sin \phi|,$$

hence

$$A(Pg) = \|\vec{x}\| \cdot \|\vec{y}\| \, |\sin \phi|.$$

On the other hand, we have by (1.7) in Proposition 1.1.3

$$\|\vec{x}\| \cdot \|\vec{y}\| \, |\sin \phi| = \|\vec{x}\| \cdot \|\vec{y}\| \sqrt{1 - \cos^2 \phi} = \sqrt{\|\vec{x}\|^2 \|\vec{y}\|^2 - (\vec{x} \bullet \vec{y})^2}$$

$$= \sqrt{(\vec{x} \times \vec{y})^2} = \|\vec{x} \times \vec{y}\|.$$

Thus the surface area of a parallelogram is the length of the vector product of the vectors along its edges,

$$A(Pg) = \|\vec{x} \times \vec{y}\| \ \ (Figure \ 1.7).$$

We may also write

$$A^2(Pg) = (\vec{x} \bullet \vec{x})(\vec{y} \bullet \vec{y}) - (\vec{x} \bullet \vec{y})^2 = \det \begin{pmatrix} \vec{x} \bullet \vec{x} & \vec{x} \bullet \vec{y} \\ \vec{x} \bullet \vec{y} & \vec{y} \bullet \vec{y} \end{pmatrix}.$$

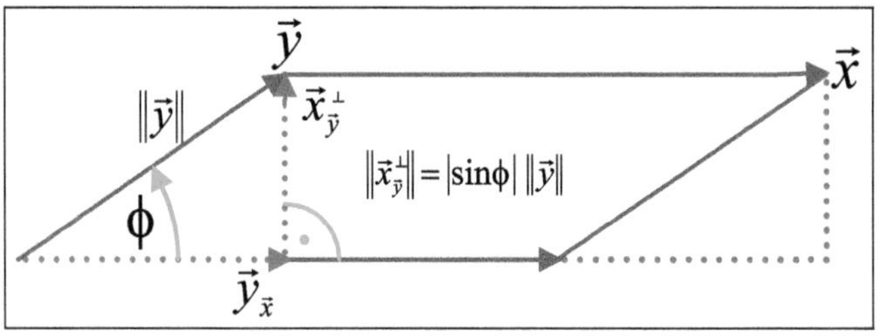

Figure 1.7 The surface area of a parallelogram

1.2 VECTOR-VALUED FUNCTIONS OF A REAL VARIABLE

In this section, we recall the definition of *vector–valued functions of a real variable* and state their well–known basic properties that are relevant in differential geometry.

Let $S \subset \mathbb{R}$ be a subset of the real line. A function $\vec{f} : S \to \mathbb{V}^n$ that assigns to every real $t \in S$ a unique vector

$$\vec{f}(t) = \{f^1(t), f^2(t), \ldots, f^n(t)\} \in \mathbb{V}^n$$

is called a *vector–valued function of a real variable*, and the real–valued functions $f^1, f^2, \ldots, f^n$ are referred to as the *component functions of $\vec{f}$*; we write

$$\vec{f} = \{f^1, f^2, \ldots, f^n\}.$$

The set S is called the *domain of $\vec{f}$*, also denoted by $D_{\vec{f}}$, and the set $\vec{f}(S) = \{f(t) : t \in S\}$ is called the *image* or *range of $\vec{f}$*.

Let X_0 be a point of the n–dimensional Euclidean space and $\vec{x}_0$ be a vector of the n–dimensional Euclidean vector space. Then, for every real $\varepsilon > 0$, the sets

$$N_\varepsilon(X_0) = \{X \in \mathbb{R}^n : d(X, X_0) < \varepsilon\}$$

and

$$N_\varepsilon(\vec{x}_0) = \{\vec{x} \in \mathbb{E}^n : \|\vec{x} - \vec{x}_0\| < \varepsilon\}$$

are called the *ε–open ball* or *ε–neighbourhood* of the point X_0 and the vector $\vec{x}_0$, and we write $\dot{N}_\varepsilon(X_0) = N_\varepsilon(X_0) \setminus \{X_0\}$ and $\dot{N}_\varepsilon(\vec{x}_0) = N_\varepsilon(\vec{x}_0) \setminus \{\vec{x}_0\}$ for the *ε–deleted neighbourhoods* of X_0 and $\vec{x}_0$, respectively.

If S is a subset of $\mathbb{R}$, then we denote the *closure of S* by $\bar{S}$. Let $\vec{a} \in \mathbb{E}^n$, $S \subset \mathbb{R}$, $t_0 \in \bar{S}$ and $\vec{f} : S \to \mathbb{E}^n$. Then the vector-valued function $\vec{f}$ is said to have a *limit $\vec{a}$ as t approaches t_0*, written as

$$\lim_{t \to t_0} \vec{f}(t) = \vec{a} \text{ or } \vec{f}(t) \to \vec{a} \ (t \to t_0),$$

if for every $\varepsilon > 0$, there is a real $\delta = \delta(\varepsilon) > 0$ such that $\vec{f}(t) \in N_\varepsilon(\vec{a})$ for all $t \in \dot{N}_\delta(t_0) \cap S$, that is, for all $t \in S$ with $0 < |t - t_0| < \delta$, or equivalently, $\vec{f}(\dot{N}_\delta(t_0) \cap S) \subset N_\varepsilon(\vec{a})$. Since

$$\max_{1 \le k \le n} |x^k| \le \|\vec{x}\| \le \sqrt{n} \max_{1 \le k \le n} |x^k| \text{ for all } \vec{x} = \{x^1, x^2, \ldots, x^n\} \in \mathbb{E}^n,$$

a vector–valued function $\vec{f} = \{f^1, f^2, \ldots, f^n\}$ has a limit $\vec{a} = \{a^1, a^2, \ldots, a^n\}$ as t approaches t_0 if and only if $f^k(t) \to a^k \ (t \to t_0)$ for all $k \in \{1, 2, \ldots, n\}$.

The following results are well known from analysis.

Proposition 1.2.1. *Let $\lambda(t) \to l_0$, $\vec{f}(t) \to \vec{a}$ and $\vec{g}(t) \to \vec{b}$ $(t \to t_0)$. Then we have*

$$\lim_{t \to t_0} (\vec{f}(t) \pm \vec{g}(t)) = \vec{a} \pm \vec{b}, \ \lim_{t \to t_0} (\lambda(t)\vec{f}(t)) = l_0\vec{a}$$

$$\lim_{t \to t_0} \|\vec{f}(t)\| = \|\vec{a}\|, \ \lim_{t \to t_0} (\vec{f}(t) \bullet \vec{g}(t)) = \vec{a} \bullet \vec{b}$$

and

$$\lim_{t \to t_0} (\vec{f}(t) \times \vec{g}(t)) = \vec{a} \times \vec{b} \text{ in } \mathbb{V}^3.$$

Furthermore, if $\lim_{s \to s_0} \tau(s) = t_0$ then $\lim_{s \to s_0} \vec{f}(\tau(s)) = \vec{a}$.

We recall that a vector–valued function $\vec{f} : S \subset \mathbb{R} \to \mathbb{E}^n$ is said to be *continuous at $t_0 \in S$* if for every $\varepsilon > 0$ there is a real $\delta = \delta(\varepsilon) > 0$ such that

$$\vec{f}(t) \in N_\varepsilon(\vec{f}(t_0)) \text{ for all } t \in N_\delta(t_0) \cap S, \text{ or equivalently, } \vec{f}(N_\delta(t_0) \cap S) \subset N_\varepsilon(\vec{f}(t_0)).$$

The function $\vec{f}$ is said to be *continuous on a subset S_0 of S* if it is continuous at all $t \in S_0$. The function $\vec{f}$ is said to be *continuous (everywhere)* if it is continuous on its domain.

Obviously $\vec{f}$ is continuous at $t_0 \in S$ if and only if

$$\lim_{t \to t_0} \vec{f}(t) = \vec{f}(t_0).$$

Furthermore, $\vec{f}$ is continuous at t_0 if and only if each of its component functions is continuous at t_0. Finally, the sum, difference, dot product, and scalar and vector products (in $\mathbb{V}^3$) of continuous functions are continuous, and the continuous function of a continuous real–valued function is a continuous function.

From now on let I always denote an interval on the real line. A function $\vec{f} : I \to \mathbb{E}^n$ is said to be *differentiable at $t_0 \in I$* if the limit

$$\vec{f}'(t_0) = \lim_{t \to t_0} \frac{\vec{f}(t) - \vec{f}(t_0)}{t - t_0}$$

exists; the vector $\vec{f}'(t_0)$ is called the *derivative of $\vec{f}$ at t_0*, and we also write

$$\vec{f}'(t_0) = \frac{d\vec{f}}{dt}(t_0).$$

The function $\vec{f}$ is said to be *differentiable on a subset S_0 of I* if it is differentiable at every $t_0 \in S_0$; it is said to be *differentiable (everywhere)* if it is differentiable on I. The vector–valued function

$$\vec{f}' : D_{\vec{f}'} = \{t \in I : \vec{f}'(t) \text{ exists}\} \to \mathbb{E}^n$$

is called the *derivative of $\vec{f}$.*

Obviously the function $\vec{f}$ is differentiable at t_0 if and only if each component function is differentiable at t_0.

The following results are well known from analysis.

Proposition 1.2.2. *(a) The function $\vec{f} : I \to \mathbb{E}^n$ is differentiable at $t_0 \in I$ if and only if there exist a vector $\vec{c}_{t_0}$ and a function $\vec{r}_{t_0}$ such that*

$$\vec{f}(t) = \vec{f}(t_0) + \vec{c}_{t_0}(t - t_0) + \vec{r}_{t_0}(t) \ \text{ and } \ \lim_{t \to t_0} \frac{\vec{r}_{t_0}(t)}{t - t_0} = \vec{0}.$$

(b) If $\vec{f}$ is differentiable at t_0 then it is continuous at t_0.

Proposition 1.2.3. *Let the functions $\vec{f}, \vec{g}$ and λ be differentiable at t_0. Then the functions $\vec{f} \pm \vec{g}$, $\lambda\vec{f}$, $\vec{f} \bullet \vec{g}$ and, in $\mathbb{E}^3$, $\vec{f} \times \vec{g}$ are differentiable at t_0 and*

$$\frac{d(\vec{f} \pm \vec{g})}{dt}(t_0) = \frac{d\vec{f}}{dt}(t_0) \pm \frac{d\vec{g}}{dt}(t_0), \tag{1.8}$$

$$\frac{d(\lambda\vec{f})}{dt}(t_0) = \frac{d\lambda}{dt}(t_0)\vec{f}(t_0) + \lambda(t_0)\frac{d\vec{f}}{dt}(t_0),$$

$$\frac{d(\vec{f} \bullet \vec{g})}{dt}(t_0) = \frac{d\vec{f}}{dt}(t_0) \bullet \vec{g}(t_0) + \vec{f}(t_0) \bullet \frac{d\vec{g}}{dt}(t_0)$$

and, in $\mathbb{E}^3$,

$$\frac{d(\vec{f} \times \vec{g})}{dt}(t_0) = \frac{d\vec{f}}{dt}(t_0) \times \vec{g}(t_0) + \vec{f}(t_0) \times \frac{d\vec{g}}{dt}(t_0). \tag{1.9}$$

Furthermore, if h is differentiable at s_0 and $\vec{f}$ differentiable at $t_0 = h(s_0)$, then $\vec{g} = \vec{f} \circ h$ is differentiable at s_0 and

$$\frac{d\vec{g}}{ds}(s_0) = \frac{d\vec{f}}{dt}(t_0)\frac{dh}{ds}(s_0) \ \text{ (chain rule)}. \tag{1.10}$$

If the vector–valued function $\vec{f}$ is differentiable on an interval I, then its derivative $\vec{f}'$ also is a vector–valued function, which again may be differentiable on I. This yields the *second-order derivative of $\vec{f}$* denoted by $\vec{f}''$. *Higher-order derivatives* are defined similarly.

If m is a nonnegative integer, then $C^m(I)$ denotes the class of all vector–valued functions that are m times continuously differentiable on the interval I, $C^0(I)$ is the class of all continuous functions on I, and $C^\infty(I)$ is the class of all functions which have continuous derivatives of all orders on I.

The following results are well known from analysis.

Proposition 1.2.4. *(a) A vector–valued function $\vec{f}$ is of class $C^m(I)$ if and only if all its component functions are of class $C^m(I)$.*
(b) If $\vec{f}, \vec{g}, h \in C^m(I)$ then $h\vec{f}, \vec{f} \pm \vec{g}, \vec{f} \bullet \vec{g} \in C^m(I)$ and, in $\mathbb{E}^3$, $\vec{f} \times \vec{g} \in C^m(I)$.

Proposition 1.2.5 (Taylor's formula). *Let $\vec{f} \in C^m(I)$. Then for every t and t_0 in I*

$$\vec{f}(t) = \sum_{k=0}^{m} \frac{1}{k!} \vec{f}^{(k)}(t_0)(t - t_0)^k + \vec{R}_{m,t_0}(t), \quad where \quad \lim_{t \to t_0} \frac{\vec{R}_{m,t_0}(t)}{(t - t_0)^m} = \vec{0}.$$

1.3 THE GENERAL CONCEPT OF CURVES

Our studies of curves take place in the three-dimensional Euclidean space with a Cartesian coordinate system. We start, however, with the more general definition of a *curve in the n–dimensional space* $\mathbb{R}^n$.

Definition 1.3.1. Let $f : \mathbb{R} \to \mathbb{R}^n$ be a continuous map on the real interval $I = [a, b]$. The point set

$$\gamma = \left\{ X \in \mathbb{R}^n : X = f(t) = \left(f^1(t), f^2(t), \ldots, f^n(t) \right) \right\} \ (t \in I) \qquad (1.11)$$

is called a *curve in* $\mathbb{R}^n$; (f, I), t and I are called a *parametric representation*, the *parameter* and the *parameter interval of* γ, respectively (Figure 1.8).

Remark 1.3.2. *(a) A curve γ may also be given in terms of the position vectors of its points*

$$\vec{x}(t) = \left\{ x^1(t), x^2(t), \ldots, x^n(t) \right\} \ (t \in I), \quad where \quad x^k(t) = f^k(t) \ for \ (k = 1, 2, \ldots, n)$$
$$(1.12)$$

and (1.12) is also referred to as a parametric representation of γ; in fact we shall rather use (1.12) than (1.11).
(b) A parametric representation induces an orientation on a curve γ; as the parameter t moves through the interval $[a, b]$ from a to b, the points $X = X(t)$ move along the curve from the point $X(a)$ to the point $X(b)$. The points $X(a)$ and $X(b)$ are called the initial and end points of the curve γ, respectively (Figure 1.8).

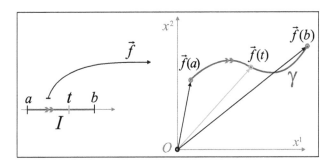

Figure 1.8 The parametric representation of a curve

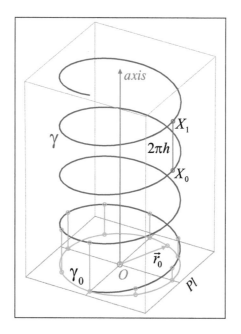

Figure 1.9 A helix

Visualization 1.3.3. *(a) A straight line through the point A with position vector $\vec{a}$ in the direction of the vector $\vec{b} \neq \vec{0}$ has a parametric representation*

$$\vec{x}(t) = \vec{a} + t\vec{b} \ (t \in \mathbb{R}).$$

(b) A parametric representation of a circle line of radius $r > 0$ and centre M in a plane spanned by the orthonormal vectors $\vec{e}^1 = \{1,0,0\}$ and $\vec{e}^2 = \{0,1,0\}$ is given by

$$\vec{x}(t) = \vec{m} + r\vec{e}^1 \cos t + r\vec{e}^2 \sin t \ (t \in [0, 2\pi]),$$

where $\vec{m}$ is the position vector of M.

(c) A parametric representation for a helix *with its axis along the x^3–axis is given by*

$$\vec{x}(t) = \{r \cos t, r \sin t, ht\} \ (t \in \mathbb{R})$$

where $r > 0$ and $h \in \mathbb{R}$ are constants (Figure 1.9).

In differential geometry, we have to assume that the function $\vec{x}$ in the parametric representation of a curve has a certain number of derivatives. Furthermore, we want to exclude trivial or pathological cases.

Definition 1.3.4. A parametric representation (1.12) is called *admissible* if the following conditions are satisfied

$$\vec{x} \in C^r(I) \text{ where } r \geq 1 \text{ is appropriately chosen} \tag{1.13}$$

and

$$\vec{x}'(t) = \left\{ \frac{dx^1}{dt}(t), \frac{dx^2}{dt}(t), \dots, \frac{dx^n}{dt}(t) \right\} \neq \vec{0} \quad \text{for all } t \in I. \tag{1.14}$$

The condition in (1.14) of Definition 1.3.4 implies that one of the equations $x^k = x^k(t)$ in (1.12) can locally be solved for t:

$$t = t(x^k) \text{ for some } k \in \{1, 2, \dots, n\}, \tag{1.15}$$

and the function in (1.15) is r times continuously differentiable on its domain.

Since, in future, we shall only deal with curves that have an admissible parametric representation, we shall omit the term *admissible*.

In general, a curve γ has various different parametric representations. We may obtain a new vector–valued function from (1.12) by introducing a new parameter t^* by

$$t = t(t^*). \tag{1.16}$$

We have to make sure that the range of the function in (1.16) contains the interval I for t, and that the conditions of admissibility also hold for the new representation.

Definition 1.3.5. A parameter transformation (1.16) is called *admissible* if the following conditions are satisfied:

$$I \subset t(I^*) \text{ for the domain } I^* \text{ of the function } t \text{ in (1.16)}, \tag{1.17}$$

$$t \in C^r(I^*), \text{ where } r \geq 1 \text{ is appropriately chosen}, \tag{1.18}$$

$$\frac{dt}{dt^*}(t^*) \neq 0 \text{ for all } t^* \in I^*. \tag{1.19}$$

Again, we shall omit the term admissible in future.

Visualization 1.3.6 (Neil's parabola).
The curve given by $\vec{x}(t) = \{t^2, t^3\}$ $(t \in \mathbb{R})$ is called Neil's parabola *(Figure 1.10).*

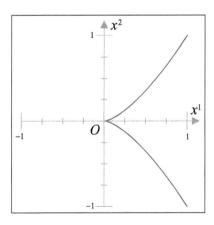

Figure 1.10 Neil's parabola

Since $x^1(t) = t^2$ and $x^2(t) = t^3$, it is obvious that $\vec{x} \in C^\infty(\mathbb{R})$. Furthermore, $\vec{x}'(t) = t\{2, 3t\} \neq \vec{0}$ for all $t \neq 0$. Thus $\vec{x}(t) = \{t^2, t^3\}$ ($t \in J$) is an admissible parametric representation for every interval J which does not contain 0.

If we consider the part of Neil's parabola in the upper half plane, that is, the part of the curve given by $\vec{x}(t) = \{t^2, t^3\}$ ($t \in I = (0, \infty)$), then we may write

$$t = \phi(t^*) = \sqrt{t^*} \quad (t^* \in I^* = (0, \infty)). \tag{1.20}$$

Then

$$I \subset \phi(I), \ \phi \in C^r(I^*) \ \text{for any } r \geq 1$$

and

$$\frac{d\phi}{dt^*}(t^*) = \frac{1}{2\sqrt{t^*}} \neq 0 \ \text{for all } t^* \in I^*,$$

and therefore, (1.20) defines an admissible parameter transformation. We obtain

$$\vec{x}^*(t^*) = \{t^*, t^{*3/2}\} \ (t^* \in (0, \infty))$$

for the part of Neil's parabola in the upper half plane; this is an admissible parametric representation.

Visualization 1.3.7 (The catenary).

The curve in which a uniform heavy cord or chain hangs is called a catenary *(Figure 1.11). It has a parametric representation of the form*

$$\vec{x}(t) = \left\{t, c \cosh\left(\frac{t}{c}\right)\right\} \ (t \in \mathbb{R}), \ \text{where } c \text{ is a constant (Figure 1.12)}. \tag{1.21}$$

Proof. We derive the parametric representation of a catenary in (1.21) (right in Figure 1.11).

Let a uniform heavy cord or chain be attached to two distinct points P_0 and P_2 at the same level, and P_1 denote the lowest point of the cord. We introduce a Cartesian coordinate system with its x–axis parallel to $\overline{P_0 P_2}$, such that $P_1 = (0, c)$ for some $c > 0$. Let P be a point on the catenary and ϕ denote the angle between the positive x–axis and the tangent to the catenary at P. We consider the equilibrium of the portion $P_1 P$ of the cord from P_1 to P. If the length of the arc from P_1 to P equals s, and w is the weight of the cord per unit length, the weight of $P_1 P$ is equal to ws. The other forces acting on $P_1 P$ are the tensions T_1 and T at P_1 and P, respectively, and the tangent at P_1 is horizontal. Resolving horizontally and vertically, we obtain

$$T \cos \phi = T_1 \tag{1.22}$$

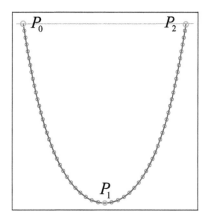 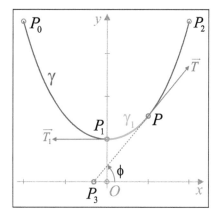

Figure 1.11 A suspended chain and a parametric representation for a catenary

and

$$T \sin \phi = ws. \tag{1.23}$$

If we put $T_1 = wc$, then dividing (1.23) by (1.22), we get

$$\tan \phi = \frac{s}{c} \text{ or } s = c \tan \phi; \tag{1.24}$$

this is called the *intrinsic equation of the catenary*. It gives the relation between the length of arc of the curve from the lowest point to any other point of the curve and the inclination of the tangent at the latter point.

We can find a parametric representation of the curve as follows (right in Figure 1.11). It follows from

$$\frac{dy}{dx} = \tan \phi \text{ and } \frac{dy}{ds} = \sin \phi$$

and (1.24) that

$$\frac{dy}{d\phi} = \frac{dy}{ds}\frac{ds}{d\phi} = \sin \phi \frac{c}{\cos^2 \phi},$$

and so

$$y = y(\phi) = \frac{c}{\cos \phi} + d, \text{ where } d \in \mathbb{R} \text{ is a constant of integration.}$$

Since $y(0) = c$, it follows that $d = 0$, hence

$$y = y(\phi) = \frac{c}{\cos \phi}. \tag{1.25}$$

We get from (1.25) and (1.24)

$$y^2 = \frac{c^2}{\cos^2 \phi} = c^2(1 + \tan^2 \phi) = c^2 + s^2, \tag{1.26}$$

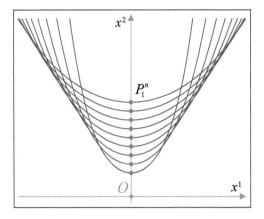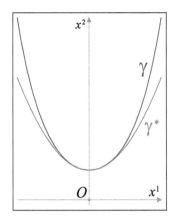

Figure 1.12 Left: catenaries for varying c. Right: a catenary γ and a quadratic parabola $\gamma*$

and then

$$\frac{dy}{dx} = \tan \phi = \frac{s}{c} = \pm \frac{\sqrt{y^2 - c^2}}{c}.$$

Taking the positive root, we have

$$\cosh^{-1}\left(\frac{y}{c}\right) = \int \frac{dy}{\sqrt{y^2 - c^2}} = \int \frac{dx}{c} - \frac{x}{c} + d,$$

where $d \in \mathbb{R}$ is a constant of integration, that is,

$$y = y(x) = c \cosh\left(\frac{x}{c} + d\right).$$

It follows from $y(0) = c$ that $d = 0$, hence

$$y = y(x) = c \cosh\left(\frac{x}{c}\right).$$

Writing $t = x$, we obtain the parametric representation in (1.21) for the catenary. □

A catenary with a parametric representation (1.21) looks very much like a quadratic parabola. The Taylor series expansion $T(f; x)$ of the function f with $f(x) = c \cosh(x/c)$ is

$$T(f; x) = c \cdot \sum_{k=0}^{\infty} \frac{(x/c)^{2k}}{(2k)!};$$

it converges to $f(x)$ on the whole real line. We consider the partial sums

$$T_{2n}(f; x) = c \cdot \sum_{k=0}^{n} \frac{(x/c)^{2k}}{(2k)!}$$

of $T(f; x)$. If $n = 1$, then $T_2(f; x)$ is a quadratic polynomial, and the quadratic parabola with a parametric representation

$$\vec{x}(t) = \{t, T_2(f; t)\} = \left\{t, c + \frac{t^2}{2c}\right\} \quad (t \in \mathbb{R})$$

is a second order approximation of the catenary for $|x|$ small (right in Figure 1.12).

1.4 SOME EXAMPLES OF PLANAR CURVES

In this section, we consider some examples of planar curves; most of them are defined by equations, such as algebraic curves. We also give the geometrical principles of their constructions.

If P and Q are given points in $\mathbb{R}^n$, then $\overline{PQ}$ denotes the straight line through P and Q. Throughout, we write $x = x^1$ and $y = x^2$.

Let n be a non–negative integer. Then the set

$$C_n = \left\{ X = (x, y) \in \mathbb{R}^2 : \sum_{0 \le k+j \le n} a_{jk} x^j y^k = 0 \ (a_{jk} \in \mathbb{R}) \right\}$$

is an *algebraic curve of order n*.

First, we consider algebraic curves of order 3.

Visualization 1.4.1 (The serpentine).
Let $S_r(M)$ be the circle line of radius $r > 0$ with its centre in M, and P_1 be a point on $S_r(M)$ such that the position vectors $\vec{m}$ and $\vec{p}_1$ of M and P_1 are not parallel. Furthermore, let P_2 be the intersection of the straight line $\overline{OP_1}$ and the straight line γ^* parallel to $\vec{m}$ at a distance a from M. Finally, let P denote the orthogonal projection of the point P_2 on the parallel of γ^* through P_1. Then the points P generate a serpentine, as the point P_1 moves along the circle line $S_r(M)$ (Figure 1.13). A serpentine $\gamma_{a,r}$ is given by the equation

$$f(x, y; a, r) = y \left(x^2 + a^2 \right) - 2arx = 0, \tag{1.27}$$

and has a parametric representation

$$\vec{x}_{a,r}(t) = \left\{ t, \frac{2art}{t^2 + a^2} \right\} \ (t \in \mathbb{R}). \tag{1.28}$$

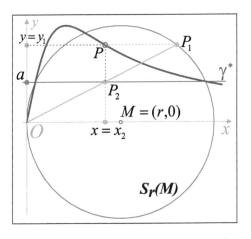
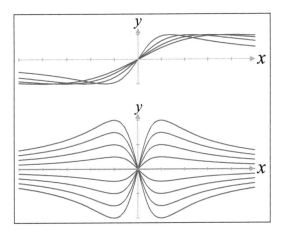

Figure 1.13 Left: the construction of a serpentine. Right: families of serpentines

Proof. We write $C = S_r(M)$, for short, and choose a Cartesian coordinate system such that $M = (r, 0)$ and γ^* is parallel to the x–axis. Let $P = (x, y)$, $P_1 = (x_1, y_1)$ and $P_2 = (x_2, y_2)$, and observe that $x = x_2$, $y_2 = a$ and $y = y_1$ by construction. It follows from elementary geometry that

$$\frac{y_1}{x_1} = \frac{y_2}{x_2}, \text{ hence } \frac{y}{x_1} = \frac{a}{x} \text{ and } x_1 = \frac{xy}{a}.$$

Since $P_1 \in C$, we have

$$(x_1 - r)^2 + y_1^2 = r^2, \text{ that is, } x_1^2 - 2rx_1 + y_1^2 = 0.$$

Substituting $x_1 = (xy)/a$ and $y_1 = y$ and multiplying by a^2, we obtain

$$x^2 y^2 - 2raxy + a^2 y^2 = 0,$$

and (1.27) follows. Writing $x = t$ and solving (1.27) for y, we obtain (1.28). □

Visualization 1.4.2 (The versiera).
Let $C = S_r(M)$ be the circle line of radius $r > 0$ with its centre in M, P_1 and P_2 be distinct points of C, γ_1^* be the straight line through M orthogonal to the vector $\overrightarrow{P_1 M}$, and P_3 be the point of intersection of γ_1^* with the straight line through P_1 and P_2. Furthermore, let P denote the intersection of the parallel γ_2^* of $\overrightarrow{P_1 M}$ through P_3 with the parallel γ_3^* of γ_1^* through P_2. Then the points P generate a versiera as the straight line through P_1 and P_2 is rotated about P_1 (Figure 1.14). A versiera γ_r is given by the equation

$$f(x, y; r) = y(x^2 + r^2) - 2r^3 = 0, \tag{1.29}$$

and has a parametric representation

$$\vec{x}_{a,r}(t) = \left\{ t, \frac{2r^3}{t^2 + r^2} \right\} \quad (t \in \mathbb{R}). \tag{1.30}$$

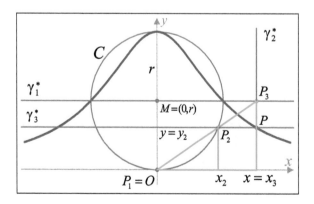 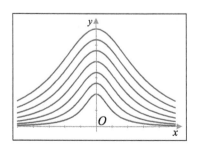

Figure 1.14 Left: The construction of a versiera. Right: families of versieras

Proof. We choose a Cartesian coordinate system with its origin in P_1 and its x–axis parallel to γ_1^*. Let $P_2 = (x_2, y_2)$, $P_3 = (x_3, y_3)$ and $P = (x, y)$. Then we have $x = x_3$, $y = y_2$ and $y_3 = r$ by construction, and it follows from elementary geometry that

$$\frac{y_2}{x_2} = \frac{y}{x_2} = \frac{y_3}{x_3} = \frac{r}{x}, \text{ hence } x_2 = \frac{yx}{r}.$$

Furthermore, $P_2 \in C$ implies

$$r^2 = x_2^2 + (y_2 - r)^2 = x_2^2 + y^2 - 2ry + r^2,$$

hence

$$x_2^2 r^2 + y^2 r^2 - 2r^3 y = y^2 x^2 + y^2 r^2 - 2r^3 y = y^2(x^2 + r^2) - 2r^3 y = 0,$$

and (1.29) follows. Writing $x = t$ and solving (1.29) for y, we obtain (1.30). □

Visualization 1.4.3 (Cissoids).
Let γ_1 and γ_2 be curves, Q be a point and γ be a straight line that intersects γ_1 in P_1 and γ_2 in P_2. A cissoid is generated by the points P for which

$$\left\| \overrightarrow{QP} \right\| = \left\| \overrightarrow{P_1 P_2} \right\|, \tag{1.31}$$

as γ is rotated about the point Q.
We consider the following special cases.

(a) Diocles's cissoid
    If Q is the origin, γ_1 the circle of radius $r > 0$ and its centre in $M = (r, 0)$ and γ_2 is the straight line through the point $(2r, 0)$ parallel to the y–axis, then we obtain Diocles's cissoid (left in Figures 1.15 and 1.16); it is given by the equation

$$f(x, y; r) = x^3 + xy^2 - 2ry^2 = 0, \tag{1.32}$$

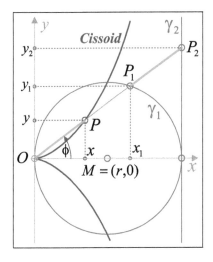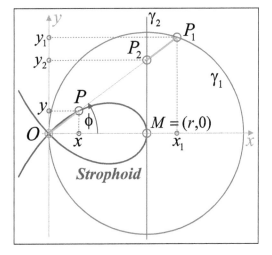

Figure 1.15 The construction of: Left: Diocles's cissoid. Right: a strophoid

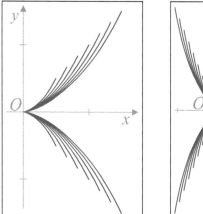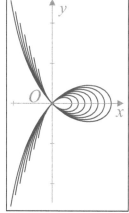

Figure 1.16 Families of Diocles's cissoids and strophoids

or in polar coordinates (ρ, ϕ) with $x = \rho \cos \phi$ and $y = \rho \sin \phi$

$$\rho(\phi) = 2r \frac{\sin^2 \phi}{\cos \phi} \quad \left(\phi \in \left(-\frac{\pi}{2}, \frac{\pi}{2} \right) \right). \tag{1.33}$$

This yields a parametric representation

$$\vec{x}(t) = \left\{ 2r \sin^2 t, 2r \sin^2 t \tan t \right\} \quad \left(t \in \left(-\frac{\pi}{2}, \frac{\pi}{2} \right) \right), \tag{1.34}$$

which is not admissible at $t = 0$.

(b) Strophoids
Choosing the point Q and γ_1 as in Part (a) and γ_2 as the straight line through the point $(r, 0)$ parallel to the y–axis, we obtain a strophoid *(right in Figures 1.15 and 1.16); it is given by the equation*

$$f(x, y; r) = x^3 + xy^2 - rx^2 + ry^2 = 0, \tag{1.35}$$

or in polar coordinates (ρ, ϕ)

$$\rho(\phi) = r \frac{\cos(2\phi)}{\cos \phi} \quad \left(\phi \in \left(-\frac{\pi}{2}, \frac{\pi}{2} \right) \right). \tag{1.36}$$

This yields a parametric representation

$$\vec{x}(t) = \{ r \cos(2t), r \cos(2t) \tan t \} \quad \left(t \in \left(-\frac{\pi}{2}, \frac{\pi}{2} \right) \right), \tag{1.37}$$

which is admissible everywhere.

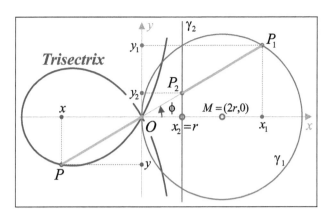
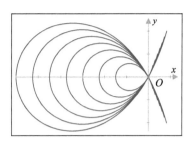

Figure 1.17 Left: The construction of MacLaurin's trisectrix. Right: A family of MacLaurin's trisectrices

(c) MacLaurin's trisectrix
Choosing $Q = (0,0)$, γ_1 as the circle of radius $2r$ and its centre in $(2r, 0)$ and γ_2 as the straight line through the point $(r, 0)$ parallel to the y–axis, we obtain MacLaurin's trisectrix (Figure 1.17); it is given by the equation

$$f(x, y; r) = x^3 + xy^2 + ry^2 - 3rx^2 = 0, \tag{1.38}$$

or in polar coordinates (ρ, ϕ)

$$\rho(\phi) = r \frac{\sin^2 \phi - 3\cos^2 \phi}{\cos \phi} \quad \left(\phi \in \left(-\frac{\pi}{2}, \frac{\pi}{2} \right) \right). \tag{1.39}$$

This yields a parametric representation

$$\vec{x}(t) = \left\{ r(\sin^2 t - 3\cos^2 t), r(\sin^2 t - 3\cos^2 t)\tan t \right\} \quad \left(t \in \left(-\frac{\pi}{2}, \frac{\pi}{2} \right) \right). \tag{1.40}$$

Proof.

(a) We write $\overrightarrow{OP} = \{x, y\}$ and obtain for $x > 0$

$$\overrightarrow{OP_2} = \frac{2r}{x} \overrightarrow{OP}, \; \overrightarrow{OP_1} = \frac{2rx}{(d(0, P))^2} \overrightarrow{OP} \text{ and } \overrightarrow{P_1 P_2} = \frac{2ry^2}{x(x^2 + y^2)} \overrightarrow{OP},$$

and (1.31) yields

$$1 = \frac{2ry^2}{x(x^2 + y^2)} \text{ or } (1.32).$$

Substituting $x = \rho \cos \phi$ and $y = \rho \sin \phi$ in (1.32), we get

$$\rho^3 \cos^3 \phi + \rho^3 \cos \phi \sin^2 \phi = \rho^3 \cos \phi \left(\cos^2 \phi + \sin^2 \phi \right) = \rho^3 \cos \phi = 2r^2 \rho^2 \sin^2 \phi$$

and (1.33) follows, since $\rho > 0$. We put $t = \phi$ and (1.34) is an immediate consequence. Finally, it follows from (1.34) that

$$\vec{x}'(t) = \left\{4r\sin t \cos t, 4r\sin t \cos t \tan t + 2r\sin^2 t \cos^2 t\right\}$$
$$= 2r\left\{\sin^2(2t), \sin^2(2t) + \tan^2(2t)\right\},$$

hence $\vec{x}'(0) = \vec{0}$, and the parametric representation (1.34) is not admissible at $t = 0$.

(b) Similarly as in Part (a), we obtain for $x > 0$

$$\overrightarrow{OP_2} = \frac{r}{x}\overrightarrow{OP}, \ \overrightarrow{OP_1} = \frac{2rx}{(d(0,P))^2}\overrightarrow{OP} \text{ and } \overrightarrow{P_1P_2} = \frac{r(x^2 - y^2)}{x(x^2 + y^2)}\overrightarrow{OP},$$

and (1.31) yields

$$1 = \frac{r(x^2 - y^2)}{x(x^2 + y^2)} \text{ or } (1.35).$$

Substituting $x = \rho\cos\phi$ and $y = \rho\sin\phi$ in (1.35), we get

$$\rho^3\cos\phi = r\rho^2\left(\cos^2\phi - \sin^2\phi\right) = r\rho^2\cos(2\phi)$$

(1.36) follows, since $\rho > 0$. We put $t = \phi$ and (1.37) is an immediate consequence. Finally, it follows from (1.37) that

$$\vec{x}'(t) = \left\{-2r\sin(2t), -2r\sin(2t)\tan t + \frac{r\cos(2t)}{\cos^2 t}\right\} \neq \vec{0} \text{ for all } t \in \left(-\frac{\pi}{2}, \frac{\pi}{2}\right).$$

(c) Part (c) follows in the same way as Parts (a) and (b).

□

Now we consider algebraic curves of order 4.

Visualization 1.4.4 (Conchoids).
The conchoid of a given curve γ is obtained by extending the position vector of every point of γ by a constant length $\pm r$. If the curve γ is given by the equation in polar coordinates

$$f_\gamma(\rho, \phi) = \rho - g(\phi) = 0$$

for some function g, then the equation for the conchoid $\gamma_{\pm r}$ of γ is

$$f_{\pm r}(\rho, \phi) = \rho - (g(\phi) \pm r) = 0.$$

We consider some special cases.

(a) Nikomedes's conchoid
Nikomedes's conchoid is the conchoid of the straight line through the point $(a, 0)$,

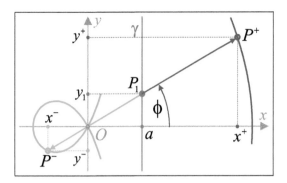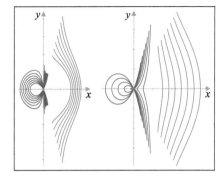

Figure 1.18 The construction (left) and families of Nikomedes's conchoids (right)

orthogonal to the x–axis (Figure 1.18). Putting $t = \phi$ where ϕ is the polar angle, we obtain the following parametric representation for Nikomedes's conchoid

$$\vec{x}(t) = \{a \pm r\cos t, a\tan t \pm r\sin t\} \quad \left(t \in \left(-\frac{\pi}{2}, \frac{\pi}{2}\right)\right), \tag{1.41}$$

or the equation in polar coordinates

$$f_{\pm r}(\rho, \phi; a) = \rho - \left(\frac{a}{\cos\phi} \pm r\right) = 0 \quad \left(\phi \in \left(-\frac{\pi}{2}, \frac{\pi}{2}\right)\right).$$

This yields the equation in Cartesian coordinates

$$f_{\pm r}^*(x, y; a) = (x - a)^2 \left(x^2 + y^2\right) - r^2 x^2 = 0.$$

(b) Pascal's snail

Pascal's snail *is the conchoid of the circle of radius $a/2 > 0$ with its centre in the point $(a/2, 0)$ (Figure 1.19). Putting $t = \phi$ where ϕ is the polar angle, we obtain the following parametric representation for Pascal's snail*

$$\vec{x}(t) = \{a\cos^2(t) \pm r\cos t, a\sin t\cos t \pm r\sin t\} \quad \left(t \in \left(-\frac{\pi}{2}, \frac{\pi}{2}\right)\right) \tag{1.42}$$

or the equation in polar coordinates

$$f_{\pm r}(\rho, \phi; a) = \rho - (a\cos\phi \pm r) = 0 \quad \left(\phi \in \left(-\frac{\pi}{2}, \frac{\pi}{2}\right)\right).$$

This yields the equation in Cartesian coordinates

$$f_{\pm r}^*(x, y; a) = \left(x^2 + y^2 - ax\right)^2 - r^2 \left(x^2 + y^2\right) = 0.$$

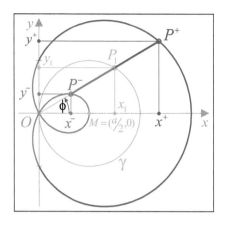 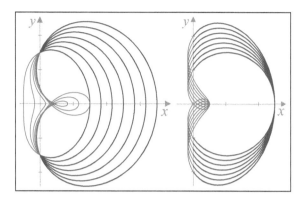

Figure 1.19 The construction (left) and families of families of Pascal's snails (right)

Visualization 1.4.5 (The κ–curve).

Let γ^* be straight line, P_0 be a point on γ^* and $\mathcal{C}$ be a family of circles lines of a fixed radius $r > 0$ with their centres on γ^*. A κ–curve is generated by the points where the tangents from P_0 to the circle lines of $\mathcal{C}$ touch the circles (Figure 1.20). A κ–curve is given by the equation

$$f(x, y; r) = y^4 + x^2 y^2 - r^2 x^2 = 0, \tag{1.43}$$

and has a parametric representation for two branches

$$\vec{x}^{(k)}(\phi) = \left\{ r \frac{|\cos \phi|}{|\sin \phi|} \cos \phi, r \frac{|\cos \phi|}{|\sin \phi|} \sin \phi \right\} \quad (\phi \in I_k)$$

$$\text{for } k = 1, 2 \text{ where } I_1 = (0, \pi) \text{ and } I_2 = (\phi, 2\pi). \tag{1.44}$$

Proof. We choose a Cartesian coordinate system with its origin in the point P_0 and its x–axis along γ^*. Let $C_a \in \mathcal{C}$ be the circle with its centre in $M_a = (a, 0)$ and

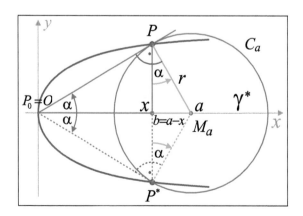 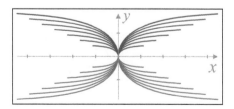

Figure 1.20 Left: The construction a κ–curve. Right: A family of κ–curves

$P = (x, y)$, $P^* = (x, y^*)$ be the points where the tangents from P_0 touch the circle, $b = a - x$, and α denote the angle between the x–axis and the tangents. Then we have $y^* = -y$. It follows that

$$\sin \alpha = \frac{b}{r} = \frac{r}{a}, \text{ that is, } b = \frac{r^2}{a} \text{ and } b(x + b) = r^2, \text{ hence } b^2 = r^2 - bx.$$

Since $y^2 = (y^*)^2 = r^2 - b^2$, we obtain

$$y^2 = (y^*)^2 = r^2 - (r^2 - bx) = bx. \tag{1.45}$$

Furthermore, the quadratic equation $b^2 + bx - r^2 = 0$ has the solutions

$$b = \frac{1}{2}\left(\pm\sqrt{x^2 + 4r^2} - x\right),$$

so (1.45) yields

$$y^2 + \frac{x^2}{2} = \pm\frac{x}{2}\sqrt{x^2 + 4r^2} \text{ and } \left(y^2 + \frac{x^2}{2}\right)^2 - \frac{x^2}{4}\left(x^2 + 4r^2\right) = 0$$

and finally

$$y^4 + x^2 y^2 - r^2 x^2 = 0.$$

We introduce polar coordinates ρ and ϕ, that is, $x = \rho\cos\phi$ and $y = \rho\sin\phi$, and obtain from (1.43)

$$\rho^4 \sin^4\phi + \rho^4 \sin^2\phi\cos^2\phi - r^2\rho^2\cos^2\phi = \rho^2\left(\rho^2\sin^2\phi(\sin^2\phi + \cos^2\phi) - r^2\cos^2\phi\right)$$
$$= \rho^2\left(\rho^2\sin^2\phi - r^2\cos^2\phi\right) = 0,$$

and, since $\rho > 0$,

$$\rho = r\frac{|\cos\phi|}{|\sin\phi|} \text{ for } \phi \neq k\pi \ (k \in \mathbb{Z}),$$

and (1.44) is an immediate consequence. □

Visualization 1.4.6 (Cassini curves and lemniscates).
Let P_1 and P_2 be distinct points and $c = (1/2)d(P_1, P_2)$.

(a) *A Cassini curve is generated by the points P with a constant product a^2 of their distances from P_1 and P_2 (Figures 1.21 and 1.22); it is given by the equation*

$$f(x, y; a, c) = \left(x^2 + y^2\right)^2 - 2c^2\left(x^2 - y^2\right) + c^4 - a^4 = 0, \tag{1.46}$$

or in polar coordinates by

$$g(\rho, \phi; a, c) = \rho^4 - 2c^2\rho^2\cos(2\phi) + c^4 - a^4 = 0. \tag{1.47}$$

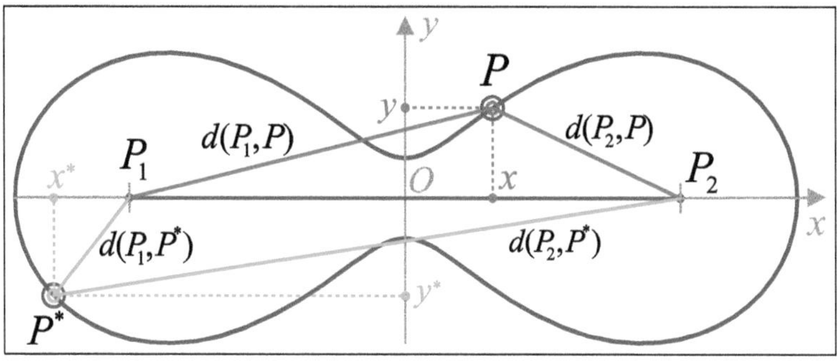

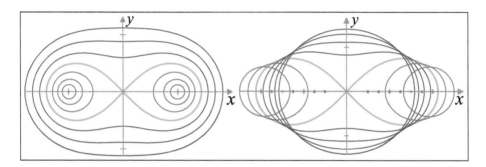

Figure 1.21 The construction of a Cassini curve

Figure 1.22 Families of Cassini curves

(b) A lemniscate *is the special case* $a = c$ *of a Cassini curve; it is given by the equation*

$$f(x, y; c) = \left(x^2 + y^2\right)^2 - 2c^2\left(x^2 - y^2\right) = 0, \tag{1.48}$$

or parametric representations in polar coordinates

$$\vec{x}^{(k)}(\phi) = \left\{\sqrt{2}c\sqrt{\cos(2\phi)}\cos\phi, \sqrt{2}c\sqrt{\cos(2\phi)}\sin\phi\right\} \ for \ \phi \in I_k \ (k = 1, 2, 3),$$

where $I_1 = (0, \pi/4)$, $I_2 = (3\pi/4, 5\pi/4)$ *and* $I_3 = (7\pi/4, 2\pi)$. (1.49)

Proof. We choose a Cartesian coordinate system with its origin in the midpoint of the straight line segment $\overline{P_1 P_2}$ and its x–axis along $\overline{P_1 P_2}$. Then $P_1 = (-c, 0)$, $P_2 = (c, 0)$ and the distances of $P = (x, y)$ from P_1 and P_2 are given by

$$d(P_1, P) = \sqrt{(x + c)^2 + y^2} \text{ and } d(P_2, P) = \sqrt{(x - c)^2 + y^2}.$$

The points P of a Cassini curve must satisfy by definition

$$a^4 = (d(P_1, P)d(P_2, P))^2 = \left((x + c)^2 + y^2\right)\left((x - c)^2 + y^2\right)$$
$$= \left(x^2 - c^2\right)^2 + \left((x + c)^2 + (x - c)^2\right)y^2 + y^4$$

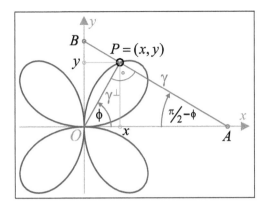 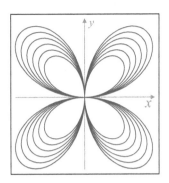

Figure 1.23 Left: the construction of a rosette. Right: a family of rosettes

$$= x^4 - 2c^2x^2 + c^4 + 2y^2\left(x^2 + c^2\right) + y^4$$
$$= \left(x^2 + y^2\right)^2 - 2c^2\left(x^2 - y^2\right) + c^4,$$

which yields (1.46). Introducing polar coordinates ρ and ϕ, we obtain with $x = \rho\cos\phi$ and $y = \rho\sin\phi$

$$g(\rho, \phi; a, c) = \rho^4 - 2c^2\rho^2\left(\cos^2\phi - \sin^2\phi\right) + c^4 - a^4$$
$$= \rho^4 - 2c^2\rho^2\cos(2\phi) + c^4 - a^4 = 0,$$

that is, (1.47).

If $a = c$ then (1.48) is an immediate consequence of (1.46), and solving (1.47) for ρ, we obtain $\rho = \sqrt{2}c\sqrt{\cos(2\phi)}$ for $\cos(2\phi) \geq 0$, which yields (1.49) and the interval I_k $(k = 1, 2, 3)$. □

Now we consider algebraic curves of order 5.

Visualization 1.4.7 (Rosettes).
Let $\gamma = \overline{AB}$ be the straight line segment of a fixed length $a > 0$ with its end points A and B on the x^1- and x^2-axes of a Cartesian coordinate system of $\mathbb{R}^2$. Furthermore, let $P = \gamma \cap \gamma^\perp$, where $\gamma^\perp$ is the straight line orthogonal to γ and through the origin O. A rosette is the curve generated by the points P, when the end points of γ move along axes of the coordinate system (Figure 1.23); it is given by the equation

$$f(x, y; a) = \left(x^2 + y^2\right)^3 - a^2\left(x^2y^2\right)^2 = 0, \tag{1.50}$$

and has a parametric representation in polar coordinates

$$\vec{x}(\phi) = \frac{a}{2}\{\sin(2\phi)\cos\phi, \sin(2\phi)\sin\phi\} \ (\phi \in (0, 2\pi)). \tag{1.51}$$

Proof. We obtain

$$d(O, B) = a\cos\phi, \ \rho = d(O, P) = d(O, B)\sin\phi = a\sin\phi\cos\phi = \frac{a}{2}\sin 2(\phi)$$

and (1.51) is an immediate consequence.

It also follows from

$$x = (a/2) \sin (2\phi) \cos \phi \text{ and } y = (a/2) \sin (2\phi) \sin \phi$$

that

$$\left(x^2 + y^2\right)^3 = \frac{a^6}{64} \sin^6 (2\phi)$$

and

$$a^2 \left(x^2 y^2\right)^2 = a^2 \left(\frac{a^2}{4} \sin^2 (2\phi) \cos \phi \sin \phi\right)^2$$

$$= \frac{a^6}{16} \sin^4 (2\phi) \left(\frac{\sin (2\phi)}{2}\right)^2 = \frac{a^6}{64} \sin^6 (2\phi),$$

and (1.50) is an immediate consequence. □

Visualization 1.4.8 (Double egg lines).
Let C be a circle line of radius $a > 0$ and its centre in the origin, and let A and B be distinct points of C. Furthermore, let F be the orthogonal projection of $\overrightarrow{OB}$ on $\overrightarrow{OA}$ and P be the orthogonal projection of F on $\overrightarrow{OB}$. A double egg line is generated by the points P, when B moves along C (Figure 1.24); it is given by the equation

$$f(x, y; a) = \left(x^2 + y^2\right)^3 - a^4 x^2 = 0, \tag{1.52}$$

and has a parametric representation

$$\vec{x}(\phi) = a\{\cos^3 \phi, \cos^2 \phi \sin \phi\} \ (\phi \in (0, 2\pi)). \tag{1.53}$$

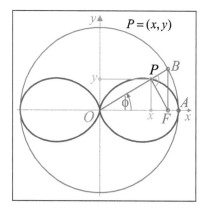 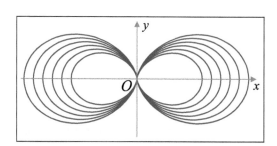

Figure 1.24 Left: the construction of a double egg line. Right: a family of double egg lines

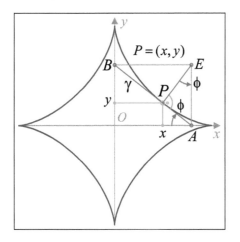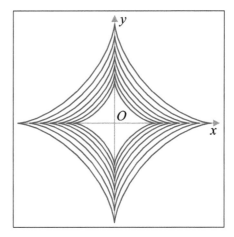

Figure 1.25 Left: the construction of an astroid. Right: a family of astroids

Proof. We choose polar coordinates (ρ, ϕ) such that $A = (0, a)$, and $B = (a \cos \phi, a \sin \phi)$. It follows that $F = (a \cos \phi, 0)$ and

$$P = (\rho \cos \phi, \rho \sin \phi), \text{ where } \rho = d(F, 0) \cos \phi = a \cos^2(\phi)$$

and (1.53) is an immediate consequence. If follows from $x = a \cos^3 \phi$ and $y = a \cos^2 \phi \sin \phi$ that

$$\left(x^2 + y^2\right)^3 - a^4 x^2 = a^6 \left(\cos^2 \phi (\cos^2 \phi + \sin^2 \phi)\right)^3 - a^6 \cos^6 \phi$$
$$= a^6 \cos^6 \phi - a^6 \cos^6 \phi = 0,$$

and (1.52) is satisfied. □

Visualization 1.4.9 (Astroids).

Let γ be the straight line segment of a fixed length $a > 0$ with its end points A and B on the x^1- and x^2–axes of a Cartesian coordinate system of $\mathbb{R}^2$. Furthermore, let P be the orthogonal projection of the point E of the rectangle $BOAE$ on γ. An astroid is generated by the points P, when the end points of γ move along the coordinate axes (Figure 1.25); it is given by the equation

$$(x^1)^{2/3} + (x^2)^{2/3} - a^{2/3} = 0. \tag{1.54}$$

Proof. We obtain

$$d(E, B) = a \cos \phi, \ d(P, B) = d(E, B) \cos \phi = a \cos^2 \phi,$$

hence

$$x = d(P, B) \cos \phi = a \cos^3 \phi \text{ and similarly } y = a \sin^3 \phi,$$

and (1.54) is an immediate consequence. □

Remark 1.4.10. *If* $(p) = (p_1, p_2, \ldots, p_n)$ *is an* n*–tuple of positive real numbers* $p_1, p_2,$ $\ldots, p_n,$ *and* $H = \max\{1, p_1, \ldots, p_n\},$ *then*

$$d_{(p)}(X, Y) = \left(\sum_{k=1}^{n} |x^k - y^k|^{p_k} \right)^{1/H} \quad (X = (x^1, \ldots, x^n), Y = (y^1, \ldots, y^n) \in \mathbb{R}^n) \tag{1.55}$$

defines a metric $d_{(p)}$ *on* $\mathbb{R}^n$.

In the special case, where $(p) = (p, p, \ldots, p)$ *for some* $p \geq 1$, *then*

$$\|\vec{x}\|_p = d_p(X, 0) = \left(\sum_{k=1}^{n} |x^k|^p \right)^{1/p}$$

is a norm on $\mathbb{R}^n$. *The sets*

$$\partial B_{(p)}(r; X_0) = \{X \in \mathbb{R}^n : d_{(p)}(X, X_0) = r\} \quad (r > 0; X_0 \in \mathbb{R}^n)$$

are the boundaries of neighbourhoods in the metric $d_{(p)}$. *Thus the astroids of Visualization 1.4.9 are a boundaries of neighbourhoods of the origin in the metric* $d_{2/3}$ *of* $\mathbb{R}^2$.

We close this section by introducing the concept of *orthogonal trajectories*. Let Γ be a family of planar curves γ_c in an open connected set $D \subset \mathbb{R}^2$ such that there is one and only one curve of Γ through every point of D. Then the curves $\gamma^\perp$ that intersect every curve of Γ at right angles are called the *orthogonal trajectories of* Γ. If the curves γ_c of the family Γ are given by the equations $f(x, y; c) = 0$, then the orthogonal trajectories of Γ are obviously given by

$$f_1(x, y; c)\frac{dy}{dx} - f_2(x, y; c) = 0, \text{ where} \tag{1.56}$$

$$f_1(x, y; c) = \frac{\partial f(x, y; c)}{\partial x} \text{ and } f_2(x, y; c) = \frac{\partial f(x, y; c)}{\partial y}.$$

Visualization 1.4.11. *Let* $\alpha, \beta \neq 2$ *be positive reals, and* $\Gamma_{\alpha,\beta}$ *be a family of curves given by the equations*

$$f(x, y; c) = x^\alpha + y^\beta - c = 0.$$

Then the orthogonal trajectories $\gamma^\perp$ *of* $\Gamma_{\alpha,\beta}$ *are given by the equations*

$$g(x, y; d) = \frac{1}{\alpha(2 - \alpha)}x^{2-\alpha} - \frac{1}{\beta(2 - \beta)}y^{2-\beta} - d = 0, \tag{1.57}$$

where $d \in \mathbb{R}$ *is a constant (Figure 1.26).*

If $\alpha = \beta = 2$, *then the orthogonal trajectories of* $\Gamma_{2,2}$ *are given by*

$$f(x, y; d) = y - dx = 0,$$

that is, the orthogonal trajectories of concentric circles with their centres in the origin are straight lines through the origin.

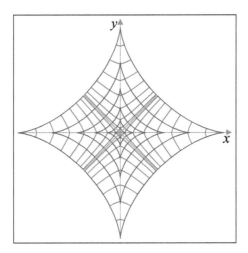

Figure 1.26 A family of astroids and their orthogonal trajectories

Proof. The orthogonal trajectories of $\Gamma_{\alpha,\beta}$ must satisfy by (1.56)

$$\alpha x^{\alpha-1}\frac{dy}{dx} - \beta y^{\beta} = 0,$$

hence for $\alpha, \beta \neq 2$

$$\frac{1}{\alpha(\alpha-2)}x^{2-\alpha} = \frac{1}{\alpha}\int x^{1-\alpha}\,dx = \frac{1}{\beta}\int y^{1-\beta}\,dy = \frac{1}{\beta(2-\beta)}y^{2-\beta} + d$$

for some constant of integration d. This yields (1.57).
If $\alpha = \beta = 2$, then we obtain from (1.56)

$$x\frac{dy}{dx} - y = 0, \text{ that is } \log|x| = \log|y| + \log\tilde{d} \text{ for some constant } \tilde{d} > 0.$$

This yields the statement for $\alpha = \beta = 2$. □

1.5 THE ARC LENGTH OF A CURVE

The *arc length of a curve* plays an eminent role as a parameter of a curve due to its significance as a *geometric invariant*.

Definition 1.5.1. Let γ be a curve with a parametric representation $\vec{x}(t)$ $(t \in I)$. Then the *arc length of γ between the points $X(t_0)$ and $X(t)$ $(t_0, t \in I)$* is given by

$$s(t) = \int_{t_0}^{t} \|\vec{x}'(\tau)\|\,d\tau. \tag{1.58}$$

We always denote the arc length of a curve by s, and use a dot to indicate differentiation with respect to s, that is, we write

$$\dot{\vec{x}}(s) = \frac{d\vec{x}}{ds}(s).$$

Proposition 1.5.2. *(a) Equation (1.58) defines an admissible parameter transformation.*
(b) The arc length of a curve is invariant under admissible parameter transformations.
(c) Every curve can be parameterized with respect to its arc length.
(d) If $\vec{x}(s)$ $(s \in I)$ is a parametric representation of a curve γ with respect to its arc length s then

$$\|\dot{\vec{x}}(s)\| = 1 \quad \text{for all } s.$$

Conversely, if $\vec{x}(t)$ $(t \in [a,b])$ is a parametric representation of a curve γ and $\|\vec{x}'(t)\| = 1$ for all $t \in [a,b]$, then the value $t - a$ is equal to the arc length of γ between its initial point $X(a)$ and the point $X(t)$.

This means that if the parameter interval I is divided into subintervals by the equidistant points s_k $(k = 0, 1, \ldots, n)$, then the arcs of the curve between $X(s_{k-1})$ and $X(s_k)$ $(k = 1, 2, \ldots, n)$ have equal lengths (Figure 1.27); or kinematically, a point moves along the curve with constant speed.

Proof.

(a) Let γ be a curve with an admissible parametric representation $\vec{x}(t)$ $(t \in I)$ and $t_0 \in I$ be given. We write $\phi : I \to \mathbb{R}$ for the function with

$$\phi(t) = s(t) = \int_{t_0}^{t} \|\vec{x}'(\tau)\| \, d\tau \text{ for all } t \in I.$$

Since $\|\vec{x}'(t)\| > 0$ for all $t \in I$ by (1.14) in Definition 1.3.4, the inverse function $\phi^{-1} : I^* = \phi(I) \to \mathbb{R}$ exists, is differentiable on I^*, and we have for every $s_0 = \phi(t_0) \in I^*$

$$\frac{d\phi^{-1}}{ds}(s_0) = \left(\frac{d\phi}{dt}(t_0)\right)^{-1} = \|\vec{x}'(t_0)\|^{-1} \neq 0.$$

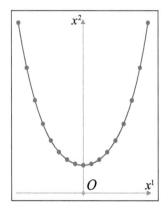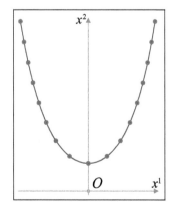

Figure 1.27 A catenary with parameters t (left) and s (right)

It also follows from this that if $\vec{x} \in C^r(I)$ for some $r \geq 1$, then $\phi \in C^r(I^*)$.

(b) Now let γ be a curve with a parametric representation as in Part (a), and $t = \psi(t^*)$ $(t^* \in I^*)$ with $\psi(I^*) = I$ be an admissible parameter transformation. Then $\vec{x}^*(t^*) = \vec{x}(\psi(t^*))$ $(t \in I^*)$ is a parametric representation of γ. Let $t_0^*, t_1^* \in I^*$ be arbitrary parameters, and $t_0 = \psi(t_0^*)$ and $t_1 = \psi(t_1^*)$. Then the rule of substitution yields the arc length $s^*(t_0^*, t_1^*)$ along γ from t_0^* to t_1^* in terms of the parameters t_0 and t_1

$$s^*(t_0^*, t_1^*) = \int_{t_0^*}^{t_1^*} \left\| \frac{d\vec{x}^*}{dt^*} \right\| dt^* = \int_{t_0^*}^{t_1^*} \left\| \frac{d\vec{x}}{d\psi}(\psi(t^*)) \frac{d\psi}{dt^*}(t^*) \right\| dt^*$$

$$= \int_{\psi(t_0^*)}^{\psi(t_1^*)} \left\| \frac{d\vec{x}}{d\psi}(\psi) \right\| d\psi = s(\psi(t_0^*), \psi(t_1^*)) = s(t_0, t_1).$$

This shows that the arc length is invariant under admissible parameter transformations.

(c) Part (c) is an immediate consequence of Part (a).

(d) If γ is a curve with a parametric representation $\vec{x}(t)$ $(t \in I)$ and t is the arc length along γ, then $s = t$ and it follows from (1.58) in Definition 1.5.1 that

$$\frac{ds}{dt}(t) = 1 = \|\vec{x}'(t)\| = \|\dot{\vec{x}}(s)\| \text{ for all } s \in I.$$

The converse part is obvious.

□

Owing to its geometric significance the parameter s is referred to as the *natural parameter*. It is convenient to use s as a parameter for theoretical purposes, since the formulae with respect to s are less complicated.

Geometric properties of a curve should not depend on some particular choice of its representation, they must be invariant under

(i) *transformations of the coordinate system,*

(ii) *admissible parameter transformations.*

Then the invariance of the conditions in (i) and (ii) is satisfied by the use of vectors and the use of the arc length of a curve as parameter, respectively.

Example 1.5.3. *(a) The arc length along the straight line in Visualization 1.3.3 (a) measured from the point A is given by*

$$s(t) = \int_0^t \|\vec{b}\| \, d\tau = \|\vec{b}\| t \ (t \in \mathbb{R}), \text{ and } t(s) = \frac{s}{\|\vec{b}\|} \ (s \in \mathbb{R}).$$

So a parametric representation of the straight line with respect to its arc length is given by

$$\vec{x}^*(s) = \vec{a} + \frac{\vec{b}}{\|\vec{b}\|} s \ (s \in \mathbb{R}).$$

(b) The arc length along the circle line in Visualization 1.3.3 (b) is given by

$$s(t) = \int_0^t r \, d\tau = rt \ (t \in [0, 2\pi]), \ and \ t(s) = \frac{s}{r} \ (s \in [0, 2\pi r]).$$

So a parametric representation of the circle line of Visualization 1.3.3 (b) with respect to the arc length is given by

$$\vec{x}^*(s) = \vec{m} + \vec{e}^1 r \cos\left(\frac{s}{r}\right) + \vec{e}^2 r \sin\left(\frac{s}{r}\right) \ (s \in [0, 2\pi r]).$$

(c) The arc length along the helix of Visualization 1.3.3 (c) is given by

$$s(t) = \int_0^t \sqrt{r^2 + h^2} \, d\tau = \sqrt{r^2 + h^2} \, t \ (t \in \mathbb{R})$$

and

$$t(s) = \frac{s}{\sqrt{r^2 + h^2}} \ (s \in \mathbb{R}).$$

We put $\omega = 1/\sqrt{r^2 + h^2}$. Then a parametric representation of the helix with respect to its arc length is given by

$$\vec{x}^*(s) = \{r \cos(\omega s), r \sin(\omega s), h\omega s\} \ (s \in \mathbb{R}).$$

(d) We consider the catenary of Visualization 1.3.7 with a parametric representation (1.21). Then it it follows from

$$\vec{x}'(t) = \left\{1, \sinh\left(\frac{t}{c}\right)\right\}$$

and

$$\|\vec{x}'(t)\|^2 = 1 + \sinh^2\left(\frac{t}{c}\right) = \cosh^2\left(\frac{t}{c}\right)$$

that

$$s(t) = \int_0^t \cosh\left(\frac{\tau}{c}\right) d\tau = c \sinh\left(\frac{t}{c}\right),$$

and so

$$t = t(s) = c \sinh^{-1}\left(\frac{s}{c}\right).$$

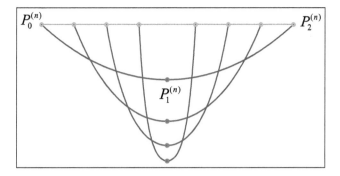

Figure 1.28 Suspended chains of constant length

Now

$$c \cosh\left(\frac{t}{c}\right) = c\sqrt{1 + \sinh^2\left(\frac{t}{c}\right)} = c\sqrt{1 + \frac{s^2}{c^2}} = \sqrt{c^2 + s^2},$$

and (1.21) yields the following parametric representation of the catenary with respect to its arc length s (Figure 1.28)

$$\vec{x}^*(s) = \left\{c\sinh^{-1}\left(\frac{s}{c}\right), \sqrt{c^2 + s^2}\right\} \quad (s \in \mathbb{R}). \tag{1.59}$$

Example 1.5.4. *We define the functions* $\phi, h : (0, \infty) \to \mathbb{R}$ *by*

$$\phi(t) = \log\left(e^t + \sqrt{e^{2t} - 1}\right) \text{ and } h(t) = \int \sqrt{1 - e^{-2t}}\, dt.$$

Let γ be the curve with a parametric representation

$$\vec{x}(t) = \left\{e^{-t} \cos \phi(t), e^{-t} \sin \phi(t), h(t)\right\} \text{ for } t > 0. \tag{1.60}$$

Then the parametric representation of γ with respect to the arc length s along γ is

$$\vec{x}^*(s) = \left\{\frac{1}{\cosh s} \cos s, \frac{1}{\cosh s} \sin s, s - \tanh s\right\} \text{ for } s > 0. \tag{1.61}$$

Proof. We write

$$\vec{y}(t) = \left\{-e^{-t} \cos \phi(t), -e^{-t} \sin \phi(t), h'(t)\right\}$$

and

$$\vec{z}(t) = e^{-t}\left\{-\sin \phi(t), \cos \phi(t), 0\right\} \phi'(t),$$

where

$$\phi'(t) = \frac{1}{e^t + \sqrt{e^{2t} - 1}}\left(e^t + \frac{e^{2t}}{\sqrt{e^{2t} - 1}}\right) = \frac{e^t}{e^t + \sqrt{e^{2t} - 1}}\left(\frac{\sqrt{e^{2t} - 1} + e^t}{\sqrt{e^{2t} - 1}}\right)$$

$$= \frac{e^t}{\sqrt{e^{2t} - 1}} = \frac{1}{\sqrt{1 - e^{-2t}}} = \frac{1}{h'(t)},$$

hence, since obviously $\vec{y}(t) \perp \vec{z}(t)$,

$$\|\vec{x}\,'(t)\|^2 = \|\vec{y}(t)\|^2 + \|\vec{z}(t)\|^2 = e^{-2t} + (h'(t))^2 + e^{-2t}(\phi'(t))^2 = 1 + \frac{e^{-2t}}{1 - e^{-2t}}$$

$$= \frac{1}{1 - e^{-2t}} = \frac{1}{(h'(t))^2}.$$

This implies

$$s(t) = \int \|\vec{x}\,'(t)\| \, dt = \int \frac{1}{h'(t)} \, dt = \int \phi'(t) \, dt = \phi(t),$$

and so

$$e^s = \exp \phi(t) = e^t + \sqrt{e^{2t} - 1},$$

hence

$$e^{2s} - 2e^{s+t} + e^{2t} = e^{2t} - 1, \text{ that is, } e^{2s} - 2e^s e^t = -1,$$

and then

$$e^t = \frac{e^{2s} + 1}{2e^s} = \cosh s.$$

Finally, we obtain

$$h(t(s)) = \int \left(\frac{dh}{dt}(t(s)) \cdot \frac{dt(s)}{ds} \right) ds = \int \left(\frac{dh(s)}{ds} \right)^2 ds$$

$$= \int \left(1 - \frac{1}{\cosh^2 s} \right) ds = s - \tanh s.$$

$\square$

Remark 1.5.5. *Of course, the statement of Example 1.5.5 could have been verified by directly showing that* $\|\dot{\vec{x}}(s)\| = 1$.

Visualization 1.5.6 (Cycloids, hypocycloids and epicycloids).
Let C be a circle line of radius $r > 0$ and centre in M, and $P \neq M$ be a point attached to C. We write $\lambda = d(M, P)/r$, that is, $\lambda < 1$, $\lambda = 1$ or $\lambda > 1$ depending on whether P is inside of C, on C or outside of C.

A cycloid *is a curve that is generated by P as C rolls along a given curve γ without sliding.*

When γ is a straight line then the cycloid is said to be an ordinary cycloid *(Figures 1.29 and 1.30).*

An epicycloid *is the special case of a cycloid where C rolls along the outside of a given circle line C_0 (Figure 1.31).*

A hypocycloid *is the special case of a cycloid where C rolls along the inside of a given circle line C_0 (Figure 1.32).*

(a) Let γ be the x^1–axis. Then the ordinary cycloid has a parametric

$$\vec{x}(t) = \{r(t - \lambda \sin t), r(1 - \lambda \cos t)\} \ (t \in \mathbb{R}). \tag{1.62}$$

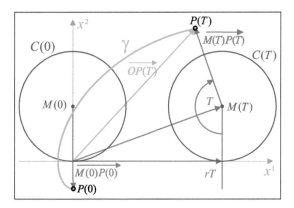

Figure 1.29 The construction of an ordinary cycloid

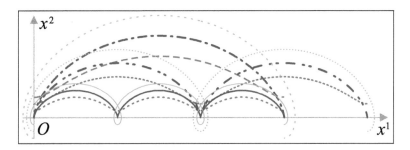

Figure 1.30 Families of ordinary cycloids

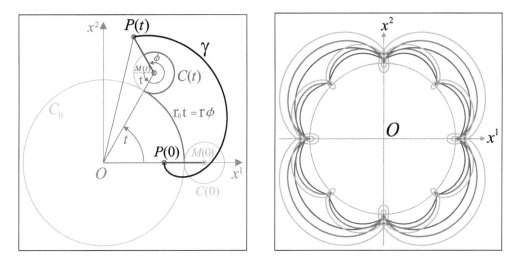

Figure 1.31 Left: the construction of an epicycloid. Right: a family of epicycloids

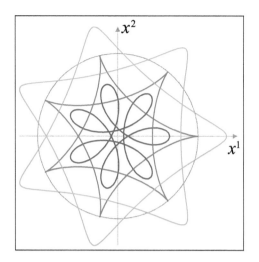

Figure 1.32 Families of hypocycloids

(b) Let γ be the circle line C_0 in the x^1x^2–plane with its centre in the origin and radius r_0. We write $R = r_0 + r$ for the epicycloid and $R = r_0 - r$ for the hypocycloid. Then the epicycloid and hypocycloid have parametric representations

$$\vec{x}(t) = \left\{ R\cos t - \lambda r \cos\left(\frac{R}{r}t\right), R\sin t - \lambda r \sin\left(\frac{R}{r}t\right) \right\} \quad (t \in \mathbb{R}) \qquad (1.63)$$

and

$$\vec{x}(t) = \left\{ R\cos t + \lambda r \cos\left(\frac{R}{r}t\right), R\sin t - \lambda r \sin\left(\frac{R}{r}t\right) \right\} \quad (t \in \mathbb{R}). \qquad (1.64)$$

(c) The arc length of the epicycloid in the special case $\lambda = 1$ is given by

$$s(t) = 8\frac{Rr}{r_0} \sin^2\left(\frac{r_0}{4r}t\right) \; for \; t \in \left(0, \frac{2\pi r}{r_0}\right). \qquad (1.65)$$

Proof.

(a) Let $P(t)$ denote the point attached to C and $M(t)$ be the centre of C, both at t. Furthermore, let the initial position of $M(0)$ and $P(0)$ be given by $M(0) = (0, r)$ and

$$P(0) = (0, r(1 - \lambda)) = \left(\cos\left(\frac{3\pi}{2}\right), r\left(1 + \lambda\sin\left(\frac{3\pi}{2}\right)\right)\right) \quad \text{(Figure 1.29)}.$$

When the circle line C has rolled the distance rt along the positive x^1–axis, we obtain

$$\overrightarrow{OP}(t) = \overrightarrow{OM}(t) + \overrightarrow{M(t)P}(t)$$
$$= \{rt, r\} + \lambda r \left\{\cos\left(\frac{3\pi}{2} - t\right), \sin\left(\frac{3\pi}{2} - t\right)\right\}$$
$$= \{rt, r\} - \lambda r \left\{\sin t, \cos t\right\} = \{r(t - \lambda\sin t), r(1 - \lambda\cos t)\}.$$

(b) Let $P(t)$ denote the point attached to C and $M(t)$ be the centre of C, both at t. Furthermore let the initial position $P(0)$ be given by $P(0) = (r_0 + (1-r)\lambda, 0)$ (left in Figure 1.31). We have

$$\overrightarrow{OP}(t) = \overrightarrow{OM}(t) + \overrightarrow{M(t)P}(t)$$

and, with

$$\phi = \frac{r_0}{r}t,$$

obviously

$$\overrightarrow{OM}(t) = (r_0 + r)\{\cos t, \sin t\}$$

and

$$\overrightarrow{M(t)P}(t) = -\lambda r\{\cos(t + \phi), \sin(t + \phi)\}.$$

Putting $\vec{x}(t) = \overrightarrow{OP}(t)$, we conclude

$$\vec{x}(t) = \left\{ R\cos t - \lambda r \cos\left(\frac{R}{r}t\right), R\sin t - \lambda r \sin\left(\frac{R}{r}t\right) \right\} \quad \text{for all } t \in \mathbb{R}.$$

The parametric representation in (1.64) for a hypocycloid is obtained analogously.

It follows from (1.63) that

$$\frac{d\vec{x}}{dt}(t) = R\left\{ -\sin(t) + \lambda \sin\left(\frac{R}{r}t\right), \cos t - \lambda \cos\left(\frac{R}{r}t\right) \right\} \quad (1.66)$$

and

$$\left\| \frac{d\vec{x}}{dt}(t) \right\|^2 = R^2 \left(\left(-\sin t + \lambda \sin\left(\frac{R}{r}t\right) \right)^2 + \left(-\cos t + \lambda \cos\left(\frac{R}{r}t\right) \right)^2 \right)$$

$$= R^2 \left(1 + \lambda^2 - 2\lambda \sin t \sin\left(\frac{R}{r}t\right) - 2\lambda \cos t \cos\left(\frac{R}{r}t\right) \right)$$

$$= R^2 \left(1 + \lambda^2 - 2\lambda \cos\left(\frac{R-r}{r}t\right) \right) = R^2 \left(1 + \lambda^2 - 2\lambda \cos\left(\frac{r_0}{r}t\right) \right).$$

If $\lambda \neq 1$, then $1 + \lambda^2 - 2\lambda \cos\left(\frac{r_0}{r}t\right) \geq (1-\lambda)^2 > 0$, and the parametric representation is admissible for all $t \in \mathbb{R}$.

If $\lambda = 1$, then $\cos\left(\frac{r_0}{r}t\right) = 1$ if and only if $t_k = \frac{2k\pi r}{r_0}$ $(k \in \mathbb{Z})$, and the parametric representation is not admissible if and only if $t = t_k$ for $k \in \mathbb{Z}$ (Figure 1.32).

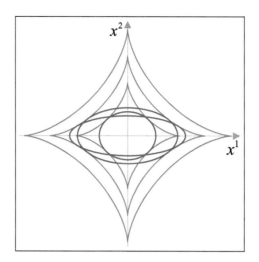

Figure 1.33 Special hypocycloids

(c) Now let $\lambda = 1$. Then we have

$$\left\|\frac{d\vec{x}}{dt}(t)\right\| = R\sqrt{2}\sqrt{1 - \cos\left(\frac{r_0}{r}t\right)} = 2R\sin\left(\frac{r_0}{2r}t\right) \text{ for } t \in \left(0, \frac{2\pi r}{r_0}\right),$$

and so

$$s(t) = \int_0^t 2R\sin\left(\frac{r_0}{2r}\tau\right) d\tau = -4\frac{Rr}{r_0}\cos\left(\frac{r_0}{2r}\tau\right)\Big|_0^t$$

$$= 4\frac{Rr}{r_0}\left(1 - \cos\left(\frac{r_0}{2r}t\right)\right) = 8\frac{Rr}{r_0}\sin^2\left(\frac{r_0}{4r}t\right). \qquad (1.67)$$

From

$$s(t) = 8R\frac{r}{r_0}\sin^2\left(\frac{r_0}{4r}t\right) \text{ for } t \in \left(0, \frac{2\pi r}{r_0}\right),$$

we obtain

$$t(s) = \frac{4r}{r_0}\sin^{-1}\sqrt{\frac{r_0}{8rR}s} \text{ for } s \in \left(0, 8R\frac{r}{r_0}\right),$$

and this yields a parametric representation of the epicycloid with respect to its arc length.

□

Remark 1.5.7. *Ellipses and astroids are special cases of hypocycloids for $r_0 = 2r$ and arbitrary λ, and for $r_0 = 4r$ and $\lambda = 1$, respectively (Figure 1.33).*

It is also interesting to combine the principles of generating ordinary cycloids and epicycloids.

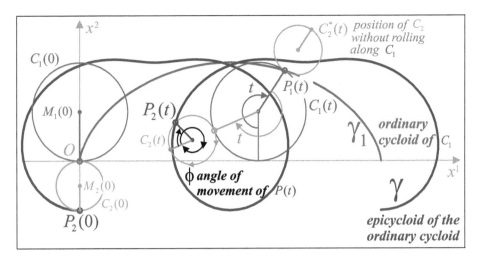

Figure 1.34 The construction of the curve of Visualization 1.5.8

Visualization 1.5.8. *Let C_1 be a circle line of radius r_1 that rolls along the x^1–axis, and C_2 be a circle line of radius r_2 that rolls along the outside of C_1 at the same time (Figure 1.34). We put $R = r_1 + 2r_2$. Then the curve γ of the point P attached to the circle line C_2 by this movement has a parametric representation*

$$\vec{x}(t) = \{r_1(t - \sin(2t)), r_1(1 - \cos(2t))\} -$$
$$- \left\{ r_2 \left(\sin(2t) + \sin\left(\frac{R}{r_2}t\right)\right), r_2\left(\cos(2t) + \cos\left(\frac{R}{r_2}t\right)\right) \right\} \quad (1.68)$$

Since it is often the case that planar curves are given in polar coordinates, it is useful to have a formula for their arc lengths.

Proposition 1.5.9. *Let γ be a planar curve with a parametric representation*

$$\vec{x}(\phi) = \{\rho(\phi)\cos\phi, \rho(\phi)\sin\phi\} \quad (\phi \in I = (\phi_0, \phi_1) \subset (0, 2\pi)).$$

Then the arc length along γ is

$$s(\phi, \phi_0) = \int_{\phi_0}^{\phi} \sqrt{\rho^2(\varphi) + (\rho'(\varphi))^2}\, d\varphi \quad (\phi \in I), \text{ where } \rho'(\phi) = \frac{d\rho(\phi)}{d\phi}. \quad (1.69)$$

Proof. We obtain

$$\vec{x}'(\phi) = \{\rho'(\phi)\cos\phi - \rho(\phi)\sin\phi, \rho'(\phi)\sin\phi + \rho(\phi)\cos\phi\}$$

and

$$\|\vec{x}'(\phi)\|^2 = (\rho'(\phi))^2 \cos^2\phi - 2\rho(\phi)\rho'(\phi)\sin\phi\cos\phi + \rho^2(\phi)\sin^2(\phi)$$
$$+ (\rho'(\phi))^2 \sin^2\phi + 2\rho(\phi)\rho'(\phi)\sin\phi\cos\phi + \rho^2(\phi)\cos^2(\phi)$$
$$= (\rho'(\phi))^2 + \rho^2(\phi)$$

and (1.69) is an immediate consequence. □

Example **1.5.10.** *(a) First we consider the rosette of Visualization 1.4.7 with a parametric representation (1.51). Then we have* $\rho(\phi) = (a/2)\sin(2\phi)$, $\rho'(\phi) = a\cos(2\phi)$ *and*

$$\rho^2(\phi) + (\rho'(\phi))^2 = \frac{a^2}{4}\left(\sin^2(2\phi) + 4\cos^2(2\phi)\right)$$
$$= \frac{a^2}{4}\left(1 + 3\cos^2(2\phi)\right).$$

Thus we obtain for the arc length along the rosette by (1.69)

$$s(\phi, 0) = \frac{a}{2}\int_0^\phi \sqrt{1 + 3\cos^2(2\varphi)}\, d\varphi \quad (\phi \in (0, 2\pi)). \tag{1.70}$$

(b) Now we consider the double egg line of Visualization 1.4.8 with a parametric representation (1.53). Then we have $\rho(\phi) = a\cos^2\phi$, $\rho'(\phi) = -2a\sin\phi\cos\phi$ *and*

$$\rho^2(\phi) + (\rho'(\phi))^2 = a^2\cos^2\phi\left(\cos^2\phi + 4\sin^2\phi\right)$$
$$= a^2\cos^2\phi\left(1 + 3\sin^2\phi\right).$$

Thus we obtain for the arc length along the double egg line by (1.69)

$$s(\phi, 0) = a\int_0^{\phi_0} |\cos\varphi|\sqrt{4 - 3\cos^2\varphi}\, d\varphi \quad (\phi \in (0, 2\pi)). \tag{1.71}$$

1.6   THE VECTORS OF THE TRIHEDRON OF A CURVE

From now on we confine our studies of curves to three-dimensional $\mathbb{E}^3$.

With every curve, we may associate the *vectors of the trihedron* or the *frame* at any of its points.

We start with the *tangent vectors of a curve*. The *tangent to a curve* γ *at a point* X is a first order approximation of γ at X by a straight line, and gives the *direction of* γ *at* X.

Let γ be a curve with a parametric representation $\vec{x}(t)$ $(t \in [a, b])$. We consider the difference vector for $t, t + h \in I$,

$$\vec{x}(t+h) - \vec{x}(t) = \left\{x^1(t+h) - x^1(t), x^2(t+h) - x^2(t), x^3(t+h) - x^3(t)\right\},$$

and divide it by h to obtain

$$\frac{\vec{x}(t+h) - \vec{x}(t)}{h} = \left\{\frac{x^1(t+h) - x^1(t)}{h}, \frac{x^2(t+h) - x^2(t)}{h}, \frac{x^3(t+h) - x^3(t)}{h}\right\}.$$

If the component functions x^k $(k = 1, 2, 3)$ are differentiable, then this yields

$$\lim_{h \to 0} \frac{\vec{x}(t+h) - \vec{x}(t)}{h} = \vec{x}'(t) = \frac{d\vec{x}}{dt}(t) = \left\{\frac{dx^1}{dt}(t), \frac{dx^2}{dt}(t), \frac{dx^3}{dt}(t)\right\}.$$

Definition 1.6.1 (Tangent, tangent vector).

Let γ be a curve in $\mathbb{R}^3$ with a parametric representation $\vec{x}(t)$ ($t \in I$) and $t_0 \in I$.
(a) If the vector

$$\vec{x}'(t_0) = \left\{ \frac{dx^1}{dt}(t_0), \frac{dx^2}{dt}(t_0), \frac{dx^3}{dt}(t_0) \right\}$$

exists and $\vec{x}'(t_0) \neq \vec{0}$, then the unit vector

$$\vec{t}(t_0) = \frac{\vec{x}'(t_0)}{\|\vec{x}'(t_0)\|}$$

is called the *tangent vector of γ at the point* $X(t_0)$.
(b) The straight line through the point $X(t_0)$ in the direction of the tangent vector $\vec{t}(t_0)$ is called the *tangent of γ at the point* $X(t_0)$; it has a parametric representation

$$\vec{x}_{t_0}(\lambda) = \vec{x}(t_0) + \lambda \vec{t}(t_0) \ (\lambda \in \mathbb{R}).$$

Visualization 1.6.2 (The tractrix).

Let γ and γ^* be a curve and a straight line in a plane that have no points of intersection. Given a point P on γ, let P^* denote the point of intersection of γ^* with the tangent to γ at P. If the distances between the points P and P^* have a constant value d, then the curve γ is called a tractrix *(Figure 1.35). We introduce a Cartesian coordinate system in the plane such that the straight line γ^* is the y–axis. Then γ has a parametric representation*

$$\vec{x}(\phi) = \left\{ d \sin \phi, d \left(\log \left(\tan \left(\frac{\phi}{2} \right) \right) + \cos \phi \right) \right\} \ (\phi \in (0, \pi/2)). \tag{1.72}$$

Proof. Let $\vec{x}(t) = \{t, y(t)\}$ be a parametric representation of γ. Then the tangent to γ at t is given by $\vec{x}(t) + \lambda \{1, y'(t)\}$ ($\lambda \in \mathbb{R}$), and so we obtain for P^*

$$\overrightarrow{OP^*} = \{0, p^*\} = \{t, y(t)\} + \lambda \{1, y'(t)\}, \text{ that is, } \lambda = -t.$$

We observe that $\gamma \cap \gamma^* = \emptyset$ implies $d > 0$, hence we have

$$d = d(P, P^*) = \|\vec{x}(t) - (\vec{x}(t) - t\{1, y'(t)\})\| = |t|\sqrt{1 + (y'(t))^2},$$

and so, since $|t| > 0$,

$$y'(t) = \pm \frac{\sqrt{d^2 - t^2}}{|t|} \text{ for } |t| < d. \tag{1.73}$$

We choose the upper sign and $t \in (0, d)$, and obtain

$$y(t) = \int \frac{\sqrt{d^2 - t^2}}{t} \, dt. \tag{1.74}$$

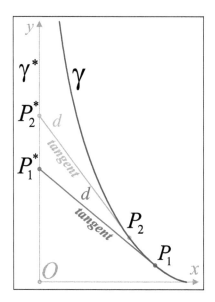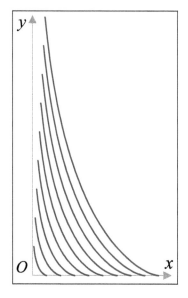

Figure 1.35 Left: the construction of a tractrix. Right: a family of tractrices

Since $t \in (0, d)$, we may put $t = d \cdot \sin \phi$ for $\phi \in (0, \pi/2)$ and this yields

$$\int \frac{\sqrt{d^2 - t^2}}{t} \, dt = \int \frac{\sqrt{d^2 - d^2 \sin^2 \phi}}{d \cdot \sin \phi} d \cdot \cos \phi \, d\phi$$

$$= d \cdot \int \frac{\cos^2 \phi}{\sin \phi} \, d\phi = d \cdot (I_1 + I_2), \qquad (1.75)$$

where

$$I_1 = \int \frac{d\phi}{\sin \phi}$$

and

$$I_2 = -\int \sin \phi \, d\phi = \cos \phi = \sqrt{1 - \sin^2 \phi} + c_2 = \frac{\sqrt{d^2 - t^2}}{d} + c_2, \qquad (1.76)$$

where c_2 is a constant of integration.

To evaluate the integral I_1, we make the substitution $z = \tan (\phi/2)$. Then $z \in (0, 1)$ and we obtain

$$\frac{d\phi}{dz} = \frac{d}{dz}(2 \tan^{-1} z) = \frac{2}{1 + z^2},$$

$$\sin \phi = 2 \sin (\phi/2) \cos (2\phi/2) = 2 \tan (\phi/2) \cos^2 (\phi/2)$$

$$= 2 \tan (\phi/2) \frac{\cos^2 (\phi/2)}{\cos^2 (\phi/2) + \sin^2 (\phi/2)}$$

$$= \frac{2 \tan (\phi/2)}{1 + \tan^2 (\phi/2)} = \frac{2z}{1 + z^2}$$

and

$$I_1 = \int \frac{1 + z^2}{2z} \frac{2}{1 + z^2} \, dz = \int \frac{dz}{z} = \log z + c_1 = \log \left(\tan \left(\frac{\phi}{2} \right) \right) + c_1, \qquad (1.77)$$

where c_1 is a constant of integration. We put $c = d \cdot (c_1 + c_2)$. Then it follows from (1.74), (1.75), (1.76) and (1.77) that

$$y(\phi) = d \cdot \left(\log \left(\tan \left(\frac{\phi}{2} \right) \right) + \cos \phi \right) + c.$$

If we choose c such that $\lim_{\phi \to \pi/2} y(\phi) = 0$, then we have $c = 0$, and (1.72) is an immediate consequence. □

We now assume that γ is a curve with a parametric representation $\vec{x}(s)$ $(s \in I)$ of class $C^r(I)$ $(r \geq 2)$. Then the derivative $\dot{\vec{x}}(s)$ is a unit vector by Proposition 1.5.2 (d), which means that the movement along the curve takes place at a constant value of speed equal to one (Figure 1.27). Since $\vec{x} \in C^r(I)$ $(r \geq 2)$, the derivative $\ddot{\vec{x}}$ is a continuous function on the interval I, and $\ddot{\vec{x}}(s)$ stands for the *speed of change in the direction of* γ *at* s. Furthermore, if $\ddot{\vec{x}}(s) \neq \vec{0}$, then we may consider the unit vectors

$$\frac{\ddot{\vec{x}}(s)}{\|\ddot{\vec{x}}(s)\|} \quad \text{and} \quad \frac{\dot{\vec{x}}(s) \times \ddot{\vec{x}}(s)}{\|\ddot{\vec{x}}(s)\|}.$$

Definition 1.6.3 (The vectors of the trihedron and curvature).
Let γ be a curve with a parametric representation $\vec{x}(s)$ $(s \in I)$ of class $C^r(I)$ $(r \geq 2)$, and $\ddot{\vec{x}}(s) \neq \vec{0}$. Then the vectors

$$\vec{v}_1(s) = \dot{\vec{x}}(s), \quad \vec{v}_2(s) = \frac{\ddot{\vec{x}}(s)}{\|\ddot{\vec{x}}(s)\|} \quad \text{and} \quad \vec{v}_3(s) = \vec{v}_1(s) \times \vec{v}_2(s)$$

are called the *tangent, principal normal* and *binormal vectors of* γ *at* s, respectively. Attached to the point $X(s)$ of the curve, the vectors $\vec{v}_k(s)$ $(k = 1, 2, 3)$ form the *trihedron* or *frame of the curve at* s (Figure 1.37).
The vector $\ddot{\vec{x}}(s)$ is called *vector of curvature*, its length $\kappa(s) = \|\ddot{\vec{x}}(s)\|$ *curvature* and $\rho(s) = 1/\kappa(s)$ for $\kappa(s) \neq 0$ *radius of curvature of the curve at* s.

Remark 1.6.4. *(a) In many textbooks, the notations* $\vec{t}$, $\vec{n}$ *and* $\vec{b}$ *are used instead of* $\vec{v}_1$, $\vec{v}_2$ *and* $\vec{v}_3$.
Obviously, we have

$$\vec{v}_i(s) \bullet \vec{v}_k(s) = \delta_{ik} = \begin{cases} 1 & (i = k) \\ 0 & (i \neq k) \end{cases} \quad \text{for all } s. \tag{1.78}$$

(b) The curvature of a curve at a point is a measure of the deviation of the curve from a straight line at that point.

Remark 1.6.5. *The trihedron of a curve is only defined at points where* $\ddot{\vec{x}}(s) \neq \vec{0}$. *We shall exclude curves or segments of curves with* $\ddot{\vec{x}} \equiv \vec{0}$ *in which case* $\vec{x}(s) = \vec{a}s + \vec{b}$ *for some constant vectors* $\vec{a}$ *and* $\vec{b}$. *The corresponding curves are straight lines.*

Visualization 1.6.6. *(a) We consider the catenary of Example 1.5.3 (d) with a parametric representation (1.59) in* $\mathbb{E}^3$

$$\vec{x}^*(s) = \left\{ c \sinh^{-1}\left(\frac{s}{c}\right), \sqrt{s^2 + c^2}, 0 \right\} \quad (s \in \mathbb{R})$$

with respect to the natural parameter s, *where* c *is a positive constant. Then we obtain*

$$\vec{v}_1^*(s) = \dot{\vec{x}}^*(s) = \left\{ \frac{c}{\sqrt{s^2 + c^2}}, \frac{s}{\sqrt{s^2 + c^2}}, 0 \right\}$$

for the tangent vectors at s *(left in Figure 1.36),*

$$\ddot{\vec{x}}^*(s) = \left\{ -\frac{cs}{(s^2 + c^2)^{3/2}}, \frac{1}{\sqrt{s^2 + c^2}} - \frac{s^2}{(s^2 + c^2)^{3/2}}, 0 \right\} = \frac{c}{(s^2 + c^2)^{3/2}} \{-s, c, 0\}$$

for the vector of curvature at s *(left in Figure 1.36),*

$$\kappa^*(s) = \|\ddot{\vec{x}}^*(s)\| = \frac{c}{(s^2 + c^2)^{3/2}} \sqrt{s^2 + c^2} = \frac{c}{s^2 + c^2}$$

for the curvature at s,

$$\vec{v}_2^*(s) = \frac{\ddot{\vec{x}}^*(s)}{\|\ddot{\vec{x}}^*(s)\|} = \frac{c}{(s^2 + c^2)^{3/2}} \frac{s^2 + c^2}{c} \{-s, c, 0\} = \frac{1}{\sqrt{s^2 + c^2}} \{-s, c, 0\}$$

for the principal normal vector at s, *and*

$$\vec{v}_3^*(s) = \vec{v}_1^*(s) \times \vec{v}_2^*(s) = \frac{1}{s^2 + c^2} \{c, s, 0\} \times \{-s, c, 0\}$$

$$= \frac{1}{s^2 + c^2} \{0, 0, s^2 + c^2\} = \vec{e}^3$$

for the binormal vector at s.

(b) We consider the tractrix of Visualization 1.6.2. Writing $c = -1/d$, *we obtain from (1.73)*

$$y'(t) = \frac{\sqrt{1 - c^2 t^2}}{|c| t} \text{ for } 0 < t < -\frac{1}{c},$$

hence

$$\|\vec{x}'(t)\|^2 = 1 + \frac{1 - c^2 t^2}{c^2 t^2} = \frac{1}{c^2 t^2},$$

and so

$$s(t) = \frac{1}{|c|} \int_{t_0}^{t} \frac{dt}{t} = \frac{1}{|c|} (\log t - \log t_0).$$

Here, we may choose $t_0 = 1$ *and then the arc length along the tractrix is given by*

$$s(t) = \frac{1}{|c|} \log t \text{ for } s < \frac{1}{|c|} \log\left(\frac{1}{|c|}\right).$$

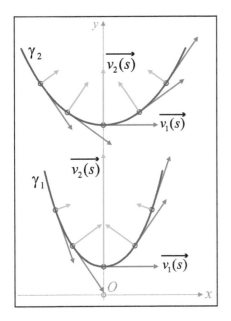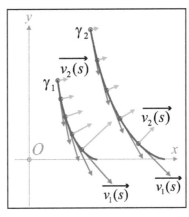

Figure 1.36 Tangent vectors and vectors of curvature of catenaries (left) and tractrices (right)

Since we have $t(s) = \exp(|c|s)$ and $(dt/ds)(s) = |c|\exp(|c|s)$, and writing $y^*(s) = y(t(s))$, we obtain

$$\dot{y}^*(s) = y'(t(s))\frac{dt}{ds}(s) = \frac{\sqrt{1 - c^2\exp(2|c|s)}}{|c|\exp(|c|s)}|c|\exp(|c|s) = \sqrt{1 - c^2\exp(2|c|s)},$$

and consequently

$$\vec{v}_1^*(s) = \left\{|c|\exp(|c|s), \sqrt{1 - c^2\exp(2|c|s)}, 0\right\}$$

for the tangent vector at s in $\mathbb{E}^3$ (right in Figure 1.36),

$$\ddot{\vec{x}}^*(s) = \left\{c^2\exp(|c|s), -\frac{c^2|c|\exp(2|c|s)}{\sqrt{1 - c^2\exp(2|c|s)}}, 0\right\}$$

$$= c^2\exp(|c|s)\left\{1, \frac{-|c|\exp(|c|s)}{\sqrt{1 - c^2\exp(2|c|s)}}, 0\right\}$$

for the vector of curvature at s (right in Figure 1.36),

$$\kappa^*(s) = \frac{c^2\exp(|c|s)\sqrt{1 - c^2\exp(2|c|s)} + c^2\exp(2|c|s)}{\sqrt{1 - c^2\exp(2|c|s)}}$$

$$= \frac{c^2\exp(|c|s)}{\sqrt{1 - c^2\exp(2|c|s)}}$$

for the curvature at s, and

$$\vec{v}_2^*(s) = \frac{\ddot{\vec{x}}(s)}{\|\ddot{\vec{x}}(s)\|} = \left\{ \sqrt{1 - c^2 \exp(2|c|s)}, -|c| \exp(|c|s), 0 \right\}$$

for the principal normal vector, and $\vec{v}_3^*(s) = \vec{e}^3(s)$ *for the binormal vector at s.*

Visualization 1.6.7. *(a) For the circle line in Example 1.5.3 (b) with a parametric representation*

$$\vec{x}(s) = \vec{m}_0 + \vec{e}^1 r \cos \frac{s}{r} + \vec{e}^2 r \sin \frac{s}{r} \quad (s \in [0, 2\pi r]),$$

we have

$$\vec{v}_1(s) = -\vec{e}^1 \sin \frac{s}{r} + \vec{e}^2 \cos \frac{s}{r} \qquad \textit{for the tangent vector,}$$

$$\vec{v}_2(s) = -\vec{e}^1 \cos \frac{s}{r} - \vec{e}^2 \sin \frac{s}{r} \qquad \textit{for the principal normal vector,}$$

$$\vec{v}_3(s) = \vec{e}^3 = \{0, 0, 1\} \qquad \textit{for the binormal vector,}$$

$$\ddot{\vec{x}}(s) = -\frac{1}{r}\vec{e}^1 \cos \frac{s}{r} - \frac{1}{r}\vec{e}^2 \sin \frac{s}{r} \quad \textit{for the vector of curvature}$$

and

$$\kappa(s) = \frac{1}{r} \qquad \textit{for the curvature.}$$

(b) For the helix in Example 1.5.3 (c) with a parametric representation

$$\vec{x}(s) = \{r \cos \omega s, r \sin \omega s, h \omega s\} \quad (s \in \mathbb{R}),$$

we have

$$\vec{v}_1(s) = \omega\{-r \sin \omega s, r \cos \omega s, h\} \quad \textit{for the tangent vector,}$$

$$\vec{v}_2(s) = -\{\cos \omega s, \sin \omega s, 0\} \qquad \textit{for the principal normal vector,}$$

$$\vec{v}_3(s) = \omega\{h \sin \omega s, -h \cos \omega s, r\} \quad \textit{for the binormal vector,}$$

$$\ddot{\vec{x}}(s) = -\omega^2 r\{\cos \omega s, \sin \omega s, 0\} \qquad \textit{for the vector of curvature}$$

and

$$\kappa(s) = r\omega^2 = \frac{r}{r^2 + h^2} \qquad \textit{for the curvature}$$

(Figure 1.37).

Visualization 1.6.8. *The vectors of the trihedra and the curvature of the curve* γ *of Example 1.5.4 with the parametric representation (1.61) are given by (Figure 1.38)*

$$\vec{v}_1(s) = -\frac{\sinh s}{\cosh^2(s)}\{\cos s, \sin s, -\sinh s + \frac{1}{\cosh s}\{-\sin s, \cos s, 0\}, \qquad (1.79)$$

$$\vec{v}_2(s) = -\frac{1}{\cosh^2(s)}\{\cos s, \sin s, -\sinh s\} - \frac{\sinh s}{\cosh s}\{-\sin s, \cos s, 0\}, \qquad (1.80)$$

$$\vec{v}_3(s) = \frac{1}{\cosh s}\{\sinh s \cos s, \sinh s \sin s, 1\} \qquad (1.81)$$

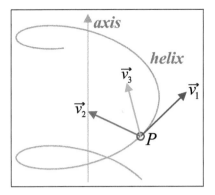

Figure 1.37 The vectors of a trihedron of a helix

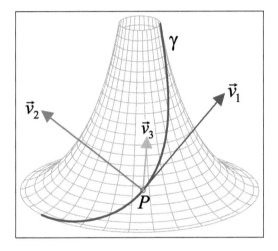

Figure 1.38 The vectors of a trihedron of the curve of Visualization 1.6.8

and

$$\kappa(s) = \frac{2}{\cosh s}. \tag{1.82}$$

Proof. We write $\phi(s) = 1/\cosh s$, omit the argument in ϕ and obtain from

$$\vec{x}(s) = \{\phi \cos s, \phi \sin s, s - \tanh s\} \text{ for } s > 0$$

that

$$\vec{v}_1(s) = \dot{\vec{x}}(s) = \{\dot{\phi} \cos s, \dot{\phi} \sin s, 1 - \phi^2\} + \{-\phi \sin s, \phi \cos s, 0\}$$

$$= -\frac{\sinh s}{\cosh^2(s)}\{\cos s, \sin s, -\sinh s\} + \frac{1}{\cosh s}\{-\sin s, \cos s, 0\}$$

which is (1.79).

This yields

$$\ddot{\vec{x}}^*(s) = \{(\ddot{\phi} - \phi)\cos s, (\ddot{\phi} - \phi)\sin s, -2\dot{\phi}\phi\} + 2\dot{\phi}\{-\sin s, \cos s, 0\}.$$

It follows from

$$\dot{\phi} = -\frac{\sinh s}{\cosh^2(s)} = -\phi^2 \sinh s$$

that

$$\ddot{\phi} = -\phi^2 \cosh s - 2\dot{\phi}\phi \sinh s = -\phi + 2\phi^3 \sinh^2 s$$
$$= -\phi + 2\phi - 2\phi^3 = \phi - 2\phi^3$$

and

$$\ddot{\phi} - \phi = -2\phi^3.$$

Therefore, we have

$$\ddot{\vec{x}}^*(s) = -\frac{2}{\cosh^3(s)}\{\cos s, \sin s, -\sinh s\} - 2\frac{\sinh s}{\cosh^2(s)}\{-\sin s, \cos s, 0\}.$$

Now we obtain for the curvature $\kappa(s)$ along γ

$$\kappa^2(s) = \|\ddot{\vec{x}}^*(s)\|^2 = 4\left(\frac{1 + \sinh^2(s)}{\cosh^6(s)} + \frac{\sinh^2(s)}{\cosh^4(s)}\right)$$
$$= 4\frac{1}{\cosh^4(s)}(1 + \sinh^2(s)) = \frac{4}{\cosh^2(s)},$$

hence

$$\kappa(s) = \frac{2}{\cosh s}$$

which is (1.82).

Now the principal normal vectors $\vec{v}_2(s)$ along γ are given by

$$\vec{v}_2(s) = \frac{1}{\kappa(s)}\ddot{\vec{x}}^*(s) = -\frac{1}{\cosh^2(s)}\{\cos s, \sin s, -\sinh s\} - \frac{\sinh s}{\cosh s}\{-\sin s, \cos s, 0\},$$

which is (1.80).

Furthermore, since

$$\vec{b}(s) = \{\cos s, \sin s, -\sinh s\} \times \{-\sin s, \cos s, 0\} = \{\sinh s \cos s, \sinh s \sin s, 1\},$$

the binormal vectors $\vec{v}_3(s)$ along γ are given by

$$\vec{v}_3(s) = \vec{v}_1(s) \times \vec{v}_2(s) = \left(\frac{\sinh^2(s)}{\cosh^3(s)} + \frac{1}{\cosh^3(s)}\right)\vec{b}$$
$$= \frac{1}{\cosh s}\{\sinh s \cos s, \sinh s \sin s, 1\}$$

which is (1.81). □

1.7 FRENET'S FORMULAE

It follows from the identities in (1.78) that if γ is a curve with a parametric representation $\vec{x}(s)$ $(s \in I)$ such that $\ddot{\vec{x}}(s) \neq \vec{0}$, then the vectors $\vec{v}_k(s)$ $(k = 1, 2, 3)$ of the trihedron of γ are a basis of $\mathbb{E}^3$ for every $s \in I$. Consequently every vector of the space $\mathbb{E}^3$ can be represented as a unique linear combination of the vectors $\vec{v}_k(s)$, in particular, this holds true for the derivatives $\dot{\vec{v}}_k(s)$ of the vectors $\vec{v}_k(s)$. The three formulae that give the vectors $\dot{\vec{v}}_j(s)$ for $j = 1, 2, 3$ as linear combinations of the vectors $\vec{v}_k(s)$ $(k = 1, 2, 3)$ are called *Frenet's formulae*. They are of the utmost importance in the theory of curves.

Theorem 1.7.1 (Frenet's formulae). *Let γ be a curve with a parametric representation $\vec{x}(s)$ $(s \in I)$ such that $\ddot{\vec{x}}(s) \neq 0$ for all $s \in I$. We put*

$$\tau(s) = \dot{\vec{v}}_2(s) \bullet \vec{v}_3(s) \ (s \in I), \tag{1.83}$$

and obtain

$$\left\{ \begin{array}{lll} \dot{\vec{v}}_1(s) & = & \kappa(s)\vec{v}_2(s) \\ \dot{\vec{v}}_2(s) & = & -\kappa(s)\vec{v}_1(s) \quad + \quad \tau(s)\vec{v}_3(s) \\ \dot{\vec{v}}_3(s) & = & -\tau(s)\vec{v}_2(s) \end{array} \right\}. \tag{1.84}$$

The formulae in (1.84) are called Frenet's formulae.

Proof. We write $\dot{\vec{v}}_k(s) = \sum_{j=1}^{3} a_{kj}(s)\vec{v}_j(s)$ $(k = 1, 2, 3)$ with the functions a_{kj} $(k, j = 1, 2, 3)$ to be determined, and omit the parameter s. First $\vec{v}_k \bullet \vec{v}_k = 1$ for $k = 1, 2, 3$ implies

$$\frac{d}{ds}(\vec{v}_k \bullet \vec{v}_k) = 2\dot{\vec{v}}_k \bullet \vec{v}_k = 0 \quad (k = 1, 2, 3)$$

and since $\vec{v}_j \bullet \vec{v}_k = 0$ for $j \neq k$, we conclude

$$0 = \dot{\vec{v}}_k \bullet \vec{v}_k = \sum_{j=1}^{3} a_{kj}\vec{v}_j \bullet \vec{v}_k = a_{kk} \text{ for } k = 1, 2, 3.$$

Furthermore, we have for $l \neq k$

$$0 = \frac{d}{ds}(\vec{v}_k \bullet \vec{v}_l) = \dot{\vec{v}}_k \bullet \vec{v}_l + \vec{v}_k \bullet \dot{\vec{v}}_l$$

$$= \sum_{j=1}^{3} a_{kj}\vec{v}_j \bullet \vec{v}_l + \sum_{j=1}^{3} a_{lj}\vec{v}_j \bullet \vec{v}_k = a_{kl} + a_{lk},$$

that is $a_{kl} = -a_{lk}$. Finally, we obtain

$$a_{12} = \dot{\vec{v}}_1 \bullet \vec{v}_2 = \ddot{\vec{x}} \bullet \vec{v}_2 = \kappa\vec{v}_2 \bullet \vec{v}_2 = \kappa, \text{ and } a_{23} = \dot{\vec{v}}_2 \bullet \vec{v}_3 = \tau \text{ by definition.}$$

□

Frenet's formulae may formally be written as

$$
\begin{pmatrix} \dot{\vec{v}}_1 \\ \dot{\vec{v}}_2 \\ \dot{\vec{v}}_3 \end{pmatrix} = \begin{pmatrix} 0 & \kappa & 0 \\ -\kappa & 0 & \tau \\ 0 & -\tau & 0 \end{pmatrix} \begin{pmatrix} \vec{v}_1 \\ \vec{v}_2 \\ \vec{v}_3 \end{pmatrix}.
$$

Now we define the *torsion* of a curve, and give some elementary results.

Definition 1.7.2 (Torsion of a curve).
Let γ be a curve with a parametric representation $\vec{x}(s)$ ($s \in I$) such that $\ddot{\vec{x}}(s) \neq \vec{0}$ for all $s \in I$. Then the value

$$
\tau(s) = \dot{\vec{v}}_2(s) \bullet \vec{v}_3(s)
$$

defined by the formula in (1.83) is called the *torsion of γ at the point $X(s)$*.

Example **1.7.3.** *We consider the helix of Visualization 1.6.7 (b). Then we have*

$$
\vec{v}_2(s) = -\{\cos \omega s, \sin \omega s, 0\},
$$
$$
\vec{v}_3(s) = \omega\{h \sin \omega s, -h \cos \omega s, r\},
$$
$$
\dot{\vec{v}}_2(s) = \omega\{\sin \omega s, -\cos \omega s, 0\}
$$

and

$$
\tau(s) = \dot{\vec{v}}_2(s) \bullet \vec{v}_3(s) = h\omega^2 = \frac{h}{r^2 + h^2} \ (s \in \mathbb{R}).
$$

for the torsion at s.

Example **1.7.4.** *The torsion of the curve of Visualization 1.6.8 is given by*

$$
\tau(s) = 1. \tag{1.85}
$$

Proof. We obtain from the definition of τ in Definition 1.7.2, and (1.80) and (1.81)

$$
\tau(s) = -\vec{v}_2(s) \bullet \dot{\vec{v}}_3(s) = \frac{1}{\cosh^4(s)}(1 + \sinh^2(s)) + \frac{\sinh^2 s}{\cosh^2 s}
$$
$$
= \frac{1}{\cosh^2 s}(1 + \sinh^2 s) = 1.
$$

$\square$

Proposition 1.7.5. *Let γ be a curve with a parametric representation $\vec{x}(s)$ ($s \in I$) such that $\ddot{\vec{x}}(s) \neq \vec{0}$ for all $s \in I$. Then the torsion τ of γ is given by*

$$
\tau(s) = \frac{\dot{\vec{x}}(s) \bullet (\ddot{\vec{x}}(s) \times \dddot{\vec{x}}(s))}{\kappa^2(s)} \quad (s \in I). \tag{1.86}
$$

Proof. We have by definition and the second identity in (1.8) of Proposition 1.2.3

$$\dot{\vec{v}}_2 = \frac{d}{ds}\left(\frac{\ddot{\vec{x}}}{\|\ddot{\vec{x}}\|}\right) = \frac{\dddot{\vec{x}}}{\|\ddot{\vec{x}}\|} + \ddot{\vec{x}}\frac{d}{ds}\left(\frac{1}{\|\ddot{\vec{x}}\|}\right), \ \ \vec{v}_3 = \vec{v}_1 \times \vec{v}_2 = \frac{\dot{\vec{x}} \times \ddot{\vec{x}}}{\|\ddot{\vec{x}}\|}$$

and

$$\tau = \dot{\vec{v}}_2 \bullet \vec{v}_3 = \frac{\dddot{\vec{x}} \bullet (\dot{\vec{x}} \times \ddot{\vec{x}})}{\|\ddot{\vec{x}}\|^2} = \frac{\dddot{\vec{x}} \bullet (\dot{\vec{x}} \times \ddot{\vec{x}})}{\kappa^2},$$

since

$$\ddot{\vec{x}}\frac{d}{ds}\left(\frac{1}{\|\ddot{\vec{x}}\|}\right) \perp (\dot{\vec{x}} \times \ddot{\vec{x}}).$$

□

In many cases, it is useful to have formulae for the vectors of the trihedron, and the curvature and torsion of a curve given by a parametric representation with respect to an arbitrary parameter.

Proposition 1.7.6. *Let γ be a curve with a parametric representation $\vec{x}(t)$ ($t \in I$) and non–vanishing vector of curvature where t is an arbitrary parameter. Then the vectors of the trihedron, and the curvature and torsion of γ are given by*

$$\vec{v}_1(t) = \frac{\vec{x}\,'(t)}{\|\vec{x}\,'(t)\|},$$

$$\vec{v}_2(t) = \frac{\vec{x}\,'(t) \times (\vec{x}\,''(t) \times \vec{x}\,'(t))}{\|\vec{x}\,'(t) \times (\vec{x}\,''(t) \times \vec{x}\,'(t))\|}$$

$$= \frac{\vec{x}\,''(t)\|\vec{x}\,'(t)\|^2 - \vec{x}\,'(t)(\vec{x}\,'(t) \bullet \vec{x}\,''(t))}{\|\vec{x}\,'(t)\| \, \|\vec{x}\,'(t) \times \vec{x}\,''(t)\|},$$

$$\vec{v}_3(t) = \frac{\vec{x}\,'(t) \times \vec{x}\,''(t)}{\|\vec{x}\,'(t) \times \vec{x}\,''(t)\|}, \tag{1.87}$$

and

$$\kappa(t) = \frac{\|\vec{x}\,'(t) \times \vec{x}\,''(t)\|}{\|\vec{x}\,'(t)\|^3} \ \ and \ \ \tau(t) = \frac{\vec{x}\,'(t) \bullet (\vec{x}\,''(t) \times \vec{x}\,'''(t))}{\|\vec{x}\,'(t) \times \vec{x}\,''(t)\|^2} \tag{1.88}$$

for all $t \in I$.

Proof. We write $\vec{x}^*(s) = \vec{x}(t(s))$ and $\vec{v}_k(t) = \vec{v}_k^*(s)$ for $k = 1, 2, 3$. Applying the chain rule in (1.10) of Proposition 1.2.3 and using the fact that $(dt/ds)(s) = \|\vec{x}\,'(t)\|$ as in the proof of Proposition 1.5.2, we first obtain

$$\vec{v}_1(t) = \vec{v}_1^*(s) = \dot{\vec{x}}^*(s) = \frac{d\vec{x}(t(s))}{ds} = \frac{d\vec{x}}{dt}(t(s))\frac{dt}{ds}(s) = \frac{\vec{x}\,'(t)}{\|\vec{x}\,'(t)\|},$$

which is the identity for $\vec{v}_1(t)$ in (1.87).

We apply the chain rule again and get, using the second identity in (1.8) of Proposition 1.2.3

$$\frac{d\vec{v}_1^*}{ds}(s) = \ddot{\vec{x}}^*(s) = \frac{d}{ds}\left(\vec{x}'(t(s))\frac{dt}{ds}(s)\right)$$

$$= \frac{d^2\vec{x}}{dt^2}(t(s))\left(\frac{ds}{dt}(s)\right)^2 + \vec{x}'(t(s))\frac{d^2s}{dt^2}(s). \tag{1.89}$$

Now we apply the chain rule and the third identity in (1.8) of Proposition 1.2.3 to obtain

$$\frac{d^2s}{dt^2}(s) = \frac{d}{dt}\left(\frac{1}{\|\vec{x}'(t(s))\|}\right) = \frac{d}{dt}\left(\frac{1}{\sqrt{\vec{x}'(t(s)) \bullet \vec{x}'(t(s))}}\right)$$

$$= -\frac{1}{2\sqrt{\vec{x}'(t(s)) \bullet \vec{x}'(t(s))}^3}\frac{d}{ds}(\vec{x}'(t(s)) \bullet \vec{x}'(t(s)))$$

$$= -\frac{\frac{d\vec{x}'}{dt}(t(s))\frac{ds}{dt}(s) \bullet \vec{x}'(t(s))}{\|\vec{x}'(t(s))\|^3} = -\frac{\vec{x}''(t(s)) \bullet \vec{x}'(t(s))}{\|\vec{x}'(t(s))\|^4}.$$

It follows from (1.89) and identity (1.5) in Proposition 1.1.3 that

$$\frac{d\vec{v}_1^*}{ds}(s) = \frac{\vec{x}''(t(s))}{\|\vec{x}'(t(s))\|^2} - \frac{\vec{x}'(t(s))(\vec{x}'(t(s)) \bullet \vec{x}''(t(s)))}{\|\vec{x}'(t(s))\|^4}$$

$$= \frac{\vec{x}''(t(s))\|\vec{x}'(t(s))\|^2 - \vec{x}'(t(s))(\vec{x}'(t(s)) \bullet \vec{x}''(t(s)))}{\|\vec{x}'(t(s))\|^4}$$

$$= \frac{\vec{x}'(t(s)) \times (\vec{x}''(t(s)) \times \vec{x}'(t(s)))}{\|\vec{x}'(t(s))\|^4}. \tag{1.90}$$

Since $\vec{v}_2^*(s)$ is a unit vector by definition, the last identity yields

$$\vec{v}_2^*(s) = \vec{v}_2(t) = \frac{\vec{x}'(t) \times (\vec{x}''(t) \times \vec{x}'(t))}{\|\vec{x}'(t) \times (\vec{x}''(t) \times \vec{x}'(t))\|},$$

which is the first identity for $\vec{v}_2(t)$ in (1.87). Furthermore, identity (1.7) in Proposition 1.1.3 yields

$$\|\vec{x}'(t) \times (\vec{x}''(t) \times \vec{x}'(t))\|^2 = \|\vec{x}'(t)\|^2\|\vec{x}''(t) \times \vec{x}'(t)\|^2 - (\vec{x}'(t) \bullet (\vec{x}''(t) \times \vec{x}'(t)))^2$$

$$= \|\vec{x}'(t)\|^2\|\vec{x}'(t) \times \vec{x}''(t)\|^2, \tag{1.91}$$

and we obtain from the third term in (1.90)

$$\vec{v}_2(t) = \frac{\vec{x}''(t)\|\vec{x}'(t)\|^2 - \vec{x}'(t)(\vec{x}'(t) \bullet \vec{x}''(t))}{\|\vec{x}'(t)\| \|\vec{x}'(t) \times \vec{x}''(t)\|},$$

which is the second term in identity (1.87) for $\vec{v}_2(t)$.

We have $\vec{v}_3(t) = \vec{v}_1^*(s) \times \vec{v}_2^*(s)$ by definition, and since $\vec{x}'(t) \times \vec{x}'(t) = \vec{0}$, the identity for $\vec{v}_3(t)$ in (1.87) immediately follows from the identity for $\vec{v}_1(t)$ and the second identity for $\vec{v}_2(t)$ in (1.87).

It follows from the definition of the curvature, the last identity in (1.90), and from (1.90) and (1.7) in Proposition 1.1.3 that

$$\kappa^2(t) = \frac{d\vec{v}_1^*}{ds}(s) \bullet \frac{d\vec{v}_1^*}{ds}(s) = \frac{\|\vec{x}'(t)\|^2 \, \|\vec{x}'(t) \times \vec{x}''(t)\|^2}{\|\vec{x}'(t)\|^8}$$
$$= \frac{\|\vec{x}'(t)\|^2 \, \|\vec{x}''(t)\|^2 - (\vec{x}'(t) \bullet \vec{x}''(t))^2}{\|\vec{x}'(t)\|^6},$$

which yields the identity for $\kappa(t)$ in (1.88).

Finally, we apply the chain rule in (1.89) and obtain

$$\dddot{\vec{x}}^*(s) = \frac{d^3\vec{x}}{dt^3}(t(s)) \left(\frac{ds}{dt}(s)\right)^3 + \alpha(t(s))\vec{x}''(t(s)) + \beta(t(s))\vec{x}'(t(s)).$$

Since the identity (1.86) for the torsion in Proposition 1.7.5 contains the expression $\ddot{\vec{x}}^*(s) \times \dddot{\vec{x}}^*(s)$, and $\dddot{\vec{x}}^*(s)$ is a linear combination of $\vec{x}'(t(s))$ and $\vec{x}''(t(s))$ by (1.89), we do not have to work out the expressions for $\alpha(t(s))$ and $\beta(t(s))$, and obtain

$$\frac{d^2\vec{x}}{dt^2}(t(s)) \times \frac{d^3\vec{x}}{dt^3}(t(s)) = \left(\frac{d^2\vec{x}}{dt^2}(t(s)) \left(\frac{dt}{ds}(s)\right)^2 + \frac{d\vec{x}}{dt}(t(s))\frac{d^2t}{ds^2}(s)\right) \times \frac{d^3\vec{x}}{dt^3}(s).$$

Thus it follows from (1.86) and the identity for $\kappa(t)$ that

$$\tau(t) = \tau^*(s) = \frac{\vec{x}'(t) \bullet (\vec{x}''(t) \times \vec{x}'''(t))}{\kappa(t)^2 \|\vec{x}(t)\|^6} = \frac{\vec{x}'(t) \bullet (\vec{x}''(t) \times \vec{x}'''(t))}{\|\vec{x}'(t) \times \vec{x}''(t)\|^2}.$$

This completes the proof. □

Visualization 1.7.7. *The curvature of the epicycloid in Visualization 1.5.6 (b) for all $\lambda > 0$ is given by (Figure 1.39)*

$$\kappa(t) = \begin{cases} \dfrac{\left|1 + \lambda^2 \frac{R}{r} - \lambda(1 + \frac{R}{r})\cos\left(\frac{r_0}{r}t\right)\right|}{R(1 + \lambda^2 - 2\lambda\cos\left(\frac{r_0}{r}t\right))^{3/2}} & \text{for all } t \in \mathbb{R} \quad \text{if } \lambda \neq 1 \\[4mm] \dfrac{2r + r_0}{4r(r + r_0)} \cdot \dfrac{1}{\left|\sin\left(\frac{r_0}{2r}t\right)\right|} & \text{for } t \neq \dfrac{2k\pi r}{r_0} \ (k \in \mathbb{Z}) \quad \text{if } \lambda = 1. \end{cases} \tag{1.92}$$

Proof. We consider the epicycloid as a curve in the $x^1 x^2$–plane of $\mathbb{E}^3$. Then we obtain from (1.66)

$$\vec{x}''(t) = R\left\{-\cos t + \frac{\lambda R}{r}\cos\left(\frac{R}{r}t\right), -\sin t + \frac{\lambda R}{r}\sin\left(\frac{R}{r}t\right), 0\right\}$$

and then

$$\vec{x}'(t) \times \vec{x}''(t) = \vec{e}^3 R^2 \left(\left(-\sin t + \lambda\sin\left(\frac{R}{r}t\right)\right)\left(-\sin t + \frac{\lambda R}{r}\sin\left(\frac{R}{r}t\right)\right) - \right.$$
$$\left. - \left(\cos t - \lambda\cos\left(\frac{R}{r}t\right)\right)\left(-\cos t + \frac{\lambda R}{r}\cos\left(\frac{R}{r}t\right)\right)\right)$$

$$= \vec{e}^3 R^2 \left(1 + \lambda^2 \frac{R}{r} - \lambda \frac{R}{r} \sin t \sin\left(\frac{R}{r}t\right) - \lambda \sin t \sin\left(\frac{R}{r}t\right) - \right.$$
$$\left. - \lambda \frac{R}{r} \cos t \cos\left(\frac{R}{r}t\right) - \lambda \cos t \cos\left(\frac{R}{r}t\right)\right)$$
$$= \vec{e}^3 R^2 \left(1 + \lambda^2 \frac{R}{r} - \lambda\left(1 + \frac{R}{r}\right) \cos\left(\frac{r_0}{r}t\right)\right).$$

Now we have by the first identity in (1.88) of Proposition 1.7.6

$$\kappa(t) = \frac{\|\vec{x}'(t) \times \vec{x}''(t)\|}{\|\vec{x}'(t)\|^3} = \frac{\left|1 + \lambda^2 \frac{R}{r} - \lambda\left(1 + \frac{R}{r}\right) \cos\left(\frac{r_0}{r}t\right)\right|}{R\left(1 + \lambda^2 - 2\lambda \cos\left(\frac{r_0}{r}t\right)\right)^{3/2}},$$

and, in the special case $\lambda = 1$,

$$\kappa(t) = \frac{\frac{r+R}{r}\left|1 - \cos\left(\frac{r_0}{r}t\right)\right|}{R \cdot 2\sqrt{2}\left(1 - \cos\left(\frac{r_0}{r}t\right)\right)^{3/2}} = \frac{2r + r_0}{r(r + r_0)}\frac{1}{2\sqrt{2}\sqrt{1 - \cos\left(\frac{r_0}{r}t\right)}}$$

$$= \frac{2r + r_0}{4r(r + r_0)}\frac{1}{\left|\sin\left(\frac{r_0}{2r}t\right)\right|} \quad \text{for } t \neq \frac{2\pi kr}{r_0} \quad (k \in \mathbb{Z}). \tag{1.93}$$

□

1.8   THE GEOMETRIC SIGNIFICANCE OF CURVATURE AND TORSION

We already know from Remark 1.6.4 (b) in Section 1.6 that the curvature of a curve is a measure of its deviation from a straight line (Figure 1.39). *The longer the vector*

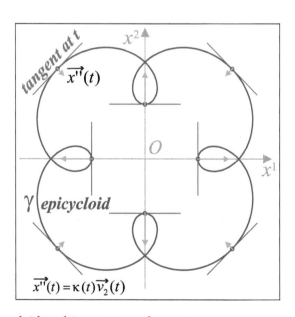

Figure 1.39   An epicycloid and its vectors of curvature

of curvature, that is, the larger the curvature, the faster the curve moves away from its tangent.

A similar geometric interpretation of the torsion can be obtained from the following theorem.

Theorem 1.8.1. *Let γ be a curve with a parametric representation $\vec{x}(s)$ ($s \in I$), and $\ddot{\vec{x}}(s) \neq \vec{0}$ for all $s \in I$. Then γ is a planar curve if and only if $\tau(s) = 0$ for all $s \in I$.*

Proof.

(i) First we assume $\tau \equiv 0$.

Then $\dot{\vec{v}}_3(s) = \vec{0}$ for all s by Frenet's third formula in (1.83), and so $\vec{v}_3 = \vec{c}$ for some constant vector $\vec{c}$. Now $\vec{v}_1 \bullet \vec{v}_3 = 0$ implies

$$\frac{d}{ds}(\vec{x} \bullet \vec{c}) = \vec{v}_1 \bullet \vec{c} = \vec{v}_1 \bullet \vec{v}_3 = 0, \text{ that is, } \vec{x} \bullet \vec{c} = const.$$

Therefore, it follows that $(\vec{x}(s) - \vec{x}_0) \bullet \vec{c} = 0$ for all s with a suitable constant vector $\vec{x}_0$. This shows that $\vec{x}(s)$ is in the plane through the point X_0 and orthogonal to $\vec{c}$.

(ii) Conversely, we assume that $\vec{x}(s)$ is a planar curve.

Then we can write

$$\vec{x}(s) = \vec{x}_0 + \alpha(s)\vec{a} + \beta(s)\vec{b},$$

where the vectors $\vec{a}$ and $\vec{b}$ are independent of s. Differentiating this identity, we see that the vectors $\vec{v}_1, \vec{v}_2, \dot{\vec{v}}_1, \dot{\vec{v}}_2$ are in the plane spanned by the vectors $\vec{a}$ and $\vec{b}$, and since $\vec{v}_3 = \vec{v}_1 \times \vec{v}_2$ by definition, it follows that $\tau(s) = \dot{\vec{v}}_2 \bullet \vec{v}_3 = 0$ for all s.

□

Thus the *torsion of a curve is a measure for its deviation from a plane* (Figure 1.40).

The longer the vector $\tau\vec{v}_3$, that is, the larger the torsion, the faster the curve moves away from the plane spanned by the vectors $\vec{v}_1$ and $\vec{v}_2$.

We shall soon restate this more precisely.

Now we apply Taylor's formula (Proposition 1.2.5) to study the shape of a curve γ in the neighbourhood of an arbitrary given point $X(s_0)$ of γ, and obtain

$$\left\{ \begin{array}{c} \vec{x}(s_0 + \Delta s) = \vec{x}(s_0) + \dot{\vec{x}}(s_0)\Delta s + \dfrac{1}{2}\ddot{\vec{x}}(s_0)(\Delta s)^2 + \dfrac{1}{6}\dddot{\vec{x}}(s_0)(\Delta s)^3 + \vec{r}_4(\Delta s), \\[2mm] \text{where } \lim\limits_{\Delta s \to 0} \dfrac{\vec{r}_4(\Delta s)}{(\Delta s)^3} = \vec{0}. \end{array} \right\}$$

(1.94)

By Definition 1.6.3, we have $\ddot{\vec{x}}(s) = \kappa(s)\vec{v}_2(s)$. Differentiating this identity and applying Frenet's second formula in (1.84) of Theorem 1.7.1, we obtain

$$\dddot{\vec{x}}(s) = \dot{\kappa}(s)\vec{v}_2(s) + \kappa(s)\dot{\vec{v}}_2(s)$$
$$= \dot{\kappa}(s)\vec{v}_2(s) + \kappa(s)\left(-\kappa(s)\right)\vec{v}_1(s) + \tau(s)\vec{v}_3(s))$$
$$= -\kappa^2(s)\vec{v}_1(s) + \dot{\kappa}(s)\vec{v}_2(s) + \kappa(s)\tau(s)\vec{v}_3(s).$$

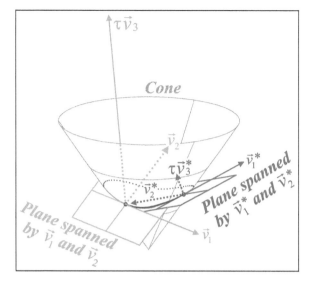

Figure 1.40 The torsion as a measure of the deviation of a curve from a plane

We substitute these expressions for $s = s_0$, and $\vec{v}_1(s_0) = \dot{\vec{x}}(s_0)$ in (1.94); this yields

$$\vec{x}(s_0 + \Delta s) - \vec{x}(s_0) = \vec{v}_1(s_0)\Delta s \left(1 - \frac{\kappa^2(s_0)}{6}(\Delta s)^2\right)$$
$$+ \vec{v}_2(s_0)(\Delta s)^2 \left(\frac{\kappa(s_0)}{2} + \frac{\dot{\kappa}(s_0)}{6}\Delta s\right)$$
$$+ \vec{v}_3(s_0)(\Delta s)^3 \cdot \frac{\kappa(s_0)\tau(s_0)}{6} + \vec{r}_4(\Delta s). \qquad (1.95)$$

Thus we have shown:

Remark 1.8.2. *Let γ be a curve with a parametric representation $\vec{x}(s)$ $(s \in I)$. Then the tangent at s is a first-order approximation of γ at s. The second-order approximation of γ at s lies in the plane spanned by the vectors $\vec{v}_1(s)$ and $\vec{v}_2(s)$; the curvature $\kappa(s)$ of γ appears in the coefficient of the second-order term and measures the deviation of γ from its tangent at s. Finally the torsion $\tau(s)$ appears in the third-order term and measures the deviation of γ from the plane spanned by the vectors $\vec{v}_1(s)$ and $\vec{v}_2(s)$.*

The so–called *osculating, normal* and *rectifying planes* can be defined at every point of a curve.

Definition 1.8.3. Let γ be a curve with a parametric representation $\vec{x}(s)$ $(s \in I)$ and $X(s_0)$ be a point on γ. Then the planes through $X(s_0)$ orthogonal to the binormal vector $\vec{v}_3(s_0)$, the tangent vector $\vec{v}_1(s_0)$, and the principal normal vector $\vec{v}_2(s_0)$ are called the *osculating, normal* and *rectifying planes of* γ *at* $X(s_0)$, respectively (Figure 1.41).

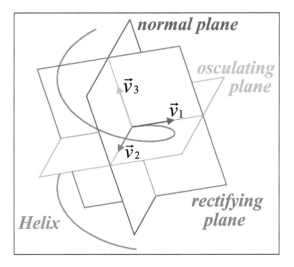

Figure 1.41 Osculating, normal and rectifying planes of a helix

We continue our studies of a curve γ with a parametric representation $\vec{x}(s)$ in a neighbourhood of an arbitrary given point $X(s_0)$ of γ, and consider certain approximations of the orthogonal projections of $\vec{x}(s)$ to the osculating, normal and rectifying planes at $X(s_0)$. We choose a coordinate system for $\mathbb{R}^3$ with its origin in the point $X(s_0)$, and the directions of the x^1–, x^2– and x^3–axes given by the vectors $\vec{v}_1(s_0)$, $\vec{v}_2(s_0)$ and $\vec{v}_3(s_0)$, respectively. Omitting the argument s_0, we obtain from (1.95)

$$x^1 = \left(1 - \frac{\kappa^2}{6}(\Delta s)^2\right)\Delta s, \ x^2 = \left(\frac{\kappa}{2} + \frac{\dot{\kappa}}{6}\Delta s\right)(\Delta s)^2 \text{ and } x^3 = \frac{\kappa\tau}{6}(\Delta s)^3.$$

Trying to eliminate the parameter Δs in these equations, we get

$$x^2 = \frac{\kappa}{2}(x^1)^2 + (\Delta s)^3(\cdots) \quad \text{in the osculating plane,}$$

$$(x^3)^2 = \frac{2\tau^2}{9\kappa}(x^2)^3 + (\Delta s)^7(\cdots) \quad \text{in the normal plane}$$

and

$$x^3 = \frac{\kappa\tau}{6}(x^1)^3 + (\Delta s)^3(\cdots) \quad \text{in the rectifying plane.}$$

Thus we obtain the following quantitative characterization of curves in a neighbourhood of any of their points. For $\kappa, \tau \neq 0$ and $\Delta s = 0$, the last three equations yield that the approximations of the projections at the point $X(s_0)$ are

    a *quadratic parabola* in the osculating plane,

    a *Neil's parabola* in the normal plane and

    a *cubic parabola* in the rectifying plane

(Figure 1.42). They are referred to as the *local canonical form of a curve*.

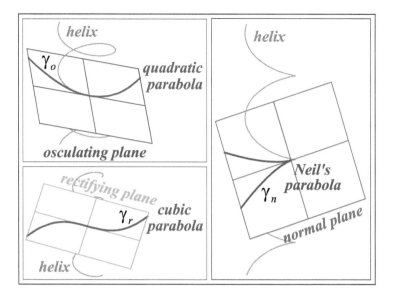

Figure 1.42 The approximations of a helix in its osculating, normal and rectifying planes

We close this section with an example.

Visualization 1.8.4. *We consider the helix with a parametric representation*

$$\vec{x}(s) = \{r\cos\omega s, r\sin\omega s, h\omega s\} \ (s \in \mathbb{R}).$$

We obtain by Visualization 1.6.7 (b),

$$\kappa(s) = r\omega^2 = \frac{r}{r^2 + h^2} \quad (s \in \mathbb{R}). \tag{1.96}$$

Furthermore it follows from

$$\vec{v}_1(s) = \omega\{-r\sin\omega s, r\cos\omega s, h\},$$
$$\vec{v}_2(s) = -\{\cos\omega s, \sin\omega s, 0\},$$
$$\vec{v}_3(s) = \omega\{h\sin\omega s, -h\cos\omega s, r\}$$

and

$$\dot{\vec{v}}_2(s) = \omega\{\sin\omega s, -\cos\omega s, 0\}$$

that

$$\tau(s) = \dot{\vec{v}}_2(s) \bullet \vec{v}_3(s) = h\omega^2 = \frac{h}{r^2 + h^2} \ (s \in \mathbb{R}). \tag{1.97}$$

Thus the approximations of the helix in an arbitrary point $X(s)$ are given by

$$x^2 = \frac{\kappa(s)}{2}(x^1)^2 = \frac{1}{2}r\omega^2(x^1)^2 \qquad \text{in the osculating plane,}$$

$$(x^3)^2 = \frac{2}{9}\frac{\tau^2(s)}{\kappa(s)}(x^2)^3 = \frac{2}{9}\frac{h^2\omega^2}{r}(x^2)^3 \qquad \textit{in the normal plane}$$

and

$$x^3 = \frac{1}{6}\kappa(s)\tau(s)(x^1)^3 = \frac{1}{6}rh\omega^4(x^1)^3 \qquad \textit{in the rectifying plane}$$

(Figure 1.42).

1.9 OSCULATING CIRCLES AND SPHERES

Osculating circles and *spheres* are used for certain approximation purposes of curves.

Definition 1.9.1. Let γ be a curve with a parametric representation $\vec{x}(s)$ $(s \in I)$ and $\ddot{\vec{x}}(s) \neq \vec{0}$ for all $s \in I$, and let $s_0 \in I$ be fixed and $X(s_0)$ be a point of γ. In the osculating plane of γ at $X(s_0)$, we consider the circle line $C(s_0)$ with radius $\rho(s_0) = 1/\kappa(s_0)$ and its centre at the point $M(s_0)$ with the position vector

$$\vec{m}(s_0) = \vec{x}(s_0) + \rho(s_0)\vec{v}_2(s_0).$$

Then $C(s_0)$ is called the *osculating circle of γ at s_0*. The point $M(s_0)$ is called the *centre of curvature of γ at s_0* (Figure 1.43).

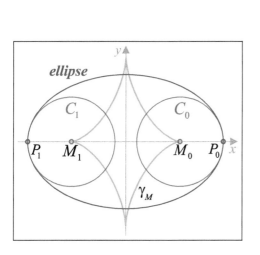

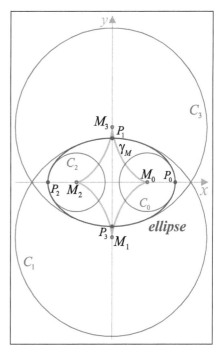

Figure 1.43 An ellipse with osculating circles and centres of curvature

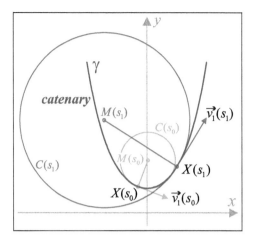

Figure 1.44 Osculating circles as second-order approximations of a curve

Remark 1.9.2. *The osculating circle $C(s_0)$ of a curve γ at s_0 is a second-order approximation of γ at a point of intersection $X(s_0) \in \gamma \cap C(s_0)$, that is, the tangents at $X(s_0)$ of γ and $C(s_0)$ coincide, and γ and $C(s_0)$ have the same curvature at $X(s_0)$ (Figure 1.44).*

Proof. The osculating circle of γ at $X(s_0)$ has a parametric representation

$$\vec{y}_{s_0}(t) = \vec{m}(s_0) + \rho(s_0)\left(\vec{v}_1(s_0) \sin t + \vec{v}_2(s_0) \cos t\right) \text{ for } t \in [0, 2\pi].$$

This yields

$$\vec{y}\,'_{s_0}(t) = \rho(s_0)(\vec{v}_1(s_0) \cos t - \vec{v}_2(s_0) \sin t) \text{ for } t \in [0, 2\pi].$$

Choosing $t = \pi$, we obtain

$$\vec{y}_{s_0}(\pi) = \vec{m}(s_0) - \rho(s_0)\vec{v}_2(s_0) = \vec{x}(s_0),$$
$$\vec{y}\,'_{s_0}(\pi) = -\rho(s_0)\vec{v}_1(s_0)$$

and

$$\frac{\vec{y}\,'_{s_0}(\pi)}{\|\vec{y}\,'_{s_0}(\pi)\|} = -\vec{v}_1(s_0).$$

The last identity shows that the tangents at $X(s_0)$ of γ and $C(s_0)$ coincide. Furthermore, the circle $C(s_0)$ has the curvature $\kappa(s_0)$ by Visualization 1.6.7 (a), and $\kappa(s_0)$ is the curvature of γ at $X(s_0)$. $\qquad\square$

Visualization 1.9.3. *The centre of curvature at s along the curve γ with a parametric representation (1.61) is given by*

$$\vec{x}_m(s) = \left\{ \frac{\cos s}{2 \cosh s}, \frac{\sin s}{2 \cosh s}, s - \frac{1}{2} \tanh s \right\} - \frac{1}{2} \sinh s\{-\sin s, \cos s, 0\}. \qquad (1.98)$$

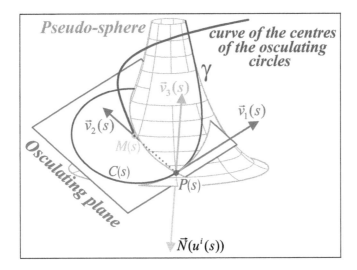

Figure 1.45 Osculating circle at a point of the curve of Visualization 1.9.3

Hence, the osculating circle of γ at s has a parametric representation (Figure 1.45)

$$\vec{y}_{m,s}(t) = \vec{x}_m(s) + \frac{\cosh s}{s}\left(\cos t\,\vec{v}_1(s) + \sin t\,\vec{v}_2(s)\right) \ \text{ for } t \in (0, 2\pi). \tag{1.99}$$

Proof. The radius of curvature of γ at s is by (1.82)

$$\rho(s) = \frac{1}{\kappa(s)} = \frac{\cosh s}{2}.$$

Therefore, we obtain from (1.61) and (1.80) for the centre of curvature of γ at s

$$\vec{x}_m(s) = \vec{x}^*(s) + \rho(s)\vec{v}_2(s) =$$
$$\left\{\frac{\cos s}{\cosh s}, \frac{\sin s}{\cosh s}, s - \frac{\sinh s}{\cosh s}\right\} - \left\{\frac{\cos s}{2\cosh s}, \frac{\sin s}{2\cosh s}, -\frac{\sinh s}{2\cosh s}\right\} -$$
$$\frac{\sinh s}{2}\{-\sin s, \cos s, 0\} =$$

$$\left\{\frac{\cos s}{2\cosh s}, \frac{\sin s}{2\cosh s}, s - \frac{1}{2}\tanh s\right\} - \frac{\sinh s}{2}\{-\sin s, \cos s, 0\}$$

which is (1.98).
Now (1.99) follows immediately from the definition of the osculating circle of a curve at s. □

Now we consider higher order approximations of curves.

Definition 1.9.4. Let γ be a curve with a parametric representation $\vec{x}(s)$ ($s \in I$) with non–vanishing curvature and torsion, and let $s_0 \in I$ be fixed. A sphere that is an approximation of highest possible order of γ at s_0 is called an *osculating sphere* of γ at s_0 (Figure 1.46).

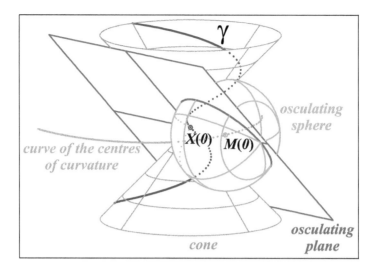

Figure 1.46 A curve, the curve of its centres of curvature, the osculating plane, circle and sphere at a point

Theorem 1.9.5. *Let γ be a curve with a parametric representation $\vec{x}(s)$ $(s \in I)$ and non–vanishing curvature and torsion, and let $s_0 \in I$ be fixed. Then the osculating sphere of γ at s_0 is uniquely determined; the position vector of its centre $M(s_0)$ and its radius $r(s_0)$ are given by*

$$\vec{m}(s_0) = \vec{x}(s_0) + \frac{1}{\kappa(s_0)}\vec{v}_2(s_0) - \frac{\dot{\kappa}(s_0)}{\tau(s_0)\kappa^2(s_0)}\vec{v}_3(s_0) \qquad (1.100)$$

and

$$r(s_0) = \frac{1}{\kappa(s_0)}\sqrt{1 + \frac{\dot{\kappa}^2(s_0)}{\kappa^2(s_0)\tau^2(s_0)}}. \qquad (1.101)$$

Proof. The sphere $C_r(M)$ with radius r and centre at M is given by the equation $(\vec{x} - \vec{m})^2 - r^2 = 0$. We define the function $f : I \to \mathbb{R}$ by $f(s) = (\vec{x}(s) - \vec{m})^2 - r^2$. If $C_r(M)$ is to be an approximation of $\vec{x}(s)$ of highest possible order at s_0, then the function f and its derivatives up to the third order should vanish at s_0. Since $\dot{\vec{x}}(s) = \vec{v}_1(s)$, $\vec{v}_1(s) \bullet \vec{v}_1(s) = 1$ and $\dot{\vec{v}}_1(s) = \kappa(s)\vec{v}_2(s)$ it follows that

$$f(s) = [\vec{x}(s) - \vec{m}]^2 - r^2 = 0, \qquad (1.102)$$

$$\dot{f}(s) = 2[\vec{x}(s) - \vec{m}] \bullet \dot{\vec{x}}(s) = 2[\vec{x}(s) - \vec{m}] \bullet \vec{v}_1(s),$$

$$\ddot{f}(s) = 2\left(\dot{\vec{x}}(s) \bullet \vec{v}_1(s) + [\vec{x}(s) - \vec{m}] \bullet \dot{\vec{v}}_1(s)\right)$$
$$= 2\left(1 + \kappa(s)[\vec{x}(s) - \vec{m}] \bullet \vec{v}_2(s)\right) \qquad (1.103)$$

and, since $\vec{v}_1(s) \bullet \vec{v}_2(s) = 0$ for all $s \in I$, we obtain, applying Frenet's second formula in (1.84) of Theorem 1.7.1

$$\dddot{f}(s) = \left(\dot{\kappa}(s)[\vec{x}(s) - \vec{m}] \bullet \vec{v}_2(s) + \kappa(s)[\vec{x}(s) - \vec{m}] \bullet \dot{\vec{v}}_2(s) \right)$$
$$= 2 \left([\vec{x}(s) - \vec{m}] \bullet [\dot{\kappa}(s)\vec{v}_2(s) - \kappa^2(s)\vec{v}_1(s) + \kappa(s)\tau(s)\vec{v}_3(s)] \right). \qquad (1.104)$$

To establish the identity in (1.100), we write

$$\vec{x}(s) - \vec{m} = a_1(s)\vec{v}_1(s) + a_2(s)\vec{v}_2(s) + a_3(s)\vec{v}_3(s) \text{ for all } s \in I,$$

and substitute this in (1.102), (1.103) and (1.104) to obtain

$$0 = \dot{f}(s_0) = 2(a_1(s_0)\vec{v}_1(s_0) + a_2(s_0)\vec{v}_2(s_0) + a_3(s_0)\vec{v}_3(s_0)) \bullet \vec{v}_1(s_0) = 2a_1(s_0)$$

which implies $a_1(s_0) = 0$,

$$0 = \ddot{f}(s_0) = 2 + 2\kappa(s_0)a_2(s_0)$$

which implies

$$a_2(s_0) = -\frac{1}{\kappa(s_0)}$$

and

$$0 = \dddot{f}(s_0) = 2 \left(-\frac{\dot{\kappa}(s_0)}{\kappa(s_0)} + \kappa(s_0)\tau(s_0)a_3(s_0) \right)$$

which implies

$$a_3(s_0) = \frac{\dot{\kappa}(s_0)}{\kappa^2(s_0)\tau(s_0)}.$$

This shows (1.100). Finally, (1.100) yields

$$r^2(s_0) = (\vec{x}(s_0) - \vec{m})^2 = \frac{1}{\kappa^2(s_0)} \left(\vec{v}_2(s_0) - \frac{\dot{\kappa}(s_0)}{\kappa(s_0)\tau(s_0)} \vec{v}_3(s_0) \right)^2$$
$$= \frac{1}{\kappa^2(s_0)} \left(1 + \frac{\dot{\kappa}^2(s_0)}{\kappa^2(s_0)\tau^2(s_0)} \right),$$

and (1.101) follows. □

Visualization 1.9.6. *(a) Let γ be the catenary of Visualization 1.6.6 (a) in the xy–plane, given by the parametric representation*

$$\vec{x}(s) = \left\{ c \sinh^{-1} \left(\frac{s}{c} \right), \sqrt{s^2 + c^2} \right\} \quad (s \in \mathbb{R}).$$

It follows from

$$\vec{v}_2(s) = \frac{1}{\sqrt{c^2 + s^2}} \{c, -s\} \text{ and } \kappa(s) = \frac{c}{c^2 + s^2} \qquad (1.105)$$

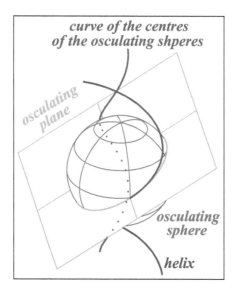

Figure 1.47 A helix, an osculating plane, circle and sphere, and the curve of its osculating spheres

that the centres of curvature of γ are given by

$$\vec{m}(s) = \vec{x}(s) + \frac{\vec{v}_2(s)}{\kappa(s)} = \left\{ c \sinh^{-1} \left(\frac{s}{c} \right) - \frac{s}{c} \sqrt{c^2 + s^2}, 2\sqrt{c^2 + s^2}, \right\} \quad (s \in \mathbb{R}).$$

(b) Let γ be the helix of Visualization 1.8.4 with a parametric representation

$$\vec{x}(s) = \{r \cos \omega s, r \sin \omega s, h\omega s\} \ (s \in \mathbb{R}) \ where \ \omega = \frac{1}{r^2 + h^2}.$$

Then we have $\kappa(s) = r\omega^2$, $\dot{\kappa}(s) = 0$ and $\tau(s) = h\omega^2$ for all s. Since $\dot{\kappa}(s) = 0$ for all s, the centres $M(s)$ of the osculating circles and spheres of the helix coincide and their radii $r(s)$ have identical values, given by

$$\vec{m}(s) = \{r \cos \omega s, r \sin \omega s, h\omega s\} + \frac{1}{r\omega^2}\{-\cos \omega s, -\sin \omega s, 0\}$$

$$= \left\{ -\frac{h^2}{r} \cos \omega s, -\frac{h^2}{r} \sin \omega s, h\omega s \right\}$$

and $r(s) = 1/(r\omega^2) = r + h^2/r \ (s \in \mathbb{R})$. Thus the centres of the osculating circles and spheres of a helix again are on a helix (Figure 1.47).

Visualization 1.9.7. *The centre and radius of the osculating sphere at s along the curve γ with a parametric representation (1.61) are given by (Figure 1.48)*

$$\vec{m}(s) = \left\{ \frac{1}{2} \cosh s \cos s, \frac{1}{2} \cosh s \sinh s, s \right\} - \frac{1}{2} \sinh s \{-\sin s, \cos s, 0\} \tag{1.106}$$

and

$$r_m(s) = \frac{1}{2}\sqrt{\cosh^2 s + \sinh^2 s}. \tag{1.107}$$

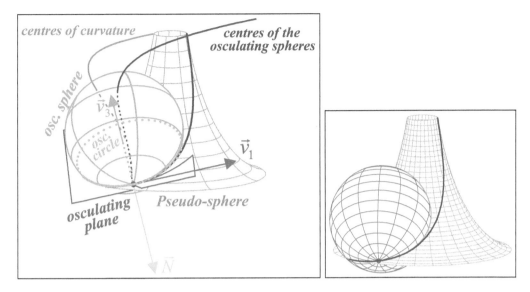

Figure 1.48   The osculating sphere at a point of the curve γ of Visualization 1.9.7

Proof. Since $\dot{\rho}(s) = \sinh(s)/2$ and $\tau(s) = 1$ by (1.85), we obtain the centre of the osculating sphere of γ from (1.81), and (1.100) in Theorem 1.9.5

$$\vec{m}(s) = \vec{x}^*(s) + \rho(s)\vec{v}_2(s) + \frac{\dot{\rho}(s)}{\tau(s)}\vec{v}_3(s) = \vec{x}_m(s) + \frac{\sinh s}{2}\vec{v}_3(s)$$

$$= \left\{ \frac{\cos s}{2\cosh s}, \frac{\sin s}{2\cosh s}, s - \frac{1}{2}\tanh s \right\} - \frac{1}{2}\sinh s\{-\sin s, \cos s\}$$

$$+ \left\{ \frac{\sinh^2 s \cos s}{2\cosh s}, \frac{\sinh^2 s \sin s}{2\cosh s}, \frac{\tanh s}{2s} \right\}$$

$$= \left\{ \frac{1}{2}\cosh s \cos s, \frac{1}{2}\cosh s \sinh s, s \right\} - \frac{1}{2}\sinh s\{-\sin s, \cos s, 0\}$$

which is (1.106). Finally, we obtain for the radius of the osculating sphere

$$r_m^2(s) = \rho^2(s) + \left(\frac{\dot{\rho}(s)}{\tau(s)}\right)^2 = \frac{1}{4}\left(\cosh^2 s + \sinh^2 s\right),$$

and (1.107) is an immediate consequence.   □

1.10   INVOLUTES AND EVOLUTES

With every curve we may associate two new curves that are closely related to one another, namely *involutes* and *evolutes*.

We start with a geometric definition of involutes.

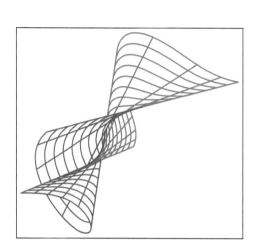 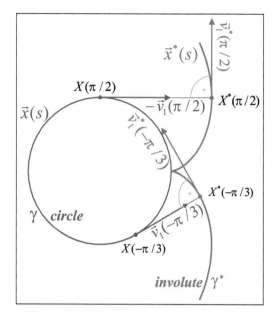

Figure 1.49 Left: Tangent surface. Right: The definition of an involute

Definition 1.10.1. Let γ be a curve with a parametric representation $\vec{x}(s)$ ($s \in I$) and $\vec{v}_1(s)$ its tangent vector at s. Then

$$\vec{y}(s,t) = \vec{x}(s) + t\vec{v}_1(s) \ ((s,t) \in I \times \mathbb{R})$$

is a parametric representation for the so–called *tangent surface of* γ (left in Figure 1.49). Curves on the tangent surface of γ which intersect the tangents to γ at right angles are called *involutes of* γ (in Figures 1.49 (right) and 1.50).

The following result gives a parametric representation for involutes.

Theorem 1.10.2. *The involutes of a curve γ with a parametric representation $\vec{x}(s)$ and tangent vectors $\vec{v}_1(s)$ ($s \in I$) have a parametric representation*

$$\vec{x}^*(s) = \vec{x}(s) + (s_0 - s)\vec{v}_1(s) \ (s \in I), \ \text{where } s_0 \in \mathbb{R} \text{ is an arbitrary constant.} \quad (1.108)$$

There is one and only one involute that corresponds to a given value of the constant s_0; thus every curve has a one–parametric collection of involutes (Figure 1.51).

Proof. A curve γ^* in the tangent surface of γ is given by $\vec{x}^*(s) = \vec{x}(s) + t(s)\vec{v}_1(s)$. By definition, γ^* is an involute of γ if

$$\frac{d\vec{x}^*}{ds}(s) \bullet \vec{v}_1(s) = 0,$$

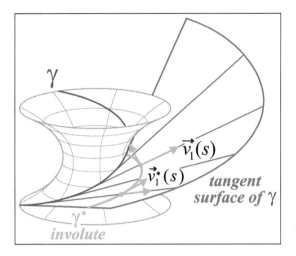

Figure 1.50 An involute in the tangent surface of a curve

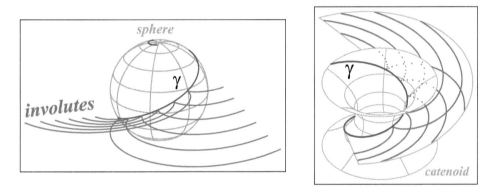

Figure 1.51 Involutes of a curve on a sphere (left) catenoid (right)

that is,

$$\left(\vec{v}_1(s) + \frac{dt}{ds}(s)\vec{v}_1(s) + t(s)\dot{\vec{v}}_1(s)\right) \bullet \vec{v}_1(s) = 1 + \frac{dt}{ds}(s) = 0.$$

This implies $t(s) = s_0 - s$ where $s_0 \in \mathbb{R}$ is a constant. □

There is a simple relation between the curvature of an involute of a curve γ and the curvature and torsion of γ.

Theorem 1.10.3. *Let γ be curve with a parametric representation $\vec{x}(s)$ ($s \in I$), non–vanishing curvature κ, and torsion τ. Furthermore, let $s_0 \in I$ be given. Then the curvature κ^* of the involute γ^* with a parametric representation (1.108) is given by*

$$\kappa^*(s) = \frac{\sqrt{\kappa^2(s) + \tau^2(s)}}{|s_0 - s|\kappa(s)} \quad \text{for all } s \in I \setminus \{s_0\}; \tag{1.109}$$

in particular, if γ is a planar curve, then

$$\kappa^*(s) = \frac{1}{|s_0 - s|} \quad \text{for all } s \in I \setminus \{s_0\}. \tag{1.110}$$

Proof. It follows from (1.108) and Frenet's first and second formulae in (1.84) of Theorem 1.7.1 that

$$\frac{d\vec{x}^*(s)}{ds} = \vec{v}_1(s) - \vec{v}_1(s) + (s_0 - s)\dot{\vec{v}}_1(s) = (s_0 - s)\kappa(s)\vec{v}_2(s) \tag{1.111}$$

and

$$\frac{d^2\vec{x}^*(s)}{ds^2} = \vec{v}_2(s)\frac{d((s_0 - s)\kappa(s))}{ds} + (s_0 - s)\kappa(s)\dot{\vec{v}}_2(s)$$

$$= \vec{v}_2(s)\frac{d((s_0 - s)\kappa(s))}{ds} + (s_0 - s)\kappa(s)(-\kappa(s)\vec{v}_1(s) + \tau(s)\vec{v}_3(s)).$$

Since $\vec{v}_j(s) \bullet \vec{v}_k(s) = \delta_{jk}$ for $j, k = 1, 2, 3$, $\vec{v}_2(s) \times \vec{v}_2(s) = \vec{0}$, $\vec{v}_2(s) \times \vec{v}_1(s) = -\vec{v}_3(s)$ and $\vec{v}_2(s) \times \vec{v}_3(s) = \vec{v}_1(s)$, we obtain

$$\left\|\frac{d\vec{x}^*(s)}{ds}\right\|^3 = |s_0 - s|^3\kappa^3(s),$$

$$\frac{d\vec{x}^*(s)}{ds} \times \frac{d^2\vec{x}^*(s)}{ds^2} = (s_0 - s)^2\kappa^2(s)\left(\kappa(s)\vec{v}_3(s) + \tau(s)\vec{v}_1(s)\right),$$

and so by (1.88) in Proposition 1.7.6

$$\kappa^*(s) = \frac{\left\|\dfrac{d\vec{x}^*(s)}{ds} \times \dfrac{d^2\vec{x}^*(s)}{ds^2}\right\|}{\left\|\dfrac{d\vec{x}^*(s)}{ds}\right\|^3} = \frac{(s_0 - s)^2\kappa^2(s)\sqrt{\kappa^2(s) + \tau^2(s)}}{|s_0 - s|^3\kappa^3(s)} = \frac{\sqrt{\kappa^2(s) + \tau^2(s)}}{|s_0 - s|\kappa(s)},$$

that is, (1.109) holds.

If γ is a planar curve, then $\tau(s) = 0$ for all $s \in I$, and (1.110) is an immediate consequence of (1.109). □

Visualization 1.10.4. *The involutes of a helix with a parametric representation*

$$\vec{x}(s) = \{r\cos\omega s, r\sin\omega s, h\omega s\} \ (s \in \mathbb{R}), \ \text{where} \ \omega = \frac{1}{r^2 + h^2},$$

are the lines of intersection of the tangents to the helix and the plane $x^3 = h\omega s_0$, where s_0 is the constant in the parametric representation (1.108) in Theorem 1.10.2 (Figure 1.52). If s^* denotes the arc length along the lines of intersection, then their curvatures are given by

$$\kappa^*(s^*) = \frac{1}{\sqrt{2r|s^*|}} \ (s^* \in \mathbb{R} \setminus \{0\}). \tag{1.112}$$

Proof. The tangent to the helix at s has a parametric representation

$$\vec{y}(s) = \vec{x}(s) + \lambda\vec{v}_1(s) \ \text{for} \ \lambda \in \mathbb{R}, \ \text{where} \ \vec{v}_1(s) = \{-\omega r\sin\omega s, \omega r\cos\omega s, h\omega\} \tag{1.113}$$

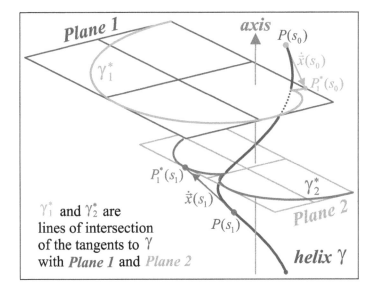

Figure 1.52 The intersections of the tangents to a helix with planes orthogonal to its axis

by Visualization 1.8.4, hence its intersection with the plane $x^3 = h\omega s_0$ is given by $h\omega s + \lambda_0 h\omega = h\omega s_0$, that is, $\lambda_0 = s_0 - s$. Substituting this in the parametric representation (1.113) of the tangent to the helix, we obtain

$$\vec{y}(s; \lambda_0) = \vec{x}(s) + (s_0 - s)\vec{v}_1(s) = x^*(s) \ (s \in \mathbb{R})$$

for the line γ^* of intersection of the tangents to the helix and the plane $x^2 = h\omega s_0$. Its curvature $\kappa^*(s)$ is given by (1.109) in Theorem 1.10.3, and by Visualization 1.8.4

$$\kappa^*(s) = \frac{\sqrt{\kappa^2(s) + \tau^2(s)}}{|s - s_0|\kappa(s)} = \frac{\sqrt{1 + \frac{h^2}{r^2}}}{|s - s_0|} = \frac{\sqrt{r^2 + h^2}}{r|s - s_0|} = \frac{1}{\sqrt{r}\sqrt{\kappa}|s - s_0|}. \tag{1.114}$$

It follows from (1.111) in the proof of Theorem 1.10.3 that the arc length s^* along γ^* is given by

$$s^* = \int |s_0 - s|\kappa(s)\,ds = \begin{cases} \kappa\dfrac{(s - s_0)^2}{2} & \text{for } s \geq s_0 \\ -\kappa\dfrac{(s - s_0)^2}{2} & \text{for } s < s_0, \end{cases}$$

that is,

$$\sqrt{\kappa}|s - s_0| = \sqrt{2|s^*|}.$$

Substituting this in (1.114), we obtain

$$\kappa^*(s^*) = \frac{1}{\sqrt{2r|s^*|}} \ (s^* \in \mathbb{R} \setminus \{0\})$$

for the curvature of γ^* with respect to its arc length, that is, (1.112) is satisfied. □

Now we consider a different type of involutes.

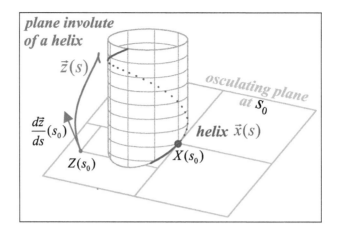

Figure 1.53 The definition of a plane involute

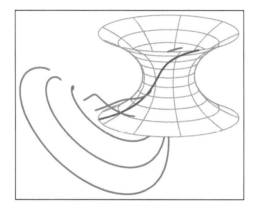

Figure 1.54 Plane involutes of a curve on a catenoid

Definition 1.10.5. Let γ be a curve. A curve that intersects the osculating planes of γ orthogonally is called a *plane involute* (Figures 1.53 and 1.54).

The next result gives a parametric representation for a plane involute.

Theorem 1.10.6. *Let γ be a curve with a parametric representation $\vec{x}(s)$ $(s \in I)$ and non–vanishing curvature κ and torsion τ. We put*

$$\psi(s) = \int_0^s \kappa(\sigma) \, d\sigma \text{ for all } s \in I.$$

Then the plane involutes of γ are given by a parametric representation

$$\vec{z}(s) = \vec{x}(s) - \left(\cos \psi(s) \left(\int \cos \psi(s) \, ds + k_1 \right) + \sin \psi(s) \left(\int \sin \psi(s) \, ds + k_2 \right) \right) \vec{v}_1(s)$$

$$+ \left(\sin \psi(s) \left(\int \cos \psi(s) \, ds + k_1 \right) - \cos \psi(s) \left(\int \sin \psi(s) \, ds + k_2 \right) \right) \vec{v}_2(s),$$

$$(1.115)$$

where $k_1, k_2 \in \mathbb{R}$ are constants of integration.

Proof. The vector $\vec{z}(s) - \vec{x}(s)$ is in the osculating plane of the curve γ by Definition 1.10.5, that is, there are functions $c_1, c_2 : I \to \mathbb{R}$ such that

$$\vec{z}(s) - \vec{x}(s) = c_1(s)\vec{v}_1(s) + c_2(s)\vec{v}_2(s) \text{ for all } s \in I. \tag{1.116}$$

Furthermore, since the plane involute intersects the osculating plane of γ orthogonally, it follows that

$$\frac{d\vec{z}(s)}{ds} \bullet \vec{v}_1(s) = \frac{d\vec{z}(s)}{ds} \bullet \vec{v}_2(s) = 0 \text{ for all } s \in I. \tag{1.117}$$

First, differentiating (1.116) and applying Frenet's first and second formulae in (1.84) of Theorem 1.7.1, we obtain, omitting the parameter s

$$\begin{aligned}
\frac{d\vec{z}}{ds} &= (1 + \dot{c}_1)\,\vec{v}_1 + c_1\dot{\vec{v}}_1 + \dot{c}_2\vec{v}_2 + c_2\dot{\vec{v}}_2 \\
&= (1 + \dot{c}_1)\,\vec{v}_1 + c_1\kappa\vec{v}_2 + \dot{c}_2\vec{v}_2 - c_2\kappa\vec{v}_1 + c_2\tau\vec{v}_3 \\
&= (1 + \dot{c}_1 - c_2\kappa)\,\vec{v}_1 + (c_1\kappa + \dot{c}_2)\,\vec{v}_2 + c_2\tau\vec{v}_3. \tag{1.118}
\end{aligned}$$

We obtain the following system of differential equations from (1.117)

$$\dot{c}_1 - c_2\kappa = -1 \text{ and } \dot{c}_2 + c_1\kappa = 0. \tag{1.119}$$

Introducing the complex function $w = c_1 + ic_2$, we can write the system in (1.119) as

$$\frac{dw}{ds} + i\kappa w + 1 = 0.$$

This differential equation has the general solution

$$w(s) = -\exp\left(-i\psi(s)\right) \int \exp\left(i\psi(s)\right) ds + k, \tag{1.120}$$

where $k = k_1 + ik_2$ is a constant of integration.

Since $\vec{z}(s)$ is a parametric representation for the plane involute, we must have $\dot{\vec{z}} \neq \vec{0}$, that is, $\tau(s) \neq 0$ for all $s \in I$, by (1.118). It follows from (1.120) that

$$\begin{aligned}
w &= -(\cos\psi - i\sin\psi)\left(\int (\cos\psi + i\sin\psi)\,ds + k_1 + ik_2\right) \\
&= -\cos\psi \int \cos\psi\,ds - k_1 - \sin\psi \int \sin\psi\,ds - k_2 \\
&\quad - i\left(\cos\psi \int \sin\psi\,ds - k_1 - \sin\psi \int \cos\psi\,ds + k_2\right),
\end{aligned}$$

and this yields

$$c_1 = -\left(\cos\psi \int \cos\psi\,ds + k_1 + \sin\psi \int \sin\psi\,ds + k_2\right)$$

and

$$c_2 = \sin\psi \int \cos\psi\,ds + k_1 - \cos\psi \int \sin\psi\,ds - k_2.$$

This establishes the parametric representation given in (1.115). $\qquad\square$

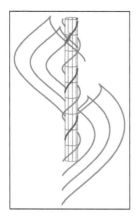 

Figure 1.55 Plane involutes of helices

Visualization 1.10.7. *We consider the helix of Visualization 1.6.7 (b) where*

$$\vec{v}_1(s) = \{-\omega r \sin \omega s, \omega r \cos \omega s, h\omega\},$$
$$\vec{v}_2(s) = \{-\cos \omega s, -\sin \omega s, 0\}$$

and

$$\kappa(s) = \omega^2 r \text{ for all } s \in \mathbb{R},$$

hence $\psi(s) = \omega^2 rs$, *and (Figure 1.55)*

$$\vec{z}(s) = \vec{x}(s) - \left[\cos\left(\omega^2 rs\right)\left(\frac{1}{\omega^2 r}\sin\left(\omega^2 rs\right) + k_1\right)\right.$$
$$\left. + \sin\left(\omega^2 rs\right)\left(-\frac{1}{\omega^2 r}\cos\left(\omega^2 rs\right) + k_2\right)\vec{v}_1(s)\right]$$
$$+ \left[\sin\left(\omega^2 rs\right)\left(\frac{1}{\omega^2 r}\sin\left(\omega^2 rs\right) + k_1\right)\right.$$
$$\left. - \cos\left(\omega^2 rs\right)\left(-\frac{1}{\omega^2 r}\cos\left(\omega^2 rs\right) + k_2\right)\vec{v}_2(s)\right]$$

$$= \vec{x}(s) - \left(k_1\cos\left(\omega^2 rs\right) + k_2\sin\left(\omega^2 rs\right)\right)\vec{v}_1(s)$$
$$+ \left(\frac{1}{\omega^2 r} + k_1\sin\left(\omega^2 rs\right) - k_2\sin\left(\omega^2 rs\right)\right)\vec{v}_2(s) \text{ for all } s \in \mathbb{R}.$$

Now we define the *evolutes of curves*.

Definition 1.10.8. A curve γ^* is called *evolute of a curve* γ if γ is an involute of $\gamma*$.

The following result gives a parametric representation for evolutes.

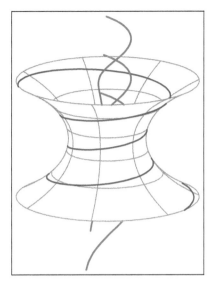

Figure 1.56 The evolutes of a curve on a catenoid

Theorem 1.10.9. *Let γ be a curve with a parametric representation $\vec{x}(s)$, non-vanishing curvature κ, and principal normal and binormal vectors $\vec{v}_2(s)$ and $\vec{v}_3(s)$ ($s \in I$). Then the evolutes of γ have parametric representations*

$$\vec{y}^*(s) = \vec{x}(s) + \frac{1}{\kappa(s)}\left(\vec{v}_2(s) + \vec{v}_3(s)\cot\alpha(s)\right) \ \text{ with } \alpha(s) = \int\limits_0^s \tau(\sigma)\,d\sigma + c,$$

where c is a constant of integration (Figure 1.56).

Proof. Let $\vec{y}^*$ be a parametric representation of an evolute γ^* of the curve γ with a parametric representation $\vec{x}(s)$. Since γ is an involute of γ^*, the tangents to γ^* intersect γ at right angles. Let $\vec{a}(s)$ be a unit vector in the normal plane of γ and orthogonal to γ; then $\vec{y}^*(s) = \vec{x}(s) + t(s)\vec{a}(s)$. Here $|t(s)|$ is the distance between the point $X(s)$ on γ and the corresponding point $Y^*(s)$ on γ^*. By the definition of an involute, $\vec{a}(s)$ has to be tangential to γ^*, hence

$$\frac{d\vec{y}^*}{ds}(s) = \lambda\vec{a}(s) = \dot{\vec{x}}(s) + \frac{dt}{ds}(s)\vec{a}(s) + t(s)\frac{d\vec{a}}{ds}(s). \tag{1.121}$$

This implies

$$\lambda = \lambda\vec{a}(s) \bullet \vec{a}(s) = \vec{v}_1(s) \bullet \vec{a}(s) + \frac{dt}{ds}(s) + t(s)\frac{d\vec{a}}{ds}(s) \bullet \vec{a}(s) = \frac{dt}{ds}(s),$$

since $\vec{a}(s) \perp \vec{v}_1(s)$, and $\|\vec{a}(s)\| = 1$ implies $\vec{a}(s) \bullet \frac{d\vec{a}}{ds}(s) = 0$. Therefore, $\lambda = \frac{dt}{ds}(s)$ and we obtain from (1.121)

$$\vec{v}_1(s) + t(s)\frac{d\vec{a}}{ds}(s) = \vec{0}. \tag{1.122}$$

Since $\vec{a}(s)$ is a unit vector in the normal plane, we may write $\vec{a}(s) = \vec{v}_2(s)\sin\alpha(s) + \vec{v}_3(s)\cos\alpha(s)$, where $\alpha(s)$ is the angle between $\vec{a}(s)$ and $\vec{v}_3(s)$. Omitting the argument s, we obtain

$$\frac{d\vec{a}}{ds} = \sin\alpha\,\dot{\vec{v}}_2 + \dot{\alpha}\cos\alpha\,\vec{v}_2 + \cos\alpha\,\dot{\vec{v}}_3 - \dot{\alpha}\sin\alpha\,\vec{v}_3$$

$$= -\kappa\sin\alpha\,\vec{v}_1 + (\dot{\alpha}\cos\alpha - \tau\cos\alpha)\vec{v}_2 + (\tau\sin\alpha - \dot{\alpha}\sin\alpha)\vec{v}_3$$

by Frenet's formulae, and substituting in (1.122), we conclude

$$(1 - \kappa t\sin\alpha)\vec{v}_1 + t\cos\alpha(\dot{\alpha} - \tau)\vec{v}_2 + t\sin\alpha(\tau - \dot{\alpha})\vec{v}_3 = \vec{0}.$$

Since the vectors $\vec{v}_1$, $\vec{v}_2$ and $\vec{v}_3$ are linearly independent and cos and sin have no common zeros, this implies

$$1 - \kappa t\sin\alpha = 0, \quad \dot{\alpha} - \tau = 0, \text{ that is, } \rho = 1/\kappa = t\sin\alpha \text{ and } \dot{\alpha} = \tau,$$

hence $\alpha(s) = \int_0^s \tau(\sigma)\,d\sigma + c$ with a constant $c \in \mathbb{R}$. Now the parametric representation for the evolute γ^* is an immediate consequence. □

Remark 1.10.10. *If γ is a planar curve with a parametric representation $\vec{x}(s)$ ($s \in I$) and non–vanishing curvature, then $\tau(s) = 0$ for all s and the parametric representation for the evolutes given in Theorem 1.10.9 reduces, for the choice $\alpha(s) = \pi/2$, to*

$$\vec{y}^*(s) = \vec{x}(s) + \frac{1}{\kappa(s)}\vec{v}_2(s) \ (s \in I).$$

Consequently, this evolute of a planar curve coincides with the centres of curvature of the curve (Figure 1.57).

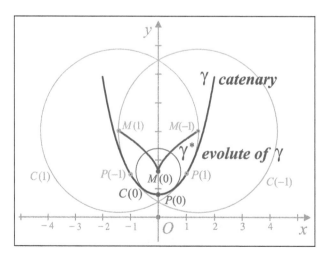

Figure 1.57 An evolute of a catenary

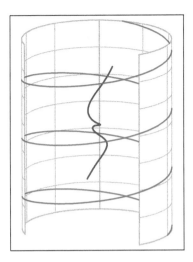

Figure 1.58 An evolute of a helix

Visualization 1.10.11. *Let γ be the helix of Visualization 1.10.7. Then, by Theorem 1.10.2, the involutes of the helix are given by*

$$\vec{x}^*(s) = \{r\cos\omega s, r\sin\omega s, h\omega s\} + (s_0 - s)\omega\{-r\sin\omega s, r\cos\omega s, h\}$$
$$= \{r(\cos\omega s - \omega(s_0 - s)\sin\omega s), r(\sin\omega s + \omega(s_0 - s)\cos\omega s), h\omega s_0\}.$$

Furthermore, since

$$\alpha(s) = \int\limits_0^s \tau(\sigma)\, d\sigma + c = h\omega^2 s + c,$$

the evolutes of the helix are given by

$$\vec{y}^*(s) = \{r\cos\omega s, r\sin\omega s, h\omega s\} + \frac{1}{r\omega^2}\{-\cos\omega s, -\sin\omega s, 0\} +$$
$$+ \frac{\cot(h\omega^2 s + c)}{r\omega}\{h\sin\omega s, -h\cos\omega s, r\} \quad \left(s \in \left(-\frac{c}{h\omega^2}, \frac{\pi - c}{h\omega^2}\right)\right),$$

by Theorem 1.10.9 (Figure 1.58).

1.11   THE FUNDAMENTAL THEOREM OF CURVES

In this section, we will show that curves are completely determined by their curvature and torsion – up to linear transformations. This result is known as the *fundamental theorem of the theory of curves*.

Theorem 1.11.1 (The fundamental theorem of curves).
Let I be an open interval containing the origin, $\bar{\kappa} \in C^1(I)$, $\bar{\tau} \in C(I)$ two functions and $\bar{\kappa}(s) > 0$ for all $s \in I$. Furthermore, let $\vec{c}_0$ be a constant vector and $\vec{c}_k$ $(k = 1, 2, 3)$ be three orthonormal vectors with $\vec{c}_1 \bullet (\vec{c}_2 \times \vec{c}_3) = 1$. Then there is one and only one curve γ with a parametric representation $\vec{x}(s)$ $(s \in I)$ that has the following properties:

(i) s *is the arc length along* γ

(ii) $\vec{x}(0) = \vec{c}_0$ *and* $\vec{v}_k(0) = \vec{c}_k$ $(k = 1, 2, 3)$ *for the vectors of the trihedron of* γ *at* 0

(iii) $\kappa(s) = \bar{\kappa}(s)$ *and* $\tau(s) = \bar{\tau}(s)$ *for the curvature* κ *and torsion* τ *of* γ.

Proof. First we find the solutions $\vec{v}_k$ $(k = 1, 2, 3)$ on I of Frenet's formulae, and then we show that

$$\vec{x}(s) = \int_0^s \vec{v}_1(\sigma)\, d\sigma + \vec{c}_0$$

is a parametric representation of a curve with the desired properties.

By the existence theorem for systems of first-order linear differential equations, the system

$$\dot{\vec{v}}_j = \sum_{k=1}^3 a_{jk}\vec{v}_k \text{ with } (a_{jk}) = \begin{pmatrix} 0 & \bar{\kappa} & 0 \\ -\bar{\kappa} & 0 & \bar{\tau} \\ 0 & -\bar{\tau} & 0 \end{pmatrix} \tag{1.123}$$

has unique solutions $\vec{v}_j$ with $\vec{v}_j(0) = \vec{c}_j$ $(j = 1, 2, 3)$. If there is a curve γ with a parametric representation $\vec{x}(s)$ $(s \in I)$ that has the desired properties, then it is given by

$$\vec{x}(s) = \vec{c}_0 + \int_0^s \vec{v}_1(\sigma)\, d\sigma. \tag{1.124}$$

This shows the uniqueness.

Now we shall prove that $\vec{x}(s)$ in (1.124) has the desired properties.

(i) *First we show* $\vec{x} \in C^3(I)$.

Since $\dot{\vec{x}} = \vec{v}_1$ by (1.124), it follows from (1.123) that $\dot{\vec{v}}_1 = \bar{\kappa}\vec{v}_2$. By hypothesis $\bar{\kappa} \in C^1(I)$ and by (1.123), we obtain $\vec{v}_2 \in C^1(I)$ and $\dot{\vec{v}}_1 \in C^1(I)$, hence $\vec{x} \in C^3(I)$.

(ii) *Now we show* $\vec{v}_j \bullet \vec{v}_k = \delta_{jk}$ $(j, k = 1, 2, 3)$ *and* $\vec{v}_1 \bullet (\vec{v}_2 \times \vec{v}_3) = 1$ *on* I.

It follows from (1.123) that

$$\frac{d}{ds}\left(\vec{v}_j \bullet \vec{v}_k\right) = \dot{\vec{v}}_j \bullet \vec{v}_k + \vec{v}_j \bullet \dot{\vec{v}}_k = \sum_{l=1}^3 a_{jl}\vec{v}_l \bullet \vec{v}_k + \sum_{l=1}^3 a_{kl}\vec{v}_l \bullet \vec{v}_j.$$

Putting $b_{jk} = \vec{v}_j \bullet \vec{v}_k$ $(j, k = 1, 2, 3)$, we obtain the following system of differential equations

$$\dot{b}_{jk} = \sum_{l=1}^3 (a_{jl}b_{lk} + a_{kl}b_{lj})$$

with $b_{lk} = b_{kl}$ and the initial conditions

$$b_{jk}(0) = \delta_{jk} \ (j, k = 1, 2, 3)$$

which has one and only one solution. Since

$$\sum_{l=1}^{3}(a_{jl}\delta_{lk} + a_{kl}\delta_{lj}) = a_{jk} + a_{kj} = 0 = \dot{\delta}_{jk} \ (j, k = 1, 2, 3),$$

the functions $b_{jk} = \delta_{jk}$ $(j, k = 1, 2, 3)$ are solutions. Consequently, we have

$$b_{jk}(s) = \vec{v}_j(s) \bullet \vec{v}_k(s) = \delta_{jk}(s) \ (j, k = 1, 2, 3) \text{ on } I.$$

Furthermore this implies

$$\vec{v}_1(s) \bullet (\vec{v}_2(s) \times \vec{v}_3(s)) = \pm 1 \text{ for all } s \in I.$$

From the initial conditions, we obtain

$$\vec{v}_1(0) \bullet (\vec{v}_2(0) \times \vec{v}_3(0)) = \vec{c}_1 \bullet (\vec{c}_2 \times \vec{c}_3) = 1,$$

and the continuity of the functions $\vec{v}_k$ $(k = 1, 2, 3)$ implies

$$\vec{v}_1(s) \bullet (\vec{v}_2(s) \times \vec{v}_3(s)) = 1 \text{ for all } s \in I.$$

(iii) *Now we show that s is the arc length of $\vec{x}$.*
By (1.124) and Part (ii), we obtain $(\dot{\vec{x}}(s))^2 = (\vec{v}_1(s))^2 = 1$, and so s is the arc length of $\vec{x}$ by Proposition 1.5.2 (d).

(iv) *Now we show that the vectors $\vec{v}_k$ $(k = 1, 2, 3)$ are the vectors of the trihedra of $\vec{x}$.*
Since $\vec{x} \in C^3(I)$ by Part (i) and

$$\ddot{\vec{x}}(s) = \dot{\vec{v}}_1(s) = \bar{\kappa}(s)\vec{v}_2(s) \neq \vec{0} \text{ for all } s \in I \text{ by Part (ii),}$$

we conclude
$$\vec{v}_1(s) = \dot{\vec{x}}(s) = \vec{v}_1(s) \text{ for all } s \in I \text{ by Part (i),}$$

and

$$\vec{v}_2(s) = \frac{\ddot{\vec{x}}(s)}{\|\ddot{\vec{x}}(s)\|} = \frac{\dot{\vec{v}}_1(s)}{\bar{\kappa}(s)} = \vec{v}_2(s) \text{ for all } s \in I.$$

Finally
$$\vec{v}_j(s) \bullet \vec{v}_k(s) = \vec{v}_j(s) \bullet \vec{v}_k(s) = \delta_{jk}(s) \ (j, k = 1, 2, 3)$$

and

$$\vec{v}_1(s) \bullet (\vec{v}_2(s) \times \vec{v}_3(s)) = \vec{v}_1(s) \bullet \left(\vec{v}_2(s) \times \vec{v}_3(s)\right) = 1 \text{ for all } s \in I \text{ by Part (ii)}$$

together imply $\vec{v}_3(s) = \vec{v}_3(s)$ for all $s \in I$.

(v) *Finally, we show that $\bar{\kappa}$ and $\bar{\tau}$ are the curvature and torsion of $\vec{x}$.*
   If we denote the curvature and torsion of $\vec{x}$ by κ and τ, then Frenet's formulae and Part (iv) together imply

$$\kappa(s) = \dot{\vec{v}}_1(s) \bullet \vec{v}_2(s) = \dot{\vec{v}}_1(s) \bullet \vec{v}_2(s) = \bar{\kappa}(s)$$

and

$$\tau(s) = -\dot{\vec{v}}_3(s) \bullet \vec{v}_2(s) = -\dot{\vec{v}}_3(s) \bullet \vec{v}_2(s) = \bar{\tau}(s) \text{ for all } s \in I.$$

$\square$

We may restate Theorem 1.11.1 as follows:

Theorem 1.11.2. *Given any two functions κ and τ that satisfy the conditions in Theorem 1.11.1 there is – up to possible translations and rotations – one and only one curve that has κ, τ and s as its curvature, torsion and arc length, respectively.*

Remark 1.11.3. *The significance of Theorem 1.11.2 is that it states that curvature and torsion are a complete system of invariants for a curve.*
By this we mean that, if $I(s)$ is any invariant, that is, a functional independent of motion

$$I(s) = J(\vec{x}(s)),$$

defined for all curves, then, by Theorem 1.11.2, the curve is completely determined by its initial point $X(0)$, the initial vectors of the trihedron and its curvature and torsion in terms of its arc length. Since by some linear transformation, the point $X(0)$ can be moved into the origin and the vectors of the trihedron can be made to coincide with the coordinate axes, the function $J(\vec{x}(s))$ in fact depends on κ, τ and s only.

In view of the fundamental theorem it is natural to define the concept of the *natural* or *intrinsic equations of a curve*.

Definition 1.11.4. The equations

$$\kappa = \kappa(s) \text{ and } \tau = \tau(s),$$

which express the curvature and torsion of a curve depending on its arc length, are called the *natural* or *intrinsic equations of the curve*.

We close this section with two examples.

Example 1.11.5. *(a) The catenary of Example 1.5.3 (d) with a parametric representation (1.59) has the curvature*

$$\kappa = \kappa(s) = \frac{a}{a^2 + s^2} \text{ for all } s \in \mathbb{R}$$

by (1.105) in Visualization 1.9.6 (a), and the torsion

$$\tau = \tau(s) = 0 \text{ for all } s \in \mathbb{R}$$

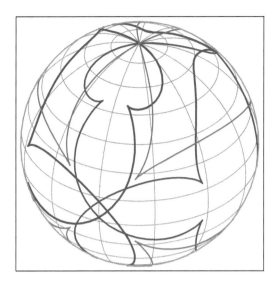

Figure 1.59 Spherical curves

by Theorem 1.8.1. Therefore, its natural equations are

$$\kappa(s) = \frac{a}{a^2 + s^2} \text{ and } \tau(s) = 0 \text{ for all } s \in \mathbb{R}.$$

(b) The natural equations for the helix of Visualization 1.8.4 with a parametric representation

$$\vec{x}(s) = \{r\cos\omega s, r\sin\omega s, h\omega s\} \ (s \in \mathbb{R}), \ where \ \omega = \frac{1}{\sqrt{r^2 + h^2}}$$

are by (1.96) and (1.97)

$$\kappa = \kappa(s) = r\omega^2 \text{ and } \tau = \tau(s) = h\omega^2 \text{ for all } s \in \mathbb{R}.$$

Visualization 1.11.6. *A curve on a sphere is called a* spherical curve *(Figure 1.59). A curve with non–vanishing torsion τ is a spherical curve if and only if its natural equations satisfy the condition*

$$\rho(s)\tau(s) + \frac{d}{ds}\left(\frac{\dot{\rho}}{\tau}\right)(s) = 0 \text{ for all } s, \tag{1.125}$$

where $\rho(s) = 1/\kappa(s)$ is the radius of curvature of the curve.

Proof.

(i) First we assume that γ is a curve with a parametric representation $\vec{x}(s)$ on a sphere S.

Then the osculating spheres of γ coincide with S, in particular their centres must be coincide. The position vector of centre of the osculating sphere at s is given by (1.100) in Theorem 1.9.5

$$\vec{m}(s) = \vec{x}(s) + \frac{1}{\kappa(s)}\vec{v}_2(s) - \frac{\dot{\kappa}(s)}{\kappa^2(s)\tau^2(s)}\vec{v}_3(s)$$

$$= \vec{x}(s) + \rho(s)\vec{v}_2(s) + \frac{\dot{\rho}(s)}{\tau(s)}\vec{v}_3(s), \tag{1.126}$$

and so by Frenet's formulae (1.84) in Theorem 1.7.1

$$\frac{d\vec{m}}{ds}(s) = \vec{v}_1(s) + \dot{\rho}(s)\vec{v}_2(s) + \rho(s)\dot{\vec{v}}_2(s) + \frac{d}{ds}\left(\frac{\dot{\rho}(s)}{\tau(s)}\right)\vec{v}_3(s) + \frac{\dot{\rho}(s)}{\tau(s)}\dot{\vec{v}}_3(s)$$

$$= (1 - \rho(s)\kappa(s))\vec{v}_1(s) + \left(\dot{\rho}(s) - \tau(s)\frac{\dot{\rho}(s)}{\tau(s)}\right)\vec{v}_2(s)$$

$$+ \left(\rho(s)\tau(s) + \frac{d}{ds}\left(\frac{\dot{\rho}(s)}{\tau(s)}\right)\right)\vec{v}_3(s)$$

$$= \left(\rho(s)\tau(s) + \frac{d}{ds}\left(\frac{\dot{\rho}(s)}{\tau(s)}\right)\right)\vec{v}_3(s) = \vec{0}, \tag{1.127}$$

which implies (1.125).

(ii)   Conversely, we assume that the condition in (1.125) is satisfied.
We define the vector–valued function $\vec{m}$ as in (1.126). Then it follows as in (1.127) that $d\vec{m}(s)/ds = \vec{0}$ for all s, hence $\vec{m}(s) = \vec{m}$ for some constant vector $\vec{m}$. Furthermore, we have

$$r^2(s) = \|\vec{x}(s) - \vec{m}\|^2 = \rho^2(s) + \left(\frac{\dot{\rho}(s)}{\tau(s)}\right)^2,$$

and so

$$\frac{dr^2}{ds}(s) = \frac{d}{ds}\left(\rho^2(s) + \left(\frac{\dot{\rho}(s)}{\tau(s)}\right)^2\right) = 2\dot{\rho}(s)\rho(s) + 2\frac{\dot{\rho}(s)}{\tau(s)}\frac{d}{ds}\left(\frac{\dot{\rho}(s)}{\tau(s)}\right)$$

$$= 2\frac{\dot{\rho}(s)}{\tau(s)}\left(\rho(s)\tau(s) + \frac{d}{ds}\left(\frac{\dot{\rho}(s)}{\tau(s)}\right)\right) = 0.$$

Thus $r(s) = r$ for some constant r, and so $\|\vec{x}(s) - \vec{m}\| = r$ for all s, that is, γ is a spherical curve.   □

1.12   LINES OF CONSTANT SLOPE

There is a class of curves for which a parametric representation can easily be derived from the natural equations.

Definition 1.12.1. A curve with a parametric representation $\vec{x}(s)$ $(s \in I)$ is called a *line of constant slope* if there is a constant c such that

$$\tau(s) = c\kappa(s) \text{ for all } s \in I \text{ (Figure 1.60)}.$$

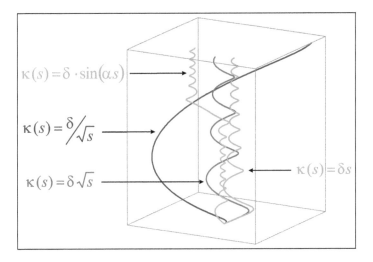

$\kappa(s) = \delta \cdot \sin(\alpha s)$

$\kappa(s) = \delta / \sqrt{s}$

$\kappa(s) = \delta \sqrt{s}$

$\kappa(s) = \delta s$

Figure 1.60 Lines of constant slope of a given curvature

Example **1.12.2.** *Helices are lines of constant slope, since it follows from Example 1.11.5 (b) that*

$$\tau(s) = \frac{r}{h}\kappa(s) \ for \ all \ s \in \mathbb{R}.$$

Now we give an explicit formula for the parametric representations of lines of constant slope.

Theorem 1.12.3. *All lines of constant slope that satisfy the natural equations* $\tau(s) = c\kappa(s)$ $(s \in I)$ *are given by*

$$\vec{x}(s) = \frac{\vec{a}}{\tilde{\omega}} \int_{s_0}^{s} \sin\left(\tilde{\omega}t(\sigma)\right) d\sigma - \frac{\vec{b}}{\tilde{\omega}} \int_{s_0}^{s} \cos\left(\tilde{\omega}t(\sigma)\right) d\sigma + \vec{c}_1 s + \vec{c}_0 \tag{1.128}$$

for $s_0 \in I$ *fixed* $(s \in I)$, *where the constant vectors* $\vec{a}$, $\vec{b}$, $\vec{c}_1$ *and* $\vec{c}_0$ *satisfy*

$$\vec{a}^2 = \vec{b}^2 = 1, \ \vec{a} \bullet \vec{b} = \vec{a} \bullet \vec{c}_1 = \vec{b} \bullet \vec{c}_1 = 0 \ and \ \vec{c}_1^2 = \frac{c^2}{1+c^2}, \tag{1.129}$$

and

$$t(s) = \int_{s_0}^{s} \kappa(\sigma) \, d\sigma \ and \ \tilde{\omega}^2 = 1 + c^2. \tag{1.130}$$

Proof. We omit the argument s in the proof.
Since $\tau = c \cdot \kappa$, Frenet's formulae (1.84) in Theorem 1.7.1 reduce to

$$\dot{\vec{v}}_1 = \kappa\vec{v}_2, \ \dot{\vec{v}}_2 = -\kappa\vec{v}_1 + c\kappa\vec{v}_3 \ and \ \dot{\vec{v}}_3 = -c\kappa\vec{v}_2.$$

We introduce a new parameter t by $dt/ds = \kappa(s)$. Then Frenet's formulae become

$$\vec{v}_1{}' = \vec{v}_2, \ \vec{v}_2{}' = -\vec{v}_1 + c\vec{v}_3 \ and \ \vec{v}_3{}' = -c\vec{v}_2. \tag{1.131}$$

Differentiating the second equation, and substituting the first into the third equation yields

$$\vec{v}_2{}'' = -\vec{v}_1{}' + c\vec{v}_3{}'$$

and

$$\vec{v}_1{}''' = \vec{v}_2{}'' = -\vec{v}_1{}' + c\vec{v}_3{}' = -\vec{v}_1{}' - c^2\vec{v}_2 = -\vec{v}_1{}'(1 + c^2),$$

hence

$$\vec{v}_1{}''' + \tilde{\omega}^2\vec{v}_1{}' = \vec{0}, \text{ where } \tilde{\omega}^2 = 1 + c^2. \tag{1.132}$$

The general solution of the system (1.132) of differential equations is

$$\vec{v}_1{}' = \vec{a}\cos\tilde{\omega}t + \vec{b}\sin\tilde{\omega}t, \tag{1.133}$$

where $\vec{a}$ and $\vec{b}$ are constant vectors. Observing that $\ddot{\vec{x}} = \dot{\vec{v}}_1 = \vec{x}'\kappa(s)$, and integrating (1.132) twice, we obtain the representation in (1.128).
Since we have

$$\dot{\vec{x}}(s) = \vec{v}_1(s) = \frac{\vec{a}}{\tilde{\omega}} \cdot \sin\tilde{\omega}t - \frac{\vec{b}}{\tilde{\omega}} \cdot \cos\tilde{\omega}t + \vec{c}_1$$

and

$$\ddot{\vec{x}}(s) = \kappa\vec{v}_2 = \kappa(\vec{a}\cos\tilde{\omega}t + \vec{b}\sin\tilde{\omega}t),$$

and $\vec{v}_1$ and $\vec{v}_2$ are orthonormal vectors, it follows, for $t = 0$, $\kappa^2\vec{a}^2 = \kappa^2$, that is, $\vec{a}^2 = 1$, and for $t = \frac{\pi}{2}$, $\kappa^2\vec{b}^2 = \kappa^2$, that is, $\vec{b}^2 = 1$. Furthermore, we have

$$\vec{v}_2^2 = \vec{a}^2\cos^2\tilde{\omega}t + 2\vec{a}\bullet\vec{b}\cos\tilde{\omega}t\sin\tilde{\omega}t + \vec{b}^2\sin^2\tilde{\omega}t$$

$$= 1 + 2\vec{a}\bullet\vec{b}\cos\tilde{\omega}t\sin\tilde{\omega}t = 1 \text{ for all } t,$$

that is, $\vec{a}\bullet\vec{b} = 0$.
The condition $\vec{v}_1 \bullet \vec{v}_2 = 0$ implies

$$(-\vec{b}/\tilde{\omega} + \vec{c}_1)\bullet\vec{a} = \vec{c}_1\bullet\vec{a} = 0 \text{ for } t = 0$$

and

$$(\vec{a}/\tilde{\omega} + \vec{c}_1)\bullet\vec{b} = \vec{c}_1\bullet\vec{b} = 0 \text{ for } t = \pi/2.$$

Finally $\vec{v}_1^2 = 1$ implies

$$1 = (\vec{a}^2/\tilde{\omega}^2)\sin^2\tilde{\omega}t + (\vec{b}^2/\tilde{\omega}^2)\cos^2\tilde{\omega}t + \vec{c}_1^2 = 1/\tilde{\omega}^2 + \vec{c}_1^2,$$

hence

$$\vec{c}_1^2 = 1 - \frac{1}{\tilde{\omega}^2} = 1 - \frac{1}{1 + c^2} = \frac{c^2}{1 + c^2}.$$

□

Proposition 1.12.4. *Every curve with constant curvature and torsion is a helix (including the special cases).*

Proof. We apply Theorem 1.12.3. Let $s_0 \in \mathbb{R}$ be given, $\kappa(s) = \kappa$ be constant, and $\tau(s) = c \cdot \kappa$ for all s. Then we have $t(s) = \kappa(s - s_0)$ by (1.130) in Theorem 1.12.3.

(i) First we assume $\kappa = 0$.

If $\kappa = 0$, then $t(s) = 0$ and it follows from (1.128) that the corresponding curve γ has a parametric representation

$$\vec{x}(s) = -\frac{\vec{b}}{\tilde{\omega}^2}(s - s_0) + \vec{c}_1 s + \vec{c}_0.$$

We put

$$\vec{b}_1 = -\left(\frac{\vec{b}}{\tilde{\omega}} - \vec{c}_1\right) \text{ and } \vec{d} = \frac{s_0}{\tilde{\omega}}\vec{b} + \vec{c}_0$$

and obtain from (1.129) and (1.130)

$$\vec{x}(s) = \vec{b}_1 s + \vec{d} \text{ with } \|\vec{b}_1\| = \sqrt{\frac{1}{\tilde{\omega}^2} + \|\vec{c}_1\|^2} = \sqrt{\frac{1}{1 + c^2} + \frac{c^2}{1 + c^2}} = 1,$$

which is a parametric representation of a straight line through the point with position vector $\vec{d}$ and in the direction of the unit vector $\vec{b}_1$.

(ii) Now we assume $\kappa \neq 0$.

If $\kappa \neq 0$, then it follows from (1.128) that the curve γ has a parametric representation

$$\vec{x}(s) = \frac{\vec{a}}{\tilde{\omega}} \int_{s_0}^{s} \sin\left(\tilde{\omega}\kappa(\sigma - s_0)\right) d\sigma - \frac{\vec{b}}{\tilde{\omega}} \int_{s_0}^{s} \cos\left(\tilde{\omega}\kappa(\sigma - s_0)\right) d\sigma + \vec{c}_1 s + \vec{c}_0$$

$$= -\frac{\vec{a}}{\tilde{\omega}^2\kappa} \cos\left(\tilde{\omega}\kappa(\sigma - s_0)\right)\Big|_{s_0}^{s} + \frac{\vec{b}}{\tilde{\omega}^2\kappa} \sin\left(\tilde{\omega}\kappa(\sigma - s_0)\right)\Big|_{s_0}^{s} + \vec{c}_1 s + \vec{c}_0$$

$$= -\frac{\vec{a}}{\tilde{\omega}^2\kappa} \cos\left(\tilde{\omega}\kappa(s - s_0)\right) + \frac{\vec{b}}{\tilde{\omega}^2\kappa} \sin\left(\tilde{\omega}\kappa(s - s_0)\right) + \vec{c}_1 s + \vec{d}, \qquad (1.134)$$

$$\text{where } \vec{d} = \vec{c}_0 + \frac{\vec{a}}{\tilde{\omega}^2\kappa}.$$

(ii.1) First we assume $c = 0$.

If $c = 0$, then $\tau = 0$, and γ is a planar curve, and we have $\vec{c}_1 = \vec{0}$ by (1.129), and it follows that

$$\|\vec{x}(s) - \vec{d}\|^2 = \left\| -\frac{\vec{a}}{\tilde{\omega}^2\kappa} \cos\left(\tilde{\omega}\kappa(s - s_0)\right) + \frac{\vec{b}}{\tilde{\omega}^2\kappa} \sin\left(\tilde{\omega}\kappa(s - s_0)\right) \right\|^2 = \frac{1}{(\tilde{\omega}^2\kappa)^2},$$

that is, γ is a circle line with radius $r = 1/(\tilde{\omega}^2\kappa)$, its centre in the point D with position vector $\vec{d}$, and in the plane through D and spanned by the vectors $\vec{a}$ and $\vec{b}$.

(ii.2) Now we assume $c \neq 0$.
If $c \neq 0$, then we put

$$\vec{b}_1 = - \left(\vec{a} \cos \left(\tilde{\omega} \kappa s_0 \right) + \vec{b} \sin \left(\tilde{\omega} \kappa s_0 \right) \right),$$
$$\vec{b}_2 = \vec{b} \cos \left(\tilde{\omega} \kappa s_0 \right) - \vec{a} \sin \left(\tilde{\omega} \kappa s_0 \right),$$
$$\vec{b}_3 = \frac{\vec{c}_1}{\|\vec{c}_1\|}, \quad r = \frac{1}{\tilde{\omega}^2 \kappa}, \quad h = cr \text{ and } \omega = \frac{1}{\sqrt{r^2 + h^2}},$$

and obtain from (1.129)

$$\|\vec{b}_1\| = \|\vec{b}_2\| = \|\vec{b}_3\| = 1, \ \vec{b}_i \bullet \vec{b}_k = 0 \text{ for } i \neq k,$$

$$\vec{c}_1 = \vec{b}_3 \|\vec{c}_1\| = \vec{b}_3 \sqrt{\frac{c^2}{1 + c^2}} = \vec{b}_3 \sqrt{\frac{\frac{h^2}{r^2}}{1 + \frac{h^2}{r^2}}} = \vec{b}_3 |h| \omega$$

and

$$\tilde{\omega} \kappa = \frac{1}{r \tilde{\omega}} = \frac{1}{r \sqrt{1 + c^2}} = \frac{1}{\sqrt{r^2 + h^2}} = \omega.$$

Thus (1.129) yields

$$\begin{aligned}
\vec{x}(s) - |h| \omega s \vec{b}_3 - \vec{d} &= - \left(\frac{\vec{a}}{\tilde{\omega}^2 \kappa} \left(\cos \left(\tilde{\omega} \kappa s \right) \cos \left(\tilde{\omega} \kappa s_0 \right) + \sin \left(\tilde{\omega} \kappa s \right) \sin \left(\tilde{\omega} \kappa s_0 \right) \right) \right) \\
&\quad + \left(\frac{\vec{b}}{\tilde{\omega}^2 \kappa} \left(\sin \left(\tilde{\omega} \kappa s \right) \cos \left(\tilde{\omega} \kappa s_0 \right) - \cos \left(\tilde{\omega} \kappa s \right) \sin \left(\tilde{\omega} \kappa s_0 \right) \right) \right) \\
&= -r \cos \left(\tilde{\omega} \kappa s \right) \left(\vec{a} \cos \left(\tilde{\omega} \kappa s_0 \right) + \vec{b} \sin \left(\tilde{\omega} \kappa s_0 \right) \right) \\
&\quad + r \sin \left(\tilde{\omega} \kappa s \right) \left(\vec{b} \cos \left(\tilde{\omega} \kappa s_0 \right) - \vec{a} \sin \left(\tilde{\omega} \kappa s_0 \right) \right) \\
&= r \cos \left(\tilde{\omega} \kappa s \right) \vec{b}_1 + r \sin \left(\tilde{\omega} \kappa s \right) \vec{b}_2 \\
&= r \cos \left(\omega s \right) \vec{b}_1 + r \sin \left(\omega s \right) \vec{b}_2.
\end{aligned}$$

This is a parametric representation of a helix with its axis through the point with position vector $\vec{d}$, and in the direction of the vector $\vec{b}_3$.

□

Proposition 1.12.5. *Let* $\delta, c > 0$, $\kappa(s) = \delta/s$ *and* $\tau(s) = c \cdot \kappa(s)$ *for all* $s > 0$. *Then all curves with curvature* κ *and torsion* τ *are given by*

$$\vec{x}(s) = \frac{s \cos \left(\tilde{\omega} y(s) \right)}{\tilde{\omega} \sqrt{(\tilde{\omega}^2 \delta^2 + 1)}} \vec{b}_1 + \frac{s \sin \left(\tilde{\omega} y(s) \right)}{\tilde{\omega} \sqrt{(\tilde{\omega}^2 \delta^2 + 1)}} \vec{b}_2 + \frac{|c|}{\tilde{\omega}} s \vec{b}_3 + \vec{d}_0, \qquad (1.135)$$

where $y(s) = \delta \log \left(\frac{s}{s_0} \right)$ *for* $s \in (0, \infty)$, *and* s_0 *is a constant, the vectors* $\vec{b}_k$ *are constant orthonormal vectors,* $\vec{d}_0$ *is a constant vector, and* $\tilde{\omega}^2 = c^2 + 1$. *Every curve is on a*

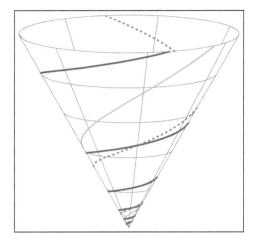

Figure 1.61 Lines of constant slope on a cone

cone with its vertex in the point with position vector $\vec{d}$, its axis along the vector $\vec{b}_3$ and an angle of

$$\beta = 2\tan^{-1}\left(\frac{1}{|c|\sqrt{(\tilde{\omega}^2\delta^2 + 1)}}\right)$$

at its vertex (Figure 1.61).

Proof. Since $\kappa(s) = \delta/s$ for $s > 0$, we obtain from (1.130) in Theorem 1.12.3

$$t(s) = \int_{s_0}^{s} \kappa(\sigma)\,d\sigma = \delta\log\left(\frac{s}{s_0}\right) \text{ for some constant } s_0 > 0.$$

We put

$$y(\sigma) = \delta\log\left(\frac{\sigma}{s_0}\right), \text{ that is, } \sigma = s_0\exp\left(y/\delta\right) \text{ and } \frac{d\sigma}{dy} = \frac{s_0}{\delta}\exp\left(y/\delta\right).$$

By (1.128) in Theorem 1.12.3, we have to evaluate the integrals

$$I_1 = \int_{s_0}^{s} \sin\left(\tilde{\omega}t(\sigma)\right)d\sigma = \frac{s_0}{\delta}\int_{0}^{y(s)} \sin\left(\tilde{\omega}y\right)\exp\left(y/\delta\right)dy$$

and

$$I_2 = \int_{s_0}^{s} \cos\left(\tilde{\omega}t(\sigma)\right)d\sigma = \frac{s_0}{\delta}\int_{0}^{y(s)} \cos\left(\tilde{\omega}y\right)\exp\left(y/\delta\right)dy.$$

Integration by parts yields

$$\tilde{I}_1 = \frac{\delta}{s_0}I_1 = \frac{\exp\left(y/\delta\right)}{\tilde{\omega}^2 + \frac{1}{\delta^2}}\left(\frac{1}{\delta}\sin\left(\tilde{\omega}y\right) - \tilde{\omega}\cos\left(\tilde{\omega}y\right)\right)\Bigg|_{0}^{y(s)}$$

$$= \frac{s}{s_0} \frac{1}{\tilde{\omega}^2 \delta^2 + 1} \left(\delta \sin(\tilde{\omega} y(s)) - \tilde{\omega} \delta^2 \cos(\tilde{\omega} y(s)) \right) + \frac{\tilde{\omega} \delta^2}{\tilde{\omega}^2 \delta^2 + 1},$$

hence

$$I_1 = \frac{s}{\tilde{\omega}^2 \delta^2 + 1} \left(\sin(\tilde{\omega} y(s)) - \tilde{\omega} \delta \cos(\tilde{\omega} y(s)) \right) + \frac{\tilde{\omega} \delta s_0}{\tilde{\omega}^2 \delta^2 + 1}, \tag{1.136}$$

and similarly

$$I_2 = \frac{s}{\tilde{\omega}^2 \delta^2 + 1} \left(\cos(\tilde{\omega} y(s)) + \tilde{\omega} \delta \sin(\tilde{\omega} y(s)) \right) - \frac{s_0}{\tilde{\omega}^2 \delta^2 + 1}. \tag{1.137}$$

We put

$$\vec{b}_1 = -\frac{\vec{b} + \tilde{\omega} \delta \vec{a}}{\sqrt{\tilde{\omega}^2 \delta^2 + 1}}, \quad \vec{b}_2 = \frac{\vec{a} - \tilde{\omega} \delta \vec{b}}{\sqrt{\tilde{\omega}^2 \delta^2 + 1}}, \quad \vec{b}_3 = \frac{\vec{c}_1}{\|\vec{c}_1\|}$$

and

$$\vec{d}_0 = \vec{c}_0 + \frac{s_0}{\tilde{\omega}(\tilde{\omega}^2 \delta^2 + 1)} (\tilde{\omega} \delta \vec{a} + \vec{b}).$$

Then it follows from (1.129) in Theorem 1.12.3 that

$$\|\vec{b}_k\| = 1 \text{ and } \vec{b}_i \bullet \vec{b}_k = 0 \text{ for } i \neq k,$$

and we obtain from (1.129) in Theorem 1.12.3, and from (1.136) and (1.137)

$$\vec{x}(s) = \vec{a} \frac{I_1}{\tilde{\omega}} - \vec{b} \frac{I_2}{\tilde{\omega}} + \vec{c}_1 s + \vec{c}_0$$

$$= \frac{s}{\tilde{\omega}(\tilde{\omega}^2 \delta^2 + 1)} \Big[\vec{a} \sin(\tilde{\omega} y(s)) - \vec{a} \tilde{\omega} \delta \cos(\tilde{\omega} y(s))$$

$$- \Big(\vec{b} \cos(\tilde{\omega} y(s)) + \vec{b} \tilde{\omega} \delta \sin(\tilde{\omega} y(s)) \Big) \Big]$$

$$+ \frac{s_0}{\tilde{\omega}(\tilde{\omega}^2 \delta^2 + 1)} (\tilde{\omega} \delta \vec{a} + \vec{b}) + \vec{c}_1 s + \vec{c}_0$$

$$= -\frac{s \cos(\tilde{\omega} y(s))}{\tilde{\omega} \sqrt{(\tilde{\omega}^2 \delta^2 + 1)}} \frac{\vec{b} + \tilde{\omega} \delta \vec{a}}{\sqrt{(\tilde{\omega}^2 \delta^2 + 1)}}$$

$$+ \frac{s \sin(\tilde{\omega} y(s))}{\tilde{\omega} \sqrt{(\tilde{\omega}^2 \delta^2 + 1)}} \frac{\vec{a} + \tilde{\omega} \delta \vec{b}}{\sqrt{(\tilde{\omega}^2 \delta^2 + 1)}} + \vec{c}_1 s + \vec{d}_0$$

$$= \frac{s \cos(\tilde{\omega} y(s))}{\tilde{\omega} \sqrt{(\tilde{\omega}^2 \delta^2 + 1)}} \vec{b}_1 + \frac{s \sin(\tilde{\omega} y(s))}{\tilde{\omega} \sqrt{(\tilde{\omega}^2 \delta^2 + 1)}} \vec{b}_2 + \frac{|c|}{\tilde{\omega}} s \vec{b}_3 + \vec{d}_0,$$

where $y(s) = y(\sigma) = \delta \log(\sigma/s_0)$. This shows (1.135). Putting

$$x^1 = \frac{s \cos(\tilde{\omega} y(s))}{\tilde{\omega} \sqrt{(\tilde{\omega}^2 \delta^2 + 1)}}, \quad x^2 = \frac{s \sin(\tilde{\omega} y(s))}{\tilde{\omega} \sqrt{(\tilde{\omega}^2 \delta^2 + 1)}}, \quad x^3 = \frac{s|c|}{\tilde{\omega}}$$

and

$$\tan \tilde{\beta} = \frac{1}{|c| \sqrt{(\tilde{\omega}^2 \delta^2 + 1)}},$$

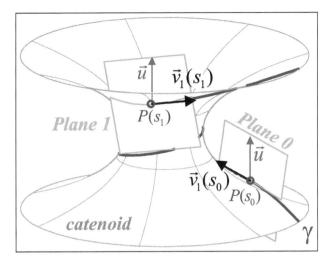

Figure 1.62 The constant angle between a line of constant slope and a direction

we see that

$$(x^1)^2 + (x^2)^2 - \tan^2 \tilde{\beta}(x^3)^2 = 0.$$

Consequently the curve is on a cone with its vertex in the point with position vector $\vec{d}_0$, its axis along the vector $\vec{b}_3$, and an angle of $\beta = 2\tilde{\beta}$ at its vertex. □

The following result explains the term *line of constant slope*.

Theorem 1.12.6. *A curve with a parametric representation $\vec{x}(s)$ $(s \in I)$ and positive curvature $\kappa(s)$ is a line of constant slope if and only if there is a unit vector $\vec{u}$ such that*

$$\vec{u} \bullet \vec{v}_1(s) = \text{ constant for all } s \in I \text{ (Figure 1.62)}. \tag{1.138}$$

If ϕ denotes the constant angle between the vectors $\vec{u}$ and $\vec{v}_1$, then we have

$$\tau(s) = \cot \phi \cdot \kappa(s) \text{ for all } s \in I.$$

Proof. (i) First we show the sufficiency of the condition in (1.138).
We assume that

$$\vec{u} \bullet \dot{\vec{x}}(s) = \vec{u} \bullet \vec{v}_1(s) = \cos \phi \text{ for all } s \in I,$$

where ϕ is the constant angle between $\vec{u}$ and the tangents of the curve. This implies $\vec{u} \bullet \ddot{\vec{x}}(s) = 0$ for all s, hence $\vec{u} \bullet \vec{v}_2(s) = 0$ for all s, since $\kappa(s) \neq 0$. Writing $\vec{u} = \sum_{k=1}^3 b_k \vec{v}_k(s)$, we obtain $b_1 = \cos \phi$, $b_2 = 0$ and $b_3 = \sin \phi$, since $\vec{u}$ is a unit vector. Hence $\vec{u} = \cos \phi \, \vec{v}_1(s) + \sin \phi \, \vec{v}_3(s)$ for all s. The first and third formulae in (1.84) together imply

$$\vec{0} = \dot{\vec{u}} = \cos \phi \, \dot{\vec{v}}_1(s) + \sin \phi \, \dot{\vec{v}}_3(s) = \kappa(s) \cos \phi \, \vec{v}_2(s) - \tau(s) \sin \phi \, \vec{v}_2(s),$$

that is,

$$0 = \kappa(s) \cos \phi - \tau(s) \sin \phi, \text{ or equivalently } \tau(s) = \kappa(s) \cot \phi \text{ for all } s \in I.$$

(ii) Now we show the necessity of the condition in (1.138).
We assume that $\tau(s) = c\kappa(s)$ for all $s \in I$, where c is a constant. Putting $c = \cot\phi$, we obtain

$$\vec{0} = (\kappa(s)\cos\phi - \tau(s)\sin\phi)\vec{v}_2(s) = \cos\phi\,\dot{\vec{v}}_1(s) + \sin\phi\,\dot{\vec{v}}_3(s)$$

and

$$\vec{u} = \vec{v}_1(s)\cos\phi + \vec{v}_3(s)\sin\phi \text{ for all } s,$$

by integration. This implies $\vec{u} \bullet \vec{v}_1(s) = const$ for all $s \in I$.

□

The following interesting result holds.

Theorem 1.12.7. *Let γ be a line of constant slope with a parametric representation $\vec{x}(s)$ ($s \in I$) and curvature $\kappa(s) \neq 0$ for all s. If φ denotes the constant angle between γ and the unit vector $\vec{u}$, then the arc length $s^{\perp}$ and the curvature $\kappa^{\perp}$ of the orthogonal projection $\gamma^{\perp}$ of γ on to a plane orthogonal to the vector $\vec{u}$ are given by*

$$s^{\perp} = (s - s_0)|\sin\varphi| \tag{1.139}$$

and

$$\kappa^{\perp}(s) = \frac{\kappa(s)}{\sin^2\varphi} \quad (\varphi \neq 0, \pi). \tag{1.140}$$

Proof. Obviously the orthogonal projection $\gamma^{\perp}$ has a parametric representation

$$\vec{x}^{\perp}(s) = \vec{x}(s) - (\vec{x}(s) \bullet \vec{u})\vec{u}.$$

First we observe that

$$\frac{d\vec{x}^{\perp}}{ds} = \vec{v}_1 - (\vec{v}_1 \bullet \vec{u})\vec{u} = \vec{v}_1 - \cos\varphi\vec{u}$$

and

$$\left\|\frac{d\vec{x}^{\perp}}{ds}\right\|^2 = 1 - 2\cos\varphi\vec{v}_1 \bullet \vec{u} + \cos^2\varphi = 1 - \cos^2\varphi = \sin^2\varphi$$

together imply (1.139) for $\varphi \neq 0, \pi$. Furthermore it follows from

$$\frac{d^2\vec{x}^{\perp}}{ds^2} = \dot{\vec{v}}_1 = \kappa\vec{v}_2,$$

$$\frac{d\vec{x}^{\perp}}{ds} \times \frac{d^2\vec{x}^{\perp}}{ds^2} = \kappa(\vec{v}_1 - \cos\varphi\vec{u}) \times \vec{v}_2 = \kappa(\vec{v}_3 - \cos\varphi(\vec{u} \times \vec{v}_2))$$

and

$$\left\|\frac{d\vec{x}^{\perp}}{ds} \times \frac{d^2\vec{x}^{\perp}}{ds^2}\right\|^2 = \kappa^2(1 - 2\vec{v}_3 \bullet (\vec{u} \times \vec{v}_2)\cos\varphi + \cos^2\varphi)$$

$$= \kappa^2(1 + \cos^2\varphi - 2\vec{u} \bullet \vec{v}_1\cos\varphi) = \kappa^2\sin^2\varphi$$

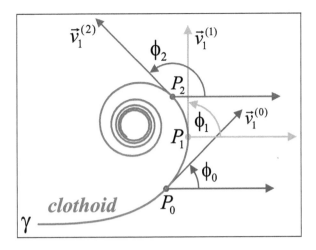

Figure 1.63 A curve with $\kappa(s) = s$, hence $\phi(s) = s^2/2$

that we have by (1.88) in Proposition 1.7.6

$$\kappa^\perp(s) = \frac{\left\| \frac{d\vec{x}^\perp}{ds} \times \frac{d^2\vec{x}^\perp}{ds^2} \right\|}{\left\| \frac{d\vec{x}^\perp}{ds} \right\|^3} = \frac{\kappa(s)|\sin\varphi|}{|\sin^3\varphi|} = \frac{\kappa(s)}{\sin^2\varphi} \quad (\varphi \neq 0, \pi).$$

This shows (1.140). □

There is a useful formula for the curvature of a planar curve.

Theorem 1.12.8. *Let γ be a planar curve with curvature $\kappa(s) \neq 0$. If $\phi(s)$ denotes the oriented angle between the positive x^1-axis and the tangent vector of γ at s, then we have*

$$\kappa(s) = \left| \frac{d\phi(s)}{ds} \right| \quad \text{for all } s \text{ (Figure 1.63)}. \tag{1.141}$$

Proof. The tangent vectors $\vec{v}_1$ of the planar curve γ satisfy

$$\vec{v}_1(\phi) = \{\cos\phi, \sin\phi\} \quad \text{and} \quad \frac{d\vec{v}_1}{d\phi}(\phi) = \{-\sin\phi, \cos\phi\},$$

hence

$$\vec{v}_1(\phi) \bullet \frac{d\vec{v}_1}{d\phi}(\phi) = 0.$$

Thus the unit vector $d\vec{v}_1/d\phi$ is orthogonal to the tangent vector at every s. Consequently we conclude

$$\frac{d\vec{v}_1}{d\phi}(\phi) = \pm\vec{v}_2(\phi). \tag{1.142}$$

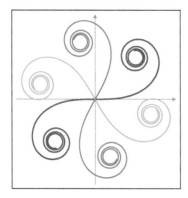 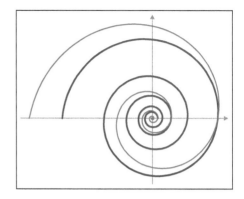

Figure 1.64 Clothoids (left) and logarithmic spirals (right)

On the other hand, it follows from Frenet's first formula (1.84) in Theorem 1.7.1 that

$$\dot{\vec{v}}_1(s) = \frac{d\vec{v}_1}{d\phi}(\phi(s))\frac{d\phi}{ds}(s) = \kappa(s)\vec{v}_2(s). \tag{1.143}$$

Since $\kappa(s) > 0$ for all s, the identity in (1.141) follows from (1.142) and (1.143). □

Remark 1.12.9. *We may choose the orientation of the curve in Theorem 1.12.8 so that*

$$\kappa(s) = \frac{d\phi(s)}{ds} \ \text{ for all } s. \tag{1.144}$$

Visualization 1.12.10. *(a) Every planar curve with $\kappa \equiv const > 0$ is a circle line of radius $r = 1/\kappa$.*
(b) Every planar curve with $\kappa(s) = c \cdot s$ ($s \in (0, \infty)$), where $c > 0$ is a constant, is a so–called clothoid *given by a parametric representation*

$$\vec{x}(\phi) = \sqrt{\frac{1}{2c}}\left\{\int \frac{\cos\phi}{\sqrt{\phi}}\,d\phi, \int \frac{\sin\phi}{\sqrt{\phi}}\,d\phi\right\} + \vec{c}_1 \ (\phi \in \mathbb{R} \setminus \{0\})$$

where $\vec{c}_1$ is a constant vector (in Figure 1.63 and left in Figure 1.64).
(c) Every planar curve with $\kappa(s) = c/s$ ($s \in (0, \infty)$), where $c > 0$ is a constant, is a logarithmic spiral *(right in Figure 1.64).*

Proof.

(a) The identity

$$\kappa(s) = \frac{d\phi(s)}{ds} = 1/r \ \text{ for all } s$$

yields

$$\vec{v}_1(s) = \{\cos\phi(s), \sin\phi(s)\} = \frac{d\vec{x}}{d\phi}(\phi(s))\frac{d\phi}{ds}(s) = \frac{1}{r}\frac{d\vec{x}}{d\phi}(\phi(s)),$$

hence
$$\vec{x}(\phi) = r\{\sin\phi, -\cos\phi\} + \vec{c}, \text{ where } \vec{c} \text{ is a constant vector.}$$

This is a parametric representation of a circle line of radius r, with its centre in the point C with the position vector $\vec{c}$.

(b) Similarly as in Part (a), the identity
$$\kappa(s) = \frac{d\phi}{ds}(s) = c \cdot s, \text{ that is, } \phi(s) = \frac{c}{2}s^2 \text{ or } s = \sqrt{\frac{2}{c}}\sqrt{\phi}$$

yields
$$\vec{v}_1(s) = \{\cos\phi(s), \sin\phi(s)\} = \frac{d\vec{x}}{d\phi}(\phi(s))\frac{d\phi}{ds}(s) = \frac{d\vec{x}}{d\phi}(\phi(s))cs = \frac{d\vec{x}(\phi)}{d\phi}\sqrt{2c\phi},$$

that is,
$$\vec{v}_1(\phi) = \sqrt{2c\phi}\frac{d\vec{x}}{d\phi}, \text{ hence } \vec{x}(\phi) = \frac{1}{\sqrt{2c}}\left\{\int\frac{\cos\phi}{\sqrt{\phi}}\,d\phi, \int\frac{\sin\phi}{\sqrt{\phi}}\,d\phi\right\} + \vec{c}_1$$

for some constant vector $\vec{c}_1$.

(c) As in Part (b), we obtain
$$\kappa(s) = \frac{d\phi}{ds}(s) = \frac{c}{s}, \text{ that is, } \phi(s) = \frac{1}{c}\log\frac{s}{s_0}$$

for some constant $s_0 > 0$, or $s = s_0 \exp(\phi/c)$, which yields
$$\vec{v}_1(s) = \{\cos\phi(s), \sin\phi(s)\} = \frac{d\vec{x}}{d\phi}(\phi(s))\frac{d\phi}{ds}(s)$$
$$= \frac{d\vec{x}}{d\phi}(\phi(s))\frac{c}{s} = \frac{d\vec{x}(\phi)}{d\phi}cs_0\exp(-\phi/c),$$

that is,
$$\vec{v}_1(\phi) = cs_0\exp(-\phi/c)\frac{d\vec{x}}{d\phi},$$

hence
$$\frac{dx^1}{d\phi} = \frac{\exp(\phi/c)}{cs_0}\cos\phi \text{ and } \frac{dx^2}{d\phi} = \frac{\exp(\phi/c)}{cs_0}\sin\phi$$

for the components of the vector $\vec{x}$. Introducing the complex number $z = x^1 + ix^2$, we obtain
$$\frac{dz}{d\phi} = \frac{\exp(\phi(i+1/c))}{cs_0} \text{ and } z = \frac{\exp(\phi(i+1/c))}{(1+ci)s_0},$$

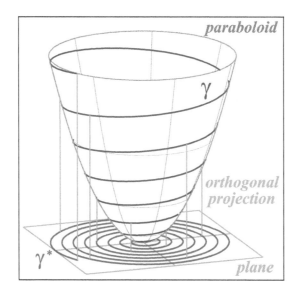

Figure 1.65 The orthogonal projection of a line of constant slope on a paraboloid

or in polar coordinates $r^2 = |z|^2$ and ϕ,

$$r^2 = \frac{\exp\left(2\phi/c\right)}{s_0^2(1+c^2)},$$

and finally

$$\vec{x}(\phi) = \frac{1}{s_0\sqrt{1+c^2}}\exp\left(\phi/c\right)\{\cos\phi, \sin\phi\} \text{ for } \phi \in \mathbb{R}.$$

$\square$

Remark 1.12.11. *We could also have used Theorem 1.12.3 to establish the statements in Visualization 1.12.10.*

Now we consider two examples that are geometrically interesting as well as instructive in the sense that most results of the theory of curves are applied.

Visualization 1.12.12. *Let γ be a curve on a paraboloid of revolution such that γ has a constant angle with the axis of the paraboloid. Then the orthogonal projection $\gamma^\perp$ of γ onto a plane orthogonal to the axis of the paraboloid is the involute of a circle line (Figure 1.65).*

Proof. Let the x^3–axis be the axis of rotation of the paraboloid. Then the paraboloid may be given by the following equation

$$(x^1)^2 + (x^2)^2 - 2ax^3 = 0, \text{ where } a \neq 0 \text{ is a constant.} \tag{1.145}$$

If $\vec{v}_k$ $(k = 1, 2, 3)$ are the vectors of the trihedra of the curve γ, then $\vec{v}_1 \bullet \vec{e}^3 = \cos \varphi$ and

$$0 = \frac{d}{ds}(\vec{v}_1 \bullet \vec{e}^3) = \dot{\vec{v}}_1 \bullet \vec{e}^3 = \kappa \vec{v}_2 \bullet \vec{e}^3 \text{ for } \kappa \neq 0.$$

This implies $\vec{v}_2 \bullet \vec{e}^3 = 0$ and consequently $\vec{v}_3 \bullet \vec{e}^3 = \sin \varphi$. The osculating plane of γ at an arbitrary point $P \in \gamma$ is given by the equation $(\overrightarrow{0X} - \overrightarrow{0P}) \bullet \vec{v}_3 = 0$. Putting

$$c(P) = \frac{1}{\sin \varphi} \overrightarrow{0P} \bullet \vec{v}_3 \quad (\varphi \neq 0, \pi),$$

we obtain

$$x^3 = -\frac{1}{\sin \varphi} \left(x^1 \vec{e}^1 \bullet \vec{v}_3 + x^2 \vec{e}^2 \bullet \vec{v}_3 \right) + c(P). \tag{1.146}$$

Let $\tilde{\gamma}$ denote the curve of intersection of the paraboloid and the osculating plane of γ at P. Then $\tilde{\gamma}$ is a second order approximation of γ. Obviously the orthogonal projection $\tilde{\gamma}^\perp$ of $\tilde{\gamma}$ on to the $x^1 x^2$–plane is also a second order approximation of $\gamma^\perp$. Eliminating x^3 from (1.145) and (1.146), we obtain an equation for $\tilde{\gamma}^\perp$

$$(x^1)^2 + (x^2)^2 + \frac{2a}{\sin \varphi} \left(x^1 \vec{e}^1 \bullet \vec{v}_3 + x^2 \vec{e}^2 \bullet \vec{v}_3 \right) = 2ac(P).$$

This is the equation of a circle line. The distance d of the centre of the circle line from the origin is given by

$$d^2 = \frac{a^2}{\sin^2 \varphi} l^2 \text{ where } l^2 = (\vec{e}^1 \bullet \vec{v}_3)^2 + (\vec{e}^2 \bullet \vec{v}_3)^2.$$

Thus $|l|$ is the length of the projection of the vector $\vec{v}_3$ on to the $x^1 x^2$–plane. Since $\vec{e}^3 \bullet \vec{v}_3 = \sin \varphi$, we have $l^2 = \cos^2 \varphi$ and consequently $d^2 = a^2 \cot^2 \varphi$. Thus the centres of curvature of $\gamma^\perp$ lie on a circle line, which is the evolute of $\gamma^\perp$. □

Visualization 1.12.13. *Let γ be a curve on a sphere of radius $R > 0$ such that γ has a constant angle with some constant unit vector $\vec{u}$. Then the orthogonal projection $\gamma^\perp$ of γ on to a plane E orthogonal to the vector $\vec{u}$ is an epicycloid with $\lambda = 1$ (Figure 1.66).*

Proof. Using the notations of Visualizations 1.5.6 and 1.7.7 with R replaced by R_0 and $\lambda = 1$, we obtain for a suitable choice s^* along the epicycloid from (1.65)

$$t(s^*) = \frac{2r}{r_0} \cos^{-1} \left(\frac{r_0}{4r R_0} s^* \right). \tag{1.147}$$

Furthermore, since we have by (1.93)

$$\kappa(t) = \frac{2r + r_0}{2\sqrt{2}r(r_0 + r)\sqrt{1 - \cos \frac{r_0}{r} t}} = \frac{2r + r_0}{4r(r_0 + r)\sqrt{1 - \cos^2 \frac{r_0}{2r} t}},$$

it follows from (1.147) that

$$\rho^*(s^*) = \frac{1}{\kappa(t(s^*))} = \frac{\sqrt{(4r R_0)^2 - r_0^2(s^*)^2}}{R_0 + r}. \tag{1.148}$$

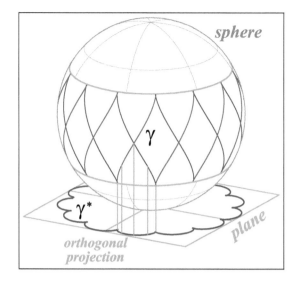

Figure 1.66 The orthogonal projection of a line of constant slope on a sphere

If S denotes a sphere of radius R, then S is the osculating sphere of any curve on S. Let ρ and τ denote the radius of curvature and torsion of γ. Then it follows by Theorem 1.9.5 that

$$R^2 = \rho^2 + \frac{\dot{\rho}^2}{\tau^2}. \tag{1.149}$$

If φ denotes the constant angle between γ and $\vec{u}$, then we have $\cos\varphi = \vec{u} \bullet \vec{v}_1$ and $0 = \vec{u} \bullet \dot{\vec{v}}_1 = \kappa \vec{u} \bullet \vec{v}_2$, and $\kappa \neq 0$ implies $0 = \vec{u} \bullet \vec{v}_2$. Since $\vec{u}$ is a unit vector orthogonal to $\vec{v}_2$, we may write

$$\vec{u} = \cos\varphi \vec{v}_1 + \sin\varphi \vec{v}_3.$$

This implies

$$\vec{0} = \dot{\vec{u}} = \cos\varphi \dot{\vec{v}}_1 + \sin\varphi \dot{\vec{v}}_3 = (\kappa\cos\varphi - \tau\sin\varphi)\vec{v}_2$$

and

$$\kappa\cos\varphi = \tau\sin\varphi \quad \text{or} \quad \tau = \frac{1}{\rho}\cot\varphi \ (\varphi \neq 0, \pi).$$

Substituting this last result into (1.149), we obtain

$$R^2 - \rho^2 = \rho^2 \dot{\rho}^2 \tan^2\varphi,$$

and for $\varphi \neq 0, \pi$

$$\dot{\rho} = \pm \frac{\sqrt{R^2 - \rho^2}}{\rho|\tan\varphi|}.$$

This yields

$$\pm \int \frac{\rho}{\sqrt{R^2 - \rho^2}} \, d\rho = \mp\sqrt{R^2 - \rho^2} = \frac{1}{|\tan\varphi|}(s - s_0).$$

We choose $s_0 = 0$ to obtain

$$\rho(s) = \sqrt{R^2 - s^2\cot^2\varphi} \quad \text{for } |s| < R|\tan\varphi|.$$

We have for the arc length $s^\perp$ and radius of curvature $\rho^\perp$ of the orthogonal projection $\gamma^\perp$ of γ onto a plane orthogonal to the vector $\vec{u}$, by (1.139) and (1.140) in Theorem 1.12.7,

$$\rho^\perp(s^\perp) = \rho^\perp(s(s^\perp)) = \rho(s(s^\perp)) \sin^2 \varphi$$

$$= \sqrt{R^2 - \frac{(s^\perp)^2}{\sin^2 \varphi} \cot^2 \varphi \sin^2 \varphi}$$

$$= \sqrt{R^2 \sin^4 \varphi - (s^\perp)^2 \cos^2 \varphi} \text{ for } |s^\perp| < R \sin \varphi |\tan \varphi|. \tag{1.150}$$

We consider the epicycloid with $r_0 = R \cos \varphi$, $r = R(1 - \cos \varphi)/2$ and $R_0 = r + r_0$. Then we obtain for the radius of curvature of this epicycloid from (1.148)

$$\rho(s^*) = \sqrt{R^2 \sin^4 \varphi - \cos^2 \varphi (s^*)^2}.$$

Thus the radii of curvature $\rho^\perp$ and ρ are the same, and the statement of the example follows from the fundamental theorem of curves. □

1.13   SPHERICAL IMAGES OF A CURVE

A good illustration can be obtained of the change in direction of the vectors of a trihedron of a curve attached to the origin of the coordinate system of three-dimensional space. Since these vectors are unit vectors this will yield three curves on the unit sphere.

Definition 1.13.1. Let γ be a curve with a parametric representation $\vec{x}(s)$ $(s \in I)$, and the vectors of the trihedra $\vec{v}_k(s)$ $(k = 1, 2, 3)$ for $s \in I$. The curves γ_1, γ_2 and γ_3 with parametric representations $\vec{v}_1(s)$, $\vec{v}_2(s)$ and $\vec{v}_3(s)$ $(s \in I)$ are called the *spherical images of the tangent, principal normal* and *binormal vectors* (Figure 1.67).

Visualization 1.13.2. *We consider the helix of Visualization 1.6.7 (b) with a parametric representation*

$$\vec{x}(s) = \{r \cos \omega s, r \sin \omega s, h \omega s\} \ (s \in \mathbb{R}).$$

Then we have

$$\vec{v}_1(s) = \{-\omega r \sin \omega s, \omega r \cos \omega s, h\omega\},$$
$$\vec{v}_2(s) = \{-\cos \omega s, -\sin \omega s, 0\}$$

and

$$\vec{v}_3(s) = \{\omega h \sin \omega s, -\omega h \cos \omega s, \omega r\} \ (s \in \mathbb{R})$$

for its spherical images (Figure 1.68).

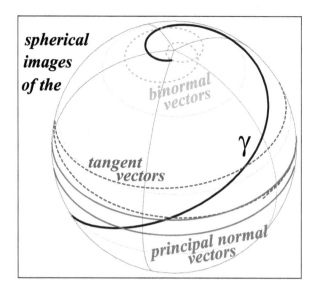

Figure 1.67 The spherical images of the tangent, principal normal and binormal vectors of γ

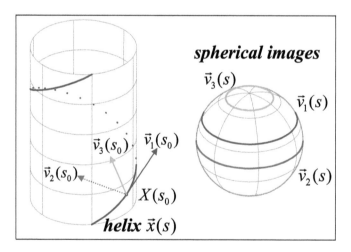

Figure 1.68 The spherical images of a helix

There are some simple relations between the arc lengths of a curve and of its spherical images.

Theorem 1.13.3. *Let γ be a curve with non–vanishing curvature κ and with torsion τ, and s_1, s_2 and s_3 denote the arc lengths along the spherical images of its tangent, principal normal and binormal vectors, respectively. Then we have*

$$\frac{ds_1}{ds} = \kappa, \ \frac{ds_2}{ds} = \sqrt{\kappa^2 + \tau^2} \ and \ \frac{ds_3}{ds} = |\tau|; \tag{1.151}$$

in particular, we have

$$\frac{ds_1}{ds} = \frac{ds_2}{ds} \tag{1.152}$$

if and only if γ is a planar curve.

Proof. Let $\vec{x}(s)$ ($s \in I$) be a parametric representation of γ, and $\vec{v}_k(s)$ denote, as usual, the tangent, principal normal and binormal vectors of γ at s. The arc lengths of the spherical images γ_k are given by

$$s_k(s) = \int_{s_0}^{s} \|\dot{\vec{v}}_k(\sigma)\| \, d\sigma$$

and so

$$\frac{ds_k}{ds}(s) = \|\dot{\vec{v}}_k(s)\| = \sqrt{\dot{\vec{v}}_k(s) \bullet \dot{\vec{v}}_k(s)} \ (k = 1, 2, 3) \text{ for all } s \in I.$$

Applying Frenet's formulae (1.84) in Theorem 1.7.1, we obtain for all $s \in I$

$$\frac{ds_1}{ds}(s) = \sqrt{\kappa(s)\vec{v}_2(s) \bullet \kappa(s)\vec{v}_2(s)} = |\kappa(s)| = \kappa(s),$$

$$\frac{ds_2}{ds}(s) = \sqrt{(-\kappa(s)\vec{v}_1(s) + \tau(s)\vec{v}_3(s)) \bullet (-\kappa(s)\vec{v}_1(s) + \tau(s)\vec{v}_3(s))} = \sqrt{\kappa^2(s) + \tau^2(s)}$$

and

$$\frac{ds_3}{ds}(s) = \sqrt{-\tau(s)\vec{v}_3(s) \bullet (-\tau(s))\vec{v}_3(s)} = |\tau(s)|.$$

Thus we have proved the identities in (1.151).
Since planar curves are characterized by the condition $\tau \equiv 0$ by Theorem 1.8.1, the equality in (1.152) is an immediate consequence of the second identity in (1.151). $\square$

The next result characterizes lines of constant slope by the spherical images of their tangent vectors.

Theorem 1.13.4. *A curve is a line of constant slope if and only if the spherical image of its tangent vectors is a circle line (Figure 1.69).*

Proof. Let γ^* denote the spherical image of the tangent vectors of a curve γ, and $\vec{x}(s)$ be a parametric representation of γ.

(i) First we assume that γ^* is a circle line.
   Then γ^* has a parametric representation

$$\vec{v}_1(s_1) = \vec{a} r \cos\left(\frac{s_1}{r}\right) + \vec{b} r \sin\left(\frac{s_1}{r}\right) + \vec{c},$$

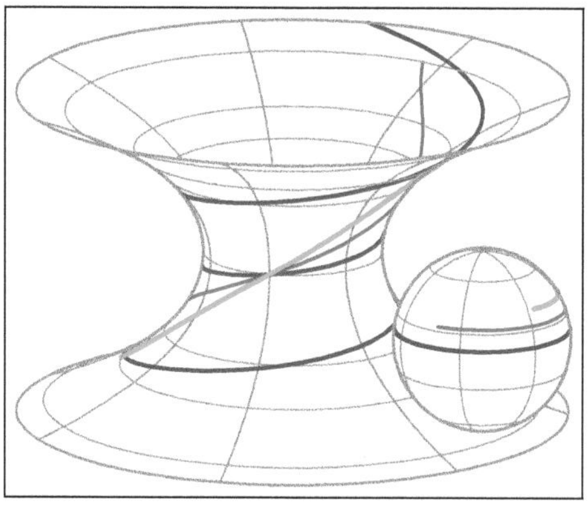

Figure 1.69 The spherical images of the tangent vectors of lines of constant slope on a catenoid

where s_1 and r are the arc length along γ^* and the radius of γ^*, and the constant vectors $\vec{a}$, $\vec{b}$ and $\vec{c}$ satisfy the conditions

$$\vec{a}^2 = \vec{b}^2 = 1,\ \vec{a} \bullet \vec{b} = \vec{a} \bullet \vec{c} = \vec{b} \cdot \vec{c} = 0. \qquad (1.153)$$

Now $\|\vec{v}_1(s_1)\| = 1$ implies $\vec{c}^2 = 1 - r^2$. If κ denotes the curvature along γ, then we have by the first identity in (1.151) of Theorem 1.13.3

$$s_1(s) = \int_{s_0}^{s} \kappa(\sigma)\,d\sigma,$$

hence

$$\vec{v}_1(s_1) = \vec{a}r\cos\left(\frac{s_1(s)}{r}\right) + \vec{b}r\sin\left(\frac{s_1(s)}{r}\right) + \vec{c},$$

and integration yields

$$\vec{x}(s) = \vec{a}r\int_{s_0}^{s}\cos\left(\frac{s_1(\sigma)}{r}\right)d\sigma + \vec{b}r\int_{s_0}^{s}\sin\left(\frac{s_1(\sigma)}{r}\right)d\sigma + \vec{c}s + \vec{c}_0,$$

where $\vec{c}_0$ is a constant vector.

This is a parametric representation of a line of constant slope by Theorem 1.12.3. Hence γ is a line of constant slope.

(ii) Now we assume that γ is a line of constant slope.

Let c be the constant such that $\tau(s) = c\kappa(s)$. Then parametric representation $\vec{x}(s)$ of γ satisfies by Theorem 1.12.3

$$\vec{v}_1(s) = \frac{\vec{a}}{\omega}\sin\left(\omega t(s)\right) - \frac{\vec{b}}{\omega}\cos\left(\omega t(s)\right) + \vec{c},$$

where the constant vectors $\vec{a}$, $\vec{b}$ and $\vec{c}$ satisfy the conditions in (1.153), and

$$\vec{c}^{\,2} = \frac{c^2}{1+c^2} \leq 1, \ \omega = 1 + c^2 \ \text{and} \ t(s) = \int_{s_0}^{s} \kappa(\sigma)\, d\sigma.$$

It follows that

$$\|\vec{v}_1(s) - \vec{c}\| = \frac{\vec{a}^{\,2}}{\omega^2} \sin^2{(\omega t(s))} + \frac{\vec{b}^{\,2}}{\omega^2} \cos^2{(\omega t(s))} = \frac{1}{\omega^2} = \frac{1}{c^2},$$

that is, $\vec{v}_2(s)$ is a parametric representation of a circle line.

□

CHAPTER 2

Surfaces in Three-Dimensional Euclidean Space

In this chapter, we deal with the local differential geometry of surfaces in three-dimensional Euclidean space $\mathbb{E}^3$. This means that we study the geometric shape of surfaces in the neighbourhood of an arbitrary one of their points. The most important concepts arising in this task are those of the *normal, principal, Gaussian* and *mean curvature* in Section 2.5.

Additional topics are

- *curves on surfaces* in Section 2.1

- *tangent planes* and *normal vectors of surfaces* in Section 2.2

- *first* and *second fundamental coefficients* in Sections 2.3 and 2.4

- *Meusnier's theorem*, Theorem 2.5.7 in Section 2.5

- *principal directions* in Section 2.5

- *Euler's formula*, (2.111) in Theorem 2.5.31 of Section 2.5

- *the local shape of surfaces* in Section 2.6

- *Dupin's indicatrix* in Section 2.7

- *lines of curvature and asymptotic lines* in Section 2.8

- *triple orthogonal systems* in Section 2.9

- *the Gauss and Weingarten equations*, (2.42) of Section 2.4 and (2.141) in Theorem 2.10.1 of Section 2.10.

DOI: 10.1201/9781003370567-2

2.1 SURFACES AND CURVES ON SURFACES

In this section, we deal with the *representation of surfaces in three-dimensional* $\mathbb{E}^3$ and with *curves on surfaces*.

Whereas a curve can be described by a parametric representation of the position vectors of its points with respect to *one real parameter, two real parameters* are needed in a parametric representation of the position vectors of the points of a surface. In general, it is not possible to globally relate all points of a surface to parameters in one region of the plane in a one–to–one way at the same time. For instance, spherical coordinates cannot be applied in the poles of a sphere. In local differential geometry, we confine our studies to sufficiently small sections of a surface.

We start with the definition of a surface in three-dimensional $\mathbb{R}^3$ and recall two notations:

An open connected subset of $\mathbb{R}^2$ is called a *domain*. By $C^r(D)$, we denote the class of all functions $f = (f^1, f^2, f^3) : D \to \mathbb{R}^3$ for which the *coordinate functions* f^k $(k = 1, 2, 3)$ *have continuous partial derivatives of order* r *on* D.

Definition 2.1.1. Let $D \subset \mathbb{R}^2$ be a domain and $f = (f^1, f^2, f^3) : D \to \mathbb{R}^3$ be a function with $f \in C^3(D)$. The point set

$$S = \left\{ X \in \mathbb{R}^3 : X = f(u^1, u^2) = (f^1(u^1, u^2), f^2(u^1, u^2), f^3(u^1, u^2)) \ \left((u^1, u^2) \in D \right) \right\} \tag{2.1}$$

is called a *surface in* $\mathbb{R}^3$, if in addition the following condition holds

$$\left\{ \frac{\partial f^1}{\partial u^1}, \frac{\partial f^2}{\partial u^1}, \frac{\partial f^3}{\partial u^1} \right\} \times \left\{ \frac{\partial f^1}{\partial u^2}, \frac{\partial f^2}{\partial u^2}, \frac{\partial f^3}{\partial u^2} \right\} \neq \vec{0} \quad \text{on } D; \tag{2.2}$$

(f, D) is called a *parametric representation of* S, and the values u^1 and u^2 are called *parameters of* S.

A surface S may also be given in terms of the position vectors of its points

$$\vec{x} = \vec{x}(u^i) = \vec{x}(u^1, u^2) = \left\{ x^1(u^1, u^2), x^2(u^1, u^2), x^3(u^1, u^2) \right\} \ \left((u^1, u^2) \in D \right), \tag{2.3}$$

where $x^k(u^i) = f^k(u^i)$ $(k = 1, 2, 3)$, and the vectors $\vec{x}_j = \dfrac{\partial \vec{x}}{\partial u^j}$ $(j = 1, 2)$ satisfy the condition

$$\vec{x}_1 \times \vec{x}_2 \neq \vec{0}. \tag{2.4}$$

As in the case of curves, (2.3) will also be referred to as a parametric representation for S; in fact we shall rather use (2.3) than (2.1).

For further studies in the theory of surfaces, the vector functions in (2.3) will often have to have higher-order continuous partial derivatives. *In general, the existence of continuous third-order partial derivatives will be sufficient.*

Curves on surfaces are of special interest.

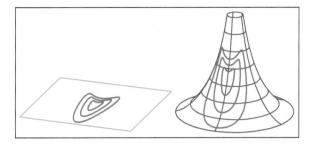

Figure 2.1 Curves on a surface (right) and in its parameter plane (left)

Definition 2.1.2. (a) A curve γ is on a surface S with a parametric representation $\vec{x}(u^i)$ $((u^1, u^2) \in D)$ if it is given by a parametric representation

$$\vec{x}(t) = \vec{x}(u^i(t)) \ (t \in I), \text{ where } u^1(I) \times u^2(I) \subset D, \tag{2.5}$$

and the functions u^k $(k = 1, 2)$ have continuous derivatives of order $(r \geq 1)$, and satisfy

$$\left|(u^1)'(t)\right| + \left|(u^2)'(t)\right| \neq 0 \quad \text{for all } t \in I.$$

The curve γ is called *curve on the surface S* (Figure 2.1).

(b) Curves for which one parameter of the surface is constant are called *parameter lines*. The curves with $u^1(t) = t$ and $u^2(t) = const$ are called u^1*-lines*, those with $u^1(t) = const$ and $u^2(t) = t$ are called u^2*-lines* (Figure 2.2).

In general, a surface S may have various parametric representations. From (2.1), we can obtain a new vector function by introducing two new parameters u^{*k} $(k = 1, 2)$ such that

$$u^i = u^i(u^{*k}) \quad (i = 1, 2). \tag{2.6}$$

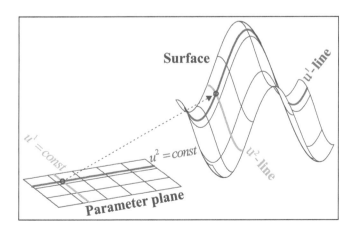

Figure 2.2 Parameter lines on a surface

We have to take care that the range of the functions defined in (2.6) contains the domain D of the parameters (u^1, u^2) and that the conditions for parametric representations of surfaces are satisfied.

Definition 2.1.3. A parameter transformation (2.6) is called *admissible* if the following three conditions hold

(i) The functions in (2.6) are defined on a domain D^* such that

$$D \subset u^1(D^*) \times u^2(D^*).$$

(ii) $u^i \in C^r(D)$ $(i = 1, 2)$ for some $r \geq 1$.

(iii) For the *Jacobian of the functions* in (2.6), we must have

$$\frac{\partial(u^i)}{\partial(u^{*k})} = \begin{vmatrix} \dfrac{\partial u^1}{\partial u^{*1}} & \dfrac{\partial u^1}{\partial u^{*2}} \\[3mm] \dfrac{\partial u^2}{\partial u^{*1}} & \dfrac{\partial u^2}{\partial u^{*2}} \end{vmatrix} \neq 0 \ \left((u^{*1}, u^{*2}) \in D^*\right). \tag{2.7}$$

The same convention as in the case of curves is used concerning the term *admissible*.

From now on we shall always make use of *Einstein's convention of summation*:

- *If the same index appears in an expression twice, once as a subscript and once as a superscript, then summation takes place with respect to this index.*

For instance, we have

$$\vec{x}_i \bullet \vec{x}_k du^i du^k = \sum_{i,k=1}^{2} \vec{x}_i \bullet \vec{x}_k du^i du^k.$$

Remark 2.1.4. *If we use (2.6) to introduce new parameters u^{*r} for a surface with a parametric representation (2.3), then we have*

$$\vec{x}(u^k) = \vec{x}(u^k(u^{*r})) = \vec{x}^*(u^{*j}).$$

This implies

$$\vec{x}_r^* = \frac{\partial \vec{x}^*}{\partial u^{*r}} = \frac{\partial \vec{x}}{\partial u^k} \frac{\partial u^k}{\partial u^{*r}} = \vec{x}_k \frac{\partial u^k}{\partial u^{*r}} \quad (r = 1, 2) \tag{2.8}$$

and

$$\vec{x}_1^* \times \vec{x}_2^* = \vec{x}_k \frac{\partial u^k}{\partial u^{*1}} \times \vec{x}_j \frac{\partial u^j}{\partial u^{*2}}$$

$$= \vec{x}_1 \frac{\partial u^1}{\partial u^{*1}} \times \vec{x}_2 \frac{\partial u^2}{\partial u^{*2}} - \vec{x}_1 \frac{\partial u^1}{\partial u^{*2}} \times \vec{x}_2 \frac{\partial u^2}{\partial u^{*1}}$$

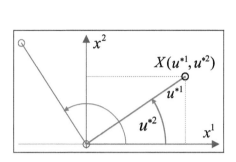 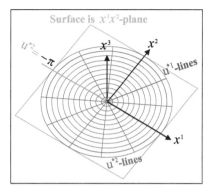

Figure 2.3 Polar coordinates in the plane (left); the $x^1 x^2$–plane with polar coordinates as parameters (right)

$$= (\vec{x}_1 \times \vec{x}_2) \left(\frac{\partial u^1}{\partial u^{*1}} \frac{\partial u^2}{\partial u^{*2}} - \frac{\partial u^1}{\partial u^{*2}} \frac{\partial u^2}{\partial u^{*1}} \right)$$

$$= (\vec{x}_1 \times \vec{x}_2) \cdot \frac{\partial(u^k)}{\partial(u^{*r})},$$

that is,

$$\vec{x}_1^* \times \vec{x}_2^* = (\vec{x}_1 \times \vec{x}_2) \cdot \frac{\partial(u^k)}{\partial(u^{*r})}. \qquad (2.9)$$

Visualization 2.1.5. *The $x^1 x^2$–plane S has a parametric representation*

$$\vec{x}(u^i) = \{u^1, u^2, 0\} \ \left((u^1, u^2) \in D = \mathbb{R}^2\right)$$

with respect to the Cartesian coordinates u^1 and u^2.
Introducing polar coordinates u^{*1} and u^{*2}, we obtain

$$\vec{x}(u^{*i}) = \{u^{*1} \cos u^{*2}, u^{*1} \sin u^{*2}, 0\} \ \left(u^{*1}, u^{*2}) \in D^* = (0, \infty) \times (-\pi, \pi)\right)$$

for S^*, which is the $x^1 x^2$–plane with the negative x^1–axis and the origin removed (Figure 2.3). It follows for S that

$$\vec{x}_1 = \{1, 0, 0\}, \ \vec{x}_2 = \{0, 1, 0\} \ and \ \vec{x}_1 \times \vec{x}_2 = \{0, 0, 1\} \neq \vec{0} \ for \ all \ (u^1, u^2) \in D,$$

and for S^* that

$$\vec{x}_1 = \{\cos u^{*2}, \sin u^{*2}, 0\}, \ \vec{x}_2 = \{-u^{*1} \sin u^{*2}, u^{*1} \cos u^{*2}, 0\},$$

and

$$\vec{x}_1 \times \vec{x}_2 = \{0, 0, u^{*1}\} \neq \vec{0} \ if \ and \ only \ if \ u^{*1} \neq 0.$$

We consider the two curves γ and γ^* in the parameter planes given by

$$u^1(t) = u^{*1}(t) = a \exp(bt), \ and \ u^2(t) = u^{*2}(t) = t \ (t \in \mathbb{R}),$$

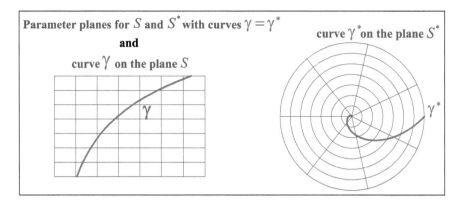

Figure 2.4 Curve on a plane with different parametric representations

where $a, b > 0$ are constants. The curve on S is given by

$$\vec{x}(u^i(t)) = \{a \exp(bt), t, 0\} \ (t \in \mathbb{R}),$$

and on S^* by

$$\vec{x}(u^{*i}(t)) = \{a \exp(bt) \cos t, a \exp(bt) \sin t, 0\} \ (t \in \mathbb{R})$$

(Figure 2.4).

Example 2.1.6. *Let $c_1 \neq c_2$. We put*

$$u^1(u^{*k}) = 2 \tan^{-1}\left(\exp\left(\frac{u^{*1} - u^{*2}}{c_1 - c_2}\right)\right) - \frac{\pi}{2}$$

and

$$u^2(u^{*k}) = \frac{c_1 u^{*1} - c_2 u^{*2}}{c_1 - c_2} \ \left((u^{*1}, u^{*2}) \in \mathbb{R}^2\right).$$

Then we have, writing $\phi(u^{*k}) = (u^{*1} - u^{*2})/(c_1 - c_2)$,

$$\frac{\partial u^1}{\partial u^{*1}} = \frac{2}{1 + (\exp(\phi(u^{*k})))^2} \cdot \exp\left(\phi(u^{*k})\right) \cdot \frac{1}{c_1 - c_2}$$

$$= \frac{1}{(c_1 - c_2) \cdot \cosh(\phi(u^{*k}))},$$

$$\frac{\partial u^1}{\partial u^{*2}} = \frac{1}{(c_1 - c_2) \cdot \cosh(\phi(u^{*k}))},$$

$$\frac{\partial u^2}{\partial u^{*1}} = \frac{c_1}{c_1 - c_2}, \ \frac{\partial u^2}{\partial u^{*2}} = -\frac{c_2}{c_1 - c_2}$$

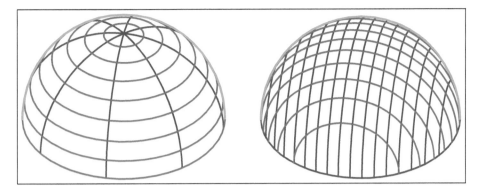

Figure 2.5 The hemispheres of Visualization 2.1.7 (a) and (b)

and

$$\frac{\partial(u^i)}{\partial(u^{*k})} = -\frac{c_2}{(c_1 - c_2)^2 \cdot \cosh(\phi(u^{*k}))} + \frac{c_1}{(c_1 - c_2) \cdot \cosh(\phi(u^{*k}))}$$

$$= \frac{1}{(c_1 - c_2) \cdot \cosh(\phi(u^{*k}))} \neq 0.$$

Therefore, the parameter transformation is admissible.

Surfaces in three-dimensional space may also be given be equations.

Visualization 2.1.7. *(a) The sphere of radius $r > 0$ centred at the origin has an equation*

$$\sum_{k=1}^{3} (x^k)^2 = r^2,$$

and we obtain a parametric representation for the upper semi–sphere with equator and north pole removed

$$\vec{x}(u^i) = \left\{ u^1, u^2, \sqrt{r^2 - (u^1)^2 - (u^2)^2} \right\} \left(0 < (u^1)^2 + (u^2)^2 < r^2 \right)$$

(right in Figure 2.5). Furthermore, it follows that

$$\vec{x}_1 = \left\{ 1, 0, \frac{-u^1}{\sqrt{r^2 - (u^1)^2 - (u^2)^2}} \right\},$$

$$\vec{x}_2 = \left\{ 0, 1, \frac{-u^2}{\sqrt{r^2 - (u^1)^2 - (u^2)^2}} \right\}$$

and

$$\vec{x}_1 \times \vec{x}_2 = \left\{ \frac{u^1}{\sqrt{r^2 - (u^1)^2 - (u^2)^2}}, \frac{u^2}{\sqrt{r^2 - (u^1)^2 - (u^2)^2}}, 1 \right\} \neq \vec{0}.$$

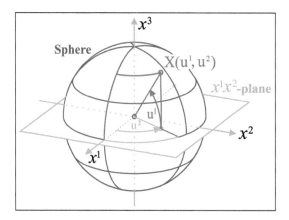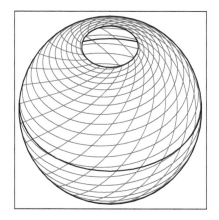

Figure 2.6 Sphere with spherical parameters (left) and the parametrization of Visualization 2.1.7 (c)

We fix c with $|c| < r$.
Then the u^1–line corresponding to $u^2 = c$ has a parametric representation

$$\vec{x}(t) = \left\{ t, c, \sqrt{r^2 - t^2 - c^2} \right\} \ \left(t \in \left(-\sqrt{r^2 - c^2}, \sqrt{r^2 - c^2}\right) \right);$$

it is a semi–circle line in the plane $x^2 = c$ with radius $\sqrt{r^2 - c^2}$ and centre in $(0, c, 0)$.
Similarly, the u^2–line corresponding to $u^1 = c$ has a parametric representation

$$\vec{x}(t) = \left\{ c, t, \sqrt{r^2 - t^2 - c^2} \right\} \ \left(t \in \left(-\sqrt{r^2 - c^2}, \sqrt{r^2 - c^2}\right) \right);$$

it is a semi–circle line in the plane $x^1 = c$ with radius $\sqrt{r^2 - c^2}$ and centre in $(c, 0, 0)$
(right in Figure 2.5).
(b) The sphere of radius $r > 0$, centred at the origin, with north and south poles
and a meridian removed has a parametric representation with respect to spherical
coordinates

$$\vec{x}(u^i) = r \left\{ \cos u^1 \cos u^2, \cos u^1 \sin u^2, \sin u^1 \right\} \ \left((u^1, u^2) \in D = \left(-\frac{\pi}{2}, \frac{\pi}{2}\right) \times (0, 2\pi) \right)$$

(Figures 2.5 and 2.6). We have

$$\vec{x}_1 = r \left\{ -\sin u^1 \cos u^2, -\sin u^1 \sin u^2, \cos u^1 \right\},$$
$$\vec{x}_2 = r \left\{ -\cos u^1 \sin u^2, \cos u^1 \cos u^2, 0 \right\}$$

and

$$\vec{x}_1 \times \vec{x}_2 = -r^2 \cos u^1 \left\{ \cos u^1 \cos u^2, \cos u^1 \sin u^2, \sin u^1 \right\}$$
$$= -r \cos u^1 \vec{x} \neq \vec{0} \ \left((u^1, u^2) \in D \right).$$

The u^1–lines are the so–called meridians, *and the u^2–lines are the so–called* parallels
(left in Figure 2.5).

(c) Let $c_1, c_2 \in \mathbb{R}$ with $c_1 \neq c_2$. We put

$$\phi = \phi(u^i) = \frac{u^1 - u^2}{c_1 - c_2}, \quad \psi = \psi(u^i) = \frac{c_1 u^1 - c_2 u^2}{c_1 - c_2}$$

and

$$\vec{x} = \vec{x}(u^i) = \frac{r}{\cosh \phi(u^i)} \{\cos \psi(u^i), \sin \psi(u^i), \sinh \phi(u^i)\} \quad \left((u^1, u^2) \in \mathbb{R}^2\right) \quad (2.10)$$

(right in Figure 2.6). This is one more parametric representation of a sphere of radius r and centre in the origin. Putting $\vec{u}(\psi) = \{\cos \psi, \sin \psi, 0\}$, we obtain

$$\vec{x} = \frac{r}{\cosh \phi}(\vec{u}(\psi) + \vec{e}^3 \cdot \sinh \phi),$$

$$\vec{x}_1 = +\frac{r}{(c_1 - c_2)\cosh \phi}\left(-\vec{u}(\psi) \cdot \tanh \phi + \vec{u}'(\psi) \cdot c_1 + \vec{e}^3 \cdot \frac{1}{\cosh \phi}\right),$$

$$\vec{x}_2 = -\frac{r}{(c_1 - c_2)\cosh \phi}\left(-\vec{u}(\psi) \cdot \tanh \phi + \vec{u}'(\psi) \cdot c_2 + \vec{e}^3 \cdot \frac{1}{\cosh \phi}\right)$$

and

$$\vec{x}_1 \times \vec{x}_2 = -\frac{r^2}{(c_1 - c_2)\cosh^2 \phi}\left(\vec{e}_3 \cdot \tanh \phi + \vec{u}(\psi) \cdot \frac{1}{\cosh \phi}\right) \neq 0.$$

The u^1–line corresponding to $u^2 = u_0^2$, and the u^2–line corresponding to $u^1 = u_0^1$ are given by

$$\vec{x}(t, u_0^2) = \frac{r}{\cosh \dfrac{t - u_0^2}{c_1 - c_2}}\left\{\cos \frac{c_1 t - c_2 u_0^2}{c_1 - c_2}, \sin \frac{c_1 t - c_2 u_0^2}{c_1 - c_2}, \sinh \frac{t - u_0^2}{c_1 - c_2}\right\} \quad (t \in \mathbb{R})$$

$$(2.11)$$

and

$$\vec{x}(t, u_0^1) = \frac{r}{\cosh \dfrac{u_0^1 - t}{c_1 - c_2}}\left\{\cos \frac{c_1 u_0^1 - c_2 t}{c_1 - c_2}, \sin \frac{c_1 u_0^1 - c_2 t}{c_1 - c_2}, \sinh \frac{u_0^1 - t}{c_1 - c_2}\right\} \quad (t \in \mathbb{R}).$$

$$(2.12)$$

2.2 THE TANGENT PLANES AND NORMAL VECTORS OF A SURFACE

In this section, we define the *tangent plane* and the *normal vector of a surface* at a given point of the surface.

The direction of a curve γ on a surface S with a parametric representation (2.3) is given by

$$\vec{x}'(u^i(t)) = \vec{x}_k(u^i(t))(u^k)'(t), \quad \text{or} \quad \vec{x}' = \vec{x}_k(u^k)' \text{ for short.} \quad (2.13)$$

The identity in (2.13) reduces to

$$\vec{x}' = \vec{x}_1 \text{ for the } u^1\text{–lines and } \vec{x}' = \vec{x}_2 \text{ for the } u^2\text{–lines.}$$

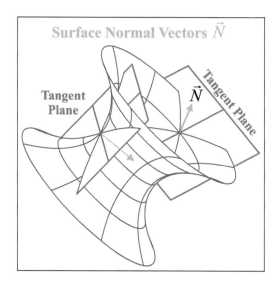

Figure 2.7   Tangent planes and normal vectors of a surface

Thus the vector $\vec{x}_k$ is a vector in the direction of the tangent to a u^k–line. Condition (2.4) means that the vectors $\vec{x}_1$ and $\vec{x}_2$ are linearly independent. They span a plane at every point of the surface S, the so–called *tangent plane*.

Definition 2.2.1. Let S be a surface with a parametric representation $\vec{x}(u^i)$ $((u^1, u^2) \in D)$ and $P = X(\underset{0}{u}^k)$ be a point of S.

(a) The plane with parametric representation

$$\vec{t}_P = \vec{x}(\underset{0}{u}^k) + \lambda^j \vec{x}_j(\underset{0}{u}^k)\ (\lambda^k \in \mathbb{R}\ (k = 1, 2))$$

is called the *tangent plane of S at P* (Figure 2.7).

(b) The vector

$$\vec{N}_P = \vec{N}(\underset{0}{u}^k) = \frac{\vec{x}_1(\underset{0}{u}^k) \times \vec{x}_2(\underset{0}{u}^k)}{\|\vec{x}_1(\underset{0}{u}^k) \times \vec{x}_2(\underset{0}{u}^k)\|}$$

is called the *surface normal vector of S at P* (Figure 2.7).

Remark 2.2.2. *(a) Obviously the tangent plane of a surface S at a point P may be given by the equation*

$$(\vec{t}_P - \vec{x}(\underset{0}{u}^k)) \bullet \vec{N}_P = 0.$$

(b) Under parameter transformations, the normal vector of a surface will at most reverse its direction; if

$$\vec{x}^*(u^{*k}) = \vec{x}(u^j(u^{*k})),$$

then we obtain from (2.9)

$$\vec{N}_P^* = \vec{N}_P \frac{\frac{\partial(u^j)}{\partial(u^{*k})}}{\left|\frac{\partial(u^j)}{\partial(u^{*k})}\right|}.$$

(c) The trihedron of a surface *spanned by the vectors $\vec{x}_1$, $\vec{x}_2$ and $\vec{N}$ is as important in the theory of surfaces as the trihedron of a curve is in the theory of curves. The vectors of the trihedron of a surface, however, are not orthonormal. We only have*

$$\vec{N}^2 = 1 \quad and \quad \vec{N} \bullet \vec{x}_k = 0 \ (k = 1, 2).$$

Example 2.2.3. *We consider the two parametric representations of the $x^1 x^2$–plane given in Visualization 2.1.5*

$$\vec{x}(u^i) = \{u^1, u^2, 0\} \ \left((u^1, u^2) \in \mathbb{R}^2 \right)$$

and

$$\vec{x}^*(u^{*i}) = \{u^{*1} \cos u^{*2}, u^{*1} \sin u^{*2}, 0\} \ \left((u^{*1}, u^{*2}) \in D = (0, \infty) \times (-\pi, \pi) \right).$$

The formulae for the parameter transformation are

$$u^1 = u^{*1} \cos u^{*2} \ and \ u^2 = u^{*1} \sin u^{*2} \ \left((u^{*1}, u^{*2}) \in D \right).$$

Now $\vec{N}(u^i) = \{0, 0, 1\}$ and $\frac{\partial(u^i)}{\partial(u^{*k})} = u^{*1}$ together imply

$$\vec{N}^*(u^{*i}) = \vec{N}(u^i(u^{*k})) \frac{\frac{\partial(u^i)}{\partial(u^{*k})}}{\left| \frac{\partial(u^i)}{\partial(u^{*k})} \right|} = \{0, 0, 1\}.$$

Example 2.2.4. *We consider the sphere of Visualization 2.1.7 (b) with a parametric representation*

$$\vec{x}(u^i) = r\{\cos u^1 \cos u^2, \cos u^1 \sin u^2, \sin u^1\} \ \left((u^1, u^2) \in D = (-\pi/2, \pi/2) \times (0, 2\pi) \right).$$

It follows for the normal vectors that

$$\vec{N}(u^i) = -\{\cos u^1 \cos u^2, \cos u^1 \sin u^2, \sin u^1\} = -\frac{1}{r} \vec{x}(u^i) \ \left((u^1, u^2) \in D \right),$$

and consequently the tangent plane to the sphere at the point $P = X(0, \pi/2) = (0, r, 0)$ is given by the equation

$$\left(\{t^1, t^2, t^3\} - \{0, r, 0\} \right) \bullet (-\{0, 1, 0\}) = -t^2 + r = 0.$$

Now let

$$\vec{x}(u^i) = \left\{ u^1, u^2, \sqrt{r^2 - (u^1)^2 - (u^2)^2} \right\}$$

be a parametric representation of the upper half of the sphere in Visualization 2.1.7 (a). Then we have

$$\left(\vec{x}_1(u^i) \right)^2 = \frac{r^2 - (u^2)^2}{r^2 - (u^1)^2 - (u^2)^2},$$

$$\left(\vec{x}_2(u^i) \right)^2 = \frac{r^2 - (u^1)^2}{r^2 - (u^1)^2 - (u^2)^2},$$

$$\vec{x}_1(u^i) \bullet \vec{x}_2(u^i) = \frac{-u^1 u^2}{r^2 - (u^1)^2 - (u^2)^2}$$

and

$$\vec{N}(u^i) = \frac{1}{r} \left\{ u^1, u^2, \sqrt{r^2 - (u^1)^2 - (u^2)^2} \right\} = \frac{1}{r} \vec{x}(u^i).$$

2.3 THE ARC LENGTH, ANGLES AND GAUSS'S FIRST FUNDAMENTAL COEFFICIENTS

We recall from the fundamental theorem of curves, Theorem 1.11.1 that a curve is uniquely determined by two local, invariant quantities, namely its curvature and torsion as functions of its arc length. Similarly, the fundamental theorem of surfaces, which we will not prove, states that a surface is uniquely determined by certain local, invariant quantities, the so–called *first and second fundamental forms*.

Lengths, arcs and surface areas can be measured when the *first fundamental coefficients* of a surface are known; this is why they are also referred to as *metric coefficients*.

Let S be a surface with a parametric representation $\vec{x}(u^j)$ and γ be a curve on S with a parametric representation $\vec{x}(t) = \vec{x}(u^j(t))$. Then the arc length along γ is given by

$$s(t) = \int_{t_0}^{t} \|\vec{x}'(\tau)\| \, d\tau. \qquad (2.14)$$

It follows from

$$\vec{x}'(t) \bullet \vec{x}'(t) = \vec{x}_i(u^j(t)) \frac{du^i(t)}{dt} \bullet \vec{x}_k(u^j(t)) \frac{du^k(t)}{dt}$$
$$= (\vec{x}_i(u^j(t)) \bullet \vec{x}_k(u^j(t))) \frac{du^i(t)}{dt} \cdot \frac{du^k(t)}{dt}$$

that we are able to compute the arc length along γ provided we know the values

$$g_{ik}(u^j) = \vec{x}_i(u^j) \bullet \vec{x}_k(u^j) \ (i, k = 1, 2).$$

Definition 2.3.1. Let S be a surface with a parametric representation $\vec{x}(u^j)$ $((u^1, u^2) \in D \subset \mathbb{R}^2)$.
(a) The functions $g_{ik} : D \to \mathbb{R}$ with

$$g_{ik} = \vec{x}_i \bullet \vec{x}_k \ (i, k = 1, 2)$$

are called the *(Gauss's) first fundamental coefficients of the surface S.*
(b) The function
$$(ds)^2 = g_{ik}(u^j) du^i du^k \qquad (2.15)$$

is called the *first fundamental form of the surface S.*
(c) The parameters of S are said to be *orthogonal*, if

$$g_{12}(u^j) = 0.$$

Remark 2.3.2. *(a) Obviously we have*

$$g_{ik} = g_{ki} \ (i, k = 1, 2).$$

(b) In many text books, the notations E, F and G are used instead of g_{11}, g_{12} and g_{22} respectively.

(c) We have

$$g_{ii} = \vec{x}_i \bullet \vec{x}_i > 0 \ (i = 1, 2)$$

and, if we put $g = \det(g_{ik})$, then

$$g = g_{11}g_{22} - g_{12}^2 = \vec{x}_1^2 \cdot \vec{x}_2^2 - (\vec{x}_1 \bullet \vec{x}_2)^2 = (\vec{x}_1 \times \vec{x}_2)^2 > 0.$$

Thus the first fundamental form (2.15) is positive definite, the inverse matrix (g^{ik}) of the matrix (g_{ik}) exists, and $g^{11} = g_{22}$, $g^{12} = -g_{12}$ and $g^{22} = g_{11}$.

(d) The first fundamental coefficients play an important role in calculating the arc length along a curve on a surface, the angle between two curves on a surface and the surface area.

Let S be a surface with a parametric representation $\vec{x}(u^j)$ and γ be a curve on S with a parametric representation $\vec{x}(t) = \vec{x}(u^j(t))$. Then the arc length along γ is given by

$$s(t) = \int_{t_0}^{t} \sqrt{g_{ik} \frac{du^i}{d\tau} \frac{du^k}{d\tau}} \, d\tau. \tag{2.16}$$

If γ^* is another curve on S with a parametric representation $\vec{x}(u^{*j}(t^*))$ and the curves γ and γ^* intersect at the point $X(\underset{0}{u^j})$ where $t = t_0$ and $t^* = t_0^*$, then the angle α between the curves γ and γ^* is defined as the angle between the tangents to the curves at the point $X(\underset{0}{u^j})$. Therefore,

$$\cos \alpha = \frac{\left. \dfrac{d\vec{x}(u^j(t))}{dt} \right|_{t=t_0} \bullet \left. \dfrac{d\vec{x}(u^{*j}(t^*))}{dt^*} \right|_{t^*=t_0^*}}{\left\| \left. \dfrac{d\vec{x}(u^j(t))}{dt} \right|_{t=t_0} \right\| \cdot \left\| \left. \dfrac{d\vec{x}(u^{*j}(t^*))}{dt^*} \right|_{t^*=t_0^*} \right\|}$$

$$= \frac{g_{ik}(\underset{0}{u^j}) \dfrac{du^i}{dt}(t_0) \dfrac{du^{*k}}{dt^*}(t_0^*)}{\sqrt{g_{ik}(\underset{0}{u^j}) g_{lm}(\underset{0}{u^j}) \dfrac{du^i}{dt}(t_0) \dfrac{du^k}{dt}(t_0) \dfrac{du^{*l}}{dt^*}(t_0^*) \dfrac{du^{*m}}{dt^*}(t_0^*)}}.$$

Finally the surface area of a region of S defined by $(u^1, u^2) \in G$ is given by

$$\iint_G \sqrt{g(u^j)} \, du^1 du^2. \tag{2.17}$$

Now we establish the transformation formulae for the first fundamental coefficients and the invariance of the arc length, surface area and angles under parameter transformations.

Remark 2.3.3. *Let S be given by the parametric representations*

$$\vec{x}(u^j) = \vec{x}(u^j(u^{*k})) = \vec{x}^*(u^{*k})$$

and the fundamental coefficients g_{ik} and g_{ik}^* ($k = 1, 2$).
(a) Then the first fundamental coefficients g_{ik} satisfy the transformation formulae

$$g_{ik}^* = g_{rs}\frac{\partial u^r}{\partial u^{*i}}\frac{\partial u^s}{\partial u^{*k}} \text{ for } i, k = 1, 2. \tag{2.18}$$

(b) The coefficients g^{ik} and g^{*ik} satisfy the transformation formulae

$$g^{*ik} = g^{jm}\frac{\partial u^{*i}}{\partial u^j}\frac{\partial u^{*k}}{\partial u^m} \text{ for } i, k = 1, 2. \tag{2.19}$$

(c) The arc length of a curve on S, angles, and the surface area of a part of S are invariant under parameter transformations.

Proof.

(a) By (2.8), the first fundamental coefficients transform as follows

$$g_{ik}^* = \vec{x}_i^* \bullet \vec{x}_k^* = \vec{x}_r \bullet \vec{x}_s \frac{\partial u^r}{\partial u^{*i}}\frac{\partial u^s}{\partial u^{*k}} = g_{rs}\frac{\partial u^r}{\partial u^{*i}}\frac{\partial u^s}{\partial u^{*k}} \text{ for } i, k = 1, 2,$$

that is, the formulae in (2.18) are satisfied.

(b) Interchanging u and u^* in (2.18), we obtain

$$g_{mr} = g_{nt}^*\frac{\partial u^{*n}}{\partial u^m}\frac{\partial u^{*t}}{\partial u^r} \text{ for } m, r = 1, 2,$$

hence by (2.8) and the interchange of the indices of summation j and m in the last step

$$\begin{aligned}
g^{*ik} &= g^{*is}\delta_s^k = g^{*is}\frac{\partial u^{*k}}{\partial u^j}\frac{\partial u^j}{\partial u^{*s}} = g^{*is}\frac{\partial u^{*k}}{\partial u^j}\frac{\partial u^r}{\partial u^{*s}}\delta_r^j\\
&= g^{*is}\frac{\partial u^{*k}}{\partial u^j}\frac{\partial u^r}{\partial u^{*s}}g_{mr}g^{mj} = g^{*is}\frac{\partial u^{*k}}{\partial u^j}\frac{\partial u^r}{\partial u^{*s}}g_{nt}^*\frac{\partial u^{*n}}{\partial u^m}\frac{\partial u^{*t}}{\partial u^r}g^{mj}\\
&= g^{*is}g_{nt}^*\frac{\partial u^{*t}}{\partial u^r}\frac{\partial u^r}{\partial u^{*s}}\frac{\partial u^{*k}}{\partial u^j}\frac{\partial u^{*n}}{\partial u^m}g^{mj} = g^{*is}g_{nt}^*\delta_s^t\frac{\partial u^{*k}}{\partial u^j}\frac{\partial u^{*n}}{\partial u^m}g^{mj}\\
&= g^{*is}g_{ns}^*g^{mj}\frac{\partial u^{*k}}{\partial u^j}\frac{\partial u^{*n}}{\partial u^m} = \delta_n^{*i}g^{mj}\frac{\partial u^{*k}}{\partial u^j}\frac{\partial u^{*n}}{\partial u^m}\\
&= g^{mj}\frac{\partial u^{*k}}{\partial u^j}\frac{\partial u^{*i}}{\partial u^m} = g^{jm}\frac{\partial u^{*i}}{\partial u^j}\frac{\partial u^{*k}}{\partial u^m} \text{ for } i, k = 1, 2.
\end{aligned}$$

Thus we have shown the formulae in (2.19).

(c) Furthermore, the arc length of a curve on S is invariant under transformations of the parameters of S, since

$$
\left(\frac{d\vec{x}^*}{dt}\right)^2 = g_{ik}^* \frac{du^{*i}}{dt}\frac{du^{*k}}{dt} = g_{rs}\frac{\partial u^r}{\partial u^{*i}}\frac{\partial u^s}{\partial u^{*k}}\frac{\partial u^{*i}}{\partial u^j}\frac{\partial u^{*k}}{\partial u^n}\frac{du^j}{dt}\frac{du^n}{dt}
$$
$$
= g_{rs}\frac{\partial u^r}{\partial u^{*i}}\frac{\partial u^{*i}}{\partial u^j}\frac{\partial u^s}{\partial u^{*k}}\frac{\partial u^{*k}}{\partial u^n}\frac{du^j}{dt}\frac{du^n}{dt} = g_{rs}\delta_j^r\delta_n^s\frac{du^j}{dt}\frac{du^n}{dt}
$$
$$
= g_{jn}\frac{du^j}{dt}\frac{du^n}{dt} = \left(\frac{d\vec{x}}{dt}\right)^2.
$$

Similarly one may show the invariance of angles and surfaces areas under parameter transformations.

□

Example **2.3.4** (**The first fundamental coefficients of explicit surfaces**).
Let $f : D \subset \mathbb{R}^2 \to \mathbb{R}$ be a function of class $C^r(D)$ for $r \geq 1$. Then the surface with a parametric representation

$$
\vec{x}(u^i) = \{u^1, u^2, f(u^1, u^2)\} \ \left((u^1, u^2) \in D\right) \tag{2.20}
$$

is called an explicit surface; *the surface S is a representation of the function f, that is, the graph of f. Then we have*

$$
\vec{x}_1(u^i) = \{1, 0, f_1(u^1, u^2)\} \ and \ \vec{x}_2(u^i) = \{0, 1, f_2(u^1, u^2)\},
$$

and the first fundamental coefficients of S are given by

$$
g_{11}(u^i) = 1 + \left(f_1(u^i)\right)^2, \ g_{12}(u^i) = f_1(u^i)f_2(u^i), \ g_{22}(u^i) = 1 + \left(f_2(u^i)\right)^2
$$

and

$$
g(u^i) = 1 + \left(f_1(u^i)\right)^2 + \left(f_2(u^i)\right)^2. \tag{2.21}
$$

Example **2.3.5.** (**The first fundamental coefficients of surfaces of revolution**)
Let γ be a curve in a plane through a point P and spanned by the unit vectors $\vec{u}_1$ and $\vec{u}_3$ in three-dimensional space. We assume that γ is given by a parametric representation

$$
\vec{x}(t) = r(t)\vec{u}_1 + h(t)\vec{u}_3 \ (t \in I),
$$

where

$$
r(t) > 0 \ and \ |r'(t)| + |h'(t)| > 0 \ for \ all \ t \in I.
$$

Then a surface of revolution SR is generated by rotating γ about the axis defined by u_3 (Figure 2.8). Putting $\vec{u}_2 = \vec{u}_3 \times \vec{u}_1$ and $u^1 = t$, and writing u^2 for the angle of revolution, we obtain the following parametric representation for SR

$$
\vec{x}(u^i) = r(u^1)\cos u^2\vec{u}_1 + r(u^1)\sin u^2\vec{u}_2 + h(u^1)\vec{u}_3 \ \left((u^1, u^2) \in D = I \times (0, 2\pi)\right). \tag{2.22}
$$

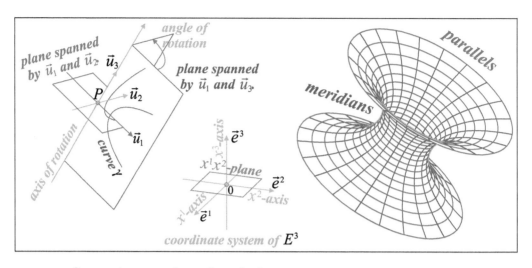

Figure 2.8 Generating a surface of revolution

We may assume that the point P is the origin of $\mathbb{E}^3$, the generating curve γ is in the x^1x^3–plane, and the axis of revolution is the x^3–axis. Then the parametric representation (2.22) becomes

$$\vec{x}(u^i) = \{r(u^1)\cos u^2, r(u^1)\sin u^2, h(u^1)\} \quad \left((u^1, u^2) \in D = I \times (0, 2\pi)\right). \quad (2.23)$$

The u^1– and u^2–lines of a surface of revolution are called meridians *and* parallels. *We obtain from (2.23)*

$$\vec{x}_1(u^i) = \{r'(u^1)\cos u^2, r'(u^1)\sin u^2, h'(u^1)\}, \ \vec{x}_2(u^i) = \{-r(u^1)\sin u^2, r(u^1)\cos u^2, 0\},$$

hence

$$g_{11}(u^i) = g_{11}(u^1) = \left(r'(u^1)\right)^2 + \left(h'(u^1)\right)^2, \ g_{12}(u^i) = g_{12}(u^1) = 0,$$
$$g_{22}(u^i) = g_{22}(u^1) = (r(u^1))^2 \quad (2.24)$$

for the first fundamental coefficients of a surface of revolution,

$$g(u^i) = g(u^1) = (r(u^1))^2 \cdot \left(\left(r'(u^1)\right)^2 + \left(h'(u^1)\right)^2\right),$$

and

$$\vec{N}(u^i) = \{-h'(u^1)\cos u^2, -h'(u^1)\sin u^2, r'(u^1)\} \cdot \frac{1}{\sqrt{(r'(u^1))^2 + (h'(u^1))^2}} \quad (2.25)$$

for the surface normal vector.

Now we consider a few special cases of surfaces of revolution.

Visualization 2.3.6 (Planes, cylinders and cones).
Let the generating curve γ be a straight line segment in the positive x^1x^3–half–plane.

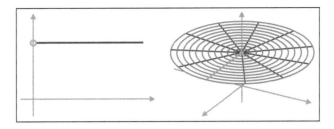

Figure 2.9 Generating a plane

(a) Planes

If γ is parallel to the x^1-axis and through a point with $x^3 = c$, then $r(u^1) = u^1$ and $h(u^1) = c$ for $u^1 \in (0, \infty)$, and the surface of revolution is the plane through the point $P = (0, 0, c)$ and parallel to the $x^1 x^2$-plane, without the point P and the points of the straight line segment parallel to the positive x^2-axis (Figure 2.9).

(b) Circular cylinders

If γ is parallel to the x^3-axis and through a point with $x^1 = r > 0$, then $r(u^1) = r > 0$ and $h(u^1) = u^1$ for $u^1 \in \mathbb{R}$, and the surface of revolution is a circular cylinder of radius r (Figure 2.10).

(c) Circular Cones

If γ has an angle $\beta \in (0, \pi) \setminus \{\pi/2\}$ with the positive x^3-axis, then $r(u^1) = u^1 \sin \beta$ and $h(u^1) = u^1 \cos \beta$ for $u^1 \in (0, \infty)$ and the surface of revolution is a circular cone with its vertex in the origin and an angle of 2β at its vertex (Figure 2.11).

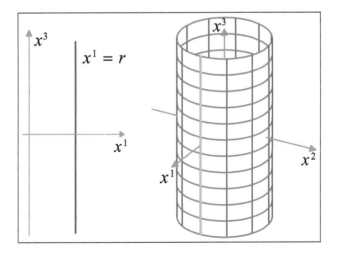

Figure 2.10 Generating a circular cylinder

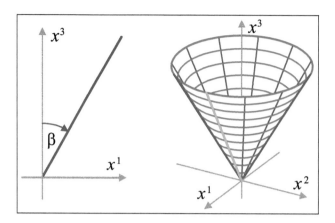

Figure 2.11   Generating a circular cone

(d)   The first fundamental coefficients of planes, circular cylinders and cones
The first fundamental coefficients are given by

$$g_{11}(u^1) = 1, \ g_{12}(u^1) = 0 \ and \ g_{22} = (u^1)^2 \tag{2.26}$$

for the plane,

$$g_{11}(u^1) = 1, \ g_{12}(u^1) = 0 \ and \ g_{22}(u^1) = r^2 \tag{2.27}$$

for the circular cylinder, and

$$g_{11}(u^1) = 1, \ g_{12}(u^1) = 0 \ and \ g_{22} = (u^1)^2 \sin^2 \beta \tag{2.28}$$

for the circular cone.

Visualization 2.3.7 (Spheres). *If the generating curve γ is the semi–circle line in the positive $x^1 x^3$–half–plane, of radius $r > 0$ and its centre in the origin, then $r(u^1) = r \cos u^1$ and $h(u^1) = r \sin(u^1)$ for $u^1 \in (-\pi/2, \pi/2)$, and the surface of revolution is a sphere with radius r, its centre in the origin, without the circle line in the $x^1 x^3$–plane, of radius r and the centre in the origin (Figure 2.12).*
The fundamental coefficients are

$$g_{11}(u^1) = r^2, \ g_{12}(u^1) = 0 \ and \ g_{22}(u^1) = r^2 \cos^2 u^1. \tag{2.29}$$

Visualization 2.3.8 (Tori). *If the generating curve γ is a circle line of radius $r_1 > 0$ and its centre in the point $(r_0, 0)$ in the $x^1 x^3$–plane, where $r_0 > r_1$, then $r(u^1) = r_0 + r_1 \cos u^1$ and $h(u^1) = r_1 \sin u^1$ for $u^1 \in (0, 2\pi)$, and the surface of revolution is a torus without the circle lines C_1 and C_2, where C_1 is in the $x^1 x^2$–plane and has the radius $r_0 + r_1$ and centre in the origin, and C_2 is in the $x^1 x^3$–plane and has the radius r_1 and centre in $(r_0, 0, 0)$ (Figure 2.13).*
The fundamental coefficients are

$$g_{11}(u^1) = r_1^2, \ g_{12}(u^1) = 0 \ and \ g_{22}(u^1) = \left(r_0 + r_1 \cos u^1 \right)^2. \tag{2.30}$$

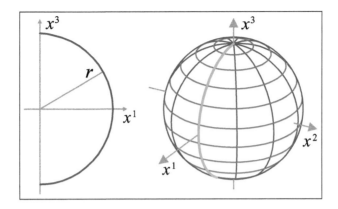

Figure 2.12 Generating a sphere

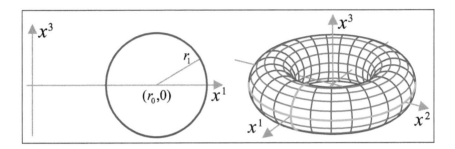

Figure 2.13 Generating a torus

Visualization 2.3.9 (Catenoids). *If the generating curve γ is a catenary with a parametric representation $x^1(t) = a \cosh(t/a)$, $x^3(t) = t$ $(t \in \mathbb{R})$, where $a > 0$ is a constant, then the surface of revolution is a* catenoid *without γ (Figure 2.14). The first fundamental coefficients are*

$$g_{11}(u^1) = \cosh^2\left(\frac{u^1}{a}\right), \ g_{12}(u^1) = 0 \ and \ g_{22}(u^1) = a^2 \cosh^2\left(\frac{u^1}{a}\right). \tag{2.31}$$

Example **2.3.10 (Some surface areas).** *Now we apply (2.17) in Remark 2.3.2 to compute the surface areas $A(G)$ of some regions G of a circular cylinder, a circular cone, a sphere, torus and catenoid.*

(a)   We obtain from (2.17) and (2.27) for the circular cylinder *and the region $G = (h_0, h_1) \times (0, 2\pi)$ with $h_1 > h_0$*

$$A(G) = \int_0^{2\pi} \int_{h_0}^{h_1} r \, du^1 du^2 = 2\pi r(h_1 - h_0).$$

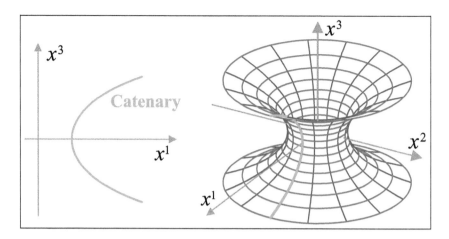

Figure 2.14   Generating a catenoid

(b)   We obtain from (2.17) and (2.28) for the circular cone and the region $G = (0, h) \times (0, 2\pi)$

$$A(G) = \int_0^{2\pi} \int_0^h |u^1 \sin \beta|\, du^1 du^2 = \pi h^2 \sin \beta.$$

(c)   We obtain from (2.17) and (2.29) for the sphere and the region $G = (-\pi/2, \pi/2) \times (0, 2\pi)$

$$A(G) = \int_0^{2\pi} \int_{-\pi/2}^{\pi/2} r^2 |\cos u^1|\, du^1 du^2 = 2\pi r^2 \sin u^1 \Big|_{-\pi/2}^{\pi/2} = 4\pi r^2.$$

(d)   We obtain from (2.17) and (2.30) for the torus and the region $G = (0, 2\pi) \times (0, 2\pi)$

$$A(G) = \int_0^{2\pi} \int_0^{2\pi} r_1 |r_0 + r_1 \cos u^1|\, du^1 du^2 = 2\pi r_1 \int_0^{2\pi} (r_0 + r_1 \cos u^1)\, du^1$$
$$= 4\pi^2 r_0 r_1.$$

(e)   We obtain from (2.17) and (2.31) for the catenoid and the region $G = (h_0, h_1) \times (0, 2\pi)$ with $h_1 > h_0$

$$A(G) = \int_0^{2\pi} \int_{h_0}^{h_1} a \cosh^2 \left(\frac{u^1}{a} \right) du^1 du^2$$

$$= 2\pi a \int_{h_0}^{h_1} \frac{1}{4} \left(\exp \left(\frac{2u^1}{a} \right) + \exp \left(-\frac{2u^1}{a} \right) + 2 \right) du^1$$

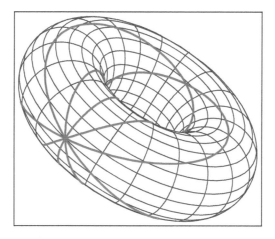

Figure 2.15 Loxodromes on a torus

$$= a\pi \int_{h_0}^{h_1} \left(\cosh\left(\frac{2u^1}{a}\right) + 1 \right) du^1$$

$$= a\pi \left(\frac{a}{2} \sinh\left(\frac{2h_1}{a}\right) - \frac{a}{2} \sinh\left(\frac{2h_0}{a}\right) + h_1 - h_0 \right).$$

Visualization 2.3.11 (Loxodromes).

A curve γ on a surface of revolution SR that intersects the meridians of SR at a constant angle $\beta \in (-\pi/2, \pi/2]$ is called a loxodrome *(Figure 2.15).*

Let $\vec{x}^*(u^i(s))$ $(s \in I)$ be the parametric representation of a curve γ on SR, and s be the arc length along γ.

If $\beta \neq \pi/2$, then the loxodrome through the point P with the parameters (u_0^1, u_0^2) is given by

$$u^2(u^1) + u_0^2 = \tan\beta \int_{u_0^1}^{u^1} \sqrt{\frac{g_{11}(u)}{g_{22}(u)}} \, du$$

$$= \tan\beta \int_{u_0^1}^{u^1} \frac{\sqrt{(r'(u))^2 + (h'(u))^2}}{r(u)} \, du. \qquad (2.32)$$

If $\beta = \pi/2$, then the loxodrome through the point P with the parameters (u_0^1, u_0^2) is the parallel corresponding to $u^1 = u_0^1$.

Proof. Since $\|\dot{\vec{x}}^*(u^i(s))\| = 1$ and

$$g_{12} = \vec{x}_1 \bullet \vec{x}_2 = 0, \ g_{11}(u^i) = g_{11}(u^1) \text{ and } g_{22}(u^i) = g_{22}(u^1)$$

for surfaces of revolution by (2.24) in Example 2.3.5, the angle $\beta(s)$ between the curve γ and the meridian at s is given by

$$\cos \beta(s) = \frac{\dot{\vec{x}}^*(u^i(s)) \bullet \vec{x}_1(u^i(s))}{\|\vec{x}_1(u^i(s))\|} = \frac{\vec{x}_1(u^i(s)) \bullet \vec{x}_1(u^i(s))}{\|\vec{x}_1(u^i(s))\|} \dot{u}^1(s) = \sqrt{g_{11}(u^1(s))} \, \dot{u}^1(s).$$
(2.33)

This and

$$\left\| \dot{\vec{x}}(u^i(s)) \right\|^2 = g_{ik}(u^1(s)) \dot{u}^i(s) \dot{u}^k(s)$$

$$= g_{11}(u^1(s)) \left(\dot{u}^1(s) \right)^2 + g_{22}(u^1(s)) \left(\dot{u}^2(s) \right)^2 = 1$$

$$= \cos^2 \beta(s) + \sin^2 \beta(s),$$

together yield

$$|\sin \beta(s)| = \left| \sqrt{g_{22}(u^1(s))} \, \dot{u}^2 \right|,$$

and, since $g_{22}(u^1(s)) > 0$ for all s, we obtain

$$\dot{u}^2(s) = \frac{\sin \beta(s)}{\sqrt{g_{22}(u^1(s))}}$$
(2.34)

for a suitable choice of the orientation of the curve.

For a loxodrome, the angle $\beta(s) = \beta$ is constant, and we obtain for $\beta \neq \pi/2$ from (2.33), (2.34) and (2.24) in Example 2.3.5

$$\frac{du^2}{du^1} = \tan \beta \sqrt{\frac{g_{11}(u^1)}{g_{22}(u^1)}},$$

and consequently, we obtain (2.32) for the loxodrome through (u_0^1, u_0^2).

If $\beta = \pi/2$, then it follows from (2.33) that $u_1 = u_0^1$, and we obtain the parallel corresponding to u_0^1. □

Visualization 2.3.12. *(a)* Loxodromes in planes

First we determine the loxodromes in the plane of Visualization 2.3.6 (a).
We obtain from (2.32) and (2.26) that the loxodromes for $\beta \in (0, \pi/2)$ are given by

$$u^2 - u_0^2 = \tan \beta \int_{u_0^1}^{u^1} \frac{du}{u} = \tan \beta \cdot \log \left(\frac{u}{u_0^1} \right)$$

for $u^1 > 0$ and $(u_0^1, u_0^2) \in (0, \infty) \times (0, 2\pi)$.
Putting $b = 1/\tan \beta$ and $a = \exp\left(-u_0^2/b\right)$, we conclude

$$u^1 = u_0^1 \cdot a \exp\left(bu^2\right),$$

that is, the loxodromes of the plane are logarithmic spirals; they intersect the straight lines through the origin, and hence the concentric circles with their centres in the origin, at a constant angle (Figure 2.16).

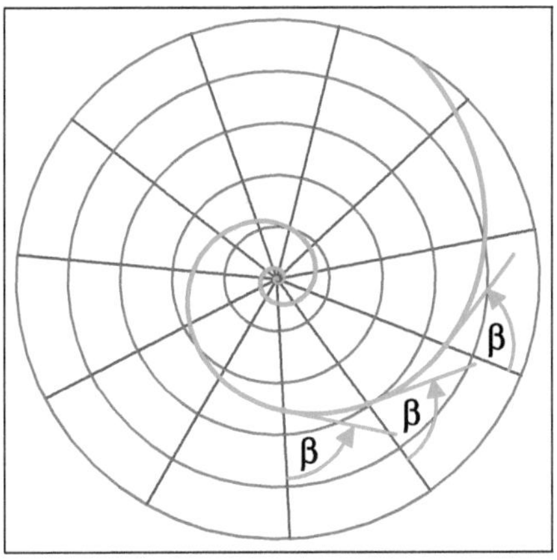

Figure 2.16 A logarithmic spiral as a loxodrome of the plane in Example 2.3.5 (a)

(b) Loxodromes on spheres

Now we determine the loxodromes on the sphere with radius $r > 0$ of Example 2.3.5 (b). For the sphere of radius $r > 0$ centred at the origin, we conclude from

$$g_{11}(u^1) = 1, \quad g_{12} = 0 \quad and \quad g_{22}(u^1) = r^2 \cos^2 u^1$$

and (2.32) that the loxodromes for $\beta \in (-\pi/2, \pi/2) \setminus \{0\}$ are given by

$$u^2 = \tan \beta \cdot \log \left(\tan \left(\frac{u_1}{2} + \frac{\pi}{4} \right) \right) + c \tag{2.35}$$

where

$$c = u_0^2 - \tan \beta \cdot \log \left(\tan \left(\frac{u_0^1}{2} + \frac{\pi}{4} \right) \right) \quad (Figure \ 2.17).$$

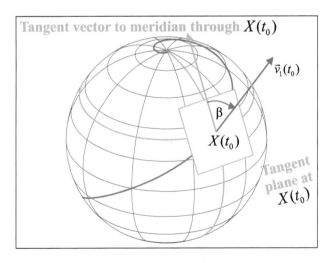

Figure 2.17 A loxodrome on a sphere

Proof of Part (b). We observe that the integral in (2.32) reduces to

$$I(u) = \int \frac{u}{\cos u} \, du.$$

The substitution $t = \tan \frac{u}{2}$ yields

$$\sin u = \frac{2t}{1+t^2}, \quad \cos u = \frac{1-t^2}{1+t^2}, \quad \frac{du}{dt} = \frac{2}{1+t^2}$$

and

$$I(u) = 2 \int \frac{dt}{1-t^2} = \int \left(\frac{1}{1-t} + \frac{1}{1+t} \right) dt = \log \frac{1+t}{1-t}$$

$$= \log \left(\frac{1 + \tan \frac{u}{2}}{1 - \tan \frac{u}{2}} \right) + \tilde{c} = \log \left(\tan \left(\frac{u}{2} + \frac{\pi}{4} \right) \right) + \tilde{c},$$

where $\tilde{c}$ is a constant of integration. □

Remark 2.3.13. *The parameter lines of the parametric representation (2.10) of the sphere in Visualization 2.1.7 (c) are loxodromes.*

Proof. We write $d = c_1 - c_2$, and put

$$t^* = 2 \cdot \tan^{-1} \left(\exp \left(\frac{t-c}{d} \right) \right) - \frac{\pi}{2}$$

to obtain

$$r \cdot \cos t^* = \frac{r}{\cosh \left(\frac{t-c}{d} \right)}, \quad \frac{c_1 - c_2}{d} = c_1 \log \left(\tan \left(\frac{t^*}{2} + \frac{\pi}{4} \right) \right) + c$$

and

$$\frac{\sin t^*}{\cos t^*} = \sinh \left(\frac{t-c}{d} \right).$$

Thus, if we write

$$u^{*1}(t^*) = t^* \text{ and } u^{*2}(t^*) = c_1 \log \left(\tan \left(\frac{t^*}{2} - \frac{\pi}{4} \right) \right) + c,$$

then identity (2.11) for the u^1–line corresponding to $u^2 = c$ yields

$$\vec{x}(t^*) = r\{\cos t^* \cos u^{*1}(t^*), \sin t^* \cos u^{*1}(t^*), \sin t^*\},$$

that is, a loxodrome by (2.35) in Visualization 2.3.11 (b) on the sphere with respect to its usual parametric representation given in Visualization 2.1.7 (b). □

Visualization 2.3.14 (Lines of constant slope on surfaces of revolution).
Now we determine all lines γ_β of constant slope on surfaces of revolution that have a constant angle $\beta \in [0, \pi)$ with the axis of revolution along $\vec{e}^{\,3}$.

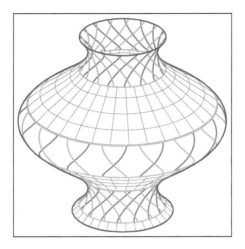

Figure 2.18 Lines of constant slope

If $\vec{x}^*(u^i(s))$ is the parametric representation of a curve γ on the surface SR of revolution given by a paramerec representation (2.23), and s is the arc length along γ, then we must have for γ_β

$$\dot{\vec{x}}^* \bullet \vec{e}^3 = h'(u^1)\dot{u}^1 = \cos\beta. \tag{2.36}$$

Let $\beta \neq \pi/2$, and I_1 denote the interval for the parameter u^1 of the parametric representation of SR. Then solutions of (2.36) only exist in subintervals $I \subset I_1$ for which $h'(u^1) \neq 0$. From

$$\|\dot{\vec{x}}^*\| = 1 \;\; and \;\; \frac{1}{(\dot{u}^1)^2} = \frac{(h'(u^1))^2}{\cos^2\beta},$$

we obtain

$$\left(\frac{du^2}{du^1}\right)^2 = \frac{\dfrac{(h'(u^1))^2}{\cos^2\beta} - g_{11}(u^1)}{g_{22}(u^1)},$$

$$\left|\frac{du^2}{du^1}\right| = \frac{1}{|\cos\beta|}\sqrt{\frac{(h'(u^1))^2 - g_{11}(u^1)\cos^2\beta}{g_{22}(u^1)}} = \frac{\sqrt{(h'(u^1))^2 \tan^2\beta - (r'(u^1))^2}}{r(u^1)}$$

and choose

$$\frac{du^2}{du^1} = \frac{\sqrt{(h'(u^1))^2 \tan^2\beta - (r'(u^1))^2}}{r(u^1)}$$

(Figures 2.18, 2.19 and 2.20).

Let I be a subinterval of I_1 for which $(r'(u^1))^2 < \tan^2\beta(h'(u^1))^2$, and $(u_0^1, u_0^2) \in I \times (0, 2\pi)$. Then γ_β is given by

$$u^1(t) = \frac{\cos\beta}{|\cos\beta|}t \;\; and \;\; u^2 - u_0^2 = \int_{u_0^1}^{u^1} \frac{\sqrt{(h'(u))^2 \tan^2\beta - (r'(u))^2}}{r(u)}\, du. \tag{2.37}$$

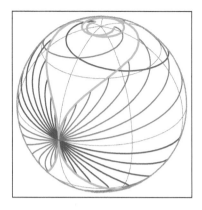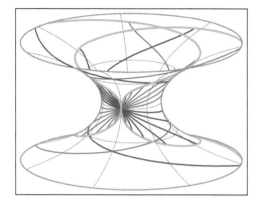

Figure 2.19 Lines of constant slope on a sphere (left) and catenoid (right)

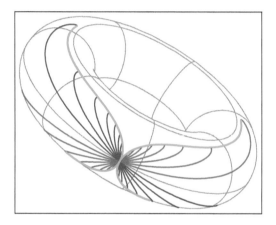

Figure 2.20 Lines of constant slope on a torus with their bounding curve

Now let $\beta = \pi/2$. In intervals with $h'(u^1) = 0$, we may choose $u^1(s) = c$, where c is a constant, and obtain from $\|x\|^2 = g_{22}(\dot{u}^2)^2 = 1$ that

$$\frac{du^2}{ds} = \pm\sqrt{\frac{1}{g_{22}(c)}} \quad and \quad u^2(s) = \sqrt{\frac{1}{g_{22}(c)}} \cdot s + \tilde{c}$$

where $\tilde{c}$ is a constant.
In intervals for which $h'(u^1) \neq 0$, we conclude from (2.36), that $\dot{u}^1 = 0$, thus we obtain u^2-lines.

2.4 THE CURVATURE OF CURVES ON SURFACES, GEODESIC AND NORMAL CURVATURE

In this section, we introduce the concepts of the *geodesic* and *normal curvature* in the study of the curvature of curves on surfaces. We will see that the geodesic curvature depends on the first fundamental coefficients, whereas the normal curvature depends on the so–called *second fundamental coefficients*.

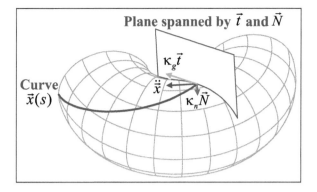

Figure 2.21 The components of a vector of curvature

Let S be a surface with a parametric representation $\vec{x}(u^j)$ and γ be a curve on S with a parametric representation $\vec{x}(s) = \vec{x}(u^j(s))$. Then we have

$$\dot{\vec{x}} = \vec{x}_j \dot{u}^j \text{ and } \ddot{\vec{x}} = \vec{x}_j \ddot{u}^j + \vec{x}_{jk} \dot{u}^j \dot{u}^k, \text{ where } \vec{x}_{jk} = \frac{\partial^2 \vec{x}}{\partial u^j \partial u^k} \ (j, k = 1, 2). \tag{2.38}$$

Furthermore, let

$$\vec{t} = \vec{N} \times \dot{\vec{x}}, \tag{2.39}$$

where $\vec{N}$ is the surface normal vector of S. Then $\vec{t}$ is a unit vector in the tangent plane of S, and the vector of curvature can be split into two components, one in the tangent plane and one along the surface normal vector (Figure 2.21),

$$\ddot{\vec{x}}(s) = \kappa_g(s)\vec{t}(s) + \kappa_n(s)\vec{N}(s). \tag{2.40}$$

Definition 2.4.1. Let S be a surface with a parametric representation $\vec{x}(u^j)$ and γ be a curve on S with a parametric representation $\vec{x}(s) = \vec{x}(u^j(s))$. Then the functions κ_g and κ_n defined in (2.40) are called the *geodesic* and *normal curvature of the curve* γ.

Figure 2.22 shows a way to represent the normal curvature of a curve on a surface. Obviously we have

$$\kappa^2(s) = \kappa_g^2(s) + \kappa_n^2(s), \text{ and } \kappa_n = \ddot{\vec{x}} \bullet \vec{N} \text{ and } \kappa_g = \ddot{\vec{x}} \bullet \vec{t} \tag{2.41}$$

for the curvature κ, and the geodesic and normal curvature κ_g and κ_n of the curve γ. In order to determine κ_n and κ_g, we put

$$\vec{x}_{ik} = \Gamma_{ik}^r \vec{x}_r + L_{ik} \vec{N} \ (i, k = 1, 2) \tag{2.42}$$

with the coefficients Γ_{ik}^r and L_{ik} yet to be determined. The identities in (2.42) are referred to as the *Gauss equations (for the derivatives)*. From (2.42), we obtain

$$L_{ik} = \vec{x}_{ik} \bullet \vec{N} \ (i, k = 1, 2).$$

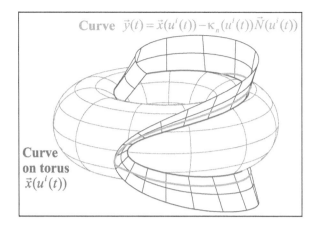

Figure 2.22 Representation of the normal curvature of a curve on a torus

Differentiating $\vec{N} \bullet \vec{x}_i = 0$ with respect to u^k, we obtain

$$\frac{\partial(\vec{N} \bullet \vec{x}_i)}{\partial u^k} = \vec{N}_k \bullet \vec{x}_i + \vec{N} \bullet \vec{x}_{ik} = 0, \text{ where } \vec{N}_k = \frac{\partial \vec{N}}{\partial u^k},$$

hence

$$L_{ik} = \vec{N} \bullet \vec{x}_{ik} = -\vec{N}_k \bullet \vec{x}_i \text{ and } L_{ik} = L_{ki} \ (i, k = 1, 2). \tag{2.43}$$

Definition 2.4.2. Let S be a surface with a parametric representation $\vec{x}(u^j)$ $((u^1, u^2) \in D \subset \mathbb{R}^2)$. The functions $L_{ik} : D \to \mathbb{R} \ (i, k = 1, 2)$ defined in (2.43) are called *second fundamental coefficients of the surface S*; we also write $L = \det(L_{ik}) = L_{11}L_{22} - L_{12}^2$.

Remark 2.4.3. *(a) In many textbooks the notations L, M and N are used instead of L_{11}, L_{12} and L_{22}.*
(b) Let S be given by the parametric representations

$$\vec{x}(u^k) = \vec{x}(u^k(u^{*r})) = \vec{x}^*(u^{*j}).$$

Then the transformation formulae for the surface normal vector and the second fundamental coefficients are

$$\vec{N}^* = \vec{N} \cdot signD, \text{ where } D = \frac{\partial(u^k)}{\partial(u^{*r})} \tag{2.44}$$

and

$$L_{ik}^* = L_{mn} \frac{\partial u^m}{\partial u^{*i}} \frac{\partial u^n}{\partial u^{*k}} \cdot \text{sign}(D) \ (i, k = 1, 2). \tag{2.45}$$

(c) The normal and geodesic curvature are given by

$$\kappa_n = L_{ik}\dot{u}^i\dot{u}^k \text{ and } \kappa_g = \left(\ddot{u}^r + \Gamma_{ik}^r \dot{u}^i \dot{u}^k\right) \vec{x}_r \bullet \vec{t}. \tag{2.46}$$

With respect to an arbitrary parameter t for the curve, we obtain

$$\kappa_n = L_{ik}\frac{du^i}{dt}\frac{du^k}{dt}\left(\frac{dt}{ds}\right)^2 = \frac{L_{ik}\frac{du^i}{dt}\frac{du^k}{dt}}{g_{ik}\frac{du^i}{dt}\frac{du^k}{dt}} \tag{2.47}$$

and

$$\kappa_g = \left(\left(\frac{d^2u^i}{dt^2} + \Gamma^i_{jk}\frac{du^i}{dt}\frac{du^k}{dt}\right)\left(\frac{dt}{ds}\right)^2 + \frac{du^i}{dt}\frac{d^2t}{ds^2}\right)\vec{x}_i \bullet \vec{t}. \tag{2.48}$$

Proof. (b) We conclude from for the surface normal vector (2.9)

$$\vec{N}^* = \frac{\vec{x}_1^* \times \vec{x}_2^*}{\|\vec{x}_1^* \times \vec{x}_2^*\|} = \frac{\vec{x}_1 \times \vec{x}_2}{\|\vec{x}_1 \times \vec{x}_2\|} \cdot \operatorname{sign}D = \vec{N} \cdot \operatorname{sign}D.$$

Furthermore,

$$\vec{x}_i^* = \vec{x}_m\frac{\partial u^m}{\partial u^{*i}} \text{ and } \vec{x}_{ik}^* = \vec{x}_{mn}\frac{\partial u^m}{\partial u^{*i}}\frac{\partial u^n}{\partial u^{*k}} + \vec{x}_m\frac{\partial^2 u^m}{\partial u^{*i}\partial u^{*k}}$$

together imply

$$L_{ik}^* = \vec{N}^* \bullet \vec{x}_{ik}^* = \vec{N} \bullet \left(\vec{x}_{mn}\frac{\partial u^m}{\partial u^{*i}}\frac{\partial u^n}{\partial u^{*k}} + \vec{x}_m\frac{\partial^2 u^m}{\partial u^{*i}\partial u^{*k}}\right)\operatorname{sign}D$$

$$= L_{mn}\frac{\partial u^m}{\partial u^{*i}}\frac{\partial u^n}{\partial u^{*k}} \cdot \operatorname{sign}D \ (i, k = 1, 2).$$

Thus we have shown (2.45).

(c) We obtain from identities (2.41) and (2.42)

$$\ddot{\vec{x}} = \vec{x}_i\ddot{u}^i + \vec{x}_{ik}\dot{u}^i\dot{u}^k = \vec{x}_i\ddot{u}^i + \left(\Gamma^r_{ik}\vec{x}_r + L_{ik}\vec{N}\right)\dot{u}^i\dot{u}^k$$

$$= \left(\ddot{u}^r + \Gamma^r_{ik}\dot{u}^i\dot{u}^k\right)\vec{x}_r + (L_{ik}\dot{u}^i\dot{u}^k)\vec{N},$$

and the identities in (2.48) are an immediate consequence.

Finally, if t is an arbitrary parameter of a curve $\vec{x}(u^i(t))$ on a surface, then writing $u^i(s) = u^i(t(s))$ for $i = 1, 2$, we obtain

$$\dot{u}^i = \frac{du^i}{dt}\frac{dt}{ds} \text{ and } \ddot{u}^i = \frac{d^2u^i}{dt^2}\left(\frac{dt}{ds}\right)^2 + \frac{du^i}{dt}\frac{d^2t}{ds^2} \text{ for } i = 1, 2,$$

and the formulae for the normal and geodesic curvature in (2.47) and (2.48) follow directly from (2.46) and (2.16).

□

Remark 2.4.4. *The coefficients L_{ik} and Γ^r_{ik} are introduced in a formal way. We want to give them a geometric meaning. The normal and geodesic curvature play as important a role in the theory of surfaces as the curvature and torsion do in the theory of curves.*

Example **2.4.5 (The second fundamental coefficients for explicit surfaces).**
Let S be an explicit surface with a parametric representation

$$\vec{x}(u^i) = \{u^1, u^2, f(u^1, u^2)\} \ ((u^1, u^2) \in D).$$

Then we have

$$L_{ik}(u^j) = \frac{f_{ik}(u^j)}{\sqrt{1 + (f_1(u^j))^2 + (f_2(u^j))^2}} \ for \ i, k = 1, 2 \tag{2.49}$$

and

$$L(u^j) = \frac{f_{11}(u^j) f_{22}(u^j) - (f_{12}(u^j))^2}{1 + (f_1(u^j))^2 + (f_2(u^j))^2}. \tag{2.50}$$

In particular, if S is a plane, then $f(u^j) = const$ for all $(u^1, u^2) \in D$ and so

$$L_{ik}(u^j) = 0 \ for \ i, k = 1, 2. \tag{2.51}$$

Proof. Since

$$\vec{x}_1 = \{1, 0, f_1(u^j)\}, \ \vec{x}_2 = \{0, 1, f_1(u^j)\} \ and \ \vec{x}_{ik}(u^j) = \{0, 0, f_{ik}(u^j)\} \ for \ i, k = 1, 2,$$

it follows from (2.43) and (1.6) the second fundamental coefficients of S are given by

$$L_{ik}(u^j) = N(\vec{u^j}) \bullet \vec{x}_{ik}(u^j) = \left(\frac{\vec{x}_1(u^j) \times \vec{x}_2(u^j)}{\|\vec{x}_1(u^j) \times \vec{x}_2(u^j)\|} \right) \bullet \vec{x}_{ik}(u^j)$$

$$\begin{vmatrix} 0 & 1 & 0 \\ 0 & 0 & 1 \\ f_{ik}(u^j) & f_1(u^j) & f_2(u^j) \end{vmatrix} \frac{1}{\sqrt{g(u^j)}}$$

$$= \frac{f_{ik}(u^j)}{\sqrt{1 + (f_1(u^j))^2 + (f_2(u^j))^2}} \ for \ i, k = 1, 2$$

and

$$L(u^j) = \frac{f_{11}(u^j) f_{22}(u^j) - (f_{12}(u^j))^2}{1 + (f_1(u^j))^2 + (f_2(u^j))^2}.$$

Thus we have shown (2.49) and (2.50).
In particular, if S is a plane, then $f(u^j) = const$ for all $(u^1, u^2) \in D$, and so (2.51) is
an immediate consequence of (2.49). □

Example **2.4.6. (The second fundamental coefficients for surfaces of revolution)** *Let SR be a surface of rotation with a parametric representation*

$$\vec{x}(u^i) = \{r(u^1) \cos u^2, r(u^1) \sin u^2, h(u^1)\}.$$

Then we have

$$L_{11}(u^i) = L_{11}(u^1) = \frac{h''(u^1)r'(u^1) - h'(u^1)r''(u^1)}{\sqrt{(r'(u^1))^2 + (h'(u^1))^2}}, \tag{2.52}$$

$$L_{12}(u^i) = L_{12}(u^1) = 0, \tag{2.53}$$

$$L_{22}(u^i) = L_{22}(u^1) = \frac{h'(u^1)r(u^1)}{\sqrt{(r'(u^1))^2 + (h'(u^1))^2}} \tag{2.54}$$

and

$$L(u^i) = L(u^1) = \frac{\left(h''(u^1)r'(u^1) - h'(u^1)r''(u^1)\right) h'(u^1)r(u^1)}{(r'(u^1))^2 + (h'(u^1))^2}. \tag{2.55}$$

Proof. We have

$$\vec{x}_1(u^i) = \{r'(u^1)\cos u^2, r'(u^1)\sin u^2, h'(u^1)\},$$
$$\vec{x}_2(u^i) = \{-r(u^1)\sin u^2, r(u^1)\cos u^2, 0\},$$
$$\vec{x}_{11}(u^i) = \{r''(u^1)\cos u^2, r''(u^1)\sin u^2, h''(u^1)\},$$
$$\vec{x}_{12}(u^i) = \{-r'(u^1)\sin u^2, r(u^1)\cos u^2, 0\},$$
$$\vec{x}_{22}(u^i) = \{-r(u^1)\cos u^2, -r(u^1)\sin u^2, 0\},$$

and so

$$L_{11}(u^i) = L_{11}(u^1)$$

$$= \begin{vmatrix} r''(u^1)\cos u^2 & r'(u^1)\cos u^2 & -\sin u^2 \\ r''(u^1)\sin u^2 & r'(u^1)\sin u^2 & \cos u^2 \\ h''(u^1) & h'(u^1) & 0 \end{vmatrix} \cdot \frac{1}{\sqrt{(r'(u^1))^2 + (h'(u^1))^2}}$$

$$= \frac{h''(u^1)r'(u^1) - h'(u^1)r''(u^1)}{\sqrt{(r'(u^1))^2 + (h'(u^1))^2}},$$

$$L_{12}(u^i) = L_{12}(u^1)$$

$$= \begin{vmatrix} -r'(u^1)\sin u^2 & r'(u^1)\cos u^2 & -\sin u^2 \\ r'(u^1)\cos u^2 & r'(u^1)\sin u^2 & \cos u^2 \\ 0 & h'(u^1) & 0 \end{vmatrix} \cdot \frac{1}{\sqrt{(r'(u^1))^2 + (h'(u^1))^2}}$$

$$= 0,$$

$$L_{22}(u^i) = L_{22}(u^1)$$

$$= \begin{vmatrix} -r(u^1)\cos u^2 & r'(u^1)\cos u^2 & -\sin u^2 \\ -r(u^1)\sin u^2 & r'(u^1)\sin u^2 & \cos u^2 \\ 0 & h'(u^1) & 0 \end{vmatrix} \cdot \frac{1}{\sqrt{(r'(u^1))^2 + (h'(u^1))^2}}$$

$$= \frac{h'(u^1)r(u^1)}{\sqrt{(r'(u^1))^2 + (h'(u^1))^2}}$$

and

$$L(u^i) = L(u^1) = L_{11}(u^1)L_{22}(u^1)$$
$$= \frac{(h''(u^1)r'(u^1) - h'(u^1)r''(u^1))\, h'(u^1)r(u^1)}{(r'(u^1))^2 + (h'(u^1))^2}.$$

□

Visualization 2.4.7. *(a)* **(Lines of vanishing normal and geodesic curvature on a hyperboloid of one sheet)**

Let $a, b, c \in \mathbb{R} \backslash \{0\}$ and H be the hyperboloid of one sheet *given by the equation*

$$\frac{(x^1)^2}{a^2} + \frac{(x^2)^2}{b^2} - \frac{(x^3)^2}{c^2} = 1.$$

Its intersection with the x^1x^2–plane is an ellipse with a parametric representation

$$\vec{x}(t) = \{a \cos t, b \sin t, 0\} \ (t \in (0, 2\pi)).$$

Given $t_0 \in (0, 2\pi)$, we consider the straight lines γ_+ and γ_- with the parametric representations

$$\vec{y}_\pm(t) = \vec{x}(t_0) + t(\pm \vec{x}'(t_0) + c\vec{e}^3) \ (t \in \mathbb{R})$$

(Figure 2.23). It follows from

$$\frac{(y_\pm^1)^2}{a^2} + \frac{(y_\pm^2)^2}{b^2} - \frac{(y_\pm^3)^2}{c^2} = (\cos t_0 - t \sin t_0)^2 + (\sin t_0 + t \cos t_0)^2 - t^2$$
$$= 1 + t^2 - t^2 = 1$$

that γ_+ and γ_- are on H. Since t_0 was arbitrary, the hyperboloid of one sheet contains two families of straight lines, which have vanishing curvature (Figure 2.24). Therefore, their geodesic and normal curvature vanish identically by (2.41).

(b) **The geodesic curvature vanishes along helices on circular cylinders**

Let $H, R > 0$, $\omega = 1/\sqrt{R^2 + H^2}$ and γ be the helix with a parametric representation

$$\vec{x}(s) = \{R \cos (\omega s), R \sin (\omega s), H \omega s\} \ (s \in \mathbb{R});$$

its vector of curvature and curvature were given in Visualization 1.6.7 (b) by

$$\ddot{\vec{x}}(s) = -\omega^2 R \{\cos \omega s, \sin \omega s, 0\} \text{ and } \kappa(s) = R\omega^2.$$

The helix is a curve on the circular cylinder of Example 2.3.5 (a) with radius $R > 0$ and axis along the x^3–axis; it is given by $u^1(s) = H\omega s$ and $u^2(s) = \omega s$ ($s \in \mathbb{R}$). By (2.25), the surface normal vectors of the cylinder along the helix are

$$\vec{N}(u^i(s)) = \{-h'(u^1(s)) \cos u^2(s), -h'(u^1(s)) \sin u^2(s), r'(u^1(s))\}.$$

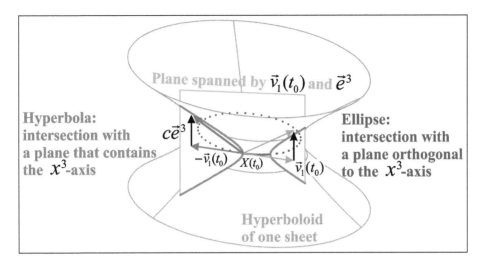

Figure 2.23 Intersections of a hyperboloid of one sheet with planes

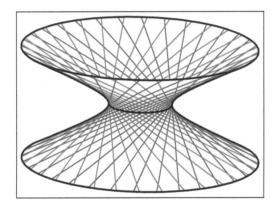

Figure 2.24 Two families of straight lines on a hyperboloid of one sheet

$$\cdot \frac{1}{\sqrt{(r'(u^1(s)))^2 + (h'(u^1(s)))^2}}$$

$$= \{-H\omega \cos \omega s, -H\omega \sin \omega s, 0\} \frac{1}{\sqrt{H^2 \omega^2}}$$

$$= -\{\cos \omega s, \sin \omega s, 0\} = \frac{1}{\omega^2 R} \ddot{\vec{x}}(s).$$

Therefore, it follows that $k_n(s) = \kappa(s)$ and $\kappa_g(s) = 0$ for all $s \in \mathbb{R}$ (Figure 2.25).

(c) **The normal and geodesic curvature along parallels on spheres**
Let S be the sphere with a parametric representation

$$\vec{x}(u^i) = r\{\cos u^1 \cos u^2, \cos u^1 \sin u^2, \sin u^1\} \left((u^1, u^2) \in (-\pi/2, \pi/2) \times (0, 2\pi)\right).$$

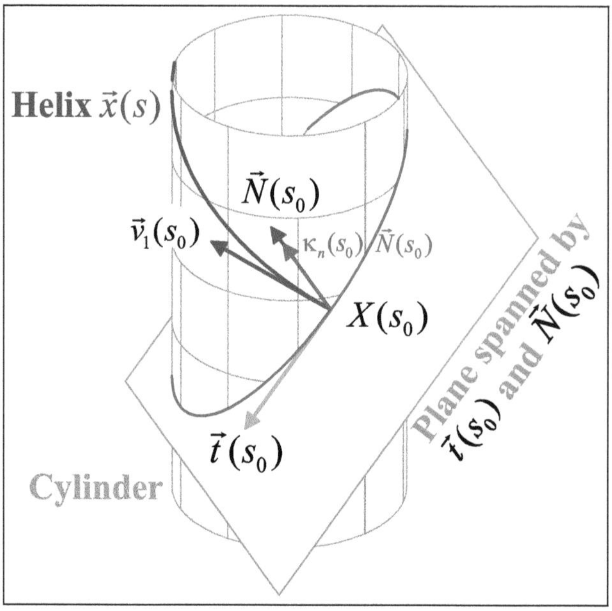

Figure 2.25 Curvature, normal and geodesic curvature along a helix on a cylinder

The u^2–line corresponding to the constant value $u^1 = \alpha \in (-\pi/2,\,\pi/2)$ has a parametric representation

$$\vec{x}(t) = \vec{x}(t, \alpha) = r\{\cos\alpha\cos t, \cos\alpha\sin t, \sin\alpha\} \quad (t \in (0, 2\pi)).$$

Obviously the curvature of this u^2–line is given by

$$\kappa = \frac{1}{r\cos\alpha}.$$

Since $\vec{N} = -\frac{1}{r}\vec{x}$, we obtain (Figure 2.26)

$$\kappa_n = \kappa\cos\alpha = \frac{1}{r}.$$

From

$$\kappa_g^2 = \kappa^2 - \kappa_n^2 = \frac{1}{r^2}\left(\frac{1}{\cos^2\alpha} - 1\right) = \frac{\tan^2\alpha}{r^2},$$

we obtain

$$|\kappa_g| = \frac{|\tan\alpha|}{r}.$$

(d) **A curve with identically vanishing normal curvature on a catenoid**
Now let $a > 0$ be given, C be the catenoid of Example 2.3.5 (b) with a parametric representation

$$\vec{x}(u^i) = \left\{a\cosh\left(\frac{u^1}{a}\right)\cos u^2, a\cosh\left(\frac{u^1}{a}\right)\sin u^2, u^1\right\} \quad \left((u^1, u^2) \in \mathbb{R} \times (0, 2\pi)\right)$$

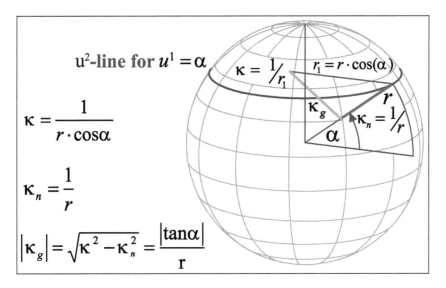

u²-line for $u^1 = \alpha$

$\kappa = \dfrac{1}{r_1}$, $r_1 = r \cdot \cos(\alpha)$

r

κ_g

$\kappa_n = \dfrac{1}{r}$

α

$$\kappa = \frac{1}{r \cdot \cos\alpha}$$

$$\kappa_n = \frac{1}{r}$$

$$|\kappa_g| = \sqrt{\kappa^2 - \kappa_n^2} = \frac{|\tan\alpha|}{r}$$

Figure 2.26 Curvature, geodesic and normal curvature at a point on the parallel of a sphere

and $\gamma_\pm$ be the curves on C given by $u^1(t) = t$ and $u^2(t) = \pm t/a$ ($t \in \mathbb{R}$). Then we have

$$r(u^1) = a\cosh\left(\frac{u^1}{a}\right), \quad r'(u^1) = \sinh\left(\frac{u^1}{a}\right), \quad r''(u^1) = \frac{1}{a}\cosh\left(\frac{u^1}{a}\right),$$

$$h(u^1) = u^1, \quad h'(u^1) = 1, \quad h''(u^1) = 0,$$

and by Example 2.4.6

$$L_{11}(u^1) = -\frac{\cosh\left(\frac{u^1}{a}\right)}{a\sqrt{\sinh^2\left(\frac{u^1}{a}\right)+1}} = -\frac{\cosh\left(\frac{u^1}{a}\right)}{a\left|\cosh\left(\frac{u^1}{a}\right)\right|} = -\frac{1}{a},$$

since $\cosh(u) > 0$ for all u, $L_{12}(u^1) = 0$ and

$$L_{22}(u^1) = \frac{a\cosh\left(\frac{u^1}{a}\right)}{\cosh\left(\frac{u^1}{a}\right)} = a.$$

Therefore, the normal curvature along $\gamma_\pm$ is by (2.47) and (2.24)

$$\kappa_n(t) = \left(L_{11}(u^1(t))\left(\frac{du^1(t)}{dt}\right)^2 + L_{22}(u^1(t))\left(\frac{du^2(t)}{dt}\right)^2\right) \cdot$$

$$\cdot \frac{1}{g_{11}(u^1(t))\left(\frac{du^1(t)}{dt}\right)^2 + g_{22}(u^1(t))\left(\frac{du^2(t)}{dt}\right)^2} =$$

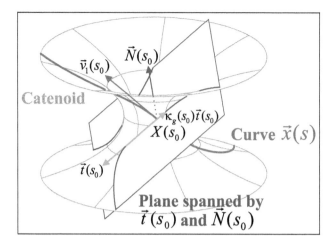

Figure 2.27 Normal and geodesic curvature $\kappa_n = 0$ and $\kappa_g = \kappa$ at a point of a curve on a catenoid

$$\left(-\frac{1}{a} + a\frac{1}{a^2}\right) \cdot \frac{1}{g_{11}(u^1(t))\left(\dfrac{du^1(t)}{dt}\right)^2 + g_{22}(u^1(t))\left(\dfrac{du^2(t)}{dt}\right)^2} = 0$$

(Figure 2.27).

2.5   THE NORMAL, PRINCIPAL, GAUSSIAN AND MEAN CURVATURE

In the previous section, we split the vector of curvature into two components the lengths of which were the geodesic and normal curvature κ_g and κ_n.

First we are going to study the normal curvature of a curve on a surface. We will see in Theorem 2.5.6 that the normal curvature of curves at a point P on a surface only depends on the direction of the curves at P. The *principal curvatures at P* are the extreme values of κ_n at P which are the real solutions of a quadratic equation (Lemma 2.5.13 and Theorem 2.5.14); their geometric and arithmetic means are called the *Gaussian* and *mean curvature* (Definition 2.5.17).

We introduce *ruled surfaces*, which among other things are useful for the visualization of the normal curvature along a given curve on a surface as in (2.58) and Figure 2.32.

Definition 2.5.1. Let γ be a curve with a parametric representation $\vec{y}(t)$ $(t \in I_1)$ and $\vec{z}(t)$ be a unit vector for every $t \in I_1$. We write $u^1 = t$. Then the surface generated by the curve γ and the family $\{\vec{z}(u^1) : u^1 \in I_1\}$ with a parametric representation

$$\vec{x}(u^i) = \vec{y}(u^1) + u^2\vec{z}(u^1) \;\left((u^1, u^2) \in D = I_1 \times I_2 \subset \mathbb{R}^2\right)$$

is called a *ruled surface*.

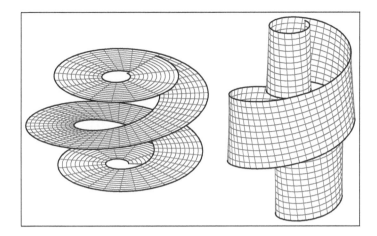

Figure 2.28 Principal normal (left) and binormal (right) surfaces of a loxodrome on a sphere

Remark 2.5.2. *(a) It is obvious that any curve on a ruled surface that intersects any vector $\vec{z}(u^1)$ exactly once with an angle unequal to zero can be chosen as a generating curve of the ruled surface.*
(b) The unit vectors $\vec{z}(u^1)$, when attached to the origin of the coordinate system, define a curve on the unit sphere, the so–called spherical indicatrix.

Visualization 2.5.3 (Some ruled surfaces).
(a) **Circular cylinders**
The circular cylinder with a parametric representation

$$\vec{x}(u^i) = \{r\cos u^1, r\sin u^1, 0\} + u^2\{0,0,1\} \ \left((u^1, u^2) \in D \subset (0, 2\pi) \times \mathbb{R}\right),$$

where $r > 0$ is a constant, is a ruled surface; its generating curve is a circle line with radius r and centre in the origin in the $x^1 x^2$–plane, here the vectors $\vec{z}(u^1) = \vec{e}^3$ are constant.

(b) **Tangent, principal normal and binormal surfaces**
Let γ be a curve with a parametric representation $\vec{y}(s)$ $(s \in I_1)$ of class C^3, where s denotes the arc length along γ. We write $u^1 = s$. If γ has non–vanishing curvature κ on I, and $\vec{v}_k$ denote the vectors of the trihedra of γ, we may consider the ruled surfaces given by the parametric representations (Figures 2.28 and 2.29)

$$\vec{x}(u^i) = \vec{y}(u^1) + u^2\vec{v}_1(u^2) \ \left((u^1, u^2) \in I_1 \times I_2\right), \quad \text{the tangent surface of } \gamma,$$

$$\vec{x}(u^i) = \vec{y}(u^1) + u^2\vec{v}_2(u^2) \ \left((u^1, u^2) \in I_1 \times I_2\right), \quad \text{the principal normal surface of } \gamma$$

and

$$\vec{x}(u^i) = \vec{y}(u^1) + u^2\vec{v}_3(u^2) \ \left((u^1, u^2) \in I_1 \times I_2)\right), \quad \text{the binormal surface of } \gamma.$$

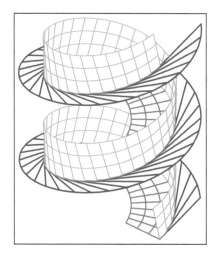

Figure 2.29 Tangent, principal normal and binormal surfaces of a helix

(c) **Conoids and helicoids**

Let $h : (0, 2\pi) \to \mathbb{R}$ be a function. Then the conoid *is a ruled surface with a parametric representation*

$$\vec{x}(u^i) = \{0, 0, h(u^1)\} + u^2\{\cos u^1, \sin u^1, 0\} \; \left((u^1, u^2) \in (0, 2\pi) \times I_2\right)$$

(left in Figure 2.30). A helicoid *is a special case of a conoid when $h(u^1) = cu^1$ for some constant $c \neq 0$; it is constructed by joining every point of the helix with the point on the x^3-axis with the same third coordinate by a straight line (right in Figure 2.30).*

We introduced conoids and helicoids as special cases of ruled surfaces. They may also be considered as special cases of the class of *general screw surfaces*.

Definition 2.5.4. Let f be a real–valued function on a domain $D = I_1 \times I_2 \subset \mathbb{R}^2$. Then a *general screw surface* is given by a parametric representation (left in Figure 2.31)

$$\vec{x}(u^i) = \{u^1 \cos u^2, u^1 \sin u^2, f(u^1, u^2)\} \; \left((u^1, u^2) \in D\right); \qquad (2.56)$$

in the special case of $f(u^i, u^2) = cu^2 + g(u^1)$, where c is a constant and g is a real valued function on I_1, we obtain the so–called *screw surface* with a parametric representation (right in Figure 2.31)

$$\vec{x}(u^i) = \{u^1 \cos u^2, u^1 \sin u^2, cu^2 + g(u^1)\} \; \left((u^1, u^2) \in D\right). \qquad (2.57)$$

Remark 2.5.5. *(a) If $c = 0$ in (2.57) then the screw surface is a surface of revolution. (b) Similarly as an explicit surface may be used as a representation of a real valued function f of two real parameters over the $x_1 x_2$-plane, where the parameters u^1 and u^2 are interpreted as the Cartesian coordinates of the plane, a general screw surface*

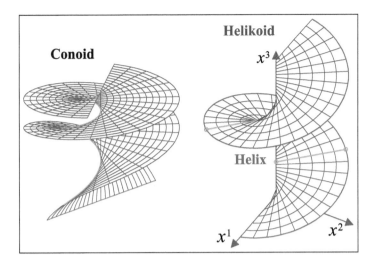

Figure 2.30   A conoid and a helicoid

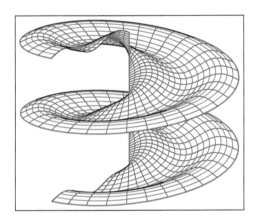
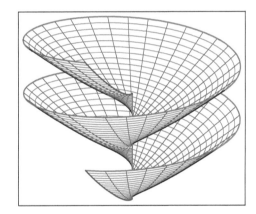

Figure 2.31   General screw surface (left) and screw surface (right)

may be used as the representation of f over the x_1x_2–plane, where the parameters u^1 and u^2 are interpreted as the polar coordinates of the plane.

(c) If we interchange u^1 and u^2 in (2.57) and put $h = g$ and $c = 0$, then we obtain a helicoid as in Visualization 2.5.3 (c).

(d) A screw surface S is generated by the simultaneous rotation of a curve about a fixed axis A and a translation along A such that the speed of translation is proportional to the speed of rotation. We choose A to be the x^3–axis and assume that the curve γ in the x^1x^3–plane is given by a parametric representation

$$\vec{x}(u^1) = \{u^1, 0, g(u^1)\} \ \text{with} \ \vec{x}(0) = \vec{0},$$

where u^1 denotes the distance between the x^3–axis and the points of γ. Let u^2 denote the angle of rotation. Since the translation of γ is parallel to the x^3–axis and

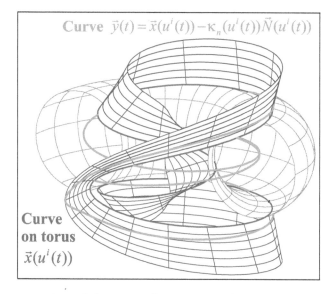

Figure 2.32 Representation of the normal curvature of a curve on a torus

proportional to u^2, we obtain the parametric representation (2.57) for S. The u^1–lines of S are referred to as the meridians *of S, and the u^2–lines are* helices.

The normal curvature along a curve on a surface can be visualized in a natural way as follows. Let S be a surface with a parametric representation $\vec{x}(u^i)$ $((u^1, u^2) \in D \subset \mathbb{R}^2)$ and γ be a curve on S with a parametric representation $\vec{x}(t) = \vec{x}(u^i(t))$ $(t \in I)$ and normal curvature $\kappa_n(t) = \kappa_n(u^i(t))$. Then we may represent $\kappa_n(t)$ by the curve γ_n with a parametric representation

$$\vec{x}^*(t) = \vec{x}(t) + \kappa_n(t)\vec{N}(t) \ (t \in I).$$

Writing $u^{*1} = t$, we see that γ_n is a curve on the ruled surface RS that has a parametric representation

$$\vec{x}^*(u^{*i}) = \vec{y}(u^{*1}) + u^{*2}\vec{z}(u^{*1}) \ \left((u^{*1}, u^{*2}) \in I \times \mathbb{R}\right),$$

$$\text{where } \vec{y}(u^{*1}) = \vec{x}(u^i(u^{*1})) \text{ and } \vec{z}(u^{*1}) = \vec{N}(u^i(u^{*1})), \quad (2.58)$$

and γ_n considered as a curve on RS is given by putting $u^{*2} = \kappa_n(u^i(u^{*1}))$ in (2.58) (Figure 2.32).

It turns out that the normal curvature of a curve at a point only depends on the direction of the curve.

Theorem 2.5.6. *All curves on a surface that intersect at a point P and have the same tangent at P, also have the same normal curvature at P (Figure 2.33).*

Proof. Since the direction of a curve with a parametric representation $\vec{x}(u^i(s))$ on a surface with a parametric representation $\vec{x}(u^i)$ is given by $\dot{u}^1(s)$ and $\dot{u}^2(s)$, the theorem is an immediate consequence of the formula for the normal curvature in (2.46). □

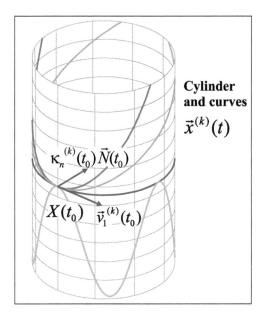

Figure 2.33 Curves on a cylinder with the same tangents and normal curvature

The following result gives a geometric interpretation of the normal curvature.

Theorem 2.5.7 (**Meusnier**). *The normal curvature of a curve on surface is equal to its curvature multiplied by the cosine of the angle between the principal normal vector of the curve and the normal vector of the surface (Figure 2.34).*

Proof. Let S be a surface with a parametric representation $\vec{x}(u^i)$ and γ be a curve on S, with a parametric representation $\vec{x}(u^i(s))$. If $\alpha(s)$ denotes the angle between the principal normal vector $\vec{v}_2(s)$ of γ at s and the surface normal vector $\vec{N}(s)$ of S at s, then it follows from (2.40) that the normal curvature of γ at s is

$$\kappa_n(s) = \left(\ddot{\vec{x}}(s) - \kappa_g(s)\vec{t}(s) \right) \bullet \vec{N}(s) = \kappa(s)\vec{v}_2(s) \bullet \vec{N}(s) = \kappa(s)\cos\alpha(s), \qquad (2.59)$$

since $\vec{t}(s) \perp \vec{N}(s)$ and $\|\vec{v}_2(s)\| = \|\vec{N}(s)\| = 1$. □

The *normal section of a surface* is closely related to the normal curvature.

Definition 2.5.8. Let S be a surface, γ be a curve on S and P be a point on the curve γ. Furthermore, let E be the plane through the point P, and spanned by the tangent vector $\vec{v}_1$ of γ at P and the normal vector of the surface at P. Then the curve of intersection of the surface S and the plane E is called the *normal section of the surface S at the point P along the direction of $\vec{v}_1$* (Figure 2.35). We will also say, the *normal section of S at P along γ.*

The next result gives a relation between the normal curvature and the curvature of the normal section.

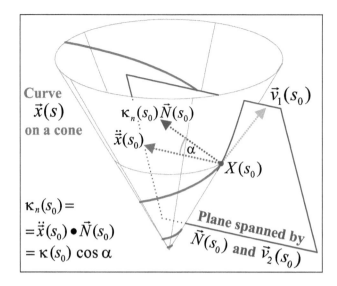

Figure 2.34 Projection of the vector of curvature of a curve on a cone to the surface normal vector of the cone

Theorem 2.5.9. *The absolute value of the normal curvature of a curve on a surface at a point P is equal to the curvature of the normal section of the surface at P along the curve (Figure 2.36).*

Proof. Let P be a point on a surface S and κ_n be the normal curvature corresponding to a given direction $\vec{d}$ at P (Theorem 2.5.6). Then the curvature κ of all curves on S through P that have a tangent in the direction of $\vec{d}$ is given by (2.59) as $\kappa_n = \kappa \cos \alpha$ where α is the angle between the principal normal vector $\vec{v}_2$ of the curves at P and the surface normal vector $\vec{N}$ of S at P. If $\alpha = 0$, then $\kappa = \kappa_n$, and if $\alpha = \pi$ then $\kappa = -\kappa_n$. Thus $|\kappa_n|$ is the curvature of the intersection of the surface S with the plane through P that contains the direction $\vec{d}$ and the surface normal vector at P. □

Visualization 2.5.10. (Normal section of a cylinder along a helix and its curvature) *Let $r, h > 0$ and Cyl be the circular cylinder with a parametric representation*

$$\vec{x}(u^i) = \{r \cos u^2, r \sin u^2, u^1\} \ \left((u^1, u^2) \in \mathbb{R} \times (0, 2\pi) \right)$$

and γ be the helix on Cyl given by

$$u^1(s) = h\omega s \ \text{and} \ u^2(s) = \omega s \ (s \in \mathbb{R}) \ \text{where} \ \omega = \frac{1}{\sqrt{r^2 + h^2}}.$$

Then the normal section of Cyl along γ at s has a parametric representation

$$\vec{y}_s(t) = \vec{x}(u^i(s)) + r(1 - \cos t)\vec{N}(u^i(s)) + \frac{1}{\omega} \sin t \, \vec{v}_1(u^i(s)) \ (t \in (0, 2\pi)) \qquad (2.60)$$

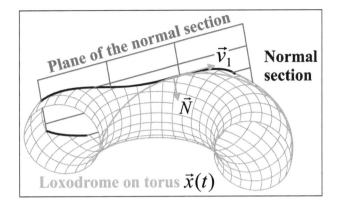

Figure 2.35 Normal section of a loxodrome on a torus

and its curvature at s is given by

$$\kappa_s(t) = \frac{r}{\omega \left(r^2 + h^2 \cos^2 t\right)^{3/2}} \quad (t \in (0, 2\pi)). \tag{2.61}$$

Proof. The tangent vector of the helix and the surface normal vector of the cylinder at any $s \in \mathbb{R}$ are by Visualizations 1.6.7 (b) and 2.4.7 (b)

$$\vec{v}_1(s) = \dot{\vec{x}}(u^i(s)) = \{-r\omega \sin (\omega s), r\omega \cos (\omega s), h\omega\}$$

and

$$\vec{N}(s) = \vec{N}(u^i(s)) = \{- \cos (\omega s), - \sin (\omega s), 0\}.$$

A parametric representation for the plane of the normal section of Cyl along γ at s is

$$\{x^1, x^2, x^3\} = \vec{x}(u^i(s)) + \lambda\vec{v}_1(s) + \mu\vec{N}(s) \text{ for all } \lambda, \mu \in \mathbb{R}.$$

The normal section of Cyl along γ at s has to satisfy the equation

$$
\begin{aligned}
r^2 &= (x^1)^2 + (x^2)^2 \\
&= ((r - \mu) \cos (\omega s) - \lambda r\omega \sin (\omega s))^2 + ((r - \mu) \sin (\omega s) + \lambda r\omega \cos (\omega s))^2 \\
&= (r - \mu)^2 + \lambda^2 r^2 \omega^2,
\end{aligned}
$$

or

$$\frac{(\mu - r)^2}{r^2} + \frac{\lambda^2}{1/\omega^2} = 1.$$

This is the equation of an ellipse in the plane spanned by the vectors $\vec{N}(s)$ and $\vec{v}_1(s)$. Hence the normal section of Cyl along γ at s has a parametric representation

$$\vec{y}_s(t) = \vec{x}(u^i(s)) + r(1 - \cos t)\vec{N}(u^i(s)) + \frac{1}{\omega} \sin t\vec{v}_1(u^i(s)) \ (t \in (0, 2\pi)).$$

Thus we have shown (2.60).
Furthermore, it follows that

$$\frac{d\vec{y}_s(t)}{dt} = r \sin t\vec{N}(u^i(s)) + \frac{1}{\omega} \cos t\vec{v}_1(u^i(s)),$$

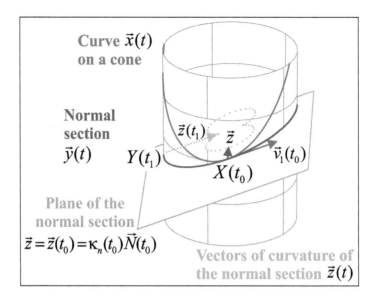

Figure 2.36 Normal curvature at a point of a curve on cylinder and curvature of the corresponding normal section

$$\frac{d^2\vec{y}_s(t)}{dt^2} = r\cos t\,\vec{N}(u^i(s)) - \frac{1}{\omega}\sin t\,\vec{v}_1(u^i(s)),$$

$$\left\|\frac{d\vec{y}_s(t)}{dt}\right\|^2 = r^2\sin^2 t + \frac{1}{\omega^2}\cos^2 t$$

and

$$\left\|\frac{d\vec{y}_s(t)}{dt} \times \frac{d^2\vec{y}_s(t)}{dt^2}\right\| = \frac{r}{\omega}\sqrt{\sin^2 t + \cos^2 t} = \frac{r}{\omega}.$$

Hence, the curvature of the normal section of Cyl along γ at s is by (1.88)

$$\kappa_s(t) = \frac{\left\|\frac{d\vec{y}_s(t)}{dt} \times \frac{d^2\vec{y}_s(t)}{dt^2}\right\|}{\left\|\frac{d\vec{y}_s(t)}{dt}\right\|^3} = \frac{r}{\omega\left(r^2\sin^2 t + \frac{1}{\omega^2}\cos^2 t\right)^{3/2}}$$

$$= \frac{r}{\omega\left(r^2 + h^2\cos^2 t\right)^{3/2}} \quad (t \in (0, 2\pi)).$$

Thus we have shown (2.61).
In particular, we have by Visualization 2.4.7 (b)

$$\kappa_s(0) = \frac{r}{\omega(r^2 + h^2)^{3/2}} = r\omega^2 = \kappa_n(s) \text{ for all } s \in \mathbb{R} \text{ (Figure 2.25)}.$$

□

Remark 2.5.11. *Meusnier's theorem may be restated as follows. The centres of curvature of all curves on a surface S that are tangent at a point P lie on a circle*

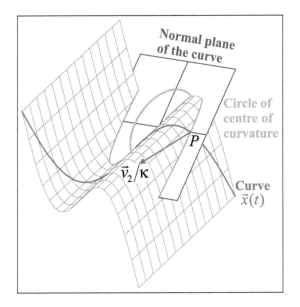

Figure 2.37 Illustration of Remark 2.5.11

line in their joint normal plane; this circle line has the diameter $1/|\kappa_n|$ and is tangent to the surface S at P (Figure 2.37).

Proof. Let $\vec{x}_0$ be the position vector of a point P on a surface S. The centres of curvature of all curves with parametric representations $\vec{x}(s)$ through P are given by the position vectors

$$\vec{m} = \vec{x}_0 + \frac{1}{\kappa}\vec{v}_2,$$

where κ and $\vec{v}_2$ are the curvature and the principal normal vectors of the curves at P. All curves that are tangent to one another at P have the same tangent vector $\vec{v}_1$ at P which is orthogonal to $\vec{v}_2$. This implies $(\vec{m} - \vec{x}_0) \bullet \vec{v}_1 = 0$, hence the vector $\vec{m}$ is in the joint normal plane of all these curves, since their normal plane is orthogonal to the vector $\vec{v}_1$. Furthermore, all these curves have the same normal curvature by Theorem 2.5.6, namely $\kappa_n = \kappa\vec{v}_2 \bullet \vec{N}$, where $\vec{N}$ is the surface normal vector of S at P. Therefore, we have

$$\left(\vec{m} - \left(\vec{x}_0 - \frac{1}{2\kappa_n}\vec{N}\right)\right)^2 = \left(\frac{1}{\kappa}\vec{v}_2 - \frac{1}{2\kappa_n}\vec{N}\right)^2 = \frac{1}{\kappa^2} - \frac{1}{\kappa\kappa_n}\vec{v}_2 \bullet \vec{N} + \frac{1}{4\kappa_n^2} = \frac{1}{4\kappa_n^2},$$

that is, $\vec{m}$ is on a circle line of radius $1/2|\kappa_n|$ with its centre given by $\vec{x}_0 - 1/(2\kappa_n)$. Obviously $\vec{x}_0$ is on this circle line, and the tangent to this circle line at P is orthogonal to $\vec{N}$. Hence this circle line is tangent to the surface S at P. □

We saw in Theorem 2.5.6 that the normal curvature of a curve on a surface at a point depends on the direction of the curve at that point only. So it is quite natural to find the directions for which the normal curvature has extreme values.

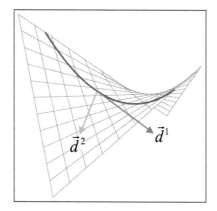

Figure 2.38 Principal directions $\vec{d}^1$ and $\vec{d}^2$ at a point of a surface

Definition 2.5.12. Let S be a surface and P a given point on S. The directions corresponding to the extreme values of the normal curvature at P are called the *principal directions at the point P* (Figure 2.38) and the extreme values of the normal curvature at P are called the *principal curvature at the point P*.

We will see that there are two orthogonal principal directions at every point of a surface. The following lemma is needed for he proof of this statement.

Lemma 2.5.13. *Let S be s surface with first and second fundamental coefficients g_{ik} and L_{ik}. Then the solutions λ of the quadratic equation*

$$\det(L_{ik} - \lambda g_{ik}) = 0 \tag{2.62}$$

are always real.

Proof. First we observe that $g = g_{11}g_{22} - g_{12}^2 > 0$ implies

$$0 \leq \left(g_{11}L_{22} - g_{22}L_{11} - \frac{2g_{12}}{g_{11}}(g_{11}L_{12} - g_{12}L_{12}) \right)^2$$

$$+ 4 \cdot \frac{g_{11}g_{22} - g_{12}^2}{g_{11}^2}(g_{11}L_{12} - g_{12}L_{11})^2$$

$$= (g_{11}L_{22} - g_{22}L_{11})^2 - 4 \cdot \frac{g_{12}}{g_{11}}(g_{11}L_{22} - g_{22}L_{11})(g_{11}L_{12} - g_{12}L_{11})$$

$$+ 4 \cdot \frac{g_{22}}{g_{11}}(g_{11}L_{12} - g_{12}L_{11})^2$$

$$= (g_{11}L_{22} - g_{22}L_{11})^2 + \frac{4}{g_{11}} \cdot (g_{11}L_{12} - g_{12}L_{11}) [g_{22}(g_{11}L_{12} - g_{12}L_{11})$$

$$- g_{12}(g_{11}L_{22} - g_{22}L_{11})]$$

$$= (g_{11}L_{22} - g_{22}L_{11})^2 + \frac{4}{g_{11}} \cdot (g_{11}L_{12} - g_{12}L_{11})(g_{11}g_{22}L_{12} - g_{11}g_{12}L_{22})$$

$$= (g_{11}L_{22} - g_{22}L_{11})^2 + 4 \cdot (g_{11}L_{12} - g_{12}L_{11})(g_{22}L_{12} - g_{12}L_{22})$$

$$= (g_{11}L_{22} + g_{22}L_{11} - 2g_{12}L_{12})^2 - 4 \cdot g_{11}g_{22}L_{11}L_{22} + 4 \cdot g_{12}L_{12}(g_{11}L_{22} + g_{22}L_{11})$$
$$- 4 \cdot g_{12}^2 L_{12}^2 + 4 \cdot (g_{11}L_{12} - g_{12}L_{11})(g_{22}L_{12} - g_{12}L_{22})$$

$$= (g_{11}L_{22} + g_{22}L_{11} - 2 \cdot g_{12}L_{12})^2 - 4 \cdot g_{11}g_{22}L_{11}L_{22} + 4 \cdot g_{11}g_{12}L_{12}L_{22}$$
$$+ 4 \cdot g_{12}g_{22}L_{11}L_{12} - 4 \cdot g_{12}^2 L_{12}^2 + 4 \cdot g_{11}g_{22}L_{12}^2 - 4 \cdot g_{11}g_{12}L_{12}L_{22}$$
$$- 4 \cdot g_{12}g_{22}L_{11}L_{12} + 4 \cdot g_{12}^2 L_{11}L_{22}$$

$$= (g_{11}L_{22} + g_{22}L_{11} - 2 \cdot g_{12}L_{12})^2$$
$$- 4 \cdot \left(g_{11}g_{22}(L_{11}L_{22} - L_{12}^2) - g_{12}^2(L_{11}L_{22} - L_{12}^2) \right)$$

$$= (g_{11}L_{22} + g_{22}L_{11} - 2 \cdot g_{12}L_{12})^2 - 4 \cdot (g_{11}g_{22} - g_{12}^2)(L_{11}L_{22} - L_{12}^2)$$
$$= (g_{11}L_{22} + g_{22}L_{11} - 2 \cdot g_{12}L_{12})^2 - 4 \cdot gL.$$

The solutions of the equation

$$0 = \det(L_{ik} - \lambda g_{ik}) = \det \begin{pmatrix} L_{11} - \lambda g_{11} & L_{12} - \lambda g_{12} \\ L_{12} - \lambda g_{12} & L_{22} - \lambda g_{22} \end{pmatrix}$$
$$= (L_{11} - \lambda g_{11})(L_{22} - \lambda g_{22}) - (L_{12} - \lambda g_{12})^2$$
$$= \lambda^2(g_{11}g_{22} - g_{12}^2) - \lambda(g_{11}L_{22} + g_{11}L_{11} - 2g_{12}L_{12}) + L_{11}L_{22} - L_{12}^2$$
$$= \lambda^2 g - \lambda(g_{11}L_{22} + g_{11}L_{11} - 2 \cdot g_{12}L_{12}) + L$$

are

$$\lambda_{1,2} = \frac{1}{2g} \left(g_{11}L_{22} + g_{22}L_{11} - 2g_{12}L_{12} \pm \sqrt{(g_{11}L_{22} + g_{22}L_{11} - 2g_{12}L_{12})^2 - 4gL} \right),$$

and

$$(g_{11}L_{22} + g_{22}L_{11} - 2 \cdot g_{12}L_{12})^2 - 4 \cdot gL \geq 0,$$

by what we have just shown.
Thus the solutions of the equation in (2.62) are always real. □

Theorem 2.5.14. *There exist two orthogonal principal directions* $\vec{d}_\mu = \xi_\mu^k \vec{x}_k$ ($\mu = 1, 2$) *at every point of a surface; they are given by the solutions of the equations*

$$(L_{ik} - \kappa_\mu g_{ik}) \, \xi_\mu^k = 0 \ (i = 1, 2) \text{ and } g_{ik}\xi_\mu^i \xi_\mu^k = 1 \text{ for } \mu = 1, 2. \tag{2.63}$$

Proof. By Theorem 2.5.6, we may assume that the directions at a point P_0 of the surface are given by ξ^1 and ξ^2 where $\xi^k \vec{x}_k = \dot{u}^k \vec{x}_k$. It follows that

$$g_{jk}\xi^j \xi^k = 1. \tag{2.64}$$

To determine the extreme values of κ_n in the point P_0, we have to find the extreme values of the function

$$f(\xi^i) = L_{jk}\xi^j \xi^k \tag{2.65}$$

under the side condition (2.64). Using the Lagrange multiplier rule, we have to solve

$$\frac{\partial}{\partial \xi^i}(L_{jk}\xi^j \xi^k - \lambda g_{jk}\xi^j \xi^k) = 0 \ (i = 1, 2), \tag{2.66}$$

or

$$(L_{ik} - \lambda g_{ik})\xi^k = 0 \ (i = 1, 2).\tag{2.67}$$

The homogeneous system (2.67) has a solution if and only if

$$\det(L_{ik} - \lambda g_{ik}) = 0.\tag{2.68}$$

The quadratic equation (2.68) for λ has two real solutions κ_1 and κ_2 by Lemma 2.5.13; we denote by ξ_1^i and ξ_2^i $(i = 1, 2)$ the corresponding solutions of (2.67) that also satisfy the side condition (2.64). Since we have by the first identity in (2.46) in Remark 2.4.3 (c)

$$\kappa_\mu = L_{ik}\xi_\mu^i\xi_\mu^k \text{ with } g_{ik}\xi_\mu^i\xi_\mu^k = 1 \text{ for } \mu = 1, 2,$$

the values κ_μ are the extreme values of the normal curvature in the directions given by ξ_μ^i. The principal directions $\vec{d}_\mu = \xi^i \vec{x}_i$ $(\mu = 1, 2)$ satisfy by (2.67)

$$\left\{ \begin{array}{l} (L_{ik} - \kappa_1 g_{ik})\xi_1^k = 0 \ (i = 1, 2), \quad g_{ik}\xi_1^i\xi_1^k = 1 \text{ and} \\ (L_{ik} - \kappa_2 g_{ik})\xi_2^k = 0 \ (i = 1, 2), \quad g_{ik}\xi_2^i\xi_2^k = 1. \end{array} \right\}\tag{2.69}$$

We multiply the first equation in (2.69) by ξ_2^i and sum with respect to i to obtain

$$(L_{ik} - \kappa_1 g_{ik})\xi_1^k\xi_2^i = 0,\tag{2.70}$$

and multiply the third equation in (2.69) by ξ_1^i and sum with respect to i to obtain

$$(L_{ik} - \kappa_2 g_{ik})\xi_1^i\xi_2^k = 0.\tag{2.71}$$

We subtract equation (2.71) from (2.70), observing the symmetries $g_{ik} = g_{ki}$ and $L_{ik} = L_{ki}$ and interchanging the order of summation in (2.71)

$$0 = (L_{ik} - \kappa_1 g_{ik})\xi_1^k\xi_2^i - (L_{ik} - \kappa_2 g_{ik})\xi_1^k\xi_2^i = (\kappa_2 - \kappa_1)g_{ik}\xi_1^k\xi_2^i$$
$$= (\kappa_1 - \kappa_2)\vec{d}_1 \bullet \vec{d}_2.$$

If $\kappa_1 \neq \kappa_2$, then this implies $\vec{d}_1 \bullet \vec{d}_2 = 0$, that is, the principal directions are orthogonal. If $\kappa_1 = \kappa_2$, then $L_{ik}\xi^i\xi^k = k_1 g_{ik}\xi^i\xi^k$ for all directions $\{\xi^1, \xi^2\}$, in particular, we obtain for $\{\xi^1, \xi^2\} = \{1, 0\}$, $\{\xi^1, \xi^2\} = \{0, 1\}$ and $\{\xi^1, \xi^2\} = 1/\sqrt{2}\{1, 1\}$ that $L_{11} = \kappa_1 g_{11}$, $L_{22} = \kappa_1 g_{22}$ and $L_{12} = \kappa_1 g_{12}$. Conversely if $L_{ik} = c \cdot g_{ik}$ $(i, k = 1, 2)$ for some constant c then

$$\kappa_n = \frac{L_{ik}\frac{du^i}{dt}\frac{du^k}{dt}}{g_{ik}\frac{du^i}{dt}\frac{du^k}{dt}} = c.$$

Thus we have shown that $\kappa_1 = \kappa_2$ if and only if

$$L_{ik} = c \cdot g_{ik} \ (i, k = 1, 2) \text{ for some constant } c \in \mathbb{R}.$$

In this case, the equations in (2.67) are satisfied for arbitrary directions $\vec{d}_1$ and $\vec{d}_2$, in particular, we may choose $\vec{d}_1$ and $\vec{d}_2$ to be orthogonal. $\quad\square$

Definition 2.5.15. A point on a surface with equal principal curvatures is called an *umbilical point*.

Remark 2.5.16. *(a) By* Vieta's formula, *the principal curvatures κ_1 and κ_2 satisfy the conditions*

$$\frac{\kappa_1 + \kappa_2}{2} = \frac{1}{2g}(g_{22}L_{11} - 2g_{12}L_{12} + g_{11}L_{22}) = \frac{1}{2}g^{ik}L_{ik} \text{ and } \kappa_1\kappa_2 = \frac{L}{g}, \qquad (2.72)$$

where g^{ik} $(i, k = 1, 2)$ are the entries of the inverse of the matrix (g_{ik}) of the first fundamental coefficients (Remark 2.3.2 (b)).
(b) We may choose any two orthogonal directions as principal directions at an umbilical point. The proof of Theorem 2.5.14 shows that points with $\kappa_1 = \kappa_2$ need a special treatment. The condition $\kappa_1 = \kappa_2$ is equivalent to

$$L_{ik} = c \cdot g_{ik} \ (i, k = 1, 2) \ \text{for some constant } c \in \mathbb{R}. \qquad (2.73)$$

Definition 2.5.17. Let g_{ik}, L_{ik}, κ_1 and κ_2 denote the first and second fundamental coefficients and the principal curvatures of a surface.
Then the functions

$$H = \frac{\kappa_1 + \kappa_2}{2} = \frac{1}{2}g^{ik}L_{ik} \text{ and } K = \kappa_1 \cdot \kappa_2 = \frac{L}{g} \qquad (2.74)$$

are called the *mean* and *Gaussian curvature of the surface.*

Remark 2.5.18. *It is clear from the definitions of the Gaussian curvature and umbilical points that a surface with negative Gaussian curvature cannot have any umbilical points.*

Visualization 2.5.19. (Gaussian and mean curvature of some surfaces of revolution) *We consider a surface of revolution with a parametric representation*

$$\vec{x}(u^i) = \{r(u^1)\cos u^2, r(u^1)\sin u^2, h(u^1)\} \ \left((u^1, u^2) \in D = I_1 \times I_2 \subset \mathbb{R} \times (0, 2\pi)\right) \qquad (2.75)$$

with the usual assumptions

$$r(u^1) > 0 \ and \ |r'(u^1)| + |h'(u^1)| > 0 \ on \ I_1.$$

Then it follows for the Gaussian and mean curvature K and H from (2.72) in Remark 2.5.16, (2.5.17) in Definition 2.5.17 and the formulae for the first and second fundamental coefficients of surfaces of revolution (2.24) in Example 2.3.5 and Example 2.4.6 that $K(u^i) = K(u^1)$, $H(u^i) = H(u^1)$,

$$K = \frac{L}{g} = \frac{(h''r' - h'r'')h'}{r\left((r')^2 + (h')^2\right)^2} \qquad (2.76)$$

and

$$H = \frac{1}{2g}(g_{11}L_{22} + g_{22}L_{11})$$

$$= \frac{1}{2(r)^2 \left((r')^2 + (h')^2\right)^{3/2}} \left(h'r \left((r')^2 + (h')^2\right) + (h''r' - h'r'') (r)^2\right)$$

$$= \frac{1}{2r \left((r')^2 + (h')^2\right)^{3/2}} \left(h' \left((r')^2 + (h')^2\right) + (h''r' - h'r'') r\right). \qquad (2.77)$$

Since the Gaussian and mean curvature K and H of a surface of revolution only depend on the parameter u^1, we may represent K and H as surfaces with parametric representations

$$\vec{x}^{(K)}(u^i) = \{u^1 \cos u^2, u^1 \sin u^1, K(u^1)\}$$

and

$$\vec{x}^{(H)}(u^i) = \{u^1 \cos u^2, u^1 \sin^{u^1}, H(u^1)\}.$$

In particular, if u^1 is the arc length along the curve that generates the surface of revolution, that is, if $(r'(u^1))^2 + (h'(u^1))^2 = 1$ on I_1, then it follows that $r'r'' + h'h'' = 0$ and, for $h' \neq 0$, $h'' = -\dfrac{r'r''}{h'}$ and so

$$h''r' - h'r'' = -\left(\frac{r''(r')^2}{h'} + h'r''\right) = -\frac{r''}{h'}\left((r')^2 + (h')^2\right) = -\frac{r''}{h'}. \qquad (2.78)$$

In this case, the formulae for the Gaussian and mean curvature reduce to

$$K = -\frac{r''}{r} \quad \text{and} \quad H = \frac{1}{2rh'}\left((h')^2 - r''r\right). \qquad (2.79)$$

(a) **Spheres.**

First we consider the sphere with radius $R > 0$. Then we have $r(u^1) = R \cos u^1$, $h(u^1) = R \sin u^1$, $(r'(u^1))^2 + (h'(u^1))^2 = R^2$, $r''(u^1) = -r(u^1)$ and $h'(u^1) = r(u^1)$, and we may apply (2.76), (2.77) and the second identity in (2.78) to obtain

$$K(u^1) = -\frac{r''(u^1)}{r(u^1)R} = \frac{1}{R}$$

and

$$H(u^1) = \frac{1}{2r(u^1)R^3}\left(R^2 h'(u^1) - R^2 \frac{r''(u^1)r(u^1)}{h'(u^1)}\right)$$

$$= \frac{1}{2Rr(u^1)}(r(u^1) + r(u^1)) = \frac{1}{R}$$

for the Gaussian and mean curvature of the sphere.

(b) **Catenoids.**

Now we consider the catenoid with $r(u^1) = a \cosh(u^1/a)$, $h(u^1) = u^1$ where $a > 0$ is a constant. Then we have

$$r'(u^1) = \sinh(u^1/a), \quad r''(u^1) = (1/a)\cosh(u^1/a),$$
$$h'(u^1) = 1, \quad h''(u^1) = 0,$$

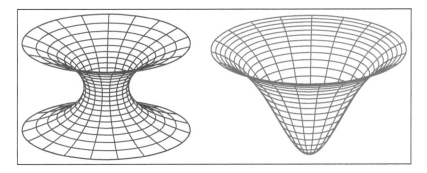

Figure 2.39 A catenoid (left) and its Gaussian curvature represented as a surface of revolution (right)

$$(r'(u^1))^2 + (h'(u^1))^2 = \sinh^2(u^1/a) + 1 = \cosh^2(u^1/a) = r''(u^1)r(u^1)$$

and it follows from (2.76) and (2.77) that

$$K(u^1) = -\frac{1}{a^2 \cosh^4\left(\frac{u^1}{a}\right)}\cosh^2\left(\frac{u^1}{a}\right) = -\frac{1}{a^2 \cosh^2\left(\frac{u^1}{a}\right)}$$

and $H(u^1) = 0$ for the Gaussian and mean curvature of the catenoid (Figure 2.39).

(c) **Pseudo–spheres.**
Finally, we consider the surface of revolution with

$$r(u^1) = \exp(-u^1) \text{ and } h(u^1) = \int \sqrt{1 - \exp(-2u^1)}\, du^1 \text{ for } u^1 > 0.$$

We have

$$r'(u^1) = -r(u^1) = -r''(u^1) = \exp(-u^1), \quad h'(u^1) = \sqrt{1 - \exp(-2u^1)}$$

and

$$(r'(u^1))^2 + (h^1(u^1))^2 = 1$$

and it follows from (2.79)

$$K(u^1) = -1 \text{ and } H(u^1) = \frac{1 - 2\exp(-2u^1)}{2\exp(-u^1)\sqrt{1 - \exp(-2u^1)}} \qquad (2.80)$$

for the Gaussian and mean curvature of the surface of revolution. A surface of revolution with constant negative Gaussian curvature is called a pseudo-sphere *(Figure 2.40).*

Now we determine surfaces of revolution for given Gaussian curvature K. The cases of given constant Gaussian curvature K are of special interest; they yield spherical and pseudo-spherical surfaces or $K > 0$ and $K < 0$, respectively. Let S

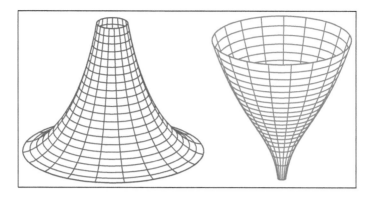

Figure 2.40 A pseudo-sphere (left) and its mean curvature represented as a surface of revolution (right)

be a surface of revolution with a parametric representation (2.75), where we may assume that $(r')^2 + (h')^2 = 1$. The Gaussian curvature of S is given by (2.79) as

$$r''(u) + K(u)r(u) = 0, \tag{2.81}$$

where we write u instead of u^1. Thus the surface with a given Gaussian curvature $K(u)$ is given by the solutions $r(u)$ of (2.81), and

$$h(u) = \pm \int \sqrt{1 - (r'(u))^2} \, du, \tag{2.82}$$

where we may choose the upper sign in (2.82) without loss of generality.

Surfaces of identically vanishing mean curvature will be studied in Section 3.11.

Visualization 2.5.20. (Surfaces of revolution with constant $K > 0$:
Spherical surfaces)
If $K(u)$ is constant and positive, we put $K = 1/c^2$ with some constant $c > 0$. Then (2.81) reduces to

$$r''(u) + \frac{1}{c^2} r(u) = 0 \tag{2.83}$$

with the general solution

$$r(u) = \lambda \cos\left(\frac{u}{c}\right) \text{ with } \lambda > 0$$

and we obtain

$$h(u) = \int \sqrt{1 - \frac{\lambda^2}{c^2} \sin^2\left(\frac{u}{c}\right)} \, du. \tag{2.84}$$

There are three different types of spherical surfaces corresponding to the cases $\lambda = c$, $\lambda > c$ or $\lambda < c$.

(i) Case 1. $\lambda = c$
Then the surface has a parametric representation

$$\vec{x}(u^i) = \left(c \cos\left(\frac{u^1}{c} \right) \cos u^2, c \cos\left(\frac{u^1}{c} \right) \sin u^2, c \sin\left(\frac{u^1}{c} \right) \right)$$

$$\left((u^1, u^2) \in (-\pi/2, \pi/2) \times (0, 2\pi) \right).$$

This is a sphere with radius c and centre in the origin.

(ii) Case 2. $\lambda > c$
The corresponding surfaces are called hyperbolic spherical surfaces. Now the integral for h in (2.84) only exists for values of u with

$$\left| \sin\left(\frac{u}{c} \right) \right| \le \frac{c}{\lambda},$$

that is,

$$u \in I_k = \left[-c \sin^{-1}\left(\frac{c}{\lambda} \right) + k\pi, c \sin^{-1}\left(\frac{c}{\lambda} \right) + k\pi \right]$$

for $k = 0, \pm 1, \pm 2, \ldots$ (left in Figure 2.41).
Every interval I_k defines a region of the surface.
The radii of the circle lines of the u^2–lines are minimal at the end points of the intervals I_k and equal to $r = \sqrt{\lambda^2 - c^2}$, whereas the maximum radius $R = \lambda$ is attained in the middle of each region (left in Figure 2.43).

(iii) Case 3. $\lambda < c$
The corresponding surfaces are called elliptic spherical surfaces (right in Figure 2.41).
Now the integral for h in (2.84) exists for all u and the radii r of the circle lines of the u^2–lines attain all values $r \le \lambda$.

Visualization 2.5.21. (Surfaces of revolution with constant $K < 0$: Pseudo–spherical surfaces)
If $K(u)$ is constant and negative, we put $K = -1/c^2$ with some constant $c > 0$. The general solution of the differential equation in (2.83) is

$$r(u) = C_1 \cosh\left(\frac{u}{c} \right) + C_2 \sinh\left(\frac{u}{c} \right)$$

with constants C_1 and C_2.
Again there are three cases.

(i) Case 1. $C_1 = -C_2 = \lambda \ne 0$
The corresponding surfaces are called parabolic pseudo-spherical surfaces (left in Figure 2.42). They have a parametric representation with

$$r(u^1) = \lambda \exp\left(-\frac{u^1}{c} \right)$$

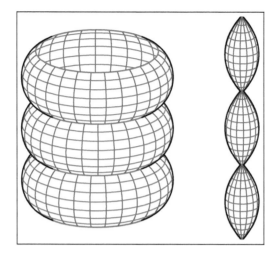

Figure 2.41 Hyperbolic (left) and elliptic (right) spherical surfaces

and

$$h(u^1) = \int \sqrt{1 - \frac{\lambda^2}{c^2} \exp\left(-\frac{2u^1}{c}\right)} \, du^1 \ \textit{for} \ u^1 > c \log\left(|\lambda|/c\right).$$

(ii)   *Case 2.* $C_2 = 0$ and $C_1 = \lambda \neq 0$

The corresponding surfaces are called hyperbolic pseudo-spherical surfaces *(middle in Figure 2.42). They have a parametric representation with*

$$r(u^1) = \lambda \cosh\left(\frac{u^1}{c}\right)$$

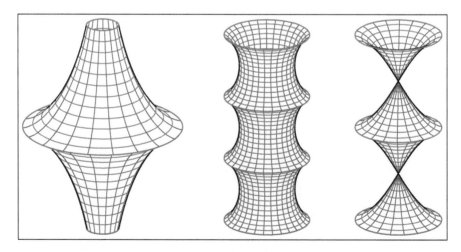

Figure 2.42 Parabolic (left), hyperbolic (middle) and elliptic (right) pseudo-spherical surfaces

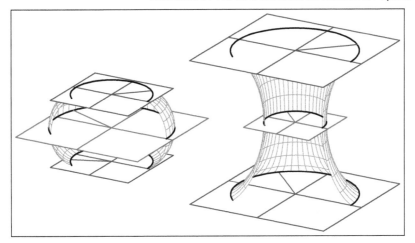

Figure 2.43 Hyperbolic spherical and pseudo-spherical surfaces, and minimal and maximal radii of u^2–lines

and

$$h(u^1) = \int \sqrt{1 - \frac{\lambda^2}{c^2} \sinh^2\left(\frac{u^1}{c^2}\right)}\, du^1$$

for

$$|u^1| \leq c \cdot \sinh^{-1}\left(\frac{c}{|\lambda|}\right) = c \log\left(\frac{c}{|\lambda|} + \sqrt{\frac{c^2}{\lambda^2} + 1}\right).$$

The radii r of the circle lines of the u^2–lines satisfy (right in Figure 2.43)

$$|\lambda| \leq r \leq \sqrt{\lambda^2 + c^2}.$$

(iii) Case 3. $C_1 = 0$ and $C_2 = \lambda \neq 0$

The corresponding surfaces are called elliptic pseudo-spherical surfaces. *They have a parametric representation with*

$$r(u^1) = \lambda \sinh\left(\frac{u^1}{c}\right)$$

$$h(u^1) = \int \sqrt{1 - \frac{\lambda^2}{c^2} \cosh^2\left(\frac{u^1}{c}\right)}\, du^1 \tag{2.85}$$

for all u^1 with

$$\cosh\left(\frac{u^1}{c}\right) \leq \frac{c}{|\lambda|};$$

(since $\cosh u^1 \geq 1$ for all u^1, we must have $|\lambda| \leq c$) (right in Figure 2.42). The integral for h in (2.85) is elliptic. The radii r of the circle lines of the u^2–lines satisfy

$$0 \leq r \leq \sqrt{c^2 - \lambda^2}.$$

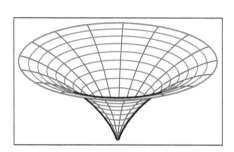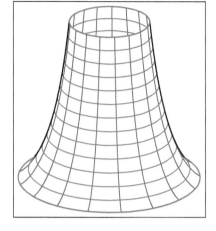

Figure 2.44 Surfaces of revolution with Gaussian curvature $K(u) = k/u^2$ for $k = -0.5$ (left) and $k = -1$ (right)

Visualization 2.5.22 (Surfaces with Gaussian curvature $K(u^1) = k/(u^1)^2$).
We determine the surfaces of revolution of Gaussian curvature $K(u^1) = k/(u^1)^2$, where $k \neq 0$ is a constant. We write $u = u^1$. Then the differential equation (2.81) reduces to

$$r''(u) + \frac{k}{u^2} \cdot r(u) = 0; \qquad (2.86)$$

which is a differential equation of Euler type. We also use (2.82).

(i) *First we consider the case $k < 1/4$ (Figure 2.44).*
We put $\gamma = \sqrt{1 - 4k}$ and obtain two linearly independent solutions

$$r_1(u) = u^{\frac{1}{2}(1+\gamma)} \text{ and } r_2(u) = u^{\frac{1}{2}(1-\gamma)}.$$

If $r(u) = cr_1(u)$ where $c \neq 0$ is a constant, then (2.82) yields

$$h(u) = \frac{1}{2} \int \sqrt{4 - (1+\gamma)^2 c^2 u^{\gamma-1}} \, du$$

for

$$u < \left(\frac{4}{(1+\gamma)^2 c^2} \right)^{1/(\gamma-1)} \text{ if } \gamma > 1, \text{ that is, } k < 0,$$

$$u > \left(\frac{(1+\gamma)^2 c^2}{4} \right)^{1/(1-\gamma)} \text{ if } 0 < \gamma < 1/4, \text{ that is, } k > 0.$$

If $r(u) = cr_2(u)$ where $c \neq 0$ is a constant, then

$$h(u) = \frac{1}{2} \int \sqrt{4 - (1-\gamma)^2 c^2 u^{-(1+\gamma)}} \, du$$

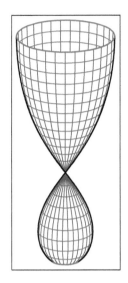
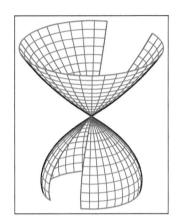

Figure 2.45 Surfaces of revolution with Gaussian curvature $K(u) = k/u^2$ for $k = 8$ (left) and $k = 2.5$ (right)

for

$$u > \left(\frac{(1-\gamma)^2 c^2}{4} \right)^{1/(1+\gamma)}.$$

(ii) *Now we consider the case $k > 1/4$ (Figure 2.45).*
We put $\delta = \sqrt{4k-1}$ and obtain two linearly independent solutions

$$r_1(u) = \sqrt{u} \cos \left(\frac{\delta}{2} \log u \right) \text{ and } r_2(u) = \sqrt{u} \sin \left(\frac{\delta}{2} \log u \right).$$

If $r(u) = cr_1(u)$ where $c \neq 0$ is a constant, then

$$r'(u) = \frac{c}{2\sqrt{u}} \left(\cos \left(\frac{\delta}{2} \log u \right) - \delta \sin \left(\frac{\delta}{2} \log u \right) \right)$$

and

$$h(u) = \int \sqrt{1 - (r'(u))^2} \, du$$

for

$$u > \frac{c^2}{4} \left(\cos \left(\frac{\delta}{2} \log u \right) - \delta \sin \left(\frac{\delta}{2} \log u \right) \right)^2.$$

If $r(u) = cr_2(u)$ where $c \neq 0$ is a constant, then

$$r'(u) = \frac{c}{2\sqrt{u}} \left(\sin \left(\frac{\delta}{2} \log u \right) + \delta \cos \left(\frac{\delta}{2} \log u \right) \right)$$

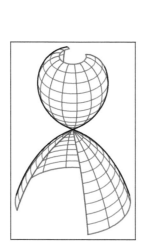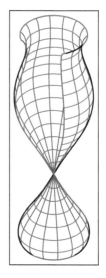

Figure 2.46 Surfaces of revolution with given Gaussian curvature $K(u)$, $K(u) = exp(u)$ (left) and $K(u) = \sin u$

and

$$h(u) = \int \sqrt{1 - (r'(u))^2}\, du$$

for

$$u > \frac{c^2}{4} \left(\sin\left(\frac{\delta}{2} \log u \right) + \delta \cos\left(\frac{\delta}{2} \log u \right) \right)^2.$$

Figure 2.46 shows surfaces of revolution with Gaussian curvature $K(u) = \exp(u)$ (left) and $K(u) = \sin u$ (right).

Visualization 2.5.23 (Gaussian and mean curvature of explicit surfaces).

(a) *Let S be the explicit surface with a parametric representation*

$$\vec{x}(u^i) = \{u^1, u^2, (u^2)^3 - a(u^1)^2 u^2\} \left((u^1, u^2) \in \mathbb{R}^2 \right),$$

where $a \in \mathbb{R}$ is a constant. Then

$$g(u^i) = 1 + 4a^2 \left(u^1 u^2 \right)^2 + \left(3(u^2)^2 - a(u^1)^2 \right)^2$$

and the Gaussian and mean curvature of S are

$$K(u^i) = \frac{L(u^i)}{g(u^i)} = -\frac{12a\left(u^2\right)^2 + 4a^2\left(u^1\right)^2}{(g(u^i))^2} \tag{2.87}$$

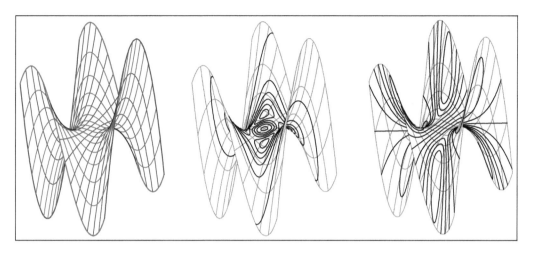

Figure 2.47 The explicit surface S of Visualization 2.5.23 (a) for $a = 3/2$ (left), lines of constant Gaussian (middle) and constant mean curvature (right)

and

$$H(u^i) = \frac{1}{(g(u^i))^{3/2}} \left[-2au^2 \left(1 + \left(3(u^2)^2 - a(u^1)^2 \right)^2 \right) \right.$$
$$\left. -8a^2 \left(u^1 \right)^2 u^2 \left(3(u^2)^2 - a(u^1)^2 \right) - 6u^2 \left(1 + 4a^2 \left(u^1 u^2 \right)^2 \right) \right], \quad (2.88)$$

where $a \in \mathbb{R}$ is a constant.

Figure 2.47 shows the surface S for $a = 3/2$, and lines of constant Gaussian and mean curvature on S.

Figure 2.48 represents the Gaussian curvature of S as an explicit surface KS with a parametric representation

$$\vec{x}(u^i) = \{u^1, u^2, K(u^1, u^2)\}$$

and the lines $K(u^i) = const$ on KS.

Figure 2.49 represents the mean curvature of S as an explicit surface HS with a parametric representation

$$\vec{x}(u^i) = \{u^1, u^2, H(u^1, u^2)\}$$

and the lines $H(u^i) = const$ on HS.

(b) *Let S be the explicit surface with a parametric representation*

$$\vec{x}(u^i) = \left\{ u^1, u^2, \left((u^1)^3 u^2 - u^1 (u^2)^3 \right) \right\} \quad \left((u^1, u^2) \in \mathbb{R}^2 \right).$$

Then

$$g(u^i) = 1 + \left((u^1)^2 + (u^2)^2 \right)^3$$

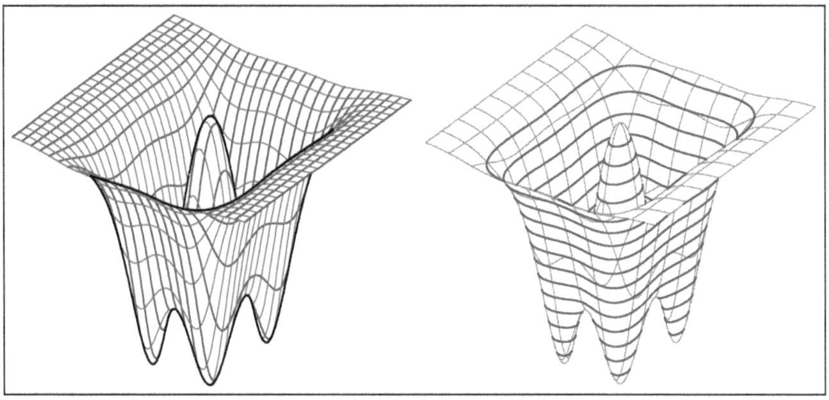

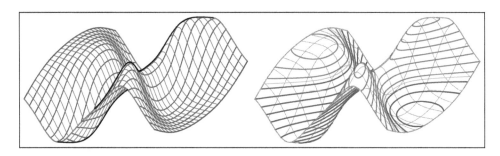

Figure 2.48 The Gaussian curvature of the explicit surface S of Visualization 2.5.23 (a) represented as an explicit surface (left) and its lines of constant Gaussian curvature (right)

Figure 2.49 The mean curvature of the explicit surface S of Visualization 2.5.23 (a) represented as an explicit surface (left) and its lines of constant mean curvature (right)

and the Gaussian and mean curvature of S are

$$K(u^i) = -\frac{9\left((u^1)^2 + (u^2)^2\right)^2}{\left(1 + ((u^1)^2 + (u^2)^2)^3\right)^2} \tag{2.89}$$

and

$$H(u^i) = -\frac{6u^1u^2\left((u^1)^4 - (u^2)^4\right)}{\left(1 + ((u^1)^2 + (u^2)^2)^3\right)^{3/2}}. \tag{2.90}$$

Figure 2.50 shows the surface S, and lines of constant Gaussian and mean curvature on S.

Figure 2.51 shows the surface S with its level lines, and its level lines in the parameter plane.

Figure 2.52 represents the Gaussian and mean curvature of S as explicit surfaces as in Part (a), and the lines $K(u^i) = const$ and $H(u^i) = const$.

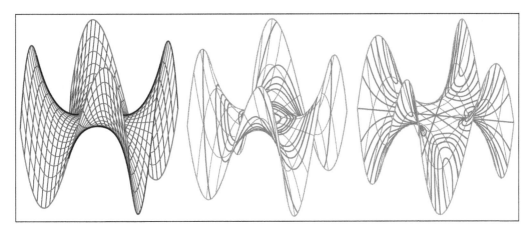

Figure 2.50 The explicit surface S of Visualization 2.5.23 (b) (left), lines of constant Gaussian (middle) and constant mean curvature (right)

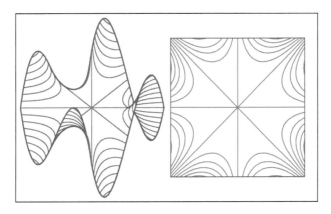

Figure 2.51 The explicit surface of Visualization 2.5.23 (b)

Figure 2.53 shows the ruled surface generated by the surface normal vectors along a curve on the surface of Part (b).

Proof. We write $x = u^1$, $y = u^2$.

(a) We put $f(x, y) = (y^2)^3 - ax^2y$.
   Then we have
   $$f_1(x, y) = -2axy, \ f_2(x, y) = 3y^2 - ax^2,$$
   $$f_{11}(x, y) = -2ay, \ f_{12}(x, y) = -2ax \text{ and } f_{22}(x, y) = 6y,$$

   and obtain for the first and second fundamental coefficients by (2.21) in Example 2.3.4 and (2.49) in Example 2.4.5

   $$g_{11}(x, y) = 1 + 4a^2 (xy)^2, \ g_{12}(x, y) = -2axy \left(3y^2 - ax^2\right),$$

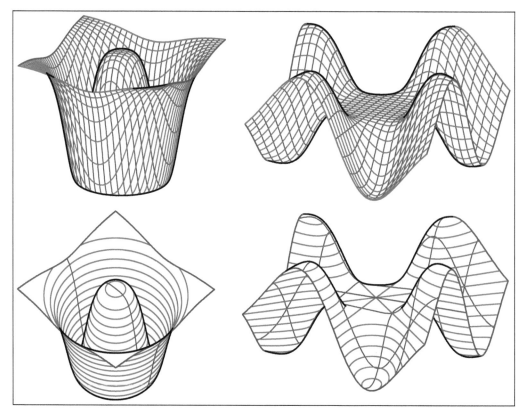

Figure 2.52 Top: the Gaussian (left) and mean curvature (right) of the surface in Visualization 2.5.23 (a) as explicit surfaces. Bottom: the lines of constant Gaussian (left) and constant mean mean curvature (right) in Part (a), and the lines $K(u^i) = const$ and $H(u^i) = const$ on the surfaces on the top.

$$g_{22}(x,y) = 1 + \left(3y^2 - ax^2\right)^2, \ g(x,y) = 1 + 4a^2(xy)^2 + \left(3y^2 - ax^2\right)^2,$$

$$L_{11}(x,y) = -\frac{2ay}{\sqrt{g(x,y)}}, \ L_{12}(x,y) = -\frac{2ax}{\sqrt{g(x,y)}}, \ L_{22}(x,y) = \frac{6y}{\sqrt{g(x,y)}}$$

and

$$L(u^i) = -\frac{12ay^2 + 4a^2x^2}{g(x,y)}.$$

This yields

$$K(x,y) = \frac{L(x,y)}{g(x,y)} = -\frac{12ay^2 + 4a^2x^2}{(g(x,y))^2},$$

that is, (2.87), and

$$H(x,y) = \frac{1}{2g(x,y)}\left(L_{11}(x,y)g_{22}(x,y) - 2L_{12}(x,y)g_{12}(x,y) + L_{22}(x,y)g_{11}(x,y)\right)$$

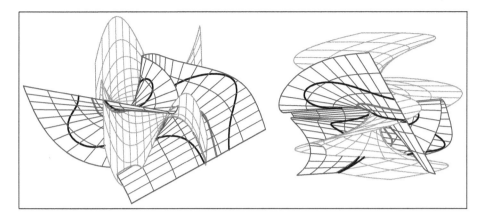

Figure 2.53 Ruled surfaces generated by the surface normal vectors along a curve on the surface of Visualization 2.5.23 (b)

$$= \frac{1}{(g(x,y))^{3/2}} \left[-2ay \left(1 + \left(3y^2 - ax^2 \right)^2 \right) \right.$$
$$\left. - 8a^2 x^2 y \left(3y^2 - ax^2 \right) - 6y \left(1 + 4a^2 (xy)^2 \right) \right],$$

that is, (2.88).

(b) We put $f(x,y) = x^3 y - xy^3$ and obtain

$$f_1(x,y) = 3x^2 y - y^3, \ f_2(x,y) = x^3 - 3xy^2,$$
$$f_{11}(x,y) = 6xy, \ f_{12}(x,y) = 3x^2 - 3y^2 \text{ and } f_{22}(x,y) = -6xy,$$

and as in Part (a)

$$g_{11}(x,y) = 1 + \left(y(3x^2 - y^2) \right)^2,$$
$$g_{12}(x,y) = xy \left(3x^2 - y^2 \right) \left(x^2 - 3y^2 \right),$$
$$g_{22}(x,y) = 1 + \left(x(x^2 - 3y^2) \right)^2,$$

$$g(x,y) = 1 + \left(y(3x^2 - y^2) \right)^2 + \left(x(x^2 - 3y^2) \right)^2$$
$$= 1 + y^2(9x^4 - 6x^2 y^2 + y^4) + x^2(9y^4 - 6x^2 y^2 + x^4)$$
$$= 1 + 9x^4 y^2 - 6x^2 y^4 + y^6 + 9x^2 y^4 - 6x^4 y^2 + x^6$$
$$= 1 + x^6 + 3x^4 y^2 + 3x^2 y^4 + y^6 = 1 + (x^2 + y^2)^3,$$

$$L_{11}(x,y) = \frac{6xy}{\sqrt{g(x,y)}} = -L_{22}(x,y), \ L_{12}(x,y) = \frac{3(x^2 - y^2)}{\sqrt{g(x,y)}}$$

and

$$L(x,y) = -9 \frac{4x^2 y^2 + (x^2 - y^2)^2}{g(x,y)} = -\frac{(x^2 + y^2)^2}{1 + (x^2 + y^2)^3}.$$

This implies

$$K(x,y) = \frac{L(x,y)}{g(x,y)} = -9\frac{(x^2+y^2)^2}{(1+(x^2+y^2)^3)^2},$$

that is, (2.89), and

$$H(x,y) = \frac{1}{2g(x,y)}(L_{11}(x,y)g_{22}(x,y) - 2L_{12}(x,y)g_{12}(x,y) + L_{22}(x,y)g_{11}(x,y))$$

$$= \frac{1}{(g(x,y))^{3/2}}\Big[6xy\left(1 + x^2(x^2 - 3y^2)^2\right)$$

$$-6xy(x^2 - y^2)(3x^2 - y^2)(x^2 - 3y^2)\Big]$$

$$- 6xy\left(1 + y^2(3x^2 - y^2)^2\right)\Big]$$

$$= \frac{3xy}{(g(x,y))^{3/2}}\Big[x^2(x^2 - 3y^2)^2 - x^2(3x^2 - y^2)(x^2 - 3y^2)$$

$$+y^2(3x^2 - y^2)(x^2 - 3y^2) - y^2(3x^2 - y^2)^2\Big]$$

$$= \frac{3xy}{(g(x,y))^{3/2}}\Big[x^2(x^2 - 3y^2)\left(x^2 - 3y^2 - (3x^2 - y^2)\right)$$

$$+y^2(3x^2 - y^2)\left(x^2 - 3y^2 - (3x^2 - y^2)\right)\Big]$$

$$= -\frac{6xy(x^2 + y^2)}{(g(x,y))^{3/2}}\left(x^2(x^2 - 3y^2) + y^2(3x^2 - y^2)\right)$$

$$= -\frac{6xy(x^2 + y^2)(x^4 - y^4)}{(g(x,y))^{3/2}} = -\frac{6xy(x^2 + y^2)(x^4 - y^4)}{(1 + (x^2 + y^2)^2)^{3/2}},$$

that is, (2.90).

□

Remark 2.5.24. *(a) The mean curvature at most changes its sign under parameter transformations.*
(b) The Gaussian curvature is invariant under parameter transformations.

Proof. Let a parameter transformation be given by $u^i = u^i(u^{*1}, u^{*2})$ for $i = 1, 2$. Then the formulae of transformations for g^{ik} and L_{ik} are by (2.19) in Remark 2.3.3 (b) and (2.45) in Remark 2.4.3 (b)

$$g^{*ik} = g^{jn}\frac{\partial u^{*i}}{\partial u^j}\frac{\partial u^{*k}}{\partial u^n} \text{ and } L_{ik}^* = \alpha L_{lm}\frac{\partial u^l}{\partial u^{*i}}\frac{\partial u^m}{\partial u^{*k}} \text{ for } i, k = 1, 2,$$

where

$$\alpha = \mathrm{sign}\left(\frac{\partial(u^i)}{\partial(u^{*k})}\right).$$

(a) We obtain for the mean curvature

$$2H^* = g^{*ik}L_{ik} = \alpha g^{jn}L_{lm}\frac{\partial u^{*i}}{\partial u^j}\frac{\partial u^{*k}}{\partial u^n}\frac{\partial u^l}{\partial u^{*i}}\frac{\partial u^m}{\partial u^{*k}}$$

$$= \alpha g^{jn}L_{lm}\frac{\partial u^l}{\partial u^{*i}}\frac{\partial u^{*i}}{\partial u^j}\frac{\partial u^m}{\partial u^{*k}}\frac{\partial u^{*k}}{\partial u^n}$$

$$= \alpha g^{jn}L_{lm}\delta_j^l\delta_n^m = \alpha g^{jn}L_{jn}\delta_j^l = \alpha 2H.$$

(b) Let A_{ik} and A_{ik}^* satisfy the formulae of transformation

$$A_{ik}^* = A_{lm}\frac{\partial u^l}{\partial u^{*i}}\frac{\partial u^m}{\partial u^{*k}} \text{ for } i,k = 1,2.$$

Writing

$$D = \begin{pmatrix} \dfrac{\partial u^1}{\partial u^{*1}} & \dfrac{\partial u^1}{\partial u^{*2}} \\ \dfrac{\partial u^2}{\partial u^{*1}} & \dfrac{\partial u^2}{\partial u^{*2}} \end{pmatrix}$$

and D^T for the transpose of D, we obtain

$$(AD)_{lk} = \sum_{m=1}^{2} A_{lm}D_{mk} = A_{lm}\frac{\partial u^m}{\partial u^{*k}} \text{ for } l,k = 1,2$$

and

$$(D^T(AD))_{ik} = \sum_{l=1}^{2} D_{il}^T(AD)_{lk} = \frac{\partial u^l}{\partial u^{*i}}A_{lm}\frac{\partial u^m}{\partial u^{*k}}$$

$$= A_{lm}\frac{\partial u^l}{\partial u^{*i}}\frac{\partial u^m}{\partial u^{*k}} = A_{ik}^* \text{ for } i,k = 1,2,$$

hence

$$\det A^* = \det(D^T A D) = (\det D)^2 \det A.$$

Therefore, we have

$$\det g^* = (\det D)^2 \det g, \ \det L^* = (\det D)^2 \det L$$

and

$$K^* = \frac{\det L^*}{\det g^*} = \frac{L}{g} = K$$

for the Gaussian curvature.

□

Example **2.5.25 (The umbilical points on spheres).**
All points of a sphere are umbilical points.

Proof. Let S be a sphere with a parametric representation

$$\vec{x}(u^i) = r\{\cos u^1 \cos u^2, \cos u^1 \sin u^2, \sin u^1\} \quad \left((u^1, u^2) \in D = (-\pi/2, \pi/2) \times (0, 2\pi)\right).$$

Then we have

$$g_{11}(u^k) = r^2, \ g_{12}(u^k) = 0, \ g_{22}(u^k) = r^2 \cos^2 u^1,$$

$$L_{11}(u^k) = r, \ L_{12}(u^k) = 0 \text{ and } L_{22}(u^k) = r \cos^2 u^1,$$

hence

$$g_{ik}(u^k) = r \cdot L_{ik}(u^k) \quad (i, k = 1, 2).$$

This is the characterization of umbilical points, by the identities in (2.73) of Remark 2.5.16 (b). □

Visualization 2.5.26. (The umbilical points on a hyperboloid of two sheets)

We determine the umbilical points on a hyperboloid of two sheets given by a parametric representation

$$\vec{x}(u^i) = \{a \sinh u^1 \cos u^2, b \sinh u^1 \sin u^2, \pm c \cosh u^1\} \quad \left((u^1, u^2) \in D \subset \mathbb{R} \times (0, 2\pi)\right),$$

where $a, b, c \in \mathbb{R} \setminus \{0\}$ are constants.
(a) If $a = b = c$, then there are no umbilical points.
(b) If $a = b > c$, then every point on the circle lines

$$\vec{x}(t) = \left\{ \frac{a^2}{c} \cos t, \frac{a^2}{c} \sin t, \pm\sqrt{c^2 + a^2} \right\}$$

is an umbilical point.
(c) If $a > b \neq c$, then there are four umbilical points at

$$X = \left(0, \pm b\sqrt{\frac{a^2 - b^2}{b^2 + c^2}}, \pm c\sqrt{\frac{c^2 + a^2}{b^2 + c^2}} \right)$$

(Figure 2.54).

Proof. We obtain

$$\vec{x}_1(u^i) = \{a \cosh u^1 \cos u^2, b \cosh u^1 \sin u^2, \pm c \sinh u^1\},$$

$$\vec{x}_2(u^i) = \sinh u^1 \{-a \sin u^2, b \cos u^2, 0\},$$

$$\vec{x}_{11}(u^i) = \vec{x}(u^i),$$

$$\vec{x}_{12}(u^i) = \cosh u^1 \{-a \sin u^2, b \cos u^2, 0\},$$

$$\vec{x}_{22}(u^i) = -\sinh u^1 \{a \cos u^2, b \cos u^2, 0\},$$

$$g_{11}(u^i) = \cosh^2 u^1 \left(a^2 \cos^2 u^2 + b^2 \sin^2 u^2\right) + c^2 \sinh^2 u^1,$$

$$g_{11}(u^i) = (b^2 - a^2) \sinh u^1 \cosh u^1 \sin u^2 \cos u^2,$$

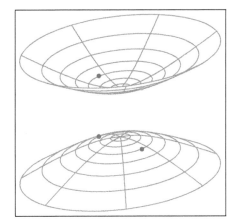

Figure 2.54 The umbilical points of a hyperboloid of two sheets for distinct a, b and c

$$g_{22}(u^i) = \sinh^2 u^1 \left(a^2 \sin^2 u^2 + b^2 \cos^2 u^2 \right),$$

$$L_{11}(u^i) = \pm \frac{abc \sinh u^1}{\sqrt{g(u^i)}}, \quad L_{12}(u^i) = 0 \text{ and } L_{22}(u^i) = \pm \frac{abc \sinh^3 u^1}{\sqrt{g(u^i)}}.$$

(a) Let $a = b = c$.

Then we have

$$g_{11}(u^i) = a^2 \left(\cosh^2 u^1 + \sinh^2 u^2 \right), \quad g_{12}(u^i) = 0, \quad g_{22}(u^i) = a^2 \sinh^2 u^1,$$

$$L_{11}(u^i) = \pm \frac{a^3 \sinh u^1}{\sqrt{g(u^i)}}, \quad L_{12}(u^i) = 0 \text{ and } L_{22}(u^i) = \pm \frac{a^3 \sinh^3 u^1}{\sqrt{g(u^i)}}.$$

The conditions in (2.73) for umbilical points now are

$$\pm \frac{a^3 \sinh u^1}{\sqrt{g(u^i)}} = ka^2 \left(\cosh^2 u^1 + \sinh^2 u^2 \right) \text{ and } \pm \frac{a^3 \sinh^3 u^1}{\sqrt{g(u^i)}} = ka^2 \sinh^2 u^1$$

for all $(u^i, u^2) \in D$ and some constant $k \in \mathbb{R}$. This implies

$$(\cosh^2 u^1 + \sinh^2 u^1) \sinh^2 u^1 = \sinh^2 u^1,$$

hence $u^1 = 0$, since $\cosh^2 u^1 + \sinh^2 u^1 > 0$ for $u^1 \neq 0$; but $\vec{x}_2(0, u^2) = \vec{0}$, and so the parametric representation is not admissible for $u^1 = 0$.

Thus there are no umbilical points.

(b) Now let $a = b > c$.

Then we have

$$g_{11}(u^i) = a^2 \cosh^2 u^1 + c^2 \sinh^2 u^2, \quad g_{12}(u^i) = 0, \quad g_{22}(u^i) = a^2 \sinh^2 u^1,$$

$$L_{11}(u^i) = \pm \frac{a^2 c \sinh u^1}{\sqrt{g(u^i)}}, \quad L_{12}(u^i) = 0 \text{ and } L_{22}(u^i) = \pm \frac{a^2 c \sinh^3 u^1}{\sqrt{g(u^i)}},$$

and the conditions for umbilical points yields

$$\left(a^2 \cosh^2 u^1 + c^2 \sinh^2 u^2 \right) \sinh^2 u^1 = a^2 \sinh^2 u^1,$$

hence $u^1 = 0$ or $a^2 - c^2 \sinh^2 u^1 = 0$, which implies $\sinh u^1 = \pm a/c$.

Thus every point on the circle lines

$$\vec{x}(t) = \left\{ \frac{a^2}{c}\cos t, \frac{a^2}{c}\sin t, \pm\sqrt{c^2 + a^2} \right\} \quad (t \in (0, 2\pi))$$

is an umbilical point.

(c) Finally, let $a > b \neq c$.
Then we have

$$g_{11}(u^i) = \cosh^2 u^1 \left(a^2 \cos^2 u^2 + b^2 \sin^2 u^2 \right) + c^2 \sinh^2 u^1,$$

$$g_{12}(u^i) = (b^2 - a^2)\sinh u^1 \cosh u^1 \sin u^2 \cos u^2,$$

$$g_{22}(u^i) = \sinh^2 u^1 \left(a^2 \sin^2 u^2 + b^2 \cos^2 u^2 \right),$$

$$L_{11}(u^i) = \pm\frac{abc \sinh u^1}{\sqrt{g(u^i)}}, \quad L_{12}(u^i) = 0 \text{ and } L_{22}(u^i) = \pm\frac{abc \sinh^3 u^1}{\sqrt{g(u^i)}}.$$

First $L_{12}(u^i) = kg_{12}(u^i)$ implies $k = 0$ or $u^2 = 0, \pi/2, \pi, 3\pi/2$, since $a^2 \neq b^2$ and $u^1 \neq 0$. If $k = 0$, then $L_{11}(u^1) = 0$ implies $u^1 = 0$, but $u^1 \neq 0$, since the parametric representation of the hyperboloid is not admissible at $u^1 = 0$. Thus k has to be unequal to zero. Furthermore, since

$$g_{ij}(u^1, 0) = g_{ij}(u^1, \pi), \quad L_{ij}(u^1, 0) = L_{ij}(u^1, \pi),$$
$$g_{ij}(u^1, \pi/2) = g_{ij}(u^1, 3\pi/2) \text{ and } L_{ij}(u^1, \pi/2) = L_{ij}(u^1, 3\pi/2)$$

for $i, j = 1, 2$, we only have to study the cases $u^2 = 0$ and $u^2 = \pi/2$.

(c.i) Let $u^2 = 0$.
Then we have

$$g_{11}(u^1, 0) = a^2 \cosh^2 u^1 + c^2 \sinh^2 u^1 \text{ and } g_{22}(u^1, 0) = b^2 \sinh^2 u^1,$$

and

$$L_{11}(u^1, 0) = \pm\frac{abc \sinh u^1}{\sqrt{g(u^1, 0)}} = kg_{11}(u^1, 0) = k(a^2 \cosh^2 u^1 + c^2 \sinh^2 u^1)$$

and

$$L_{22}(u^1, 0) = \pm\frac{abc \sinh^3 u^1}{\sqrt{g(u^1, 0)}} = kg_{22}(u^1, 0) = kb^2 \sinh^2 u^1$$

imply $a^2 \cosh^2 u^1 + c^2 \sinh^2 u^1 = b^2$, since $u^1 \neq 0$. But $a^2 \cosh^2 u^1 - b^2 > 0$ for all u^1, and so there is no umbilical point with $u^2 = 0$.

(c.ii)   Finally, let $u^2 = \pi/2$.
Then we have

$$g_{11}(u^1, \pi/2) = b^2 \cosh^2 u^1 + c^2 \sinh^2 u^1 \text{ and } g_{22}(u^1, \pi/2) = a^2 \sinh^2 u^1,$$

and

$$L_{11}(u^1, \pi/2) = \pm \frac{abc \sinh u^1}{\sqrt{g(u^1, \pi/2)}} = kg_{11}(u^1, \pi/2)$$
$$= k(b^2 \cosh^2 u^1 + c^2 \sinh^2 u^1)$$

and

$$L_{22}(u^1, \pi/2) = \pm \frac{abc \sinh^3 u^1}{\sqrt{g(u^1, 0)}} = kg_{22}(u^1, \pi/2) = ka^2 \sinh^2 u^1$$

imply $b^2 \cosh^2 u^1 + c^2 \sinh^2 u^1 = a^2$, that is, $(b^2 + c^2) \sinh^2 u^1 = a^2 - b^2 > 0$
or

$$\sinh u^1 = \pm \sqrt{\frac{a^2 - b^2}{b^2 + c^2}} \text{ and } \cosh u^1 = \pm \sqrt{\frac{c^2 + a^2}{b^2 + c^2}}.$$

Therefore, there are four umbilical points on the hyperboloid of two sheets, at

$$X = \left(0, \pm b \sqrt{\frac{a^2 - b^2}{b^2 + c^2}}, \pm c \sqrt{\frac{c^2 + a^2}{b^2 + c^2}}\right).$$

□

Visualization 2.5.27 (Principal curvature on a hyperbolic paraboloid).

We consider the hyperbolic paraboloid *given by a parametric representation*

$$\vec{x}(u^i) = \{u^1, u^2, u^1 u^2\} \quad \left((u^1, u^2) \in \mathbb{R}^2\right).$$

Then it is easy to see that

$$\vec{x}_1(u^k) = \{1, 0, u^2\}, \ \vec{x}_2(u^k) = \{0, 1, u^1\},$$
$$g_{11}(u^k) = 1 + (u^2)^2, \ g_{12}(u^k) = u^1 u^2,$$
$$g_{22}(u^k) = 1 + (u^1)^2, \ g(u^k) = 1 + (u^1)^2 + (u^2)^2,$$
$$L_{11}(u^k) = L_{22}(u^k) = 0, \ L_{12}(u^k) = \frac{1}{\sqrt{g(u^k)}} \text{ and } L(u^k) = -\frac{1}{g(u^k)}.$$

The principal curvature is given by

$$\kappa_{1,2}(u^k) = \frac{-u^1 u^2 \pm \sqrt{(1 + (u^1)^2)(1 + (u^2)^2)}}{(1 + (u^1)^2 + (u^2)^2)^{3/2}} = \frac{-g_{12}(u^k) \pm \sqrt{g_{11}(u^k)g_{22}(u^k)}}{(g(u^k))^{3/2}}.$$

$$(2.91)$$

There are no umbilical points, since $\kappa_1(u^k) \neq \kappa_2(u^k)$ for all $(u^1, u^2) \in \mathbb{R}^2$.
The mean and Gaussian curvature are given by

$$H(u^k) = -\frac{u^1 u^2}{(1 + (u^1)^2 + (u^2)^2)^{3/2}} \quad \text{and} \quad K(u^k) = -\frac{1}{(1 + (u^1)^2 + (u^2)^2)^2}.$$

(a) *First we find the principal curvature along a curve on the hyperbolic paraboloid.*
We omit the parameters u^1 and u^2 and obtain from (2.91)

$$\kappa_{1,2} = \frac{-g_{12} \pm \sqrt{g_{11} g_{22}}}{g^{3/2}}.$$

By (2.63), the principal direction corresponding to κ_1 is given by

$$\left\{ \begin{array}{c} \kappa_1 g_{11} \xi_1^1 = (L_{12} - \kappa_1 g_{12}) \xi_1^2 \\ (L_{12} - \kappa_1 g_{12}) \xi_1^1 = \kappa_1 g_{22} \xi_1^2 \\ \text{and} \\ g_{11}(\xi_1^1)^2 + 2g_{12}\xi_1^1 \xi_1^2 + g_{22}(\xi_1^2)^2 = 1. \end{array} \right\} \tag{2.92}$$

It follows from the first equation on the left hand side of (2.92) that

$$\xi_1^1 = \frac{L_{12} - \kappa_1 g_{12}}{\kappa_1 g_{11}} \xi_1^2 = \frac{\dfrac{1}{\sqrt{g}} + \dfrac{g_{12}^2 - g_{12}\sqrt{g_{11}g_{22}}}{g^{3/2}}}{g_{11} \dfrac{-g_{12} + \sqrt{g_{11}g_{22}}}{g^{3/2}}} \xi_1^2$$

$$= \frac{g + g_{12}^2 - g_{12}\sqrt{g_{11}g_{22}}}{g_{11}\left(-g_{12} + \sqrt{g_{11}g_{22}}\right)} \xi_1^2 = \frac{g_{11}g_{22} - g_{12}\sqrt{g_{11}g_{22}}}{g_{11}\left(\sqrt{g_{11}g_{22}} - g_{12}\right)} \xi_1^2$$

$$= \frac{\sqrt{g_{11}g_{22}}\left(\sqrt{g_{11}g_{22}} - g_{12}\right)}{g_{11}\left(\sqrt{g_{11}g_{22}} - g_{12}\right)} \xi_1^2 = \sqrt{\frac{g_{22}}{g_{11}}} \xi_1^2.$$

Substituting this in the equation on the right hand side of (2.92), we obtain

$$g_{11}(\xi_1^1)^2 + 2g_{12}\xi_1^1\xi_1^2 + g_{22}(\xi_1^2)^2 = g_{22}(\xi_1^2)^2 + 2g_{12}\sqrt{\frac{g_{22}}{g_{11}}}(\xi_1^2)^2 + g_{22}(\xi_1^2)^2$$

$$= 2\left(g_{22} + g_{12}\sqrt{\frac{g_{22}}{g_{11}}}\right)(\xi_1^2)^2 = 1,$$

and we may choose

$$\xi_1^2 = \frac{1}{\sqrt{2}\sqrt{g_{22} + g_{12}\sqrt{\dfrac{g_{22}}{g_{11}}}}} = \frac{1}{\sqrt{2}\sqrt{\sqrt{\dfrac{g_{22}}{g_{11}}}\left(\sqrt{g_{11}g_{22}} + g_{12}\right)}}$$

and

$$\xi_1^1 = \sqrt{\frac{g_{22}}{g_{11}}} \frac{1}{\sqrt{2}\sqrt{g_{22} + g_{12}\sqrt{\frac{g_{22}}{g_{11}}}}} = \frac{1}{\sqrt{2}\sqrt{g_{11} + g_{12}\sqrt{\frac{g_{11}}{g_{22}}}}}$$

$$= \frac{1}{\sqrt{2}\sqrt{\sqrt{\frac{g_{11}}{g_{22}}}\left(\sqrt{g_{11}g_{22}} + g_{12}\right)}}$$

as the principal direction corresponding to κ_1.
Similarly we may choose

$$\xi_2^2 = \frac{1}{\sqrt{2}\sqrt{g_{22} - g_{12}\sqrt{\frac{g_{22}}{g_{11}}}}} = \frac{1}{\sqrt{2}\sqrt{\sqrt{\frac{g_{22}}{g_{11}}}\left(\sqrt{g_{11}g_{22}} - g_{12}\right)}}$$

and

$$\xi_2^1 = -\sqrt{\frac{g_{22}}{g_{11}}} \frac{1}{\sqrt{2}\sqrt{g_{22} - g_{12}\sqrt{\frac{g_{22}}{g_{11}}}}} = -\frac{1}{\sqrt{2}\sqrt{g_{11} - g_{12}\sqrt{\frac{g_{11}}{g_{22}}}}}$$

$$= -\frac{1}{\sqrt{2}\sqrt{\sqrt{\frac{g_{11}}{g_{22}}}\left(\sqrt{g_{11}g_{22}} - g_{12}\right)}}$$

as the principal direction corresponding to κ_2; we observe that since $g_{11}g_{22} > g_{12}^2$, $g_{11}, g_{22} > 0$ and $g_{12}^2 \geq 0$ for any surface, the expressions above are well defined and real.

(b) *We may represent the principal curvature along a given curve γ on the hyperbolic paraboloid as a curve γ^* on the ruled surface generated by γ and the principal direction corresponding to the principal curvature as follows. If γ is given by $(u^1(t), u^2(t))$, and $\kappa_\mu(t) = \kappa(u^i(t))$ and $\vec{d}_\mu(t) = \vec{d}_\mu(u^i(t))$ $(\mu = 1, 2)$ denote the principal curvature and corresponding principal direction, then we write $u^{*1} = t$ and consider the ruled surfaces RS^μ $(\mu = 1, 2)$ given by the parametric representations*

$$\vec{x}^{*(\mu)}(u^{*i}) = \vec{x}((u^1(u^{*1}), u^2(u^{*1}))) + u^{*2}\vec{d}_\mu(u^1(u^{*1}), u^2(u^{*1})) \text{ for } \mu = 1, 2,$$
$$(2.93)$$

and represent $\kappa_\mu(u^i(t))$ as a curve γ_μ^* on RS^μ by substituting $u^{*2(\mu)} = u^{*2}(t) = \kappa_\mu(u^i(t))$ in (2.93). Thus each of the curves γ_μ^* for $\mu = 1, 2$ is given by a parametric representation

$$\vec{x}^{*(\mu)}(t) = \vec{x}^{*(\mu)}(u^{*i}(t))$$
$$= \vec{x}(u^1(u^1(u^{*1}), u^2(u^{*2}))) + \kappa_\mu(u^1(t), u^2(t))\vec{d}_\mu(u^1(u^{*1}), u^2(u^{*1})).$$
$$(2.94)$$

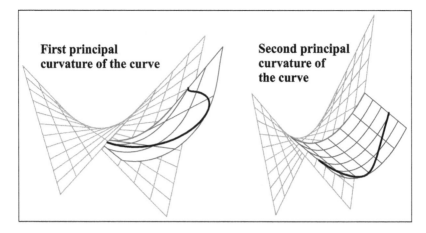

Figure 2.55 Representation of the first and second principal curvature along a curve on a hyperbolic paraboloid

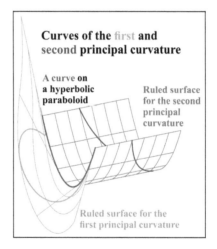

Figure 2.56 Representation of the first and second principal curvature on the corresponding ruled surfaces

Figures 2.55 and 2.56 show the representations of the first and second principal curvature along the curve γ given by $u^1(t) = t$ and $u^2(t) = t$.

(c) *Figure 2.57 shows the representations of the Gaussian and mean curvature along γ as curves on the ruled surface generated by γ and the surface normal vectors of the hyperbolic paraboloid along γ.*

The next result is useful to find the normal, principal, Gaussian and mean curvature of a surfaces that are given by an equation.

Theorem 2.5.28. *Let S be a surface which is given by an equation $F(x^1, x^2, x^3) = 0$.*
(a) Then the values of the principal curvature are

$$\kappa_i = -\frac{\lambda_i}{F_1^2 + F_2^2 + F_3^2} \; for \; i = 1, 2, \tag{2.95}$$

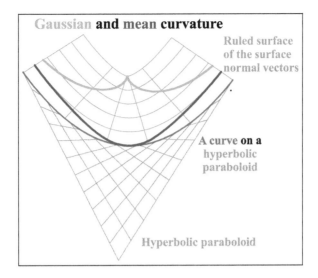

Figure 2.57 Representation of the Gaussian and mean curvature of a curve on a hyperbolic paraboloid

where the values λ_i are the solutions of the quadratic equation for λ

$$\det \begin{pmatrix} F_{11} - \lambda & F_{12} & F_{13} & F_1 \\ F_{21} & F_{22} - \lambda & F_{23} & F_2 \\ F_{31} & F_{32} & F_{33} - \lambda & F_3 \\ F_1 & F_2 & F_3 & 0 \end{pmatrix} = 0. \tag{2.96}$$

(b) A direction given by the unit vector $\vec{d} = \{\xi^1, \xi^2, \xi^3\}$ is a principal direction if and only if

$$\det \begin{pmatrix} F_1 & \xi^k F_{1k} & \xi^1 \\ F_2 & \xi^k F_{2k} & \xi^2 \\ F_3 & \xi^k F_{3k} & \xi^3 \end{pmatrix} = 0 \text{ and } \xi^k F_k = 0. \tag{2.97}$$

(c) Let A, B and C denote the coefficients of λ^2, λ and the constant term in the quadratic equation (2.96). Then the Gaussian and mean curvature are given by

$$K = \frac{C}{A} \cdot \frac{1}{F_1^2 + F_2^2 + F_3^2} \text{ and } H = \frac{-B}{2A} \cdot \frac{1}{\sqrt{F_1^2 + F_2^2 + F_3^2}}. \tag{2.98}$$

Proof. The normal curvature of a curve γ with a parametric representation $\vec{x}(s)$ on a surface is given by $\kappa_n(s) = \ddot{\vec{x}}(s) \bullet \vec{N}(s)$, and $\dot{\vec{x}}(s) \bullet \vec{N}(s) = 0$ implies $\ddot{\vec{x}}(s) \bullet \vec{N}(s) + \dot{\vec{x}}(s) \bullet \dot{\vec{N}}(s) = 0$, hence $\kappa_n(s) = -\dot{\vec{x}}(s) \bullet \dot{\vec{N}}(s)$. If the surface S is given by an equation $F(x^1, x^2, x^3) = 0$, then the gradient of F and the surface normal vector of S at s are linearly dependent, that is,

$$\frac{\mathrm{grad} F(s)}{\|\mathrm{grad} F(s)\|} = \pm \vec{N}(s).$$

This implies

$$\pm \dot{\vec{N}}(s) = \frac{d}{ds}\left(\frac{\operatorname{grad}F(s)}{\|\operatorname{grad}F(s)\|}\right)$$

$$= \frac{\partial \operatorname{grad}F(s)}{\partial x^k}\dot{x}^k(s)\frac{1}{\|\operatorname{grad}F(s)\|} + \operatorname{grad}F(s)\frac{d}{ds}\left(\frac{1}{\|\operatorname{grad}F(s)\|}\right),$$

and

$$\dot{\vec{x}}(s) \bullet \dot{\vec{N}}(s) = \pm\frac{1}{\|\operatorname{grad}F(s)\|}\dot{\vec{x}}(s) \bullet \dot{x}^k(s)\frac{\partial \operatorname{grad}F(s)}{\partial x^k}. \tag{2.99}$$

(a) First we show (2.95).
Let $\vec{d} = \{\xi^1, \xi^2, \xi^3\}$ be the direction of a tangent, that is, $\|\vec{d}\| = 1$ and $\operatorname{grad}F \bullet \vec{d} = \xi^k F_k = 0$. To find the principal directions, by (2.99), we have to find the extreme values of the function

$$G(\xi^1, \xi^2, \xi^3) = \xi^k\xi^i F_{ik} \text{ on the set}$$
$$M = \left\{\{\xi^1, \xi^2, \xi^3\} : \xi^k F_k = 0 \text{ and } \|\{\xi^1, \xi^2, \xi^3\}\| = 1\right\}.$$

Using Lagrange's multipliers, we obtain

$$\frac{\partial G(\xi^j)}{\partial \xi^m} = \frac{\partial(\xi^i\xi^k F_{ik})}{\partial \xi^m} = \lambda\frac{\partial\left(\sum_{i=1}^{3}(\xi^i)^2\right)}{\partial \xi^m} + \mu\frac{\partial\left(\xi^i F_i\right)}{\partial \xi^m},$$

that is,

$$\xi^k F_{mk} = \lambda\xi^m + \frac{\mu}{2}F_m \text{ for } m = 1, 2, 3. \tag{2.100}$$

We multiply (2.100) by ξ^m and sum to obtain

$$\xi^m\xi^k F_{mk} = \lambda\sum_{m=1}^{3}(\xi^m)^2 + \frac{\mu}{2}\xi^m F_m = \lambda.$$

Putting $t = \mu/2$, we can rewrite the last equation as

$$\begin{pmatrix} F_{11} - \lambda & F_{12} & F_{13} & F_1 \\ F_{21} & F_{22} - \lambda & F_{23} & F_2 \\ F_{31} & F_{32} & F_{33} - \lambda & F_3 \\ F_1 & F_2 & F_3 & 0 \end{pmatrix}\begin{pmatrix} \xi^1 \\ \xi^2 \\ \xi^3 \\ t \end{pmatrix} = 0, \tag{2.101}$$

and $\{\xi^1, \xi^2, \xi^3, t\}$ is a solution of (2.101) if and only if (2.96) holds; (2.96) is a quadratic equation for λ and, by (2.99), the values of the principal curvature are as stated in (2.95).

(b) Now we show (2.97).
We eliminate λ and t in (2.101). First, we multiply the first equation in (2.101) by F_2 and the second equation in (2.101) by F_1 and obtain

$$\xi^k F_{1k}F_2 - \lambda\xi^1 F_2 + tF_1 F_2 = 0 \text{ and } \xi^k F_{2k}F_1 - \lambda\xi^2 F_1 + tF_1 F_2 = 0;$$

subtraction of the second from the first equation now yields

$$\xi^k(F_{1k}F_2 - F_{2k}F_1) - \lambda(\xi^1 F_2 - \xi^2 F_1) = 0. \tag{2.102}$$

Similarly, multiplication of the first and second equations in (2.101) by ξ^2 and ξ^1, respectively, followed by subtraction, yields

$$\xi^1 \xi^k F_{2k} - \xi^2 \xi^k F_{1k} + t(\xi^1 F_2 - \xi^2 F_1) = 0. \tag{2.103}$$

Now we multiply the third equation in (2.101) by $\xi^1 F_2 - \xi^2 F_1$ and substitute (2.102) in (2.103) to obtain

$$(\xi^1 F_2 - \xi^2 F_1)\xi^k F_{3k} + \xi^k(F_{2k}F_1 - F_{1k}F_2) + (\xi^2 \xi^k F_{1k} - \xi^1 \xi^k F_{2k})$$

$$= \xi^1(\xi^k F_{3k}F_2 - \xi^k F_{2k}F_3) - \xi^2(\xi^k F_{3k}F_1 - \xi^k F_{1k}F_3) + \xi^3(\xi^k F_{2k}F_1 - \xi^k F_{12}F_2)$$

$$= -\det \begin{pmatrix} \xi^1 & \xi^k F_{1k} & F_1 \\ \xi^2 & \xi^k F_{2k} & F_2 \\ \xi^3 & \xi^k F_{3k} & F_3 \end{pmatrix} = 0.$$

This and the last equation in (2.101) are equivalent to (2.97).

(c) Finally, we prove the formulae for K and H.
Writing (2.96) as $A\lambda^2 + B\lambda + C = 0$, we obtain from Vieta's rule

$$\lambda_1 \lambda_2 = \frac{C}{A} \quad \text{and} \quad \lambda_1 + \lambda_2 = -\frac{B}{A},$$

hence

$$K = \kappa_1 \kappa_2 = \frac{C}{A} \cdot \frac{1}{\|\{F_1, F_2, F_3\}\|^2} = \frac{C}{A} \cdot \frac{1}{F_1^2 + F_2^2 + F_3^2}$$

and

$$H = \frac{\kappa_1 + \kappa_2}{2} = -\frac{B}{2A} \cdot \frac{1}{\sqrt{F_1^2 + F_2^2 + F_3^2}}.$$

□

Visualization 2.5.29 (The umbilical points on ellipsoids).
The umbilical points of an ellipsoid *defined by the equation*

$$\frac{(x^1)^2}{a^2} + \frac{(x^2)^2}{b^2} + \frac{(x^3)^2}{c^2} = 1,$$

where $a > b > c > 0$ are constants, are given by

$$X = \left\{ \pm a\sqrt{\frac{a^2 - b^2}{a^2 - c^2}}, 0, \pm c\sqrt{\frac{b^2 - c^2}{a^2 - c^2}} \right\}. \tag{2.104}$$

Proof. We apply Theorem 2.5.28 to find the umbilical points of the ellipsoid. We have

$$F_1(x^i) = \frac{2x^1}{a^2}, \quad F_2(x^i) = \frac{2x^2}{b^2}, \quad F_3(x^i) = \frac{2x^3}{c^2},$$

$$F_{11}(x^i) = \frac{2}{a^2}, \quad F_{22}(x^i) = \frac{2}{b^2}, \quad F_{33}(x^i) = \frac{2}{c^2}$$

and

$$F_{jk}(x^i) = 0 \text{ for } j \neq k.$$

A direction $\vec{d} = \{\xi^1, \xi^2, \xi^3\}$ is a principal direction by (2.97) in Theorem 2.5.28 if and only if

$$D = \det \begin{pmatrix} \dfrac{2x^1}{a^2} & \dfrac{2\xi^1}{a^2} & \xi^1 \\ \dfrac{2x^2}{b^2} & \dfrac{2\xi^2}{b^2} & \xi^2 \\ \dfrac{2x^3}{c^2} & \dfrac{2\xi^3}{c^2} & \xi^3 \end{pmatrix} = 0 \text{ for } \frac{\xi^1 x^1}{a^2} + \frac{\xi^2 x^2}{b^2} + \frac{\xi^3 x^3}{b^3} = 0.$$

We have

$$D - 4\det \begin{pmatrix} \dfrac{x^1}{a^2} & \dfrac{\xi^1}{a^2} & \xi^1 \\ \dfrac{x^2}{b^2} & \dfrac{\xi^2}{b^2} & \xi^2 \\ \dfrac{x^3}{c^2} & \dfrac{\xi^3}{c^2} & \xi^3 \end{pmatrix}$$

$$= 4\left(\frac{x^1}{a^2}\left(\xi^2\xi^3\left(\frac{1}{b^2} - \frac{1}{c^2}\right)\right) - \frac{x^2}{b^2}\left(\xi^1\xi^3\left(\frac{1}{a^2} - \frac{1}{c^2}\right)\right) + \frac{x^3}{c^2}\left(\xi^1\xi^2\left(\frac{1}{a^2} - \frac{1}{b^2}\right)\right) \right)$$

$$= 0$$

if and only if

$$x^1\xi^2\xi^3(b^2 - c^2) + x^2\xi^1\xi^3(c^2 - a^2) + x^3\xi^1\xi^2(a^2 - b^2) = 0. \tag{2.105}$$

We saw in Remark 2.5.16 (b) and at the end of the proof of Theorem 2.5.14 that any direction can be chosen as a principal direction at an umbilical point. Therefore, the equation in (2.105) has to be satisfied at an umbilical point for all directions $\vec{d} = \{\xi^1, \xi^2, \xi^3\}$ with

$$\frac{\xi^1 x^1}{a^2} + \frac{\xi^2 x^2}{b^2} + \frac{\xi^3 x^3}{b^3} = 0. \tag{2.106}$$

(i) First we assume $x^2 \neq 0$.

(i.1) If $x^1 = 0$, then the equations in (2.105) and (2.106) reduce to

$$(c^2 - a^2)x^2\xi^1\xi^3 + (a^2 - b^2)x^3\xi^1\xi^2 = 0 \text{ for all } \vec{d} = \{\xi^1, \xi^2, \xi^3\}$$

$$\text{with } \frac{x^2\xi^2}{b^2} + \frac{x^3\xi^3}{c^2} = 0. \tag{2.107}$$

If we choose

$$\xi^1, \xi^3 \neq 0 \text{ and } \xi^2 = -\frac{x^3 \xi^3 b^2}{x^2 c^2}$$

then the first equation in (2.107) is not satisfied, since

$$(c^2 - a^2)(x^2)^2 \xi^2 - (a^2 - b^2)(x^3)^2 \frac{b^2}{c^2} \xi^3 \neq 0.$$

Thus we cannot have $x^1 = 0$.

(i.2) It can be shown as in Part (i.1) that we cannot have $x^3 = 0$.

(i.3) Thus we assume $x^1, x^3 \neq 0$.

We choose $\xi^2 = 0$. Then the equation in (2.105) reduces to

$$(c^2 - a^2)x^2 \xi^1 \xi^3 = 0,$$

which implies $\xi^1 = 0$ or $\xi^2 = 0$.

If $\xi^3 = 0$, then the equation in (2.106) reduces to $x^3 \xi^3 = 0$. But $(\xi^1)^2 + (\xi^2)^2 + (\xi^3)^2) = 1$ implies $\xi^3 = \pm 1$, hence $x^3 \xi^3 = 0$ would yield $x^3 = 0$, which is a contradiction to the assumption $x^3 \neq 0$.

Similarly it can be shown that $\xi^3 = 0$ cannot hold.

Thus there cannot be umbilical points with $x^2 \neq 0$.

(ii) Now let $x^2 = 0$.

Then we must have

$$(b^2 - c^2)x^1 \xi^2 \xi^3 + (a^2 - b^2)x^3 \xi^1 \xi^2 = 0 \tag{2.108}$$

for all directions $\vec{d} = \{\xi^2, \xi^2, \xi^3\}$ with

$$\frac{x^1 \xi^1}{a^2} + \frac{x^3 \xi^3}{c^2} = 0. \tag{2.109}$$

Since the equation in (2.108) has to hold for all directions, it follows that $x^1 \neq 0$, for otherwise x^3 would also be equal to zero. But the point $(0, 0, 0)$ is not on the ellipsoid.

If $\xi^1, \xi^2, \xi^3 \neq 0$, then the equation in (2.108) implies

$$\frac{b^2 - c^2}{b^2 - a^2} = \frac{x^3}{x^1} \cdot \frac{\xi^1}{\xi^3},$$

and the equation in (2.109) implies

$$\frac{b^2 - c^2}{a^2 - b^2} = \left(\frac{x^3}{x^1}\right)^2 \frac{a^2}{c^2} \text{ or } (b^2 - c^2)c^2(x^1)^2 = (a^2 - b^2)a^2(x^3)^2. \tag{2.110}$$

Since $a^2(x^3)^2 + c^2(x^1)^2 = a^2 c^2$, the equation in (2.110) yields

$$(b^2 - c^2)(x^1)^2 c^2 = (a^2 - b^2)c^2(a^2 - (x^1)^2)$$

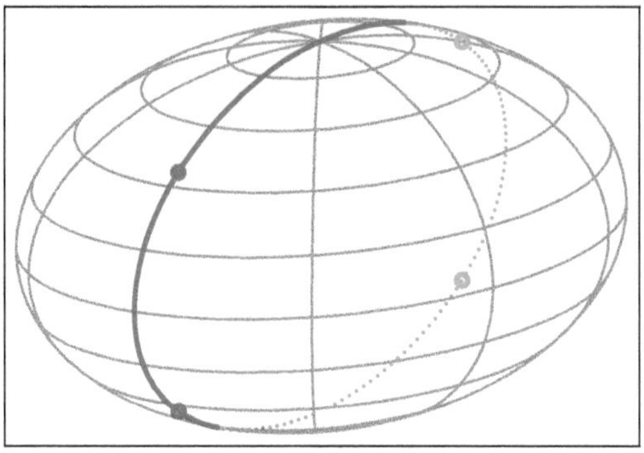

Figure 2.58 The umbilical points of an ellipsoid

and

$$(x^1)^2(b^2 - c^2 + a^2 - b^2) = (a^2 - b^2)a^2,$$

hence

$$x^1 = \pm a\sqrt{\frac{a^2 - b^2}{a^2 - c^2}}.$$

Now the first identity in (2.110) yields

$$x^3 = \pm\frac{c}{a}\sqrt{\frac{b^2 - c^2}{a^2 - b^2}}a\sqrt{\frac{a^2 - b^2}{a^2 - c^2}} == \pm c\sqrt{\frac{b^2 - c^2}{a^2 - c^2}}.$$

Finally, since

$$\frac{(x^1)^2}{a^2} + \frac{(x^3)^2}{c^2} = \frac{a^2 - b^2}{a^2 - c^2} + \frac{b^2 - c^2}{a^2 - c^2} = \frac{a^2 - c^2}{a^2 - c^2} = 1,$$

the points

$$X = \left\{\pm a\sqrt{\frac{a^2 - b^2}{a^2 - c^2}}, 0, \pm c\sqrt{\frac{b^2 - c^2}{a^2 - c^2}}\right\}$$

are the umbilical points of the ellipsoid (Figure 2.58).

□

Remark 2.5.30. *The second fundamental coefficients of a hyperboloid of one sheet with a parametric representation*

$$\vec{x}(u^i) = \{a\cosh u^1 \cos u^2, b\cosh u^1 \sin u^2, c\sinh u^1\}\left((u^1, u^2) \in D \subset \mathbb{R} \times (0, 2\pi)\right),$$

where $a, b, c \in \mathbb{R} \setminus \{0\}$, are

$$L_{11}(u^i) = -\frac{abc\cosh u^1}{\sqrt{g(u^i)}}, \ L_{12}(u^i) = 0 \ and \ L_{22}(u^i) = \frac{abc\cosh^3 u^1}{\sqrt{g(u^i)}},$$

hence its Gaussian curvature is negative at every point. Consequently there are no umbilical points on hyperboloids of one sheet by Remark 2.5.18.

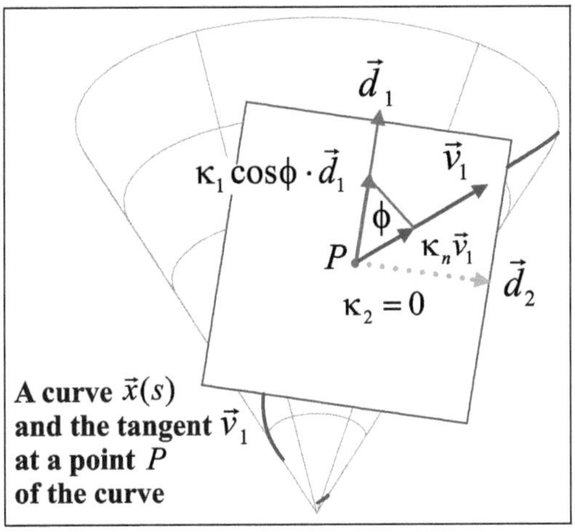

Figure 2.59 Euler's theorem

The next result gives the normal curvature in terms of the principal curvature.

Theorem 2.5.31 (Euler). *Let κ_1 and κ_2 denote the principal curvatures. Then the normal curvature κ_n of a curve on a surface and the angle ϕ between the principal direction of κ_1 and the curve satisfy the relation (Figure 2.59)*

$$\kappa_n = \kappa_1 \cos^2 \varphi + \kappa_2 \sin^2 \varphi. \tag{2.111}$$

Proof. Let S be a surface with a parametric representation $\vec{x}(u^i)$, and γ be a curve on S with a parametric representation $\vec{x}(s) = \vec{x}(u^i(s))$. The normal curvature of γ at s is given by

$$\kappa_n(s) = \vec{v}_1(s) \bullet \vec{N}(s) \text{ and } \vec{v}_1(s) \text{ is the corresponding direction.} \tag{2.112}$$

If $\vec{d}_\mu$ ($\mu = 1, 2$) are the principal directions at the point $X(s)$, then we have

$$\vec{v}_1(s) = \vec{d}_1 \cos \varphi(s) + \vec{d}_2 \sin \varphi(s). \tag{2.113}$$

If γ_μ ($\mu = 1, 2$) are curves on S, given by the parametric representations $\vec{y}_\mu(s) = \vec{y}_\mu(u^i(s))$ such that

$$\frac{d\vec{y}_\mu(s)}{ds} = \vec{d}_\mu(s) \cos \varphi(s) \text{ for } \mu = 1, 2,$$

and if s_μ denotes the arc length along γ_μ, then we obtain

$$\frac{ds_1}{ds} = \cos \varphi(s) \text{ and } \frac{ds_2}{ds} = \sin \varphi(s) \tag{2.114}$$

for suitable orientations of s_1 and s_2. Writing

$$\vec{d}_\mu^*(s_\mu) = \vec{d}_\mu^*(s_\mu(s)) = \vec{d}_\mu(s) \text{ for } \mu = 1, 2, \tag{2.115}$$

we obtain

$$\kappa_\mu(s_\mu) = \left(\frac{d\vec{d}_\mu^*(s_\mu)}{ds_\mu}\right) \bullet \vec{N}(s_\mu), \tag{2.116}$$

and it follows from (2.112), (2.113), (2.114), (2.115) and (2.116) that

$$\kappa_n(s) = \left(\left(\frac{d\vec{d}_1(s)}{ds}\right)\cos\varphi(s) - \left(\sin\varphi(s)\frac{d\varphi(s)}{ds}\right)\vec{d}_1(s)\right.$$
$$\left. + \left(\frac{d\vec{d}_2(s)}{ds}\right)\sin\varphi(s) + \left(\cos\varphi(s)\frac{d\varphi(s)}{ds}\right)\vec{d}_2(s)\right) \bullet \vec{N}(s)$$

$$= \cos\varphi(s)\frac{ds_1}{ds}\frac{d\vec{d}_1^*(s_1)}{ds_1} \bullet \vec{N}(s) + \sin\varphi(s)\frac{ds_2}{ds}\frac{d\vec{d}_2^*(s_2)}{ds_2} \bullet \vec{N}(s)$$

$$= \cos^2\varphi(s)\kappa_1(s) + \sin^2\varphi(s)\kappa_2(s),$$

since $\vec{d}_\mu(s) \bullet \vec{N}(s) = \kappa_\nu(s)$ for $\mu = 1, 2$. □

Theorem 2.5.32. *Let P be a point on a surface S.*
(a) If ϕ denotes the angle between the principal direction corresponding to κ_1 and the direction corresponding to κ_n, then the mean curvature H at P is given by

$$H = \int_0^{2\pi} \kappa_n \, d\phi.$$

(b) Let κ_n and $\kappa_n^\perp$ be the values of the normal curvatures at P corresponding to two orthogonal directions. Then the mean curvature H at P is given by

$$H = \frac{1}{2}(\kappa_n + \kappa_n^\perp).$$

Proof. (a) Applying (2.111) in Euler's Theorem 2.5.31, we obtain

$$\frac{1}{2\pi}\int_0^{2\pi} \kappa_n \, d\phi = \frac{1}{2\pi}\int_0^{2\pi}\left(\kappa_1\cos^2\phi + \kappa_2\sin^2\phi\right)d\phi$$

$$= \frac{\kappa_1}{2\pi}\int_0^{2\pi}\frac{1}{2}(1 + \cos 2\phi)\,d\phi + \frac{\kappa_2}{2\pi}\int_0^{2\pi}\frac{1}{2}(1 - \cos 2\phi)\,d\phi$$

$$= \frac{\kappa_1 + \kappa_2}{2} = H.$$

(b) Let ϕ and $\phi^\perp$ denote the angles between the principal direction corresponding to κ_1 and the directions $\vec{d}$ and $\vec{d}^\perp$ corresponding to κ_n and $\kappa_n^\perp$, respectively. Then it follows from (2.111) in Euler's theorem, Theorem 2.5.31, that

$$\kappa_n = \kappa_1\cos^2\phi + \kappa_2\sin^2\phi \text{ and } \kappa_n^\perp = \kappa_1\cos^2\phi^\perp + \kappa_2\sin^2\phi^\perp. \tag{2.117}$$

Since the directions $\vec{d}$ and $\vec{d}^\perp$ are orthogonal, we have $\cos^2 \phi^\perp = \sin^2 \phi$ and $\sin^2 \phi^\perp = \cos^2 \phi$, and obtain from (2.117)

$$\kappa_n + \kappa_n^\perp = \kappa_1 + \kappa_2 = 2H.$$

□

Example 2.5.33 (An application of Euler's theorem).

We consider the cone given by a parametric representation

$$\vec{x}(u^i) = \left\{ \frac{1}{\sqrt{a}} u^1 \cos u^2, \frac{1}{\sqrt{a}} u^1 \sin u^2, u^1 \right\} \ \left((u^1, u^2) \in D = (0, \infty) \times (0, 2\pi) \right),$$

where $a \neq 0$ is a positive constant. Omitting the arguments (u^k), we obtain

$$g_{11} = \frac{a}{a+1}, \ g_{12} = 0, \ g_{22} = \frac{(u^1)^2}{a},$$

$$L_{11} = L_{12} = 0 \ and \ L_{22} = \frac{u^1}{\sqrt{a+1}}.$$

The two values of the principal curvature are given by

$$\kappa_1 = \frac{L_{22}}{g_{22}} = \frac{a}{u^1 \sqrt{a+1}} \ and \ \kappa_2 = 0.$$

We obtain the principal direction corresponding to $\kappa_2 = 0$ from

$$\left\{ \begin{array}{l} L_{11}\xi_2^1 + L_{12}\xi_2^2 = 0 \cdot \xi_2^1 + 0 \cdot \xi_2^2 = 0 \\ L_{12}\xi_2^1 + L_{22}\xi_2^2 = 0 \cdot \xi_2^1 + L_{22}\xi_2^2 = 0 \end{array} \right\}, \ that \ is, \ \xi_2^2 = 0.$$

Consequently, we may choose the principal directions corresponding to κ_2 and κ_1 as

$$\vec{d}_2 = \frac{\vec{x}_1}{\|\vec{x}_1\|} \ and \ \vec{d}_1 = \frac{\vec{x}_2}{\|\vec{x}_2\|}.$$

Now let $c > 0$ be given. Putting

$$\beta = \sqrt{\frac{a+1}{a}} \sqrt{1+c^2} = \sqrt{g_{11}} \sqrt{1+c^2},$$

we consider the curve γ on the cone given by

$$u^1(s) = \frac{s}{\beta} \ and \ u^2(s) = c\sqrt{a+1} \log \frac{s}{\beta} \ \ (s > 0).$$

Since

$$g_{ik} \dot{u}^i \dot{u}^k = g_{11}(\dot{u}^1)^2 + g_{22}(\dot{u}^2)^2 = \frac{(a+1)(1+c^2)}{a\beta^2} = 1$$

and

$$L_{ik} \dot{u}^i \dot{u}^k = L_{22}(\dot{u}^2)^2 = \frac{c^2 \sqrt{a+1}}{\beta s},$$

the normal curvature is given by

$$\kappa_n = \frac{c^2\sqrt{a+1}}{\beta s}.$$

We have for the angle φ between the curve γ and the principal direction given by $\vec{d}_1$

$$\cos\varphi = \vec{x}_k\dot{u}^k \bullet \frac{\vec{x}_2}{\|\vec{x}_2\|} = \dot{u}^2\sqrt{g_{22}},$$

and consequently Euler's formula yields

$$\kappa_n = \cos^2\varphi\kappa_1 + \sin^2\varphi\kappa_2 = \frac{(\dot{u}^2)^2 g_{22} L_{22}}{g_{22}} = \frac{(a+1)c^2 s}{s^2\beta\sqrt{a+1}} = \frac{c^2\sqrt{a+1}}{\beta s}.$$

2.6   THE SHAPE OF A SURFACE IN THE NEIGHBOURHOOD OF A POINT

We know from Theorem 2.5.9 that the absolute value of the normal curvature of a curve at a point is equal to the curvature of the corresponding normal section. The study of how the change of direction of the tangent to the normal section affects the normal curvature leads to three different cases for the shape of the surface in the neighbourhood of a point with non–vanishing *second fundamental form* $L_{ik}du^i du^k$ depending on the sign of $L = \det(L_{ik})$.

Definition 2.6.1. Let S be a surface with a parametric representation $\vec{x}(u^i)$ $((u^1, u^2) \in D \subset \mathbb{R}^2)$ and second fundamental coefficients L_{ik}. Then a point $P = X(u_0^1, u_0^2)$ on S is called an *elliptic point* if $L(u_0^1, u_0^2) > 0$, a *hyperbolic point* if $L(u_0^1, u_0^2) < 0$, and a *parabolic point* if $L(u_0^1, u_0^2) = 0$.

Remark 2.6.2. *(a) Since the first fundamental form is positive definite, the sign of κ_n depends on the second fundamental form only.*
We will see in Theorem 2.6.6:
If P is an elliptic point, that is, $L > 0$ at P, then κ_n has the same sign at P for all directions of the normal sections, and consequently the centres of curvature of all normal sections are on one side of the surface.
If P is a parabolic point, that is, $L = 0$ at P, then κ_n does not change sign, there is, however, one and only one direction for which $\kappa_n = 0$.
If P is a hyperbolic point, that is, $L < 0$ at P, then κ_n changes sign. Part of the centres of curvature of the normal sections are on one side of the surface and part on the other. There are two directions for which $\kappa_n = 0$, and the normal curvature has the same sign in a sector bounded by these two directions. The normal curvature changes sign when the direction of the tangent moves across the direction corresponding to $\kappa_n = 0$.
Summarizing we have

	Property of point	Number of directions of $\kappa_n = 0$
$L > 0$	*elliptic*	*0*
$L = 0$	*parabolic*	*1*
$L < 0$	*hyperbolic*	*2*

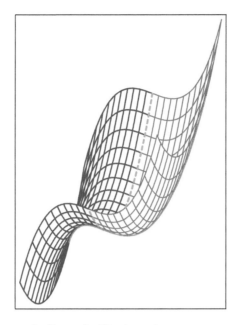

Figure 2.60 Hyperbolic, parabolic and elliptic points on an explicit surface

(b) The elliptic, parabolic or hyperbolic property of a point is invariant with respect to parameter transformations by Remark 2.5.24 (b).
(c) The use of the terms elliptic, parabolic *and* hyperbolic points *will become clear in Remark 2.7.2, when we consider* Dupin's indicatrix *(Definition 2.7.1).*

Visualization 2.6.3. *We consider the explicit surface ES given by the function*

$$f(u^j) = (u^1)^2 + (u^2)^3 \ \left((u^1, u^2) \in \mathbb{R}^2\right).$$

Then all the points of ES with $u^2 > 0$ are elliptic, with $u^2 = 0$ parabolic, and with $u^2 < 0$ hyperbolic (Figure 2.60).

Proof. We have

$$f_1(u^j) = 2u^1, \ f_2 = 3(u^2)^2, \ f_{11}(u^j) = 2, \ f_{12}(u^j) = 0 \text{ and } f_{22}(u^j) = 6u^2,$$

hence by (2.49) in Example 2.4.5

$$L_{11}(u^j) = \frac{2}{\sqrt{1 + 4(u^1)^2 + 9(u^2)^4}}, \ L_{12}(u^j) = 0,$$

$$L_{22}(u^j) = \frac{6u^2}{\sqrt{1 + 4(u^1)^2 + 9(u^2)^4}} \text{ and } L(u^j) = \frac{12u^2}{1 + 4(u^1)^2 + 9(u^2)^4}.$$

□

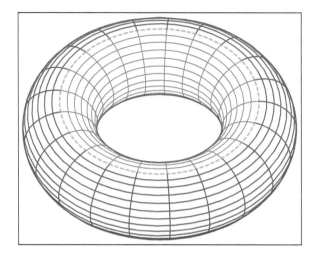

Figure 2.61 A torus with elliptic (outside), parabolic (dashed) and hyperbolic (inside) points

Visualization 2.6.4. *(a) Every point of an ellipsoid with a parametric representation*

$$\vec{x}(u^i) = \{a \cos u^1 \cos u^2, b \cos u^1 \sin u^2, c \sin u^1\} \ \left((u^1, u^2) \in (-\pi/2, \pi/2) \times (0, 2\pi)\right),$$

where a, b and c are distinct positive constants, is an elliptic point.
(b) Every point of an elliptic cone is a parabolic point.
(c) Every point of the hyperbolic paraboloid of Visualization 2.5.27 is a hyperbolic point.
(d) We consider the torus with a parametric representation

$$\vec{x}(u^i) = \{(r_0+r_1 \cos u^1) \cos u^2, (r_0+r_1 \cos u^1) \sin u^2, r_1 \sin u^1\} \ \left((u^1, u^2) \in D = (0, 2\pi)^2\right),$$

where r_0 and r_1 are constants with $0 < r_1 < r_0$. Then all points with $u^1 \in (0, \pi/2) \cup (3\pi/2, 2\pi)$ are elliptic points, all points with $u^1 = \pi/2$ or $u^1 = 3\pi/2$ are parabolic points, and all points with $u^1 \in (\pi/2, 3\pi/2)$ are hyperbolic points (Figure 2.61).

Proof. (a) It follows that

$$\vec{x}_1(u^i) = \{-a \sin u^1 \cos u^2, -b \sin u^1 \sin u^2, c \cos u^1\},$$
$$\vec{x}_2(u^i) = \{-a \cos u^1 \sin u^2, b \cos u^1 \cos u^2, 0\},$$
$$\vec{x}_{11}(u^i) = \{-a \cos u^1 \cos u^2, -b \cos u^1 \sin u^2, -c \sin u^1\},$$
$$\vec{x}_{12}(u^i) = \{a \sin u^1 \sin u^2, -b \sin u^1 \cos u^2, 0\},$$
$$\vec{x}_{22}(u^i) = \{-a \cos u^1 \cos u^2, -b \cos u^1 \sin u^2, 0\},$$

$$L_{11}(u^i) = \frac{1}{\sqrt{g}} \begin{vmatrix} -a \cos u^1 \cos u^2 & -a \sin u^1 \cos u^2 & -a \sin u^2 \\ -b \cos u^1 \sin u^2 & -b \sin u^1 \sin u^2 & b \cos u^2 \\ -c \sin u^1 & c \cos u^1 & 0 \end{vmatrix} \cos u^1$$

$$= \frac{abc}{\sqrt{g}} \cos u^1 (-\sin u^1 (-\sin u^1) - \cos u^1 (-\cos u^1)) = \frac{abc}{\sqrt{g}} \cos u^1,$$

$$L_{12}(u^i) = \frac{abc \sin u^1 \cos u^1}{\sqrt{g}} \begin{vmatrix} \sin u^2 & -\sin u^1 \cos u^2 & -\sin u^2 \\ -\cos u^2 & -\sin u^1 \sin u^2 & \cos u^2 \\ 0 & \cos u^1 & 0 \end{vmatrix}$$

$$= \frac{abc \sin u^1 \cos^2 u^1}{\sqrt{g}} (\sin u^2 \cos u^2 - \sin u^2 \cos u^2) = 0,$$

$$L_{22}(u^i) = \frac{abc \cos^2 u^1}{\sqrt{g}} \begin{vmatrix} -\cos u^2 & -\sin u^1 \cos u^2 & -\sin u^2 \\ -\sin u^1 & -\sin u^1 \sin u^2 & \cos u^2 \\ 0 & \cos u^1 & 0 \end{vmatrix}$$

$$= \frac{abc \cos^3 u^1}{\sqrt{g}} (\cos^2 u^2 + \sin^2 u^2) = \frac{abc \cos^3 u^1}{\sqrt{g}}$$

and so

$$L(u^i) = L_{11}(u^i) L_{22}(u^i) - L_{12}^2(u^i) = \frac{a^2 b^2 c^2 \cos^4 u^1}{g} > 0 \text{ for all } u^1 \in (-\pi/2, \pi/2).$$

(b) Let $a_1, a_2 > 0$ and the elliptic cone be given by a parametric representation

$$\vec{x}(u^i) = \left\{ \frac{u^1}{\sqrt{a_1}} \cos u^2, \frac{u^1}{\sqrt{a_2}} \sin u^2, u^1 \right\} \quad \left((u^1, u^2) \in D \subset (0, \infty) \times (0, 2\pi) \right).$$

Then we have

$$\vec{x}_1(u^i) = \left\{ \frac{1}{\sqrt{a_1}} \cos u^2, \frac{1}{\sqrt{a_2}} \sin u^2, 1 \right\},$$

$$\vec{x}_2(u^i) = \left\{ -\frac{u^1}{\sqrt{a_1}} \sin u^2, \frac{u^1}{\sqrt{a_2}} \cos u^2, 0 \right\},$$

$$g_{11}(u^i) = \frac{\cos^2 u^2}{a_1} + \frac{\sin^2 u^2}{a_2} + 1, \quad g_{12}(u^i) = \left(\frac{1}{a_2} - \frac{1}{a_1} \right) u^1 \sin u^2 \cos u^2,$$

$$g_{22}(u^i) = \frac{(u^1)^2}{a_1} \sin^2 u^2 + \frac{(u^1)^2}{a_2} \cos^2 u^2,$$

$$\vec{N}(u^i) = \frac{1}{\sqrt{a_1 \cos^2 u^2 + a_2 \sin^2 u^2 + 1}} \left\{ -\sqrt{a_1} \cos u^2, -\sqrt{a_2} \sin u^2, 1 \right\},$$

$$\vec{x}_{11}(u^i) = \vec{0}, \quad \vec{x}_{12}(u^i) = \left\{ -\frac{1}{\sqrt{a_1}} \sin u^2, \frac{1}{\sqrt{a_2}} \cos u^2, 0 \right\},$$

$$\vec{x}_{22}(u^i) = \left\{ -\frac{u^1}{\sqrt{a_1}} \cos u^2, -\frac{u^1}{\sqrt{a_2}} \sin u^2, 0 \right\},$$

$$L_{11}(u^i) = L_{12}(u^i) = 0 \text{ and } L_{22}(u^i) = \frac{u^1}{\sqrt{a_1 \cos^2 u^2 + a_2 \sin^2 u^2 + 1}},$$

hence $L(u^i) = 0$ on D.

(c)   We have by Visualization 2.5.27, $L(u^i) = -1/g(u^i) < 0$ for all u^1 and u^2.

(d)   Since $r'(u^1) = -r_1 \sin u^1$, $h'(u^1) = r_1 \cos u^1$, $r'(u^1(u^1))^2 + (h'(u^1))^2 = r_1^2$ and $r''(u^1) = -r_1 \cos u^1$, it follows from (2.76) in Visualization 2.5.19 as in the proof of (2.78) that

$$K(u^1) = -\frac{r''(u^1)r_1^2}{r(u^1)r_1^4} = \frac{\cos u^1}{r_1(r_0 + r_1 \cos u^1)}, \qquad (2.118)$$

and the conclusion is an immediate consequence of (2.118).

$\square$

The following result gives a geometric interpretation of the second fundamental form.

Theorem 2.6.5. *Let S be a surface with a parametric representation $\vec{x}(u^i)$ $((u^1, u^2) \in D)$ and X_0 be a point on S with position vector $\vec{x}(u_0^k)$. If the second fundamental form of S does not vanish identically at (u_0^1, u_0^2), then – up to terms in $((du^1)^2 + (du^2)^2)^{1/2}$ of order higher than two – it is equal to twice the distance (including the sign) of a point $Q \in S$ with position vector $\vec{x}(u_0^k + du^k)$ from the tangent plane of S at X_0.*

Proof. We obtain for the point Q by Taylor's formula

$$\vec{x}(u_0^i + du^i) = \vec{x}(u_0^i) + du^k \vec{x}_k(u_0^i) + \frac{1}{2} du^k du^j \vec{x}_{kj}(u_0^i) + \vec{o}\left((du^1)^2 + (du^2)^2\right)$$

and the distance d of Q from the tangent plane to S at the point X_0 is given by

$$d = \left(\vec{x}(u_0^i + du^i) - \vec{x}(u_0^i)\right) \bullet \vec{N}(u_0^i)$$
$$= \frac{1}{2} du^k du^j \vec{x}_{kj}(u_0^i) \bullet \vec{N}(u_0^i) + o\left((du^1)^2 + (du^2)^2\right)$$
$$= \frac{1}{2} L_{kj}(u_0^i) du^k du^j + o\left((du^1)^2 + (du^2)^2\right).$$

The sign of the distance depends on which side of the tangent plane the point Q is, hence depends on the orientation. $\square$

The next result is an immediate consequence of Theorem 2.6.5.

Theorem 2.6.6. *Let P be a point of a surface S and $T(P)$ denote the tangent plane of S at P.*
If P is an elliptic or parabolic point, then a sufficiently small neighbourhood of P in $T(P)$ lies completely on one side of S (Figure 2.62).
If P is a hyperbolic point, then every neighbourhood of P in $T(P)$ contains points on either side of S (Figure 2.63).

Tangent planes at elliptic, parabolic and hyperbolic points of a torus are shown in Figure 2.64.

Since g is always positive and the Gaussian curvature is given by $K = L/g$, the sign of K is equal to that of L. Thus the following result holds.

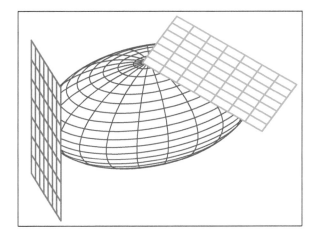

Figure 2.62 Tangent planes of an ellipsoid

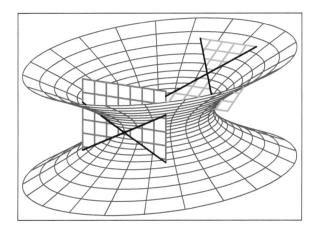

Figure 2.63 Tangent planes of a hyperbolid of one sheet

Theorem 2.6.7. *The Gaussian curvature at a point P of a surface is positive, if P is an elliptic point, negative, if P is a hyperbolic point and it vanishes, if P is a parabolic point.*

Remark 2.6.8. *We already know from (2.73) in Remark 2.5.16 (b) that $L = c^2 g$ in an umbilical point, for some constant c. Since $g > 0$ and $c^2 \geq 0$, it follows that there are only elliptic umbilical points for $c \neq 0$, and parabolic umbilical points for $c = 0$. A parabolic umbilical point is called* planar *or* flat *point.*

Example **2.6.9.** *(a) Every point of a sphere is an elliptic umbilical point.*
(b) Every point in a plane is a planar point.

Proof. (a) Every point of a sphere of radius $r > 0$ is an umbilical point by Example 2.5.25. Since $L_{11}(u^i) = r$, $L_{12}(u^i) = 0$ and $L_{22}(u^i) = r \cos^2 u^2$, we have $L(u^i) = r^2 \cos^2 u^2 > 0$ for all $u^2 \in (-\pi/2, \pi/2)$.

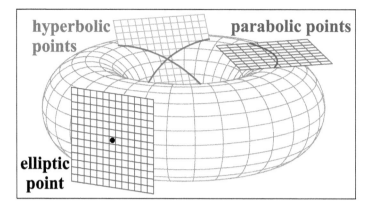

Figure 2.64 Tangent planes of a torus at elliptic, parabolic and hyperbolic points

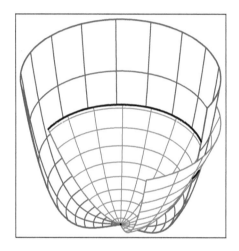

Figure 2.65 A paraboloid of revolution and the surface of Visualization 2.6.10

(b) Let the plane be given by a parametric representation $\vec{x}(u^i) = \{u^1, u^2, 0\}$ $((u^1, u^2) \in D \subset \mathbb{R}^2)$. Then it is easy to see that $g_{11}(u^i) = g_{22}(u^i) = 1$, $g_{12}(u^i) = 0$, $L_{11}(u^i) = L_{12}(u^i) = L_{22}(u^i) = 0$ for all $(u^1, u^2) \in D$, and consequently every point of the plane is a flat point. ☐

Visualization 2.6.10. (A flat point on a biquadratic parabola of revolution)
Let RS be the surface of revolution generated by the rotation about the x^3-axis of the curve with the equation $x^3 = (x^1)^4$ in the x^1x^3-plane (Figure 2.65). Then RS is given by the parametric representation

$$\vec{x}(u^i) = \left\{u^1, u^2, \left((u^1)^2 + (u^2)^2\right)^2\right\} \ \left((u^1, u^2) \in \mathbb{R}^2\right).$$

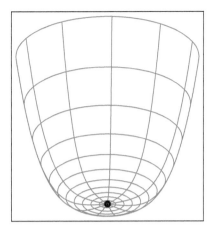

Figure 2.66 A flat point on the surface of Visualization 2.6.10

Then

$$g_{11}(u^i) = 1 + 16(u^1)^2 \left((u^1)^2 + (u^2)^2\right)^2, \ g_{12}(u^i) = 16u^1u^2 \left((u^1)^2 + (u^2)^2\right)^2,$$

$$g_{22}(u^i) = 1 + 16(u^2)^2 \left((u^1)^2 + (u^2)^2\right)^2; \ L_{11}(u^i) = \frac{12(u^1)^2 + 4(u^2)^2}{\sqrt{g}},$$

$$L_{12}(u^i) = \frac{8u^1u^2}{\sqrt{g}}, \qquad\qquad L_{22}(u^i) = \frac{12(u^2)^2 + 4(u^1)^2}{\sqrt{g}},$$

and so

$$g(u^i) = 1 + 16 \left((u^1)^2 + (u^2)^2\right)^3 \ and \ L(u^i) = \frac{48 \left((u^1)^2 + (u^2)^2\right)^2}{g}.$$

Thus $L_{ik}(u^1, u^2) = 0$ for all $i, k = 1, 2$ if and only if $(u^1, u^2) = (0, 0)$. Therefore, the surface has a plane point in the origin (Figure 2.66), and all other points are elliptic points.

Visualization 2.6.11 (The monkey saddle).
Let S be the so–called monkey saddle *with a parametric representation (Figure 2.67)*

$$\vec{x}(u^i) = \left\{u^1, u^2, u^2(u^2 - \sqrt{3}u^1)(u^2 + \sqrt{3}u^1)\right\} \left((u^1, u^2) \in \mathbb{R}^2\right).$$

Then

$$g_{11}(u^1, u^2) = 1 + 36(u^1u^2)^2, \ g_{12}(u^1, u^2) = 18u^1u^2 \left((u^2)^2 - (u^1)^2\right),$$

$$g_{22}(u^1, u^2) = 1 + 9 \left((u^1)^2 - (u^2)^2\right)^2, \ g(u^1, u^2) = 1 + 9 \left((u^1)^2 + (u^2)^2\right)^2,$$

$$L_{11}(u^1, u^2) = -\frac{6u^2}{\sqrt{g}}, \ L_{12}(u^1, u^2) = -\frac{6u^1}{\sqrt{g}},$$

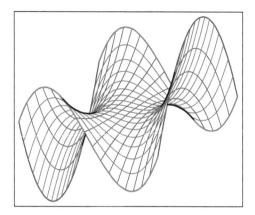

Figure 2.67 The monkey saddle

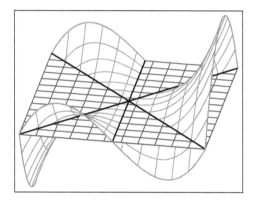

Figure 2.68 The monkey saddle and its intersection with the x^1x^2–plane

$$L_{22}(u^1, u^2) = \frac{6u^2}{\sqrt{g}} \text{ and } L(u^1, u^2) = -\frac{36\left((u^1)^2 + (u^2)^2\right)}{g}.$$

The second fundamental form vanishes identically at $(0,0)$. Thus the monkey saddle has a planar point in the origin. Since $\vec{x}_1(0,0) = \vec{e}^1$ and $\vec{x}_2(0,0) = \vec{e}^2$, the x^1x^2–plane is the tangent plane of the monkey saddle at the origin. It intersects the monkey saddle in three straight lines that are given by the following equations in the x^1x^2–plane (Figure 2.68)

$$x^2 = 0, \quad x^2 = \sqrt{3}x^1 \text{ and } x^3 = -\sqrt{3}x^1. \tag{2.119}$$

Finally we consider a normal section at the origin, different from the three lines of intersection given in (2.119). If the tangent to the curve has direction $\vec{v} = \{\xi^1, \xi^2\}$, that is, if

$$\vec{x}' = \xi^1\vec{x}_1(0,0) + \xi^2\vec{x}_2(0,0) = \xi^1\vec{e}^1 + \xi^2\vec{e}^2,$$

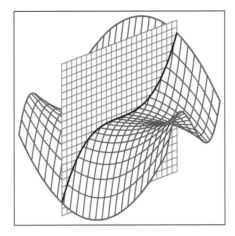

Figure 2.69 The monkey saddle and a normal section at $(u^1, u^2) = (0,0)$

then a parametric representation of the plane of the corresponding normal section is given by

$$\overrightarrow{OX} = \lambda \vec{x}'(0,0) + \mu \vec{N}(0,0) = \lambda \xi^1 \vec{e}^1 + \lambda \xi^2 \vec{e}^2 + \mu \vec{e}^3 \ (\lambda, \mu \in \mathbb{R}).$$

The normal section is given by the following equation in the plane of the normal section $\mu = \lambda^3(\xi^2((\xi^2)^2 - 3(\xi^1)^2)))$. If $\xi^2 \neq 0$ and $\xi^2 \neq \pm\sqrt{3}\xi^1$, then putting $a = \xi^2((\xi^2)^2 - 3(\xi^1)^2)) \neq 0$, we obtain the cubic parabola $\mu = a\lambda^3$ (Figure 2.69).

2.7 DUPIN'S INDICATRIX

In this section, we explain the geometric meaning of elliptic, parabolic and hyperbolic points, of the principal directions and the directions for which $\kappa_n = 0$, the so–called *asymptotic directions*. This leads to *Dupin's indicatrix*.

Meusnier and Euler's theorems (Theorems 2.5.7 and 2.5.31) give complete information on the curvature of any curve on a surface through a given point.

Definition 2.7.1. Let P be a point on a surface S, and we assume that P is not an umbilical point. We introduce a Cartesian coordinate system in the tangent plane $T(P)$ of S at P with its origin at P and its x^1– and x^2–axes along the principal directions corresponding to the principal curvatures κ_1 and κ_2. If φ denotes the angle between the positive x^1–axis and direction that corresponds to the normal curvature κ_n, then the point set

$$DI = \left\{ X = (x^1, x^2) \in T(P) : \ x^1 = \sqrt{\frac{1}{|\kappa_n|}} \cos \varphi, x^2 = \sqrt{\frac{1}{|\kappa_n|}} \sin \varphi \ (\varphi \in [0, 2\pi]) \right\}$$

is called *Dupin's indicatrix of the surface S at the point P.*

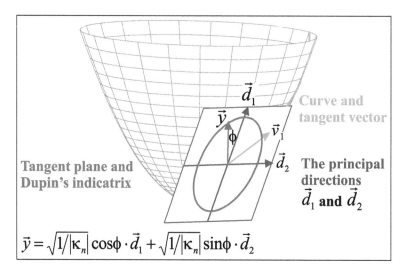

Figure 2.70 Dupin's indicatrix at an elliptic point

Remark 2.7.2. *(a) If we multiply equation (2.111) in Euler's theorem, Theorem 2.5.31, by $1/|\kappa_n|$, then we obtain, since*

$$x^1 = \sqrt{\frac{1}{|\kappa_n|}} \cos\varphi \ \text{and} \ x^2 = \sqrt{\frac{1}{|\kappa_n|}} \sin\varphi,$$

that

$$\kappa_1(x^1)^2 + \kappa_2(x^2)^2 = \frac{\kappa_1}{|\kappa_n|}\cos^2\varphi + \frac{\kappa_2}{|\kappa_n|}\sin^2\varphi = \pm 1.$$

If P is an elliptic point, then the principal curvatures κ_1 and κ_2 have the same sign, and consequently Dupin's indicatrix is an ellipse (Figure 2.70).

If P is a hyperbolic point, then κ_1 and κ_2 have different signs, and consequently Dupin's indicatrix consists of two hyperbolas with a joint pair of asymptotes. The so-called asymptotic directions *on the surface, that is, the directions for which $\kappa_n = 0$, coincide with the directions of the asymptotes of the hyperbolas.*

This explains the terms elliptic *and* hyperbolic *points and* asymptotic *directions (right in Figure 2.71).*

If P is a parabolic point, then $\kappa_1 = 0$, say, and the equation for Dupin's indicatrix reduces to

$$(x^2)^2 = \pm\frac{1}{\kappa_2},$$

which is the equation for a pair of straight lines parallel to the x^1-axis, which intersect the x^2-axis at the points $\pm\frac{1}{\sqrt{|\kappa_2|}}$ (left in Figure 2.71).

(b) The principal directions bisect the angle between the asymptotic directions. The directions of equal normal curvatures are symmetric with respect to the principal directions.

(c) Dupin's indicatrix at a point P of a surface S is closely related to the lines of intersection of S with planes parallel to the tangent plane $T(P)$ of S at P. If we choose two planes parallel to $T(P)$ at a distance ε on either side of S, then an equation of the

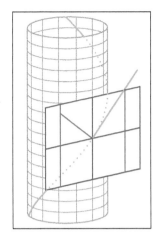
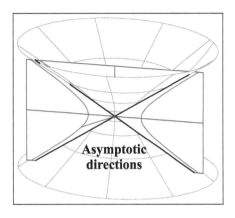

Figure 2.71 Dupin's indicatrix at a parabolic point (left) and at a hyperbolic point with asymptotic directions (right)

line of intersection – up to terms of higher than second order in $((du^1)^2 + (du^2)^2)^{1/2}$ – is given by Theorem 2.6.5

$$\frac{1}{2}L_{ik}du^i du^k = \pm\varepsilon.$$

Choosing the parameters such that for each parameter line its tangents at each point coincide with one principal direction, we obtain

$$L_{11}(du^1)^2 + L_{22}(du^2)^2 = \pm 2\varepsilon,$$

as we shall see in Theorem 2.8.6. Finally

$$\kappa_1 = \frac{L_{11}}{g_{11}} \text{ and } \kappa_2 = \frac{L_{22}}{g_{22}}$$

together imply

$$\kappa_1 g_{11}(du^1)^2 + \kappa_2 g_{22}(du^2)^2 = \pm 2\varepsilon \text{ (Figure 2.72)}.$$

2.8 LINES OF CURVATURE AND ASYMPTOTIC LINES

In view of the results of the previous sections, it is interesting to find those curves on surfaces the directions of which coincide with the principal or *asymptotic* directions.

Definition 2.8.1. (a) A curve on a surface such that the direction of its tangent at any point coincides with one principal direction is called a *line of curvature*.
(b) *Asymptotic directions* are directions for which the normal curvature vanishes. A curve on a surface such that the direction of its tangent at any point coincides with an asymptotic direction is called *asymptotic line*.

The next result gives the differential equations for lines of curvature and asymptotic lines.

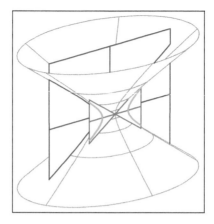

Figure 2.72 Visualization of Remark 2.7.2

Theorem 2.8.2. *(a) There are two lines of curvature through every point of a surface that is not an umbilical point (Theorem 2.5.14). They are the solutions of the differential equations*

$$\det \begin{pmatrix} L_{1k}du^k & g_{1k}du^k \\ L_{2k}du^k & g_{2k}du^k \end{pmatrix} = 0. \tag{2.120}$$

(b) Since

$$\kappa_n = \frac{L_{ik}u^{i'}u^{k'}}{g_{ik}u^{i'}u^{k'}}$$

for the normal curvature, a direction given by $\{\xi^1, \xi^2\}$ is an asymptotic direction, if and only if

$$L_{ik}\xi^i\xi^k = 0.$$

Thus the differential equation for an asymptotic line is

$$L_{ik}u^{i'}u^{k'} = 0. \tag{2.121}$$

The existence of a (real) asymptotic line depends on the shape of the surface.

If a point P on a surface S is an elliptic point, then $L > 0$ in some neighbourhood of P and the second fundamental $L_{ik}du^i du^k$ is positive definite. Consequently there exists no (real) asymptotic line through P.

If P is a hyperbolic point, then $L < 0$ in some neighbourhood of P and there are two asymptotic directions at P. Consequently there are two asymptotic lines through P in a neighbourhood of P.

If P is a parabolic point, then $L = 0$ at P and there is one and only one asymptotic direction at P.

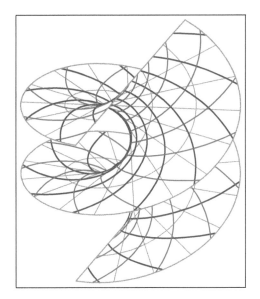

Figure 2.73 Lines of curvature on a helicoid

Visualization 2.8.3 (Lines of curvature on a helicoid).
The lines of curvature on the helicoid of Visualization 2.5.3 with a parametric representation

$$\vec{x}(u^i) = \{u^2 \cos u^1, u^2 \sin u^1, cu^1\} \ \left((u^1, u^2) \in D\right),$$

where $c \neq 0$ is a constant, are given by (Figure 2.73)

$$\pm(u^1 + d) = \sinh^{-1} \frac{u^2}{c}, \ or \ u^2(u^1) = \pm c \sinh (u^1 + d), \qquad (2.122)$$

where d is a constant.

Proof. We obtain

$$\vec{x}_1(u^i) = \{-u^2 \sin u^1, u^2 \cos u^1, c\}, \ \vec{x}_2(u^i) = \{\cos u^1, \sin u^1, 0\},$$

$$\vec{x}_{11}(u^i) = \{-u^2 \cos u^1, -u^2 \sin u^1, 0\}, \ \vec{x}_{12}(u^i) = \{-\sin u^1, \cos u^1, 0\}, \ \vec{x}_{22}(u^i) = \vec{0},$$

$$g_{11}(u^i) = (u^2)^2 + c^2, \ g_{12}(u^i) = 0, \ g_{22}(u^i) = 1, \ g(u^i) = (u^2)^2 + c^2,$$

$$\sqrt{g(u^i)} L_{11}(u^i) = -u^2 \begin{vmatrix} \cos u^1 & -u^2 \sin u^1 & \cos u^1 \\ \sin u^1 & u^2 \cos u^1 & \sin u^1 \\ 0 & c & 0 \end{vmatrix} = 0,$$

$$\sqrt{g(u^i)} L_{12}(u^i) = \begin{vmatrix} -\sin u^1 & -u^2 \sin u^1 & \cos u^1 \\ -\cos u^1 & u^2 \cos u^1 & \sin u^1 \\ 0 & c & 0 \end{vmatrix} = \sqrt{g(u^i)}c$$

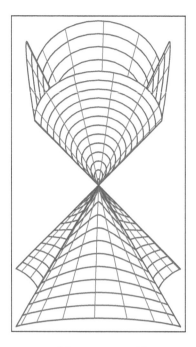

Figure 2.74 A general cone generated by an astroid

and

$$L_{22}(u^i) = 0.$$

Thus the differential equation (2.120) for the lines of curvature on the helicoid reduce to

$$c\left(du^2\right)^2 - c\left((u^2)^2 + c^2\right)\left(du^1\right)^2 = 0$$

and since $c \neq 0$, to

$$\frac{du^2}{\sqrt{((u^2)^2 + c^2)}} = \pm du^1$$

with the solutions in (2.122). □

Visualization 2.8.4 (Lines of curvature on a general cone).
Let γ be a curve of non–vanishing curvature in the x^1x^2–plane, and $\vec{y}(s)$ ($s \in I$) be a parametric representation of γ where s denotes the arc length along γ. A general cone GC is generated by the straight lines through the point $(0,0,1)$ and the points of γ (Figure 2.74).

Writing $u^1 = s$, we obtain the following parametric representation for GC

$$\vec{x}(u^i) = u^2\vec{y}(u^1) + (1 - u^2)\vec{e}^3 \ \left((u^1, u^2) \in I \times (0, \infty)\right).$$

The lines of curvature on GC are given by

$$u^2(u^1) = \frac{c^2}{\sqrt{(\vec{y}(u^1))^2 + 1}} \tag{2.123}$$

for some constant $c \neq 0$ (Figure 2.75).

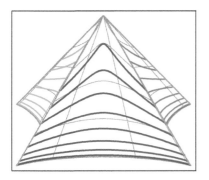 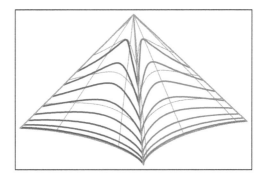

Figure 2.75 The second family of lines of curvature on the general cone of Figure 2.74

Proof. We obtain

$$\vec{x}_1(u^i) = u^2\vec{y}'(u^1), \ \vec{x}_2(u^i) = \vec{y}(u^1) - \vec{e}^3,$$
$$\vec{x}_{11}(u^i) = u^2\vec{y}''(u^1), \ \vec{x}_{12}(u^i) = \vec{y}'(u^1), \ \vec{x}_{22}(u^i) = \vec{0},$$

$$\vec{x}_1(u^i) \times \vec{x}_2(u^i) = u^2\vec{y}'(u^1) \times (\vec{y}(u^1) - \vec{e}^3) \neq \vec{0},$$
$$g_{11}(u^i) = (u^2)^2, \ \|\vec{y}'(u^1)\| = 1, \text{ since } u^1 \text{ is the arc length along } \gamma,$$
$$g_{12}(u^i) = u^2\vec{y}'(u^1) \bullet \vec{y}(u^1), \ g_{22}(u^i) = \left(\vec{y}(u^1)\right)^2 + 1,$$

$$\|\vec{x}_1(u^i) \times \vec{x}_2(u^i)\|L_{11}(u^1) = (u^2)^2\vec{y}''(u^1) \bullet \left(\vec{y}'(u^1) \times (\vec{y}(u^1) - \vec{e}^3)\right),$$
$$L_{12}(u^i) = L_{22}(u^i) = 0.$$

Since $\|\vec{y}'\| = 1$ implies $\vec{y}' \perp \vec{y}''$ in the x^1x^2–plane, we conclude that $\vec{y}' \times \vec{y}''$ is in the direction of $\vec{e}^3$ and $\vec{y}' - e^3$ is not orthogonal to $\vec{e}^3$. Thus we have $L_{11}(u^i) \neq 0$ for $u^2 > 0$. The differential equation for the lines of curvature (2.120) on GC reduces to

$$L_{11}(u^i)du^1 \left(g_{12}(u^i)du^1 + g_{22}(u^2)du^2\right) = 0.$$

It follows from $L_{11}(u^i) \neq 0$ that the u^2–lines are lines of curvature and the second family of lines of curvature is given by

$$u^2\vec{y}'(u^1) \bullet \vec{y}(u^1)du^1 = -\left((\vec{y}(u^1))^2 + 1\right)du^2,$$

that is,

$$\log u^2 = -\int \frac{\vec{y}'(u^1) \bullet \vec{y}(u^1)}{(\vec{y}(u^1))^2 + 1} du^1 = -\frac{1}{2}\log\left((\vec{y}(u^1))^2 + 1\right) + \log k$$

for some constant $k > 0$, hence (2.123) follows. □

Visualization 2.8.5 (Asymptotic lines on ruled surfaces).

We consider the ruled surface RulS with a parametric representation

$$\vec{x}(u^i) = \vec{y}(u^1) + u^2\vec{z}(u^1) \ \left((u^1, u^2) \in D\right)$$

with $\|\vec{z}(u^1)\| = 1$ and $L_{11}(u^i) \neq 0$. Omitting the parameter u^1, we obtain

$$\vec{x}_1 = \vec{y}' + u^2\vec{z}', \ \vec{x}_2 = \vec{z}, \ \vec{x}_{11} = \vec{y}'' + u^2\vec{z}'', \ \vec{x}_{12} = \vec{z}', \ \vec{x}_{22} = \vec{0},$$

$$L_{11}(u^i) = \vec{x}_{11} \bullet \frac{\vec{x}_1 \times \vec{x}_2}{\sqrt{g(u^i)}} = \frac{1}{\sqrt{g(u^i)}}(\vec{y}'' + u^2\vec{z}'') \bullet \left((\vec{y}' + u^2\vec{z}') \times \vec{z}\right),$$

$$L_{12}(u^i) = \vec{x}_{12} \bullet \frac{\vec{x}_1 \times \vec{x}_2}{\sqrt{g(u^i)}} = \frac{1}{\sqrt{g(u^i)}}\vec{z}' \bullet (\vec{y}' \times \vec{z}) \ and \ L_{22}(u^i) = 0.$$

Thus the differential equation (2.121) for the asymptotic lines on RulS reduces to

$$du^1 \left(L_{11}(u^i)du^1 + 2L_{12}(u^i)du^2\right) = 0,$$

and the u^2–lines are one family of asymptotic lines. Further solutions are obtained from

$$2\vec{z}' \bullet (\vec{y}' \times \vec{z})\frac{du^2}{du^1} + (\vec{y}'' + u^2\vec{z}'') \bullet \left((\vec{y}' + u^2\vec{z}') \times \vec{z}\right) = 0. \qquad (2.124)$$

If $\vec{z}' \bullet (\vec{y}' \times \vec{z}) = 0$, then there are no more solutions.
We assume $\vec{z}' \bullet (\vec{y}' \times \vec{z}) \neq 0$ and put

$$a(u^1) = 2\vec{z}' \bullet (\vec{y}' \times \vec{z}), \ f(u^1) = -\frac{\vec{y}'' \bullet (\vec{y}' \times \vec{z})}{a(u^1)},$$

$$g(u^1) = -\frac{\vec{y}''(\vec{z}' \times \vec{z}) + \vec{z}'' \bullet (\vec{y}' \times \vec{z})}{a(u^1)}$$

and

$$h(u^1) = -\frac{\vec{z}'' \bullet (\vec{z}' \times \vec{z})}{a(u^1)}.$$

Then we see that (2.124) is a differential equation of Riccati type

$$\frac{du^2}{du^1} = f(u^1) + u^2 g(u^1) + (u^2)^2 h(u^1), \qquad (2.125)$$

and we obtain one more family of asymptotic lines from the solutions of (2.125).

(a) *First we consider the* conoid *of Visualization 2.5.3 (c) with*

$$\vec{y}(u^1) = \{0, 0, \varphi(u^1)\}, \ \varphi'(u^1) \neq 0 \ and \ z(u^1) = \{\cos u^1, \sin u^1, 0\}.$$

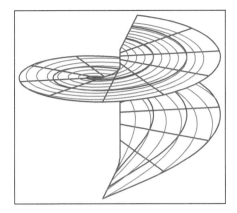

Figure 2.76 Asymptotic lines on the conoid with $\varphi(u^1) = \log(1 + u^1)$

Then we obtain

$$\vec{y}'(u^1) = \{0, 0, \varphi'(u^1)\}, \ \vec{y}''(u^1) = \{0, 0, \varphi''(u^{t1})\},$$
$$\vec{z}'(u^1) = \{-\sin u^1, \cos u^1, 0\} \ \text{and} \ \vec{z}''(u^1) = -\vec{z}(u^1).$$

Therefore, we have

$$\vec{z}''(u^1) \bullet \left(\vec{y}'(u^1) \times \vec{z}(u^1)\right) = \vec{z}''(u^1) \bullet \left((\vec{z}'(u^1) \times \vec{z}(u^1)\right)$$
$$= \vec{y}''(u^1) \bullet \left(\vec{y}'(u^1) \times \vec{z}(u^1)\right) = 0,$$

$$a(u^1) = -2\varphi(u^1), \ f(u^1) = 0,$$

$$g(u^1) = -\frac{\vec{y}(u^1)'' \bullet (\vec{z}'(u^1) \times \vec{z}(u^1))}{a(u^1)} = \frac{\varphi''(u^1)}{2\varphi'(u^1)} \ \text{and} \ h(u^1) = 0.$$

Hence the differential equation (2.125) reduces to

$$\frac{du^2}{du^1} = \frac{u^2 \varphi''(u^1)}{2\varphi'(u^1)}$$

with the solutions

$$\log|u^2| = \frac{1}{2} \int \frac{\varphi''(u^1)}{\varphi'(u^1)} \, du^1$$

for $u^2 \neq 0$, that is,

$$u^2 = c \exp\left(\frac{1}{2} \int \frac{\varphi''(u^1)}{\varphi'(u^1)} \, du^1\right) = c \exp\left(\frac{1}{2} \log|\varphi'(u^1)|\right) = c\sqrt{|\varphi'(u^1)|}$$

for some constant $c \neq 0$ (Figure 2.76).

(b) Now we consider the ruled surface with

$$\vec{y}(u^1) = r\{\cos u^1, \sin u^1, 0\} \ \text{and} \ \vec{z}(u^1) = \{\cos u^1, 0, \sin u^1\}$$

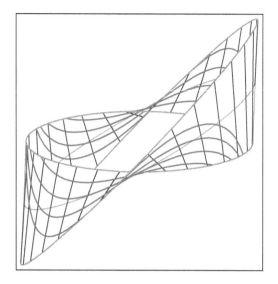

Figure 2.77 Lines of curvature on a Moebius strip

for some constant $r > 0$, the so–called Moebius strip *or* Moebius band *(Figure 2.77).*
Then we have

$$\vec{y}'(u^1) = r\{-\sin u^1, \cos u^1, 0\}, \ \vec{y}''(u^1) = -\vec{y}(u^1),$$
$$\vec{z}'(u^1) = r\{-\sin u^1, 0, \cos u^1\}, \ \vec{z}''(u^1) = -\vec{z}(u^1),$$

$$a(u^1) = 2r(-\sin^2 u^1 \cos u^1 - \cos^3 u^1) = -2r \cos u^1 \neq 0 \ for \ u^1 \neq \frac{\pi}{2}, \frac{3\pi}{2},$$

$$\vec{y}''(u^1) \bullet (\vec{y}'(u^1) \times \vec{z}(u^1)) = -r^2 \sin u^1,$$
$$\vec{y}''(u^1) \bullet (\vec{z}'(u^1) \times \vec{z}(u^1)) = -r \sin u^1,$$
$$\vec{z}''(u^1) \bullet (\vec{y}'(u^1) \times \vec{z}(u^1)) = \vec{z}''(u^1) \bullet (\vec{z}'(u^1) \times \vec{z}(u^1)) = 0,$$

$$f(u^1) = -\frac{r}{2} \tan u^1, \ g(u^1) = -\frac{1}{2} \tan u^1 \ and \ h(u^1) = 0 \ for \ u^1 \neq \frac{\pi}{2}, \frac{3\pi}{2}.$$

Now (2.125) reduces to

$$\frac{du^2}{du^1} = -\frac{1}{2} \tan u^1 (r + u^2)$$

with the solutions

$$u^2(u^1) = c\sqrt{|\cos u^1|} - r$$

for some constant $c > 0$.

The differential equations for lines of curvature and asymptotic lines reduce considerably for suitable choices of the parameters of the surface.

Theorem 2.8.6. *(a) The parameter lines of a surface are lines of curvature if and only if $g_{12} = L_{12} = 0$.*
(b) The parameter lines of a surface are asymptotic lines if and only if $L_{11} = L_{22} = 0$ and $L_{12} \neq 0$.

Proof. (a) First we prove Part (a).

(a.i) We prove the necessity of the condition $g_{12} = L_{12} = 0$.
We assume that the parameter lines of the surface S are lines of curvature. Then it follows that $g_{12} = 0$, since the principal directions are orthogonal by Theorem 2.5.14. Furthermore, since $du^1 = 0$ for the u^2–lines, the differential equations (2.120) for the lines of curvature reduce to

$$\det \begin{pmatrix} L_{12}du^2 & 0 \\ L_{22}du^2 & g_{22}du^2 \end{pmatrix} = L_{12}g_{22}(du^2)^2 = 0.$$

This implies $L_{12} = 0$, since $g_{22} \neq 0$.

(a.ii) Now we show the sufficiency of the condition $L_{12} = g_{12} = 0$.
We assume that $L_{12} = g_{12} = 0$. Then the differential equations (2.120) for the lines of curvature on S reduce to

$$\det \begin{pmatrix} L_{11}du^1 & g_{11}du^1 \\ L_{22}du^2 & g_{22}du^2 \end{pmatrix} = (L_{11}g_{22} - L_{22}g_{11})du^1 du^2 = 0$$

and the parameter lines of S are lines of curvature.

(b) Now we prove Part (b).

(b.1) First, we show the necessity of the condition $L_{11} = L_{22} = 0$ and $L_{12} \neq 0$.
We assume that the parameter lines of the surface S are asymptotic lines. Since $du^2 = 0$ for the u^1–lines, the differential equations (2.121) for the asymptotic lines reduce to $L_{11}(du^1)^2 = 0$; this implies $L_{11} = 0$. Similarly the differential equations (2.121) reduce to $L_{22}(du^2)^2 = 0$ for the u^2–lines, which implies $L_{22} = 0$.

(b.2) Finally, we show the sufficiency of the conditions $L_{11} = L_{22} = 0$ and $L_{12} \neq 0$.
If $L_{11} = L_{22} = 0$, then the differential equations (2.121) for the asymptotic lines on S reduce to $L_{12}du^1 du^2 = 0$ and the parameter lines of S are asymptotic lines, since $L_{12} \neq 0$.

□

Theorem 2.8.7. *At every point of an asymptotic line with non–vanishing curvature κ, its binormal vector and the surface normal vector coincide, or equivalently, the osculating planes of an asymptotic line coincide with the tangent planes of the surface (Figure 2.78).*

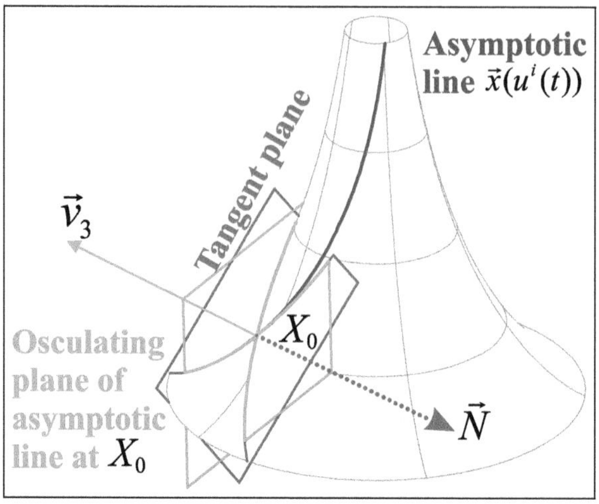

Figure 2.78 Osculating and tangent planes at a point of an asymptotic line

Proof. The normal curvature of a curve is given by (2.40). Thus we have for an asymptotic line

$$0 = \kappa_n = \kappa \vec{v}_2 \bullet \vec{N}$$

and $\kappa \neq 0$ implies that $\vec{v}_2$ is orthogonal to $\vec{N}$. Since $\vec{v}_3 \perp \vec{v}_1, \vec{v}_2$ and $\vec{N} \perp \vec{v}_1$ it follows that $\vec{v}_3 = \pm \vec{N}$. □

Visualization 2.8.8 (Lines of curvature on a hyperbolic paraboloid).

We consider the hyperbolic paraboloid of Visualization 2.5.27 with a parametric representation

$$\vec{x}(u^i) = \{u^1, u^2, u^1 u^2\} \ \left((u^1, u^2) \in \mathbb{R}^2\right).$$

The lines of curvature on the hyperbolic paraboloid are the u^2–lines and the curves given by

$$u^1 = \sinh\left(\pm \sinh^{-1} u^2\right), \tag{2.126}$$

where $c \in \mathbb{R}$ is a constant (Figure 2.79).

Proof. The differential equations (2.120) for the lines of curvature reduce to

$$\det \begin{pmatrix} L_{12}du^2 & g_{11}du^1 + g_{12}du^2 \\ L_{12}du^1 & g_{12}du^1 + g_{22}du^2 \end{pmatrix} = 0,$$

or equivalently

$$g_{22}\left(du^2\right)^2 - g_{11}\left(du^1\right)^2 = 0,$$

hence

$$\frac{du^2}{\sqrt{1 + (u^2)^2}} = \pm \frac{du^1}{\sqrt{1 + (u^1)^2}}.$$

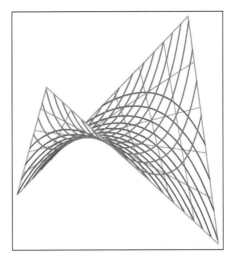

Figure 2.79 Lines of curvature on a hyperbolic paraboloid

This implies

$$\sinh^{-1} u^1 = \pm \sinh^{-1} u^2 + c,$$

where $c \in \mathbb{R}$ is a constant, or (2.126). □

Remark 2.8.9. *Alternatively, since $L_{11} = L_{22} = 0$ and $L_{12} \neq 0$ for the hyperbolic paraboloid of Visualization 2.8.8, the asymptotic lines are the parameter lines by Theorem 2.8.6 (b).*

Visualization 2.8.10 (Lines of curvature on an elliptic cone).
Let a_1, a_2 be two positive constants. We consider the elliptic cone of Visualization 2.6.4 with a parametric representation

$$\vec{x}(u^i) = \left\{ \frac{1}{\sqrt{a_1}} u^1 \cos u^2, \frac{1}{\sqrt{a_2}} u^1 \sin u^2, u^1 \right\} \ \left((u^1, u^2) \in (0, \infty) \times (0, 2\pi) \right).$$

Then the lines of curvature are the u^1–lines and the curves given by (Figure 2.80)

$$u^1(u^2) = \frac{k}{\sqrt{a_2 \cos^2 u^2 + a_1 \sin^2 u^2 + a_1 a_2}}, \qquad (2.127)$$

where $k > 0$ is a constant.

Proof. We have $L_{12} = g_{12} = 0$ if and only if $a_1 = a_2$ and consequently the parameter lines are lines of curvature by Theorem 2.8.6 (a) if and only if the cone is circular. Now let $a_1 \neq a_2$. Then we have $L_{11}(u^k) = L_{12}(u^k) = 0$ by Visualization 2.6.4 (b) and the differential equations (2.120) for the lines of curvature reduce to

$$L_{22}(u^k)du^2 \left(g_{11}(u^k)du^1 + g_{12}(u^k)du^2 \right) = 0.$$

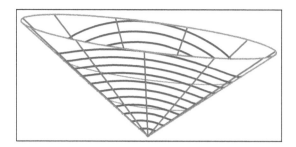

Figure 2.80 Lines of curvature on an elliptic cone

The u^1–lines are one family of lines of curvature. Furthermore, $L_{22}(u^k) \neq 0$ implies

$$\frac{du^1}{du^2} = -\frac{g_{12}(u^k)}{g_{11}(u^1)} = -\frac{u^1}{2g_{11}(u^1)}\frac{dg_{11}(u^1)}{du^2}$$

and consequently

$$\log u^1 = -\frac{1}{2}\log g_{11}(u^1) + c,$$

where $c \in \mathbb{R}$ is a constant.

Since, by the proof of Visualization 2.6.4 (b),

$$g_{11}(u^1) = \frac{1}{a_1 a_2}\left(a_2\cos^2 u^2 + a_1\sin u^2 + a_1 a_2\right),$$

the second family of lines of curvature is given by (2.127). □

Visualization 2.8.11 (Asymptotic lines on surfaces of rotation).
Let RS be a surface of rotation with a parametric representation

$$\vec{x}(u^i) = \{r(u^1)\cos u^2, r(u^1)\sin u^2, h(u^1)\}\ \left((u^1,u^2)\in D\right)$$

with $r(u^1) > 0$ and $|r'(u^1)| + |h'(u^1)| > 0$ for all u^1.
Then we have by Example 2.4.6

$$L_{11}(u^i) = L_{11}(u^1) = \frac{h''(u^1)r'(u^1) - h'(u^1)r''(u^1)}{\sqrt{(r'(u^1))^2 + (h'(u^1))^2}},\ L_{12}(u^i) = 0$$

and

$$L_{22}(u^i) = L_{22}(u^1) = \frac{h'(u^1)r(u^1)}{\sqrt{(r'(u^1))^2 + (h'(u^1))^2}}.$$

The differential equations for asymptotic lines (2.121) reduce to

$$\frac{du^2}{du^1} = \pm\sqrt{-\frac{L_{11}(u^1)}{L_{22}(u^1)}} = \pm\sqrt{\frac{h'(u^1)r''(u^1) - h''(u^1)r'(u^1)}{h'(u^1)r(u^1)}}, \qquad (2.128)$$

which have real solutions for those u^1 for which

$$\frac{h'(u^1)r''(u^1) - h''(u^1)r'(u^1)}{h'(u^1)r(u^1)} \geq 0.$$

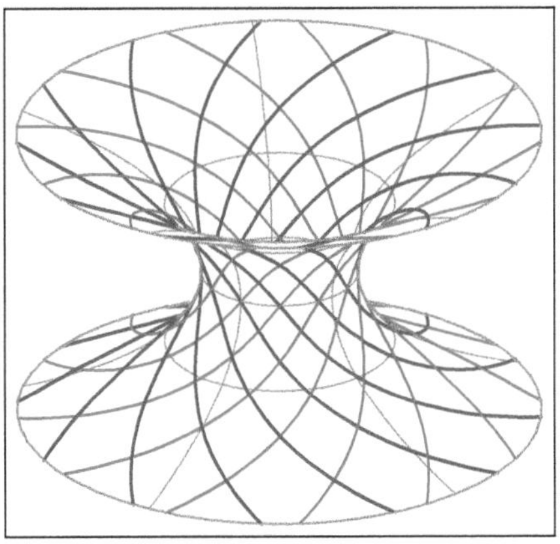

Figure 2.81 Asymptotic lines on a catenoid

(a) *Asymptotic lines on the catenoid.*
    Let $a > 0$ be a constant. We consider the catenoid with

$$r(u^1) = a \cosh \frac{u^1}{a} \text{ and } h(u^1) = u^1 \text{ for } u^1 \in \mathbb{R}.$$

Since $h''(u^1) = 0$, (2.128) reduces to

$$\frac{du^2}{du^1} = \pm \sqrt{\frac{r''(u^1)}{r(u^1)}} = \pm \frac{1}{a} \text{ for all } u^1 \in \mathbb{R}$$

with the solutions

$$u^2(u^1) = \pm \frac{u^1}{a} + c,$$

where $c \in \mathbb{R}$ is a constant (Figure 2.81).

(b) *We consider the surface of rotation with*

$$r(u^1) = u^1 \text{ and } h(u^1) = \log u^1 \text{ for } u^1 \in (0, \infty).$$

Since $r''(u^1) = 0$, (2.128) reduces to

$$\frac{du^2}{du^1} = \pm \sqrt{-\frac{h''(u^1)}{h'(u^1)r(u^1)}} = \pm \frac{1}{u^1} \text{ for all } u^1 \in (0, \infty)$$

with the solutions $u^2(u^1) = \pm \log u^1 + \log c$, where $c > 0$ is a constant (Figure 2.82).

(c) *Now we consider the torus with*

$$r(u^1) = r_0 + r_1 \cos u^1 \text{ and } h(u^1) = r_1 \sin u^1 \text{ for } u^1 \in (0, 2\pi),$$

where $r_0 > r_1$.
    Then (2.128) reduces to (Figure 2.83)

$$\frac{du^2}{du^1} = \pm \sqrt{-\frac{r_1}{\cos u^1 (r_0 + r_1 \cos u^1)}} \text{ for } u^1 \in \left(\frac{\pi}{2}, \frac{3\pi}{2} \right).$$

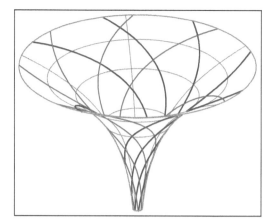

Figure 2.82 Asymptotic lines on the surface of rotation with $r(u^1) = 1$ and $h(u^1) = \log u^1$

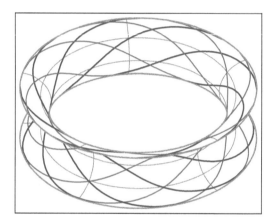

Figure 2.83 Asymptotic lines on the part of $u^1 \in (\pi/2, 3\pi/2)$ of a torus

Visualization 2.8.12 (Asymptotic lines on a pseudo-sphere).
The asymptotic lines on the pseudo-sphere of Visualization 2.5.19 (c) with its parametric representation defined by

$$r(u^1) = \exp\left(-u^1\right) \text{ and } h(u^1) = \int \sqrt{1 - \exp\left(-2u^1\right)}\, du^1 \text{for } u^1 > 0$$

are given by (Figure 2.84)

$$u_1(t) = t \text{ and } u_2(t) = \pm \log\left(e^t + \sqrt{e^{2t} - 1}\right) + c \text{ for all } t > 0, \qquad (2.129)$$

where $c \in \mathbb{R}$ is a constant of intgration.

Proof. We write $u = u^1$, and observe that

$$r(u) = e^{-u}, \; h'(u) = \sqrt{1 - e^{-2u}} \text{ and } r''(u) = r(u).$$

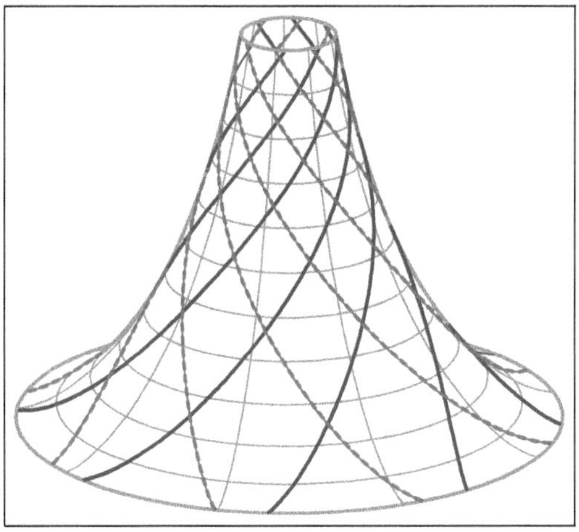

Figure 2.84 Families of asymptotic lines on the pseudo-sphere of Visualization 2.8.12

Hence the differential equations (2.121) for the asymptotic lines reduce to

$$\frac{du^2}{du^1} = \pm \frac{1}{|h'(u)|} = \pm \frac{1}{h'(u)},$$

since $h'(u) > 0$ for all $u > 0$. Therefore, we have to solve the integral

$$I = \int \frac{du}{h'(u)} = \int \frac{du}{\sqrt{1 - e^{-2u}}} = \int \frac{e^u}{\sqrt{e^{2u} - 1}} \, du.$$

The substitution $t = e^u > 1$ yields

$$I = \int \frac{dt}{\sqrt{t^2 - 1}} = \log\left(t + \sqrt{t^2 - 1}\right) = \log\left(e^u + \sqrt{e^{2u} - 1}\right).$$

We choose the upper sign and $c = 0$ in (2.129), that is, we consider the asymptotic line given by

$$u_1(t) = t \text{ and } u_2(t) = \log\left(e^t + \sqrt{e^{2t} - 1}\right) \text{ for all } t > 0. \tag{2.130}$$

□

Visualization 2.8.13. (Asymptotic lines on the explicit surface of Visualization 2.6.3) *The differential equations for the asymptotic lines on the explicit surface of Visualization 2.6.3 reduce to*

$$2(du^1)^2 + 6u^2(du^2)^2 = 0$$

with the real solutions (Figure 2.85)

$$u^1(u^2) = \pm \frac{2}{\sqrt{3}} \left(\sqrt{-u^2}\right)^3 + u_0^1$$

for $u^2 < 0$, where u_0^1 is a constant.

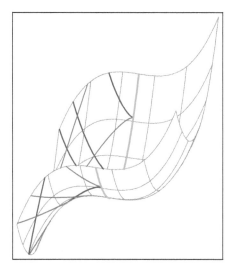

Figure 2.85 Asymptotic lines on the explicit surface with $f(u^i) = (u^1)^2 + (u^2)^3$

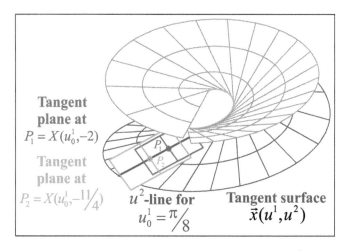

Figure 2.86 A developable surface and tangent planes along a u^2–line

Definition 2.8.14. A ruled surface with a parametric representation

$$\vec{x}(u^i) = \vec{y}(u^1) + u^2 \vec{z}(u^1) \; ((u^1, u^2) \in D \subset \mathbb{R}^2) \tag{2.131}$$

is said to be a *torse*, or *developable*, if all the tangent planes along a u^2–line coincide (Figures 2.86 and 2.87).

First we give a characterization of developable surfaces.

Theorem 2.8.15. *A ruled surface is a developable if and only if*

$$\vec{y}'(u^1) \bullet (\vec{z}(u^1) \times \vec{z}'(u^1)) = 0 \; \text{for all } u^1. \tag{2.132}$$

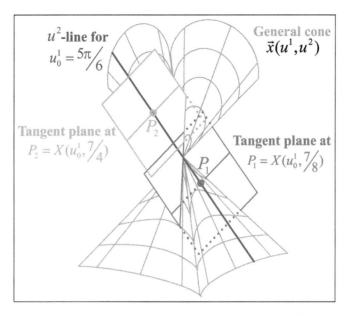

Figure 2.87 A developable surface and tangent planes along a u^2–line

Proof. A developable surface is characterized by the fact that the normal vectors along a u^2–line are constant, that is, $\vec{N}_2 = \vec{0}$. Now $\vec{N}^2 = 1$ implies $\vec{N} \bullet \vec{N}_2 = 0$, hence $\vec{N}_2 = \vec{0}$ and $\vec{N} \bullet \vec{N}_2$ are equivalent conditions. We have $\vec{x}_1 = \vec{y}' + u^2\vec{z}'$ and $\vec{x}_2 = \vec{z}$, hence

$$\vec{N} = \frac{\vec{x}_1 \times \vec{x}_2}{\|\vec{x}_1 \times \vec{x}_2\|} = \frac{1}{\sqrt{g}}(\vec{y}' + u^2\vec{z}') \times \vec{z},$$

and

$$\vec{N}_2 = \frac{1}{\sqrt{g}}(\vec{z}' \times \vec{z}) + \vec{N}\sqrt{g}\frac{\partial}{\partial u^2}\left(\frac{1}{\sqrt{g}}\right),$$

hence by the Grassmann identity (1.5) and the identities in Remark 1.1.4

$$\vec{N} \times \vec{N}_2 = \frac{1}{g}(\vec{y}' \times \vec{z}) \times (\vec{z}' \times \vec{z})$$

$$= \frac{1}{g}\left(((\vec{y}' \times \vec{z}) \bullet \vec{z})\vec{z}' - ((\vec{y}' \times \vec{z}) \bullet \vec{z}')\vec{z}\right)$$

$$= -\frac{1}{g}((\vec{y}' \times \vec{z}) \bullet \vec{z}')\vec{z} = -\frac{1}{g}(\vec{y}' \bullet (\vec{z} \times \vec{z}'))\vec{z}.$$

Thus $\vec{N} \times \vec{N}_2 = \vec{0}$ if and only if the condition in (2.132) holds. □

Developable surfaces can be characterized explicitly.

Theorem 2.8.16. *A surface is developable if and only if it is a cylinder, cone or tangent surface (Figure 2.88).*

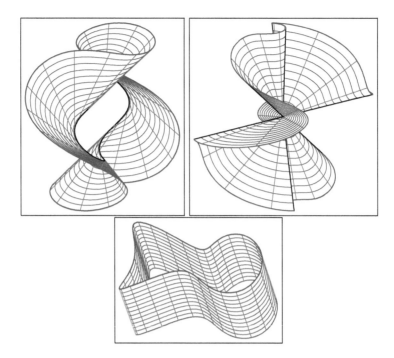

Figure 2.88 Developable surfaces: Top left tangent surface, right cone. Bottom cylinder

Proof.

(i) First we show that cylinders, cones and tangent surfaces are developable surfaces.

Let γ be a curve with a parametric representation $\vec{x}^*(t)$ $(t \in I_1)$. We put $u^1 = t$ and $\vec{y}(u^1) = \vec{x}^*(u^1)$.

(i.1) Then the *general cylinder* generated by γ has a parametric representation

$$\vec{x}(u^i) = \vec{y}(u^1) + u^2 \vec{z} \ \left((u^1, u^2) \in D\right),$$

where $\vec{z}$ is a constant vector.

Since $\vec{z}' = \vec{0}$, the condition in (2.132) for a developable surface is satisfied.

(i.2) The tangent surface generated by γ has a parametric representation

$$\vec{x}(u^i) = \vec{y}(u^1) + u^2 \vec{y}'(u^1) \ ((u^1, u^2) \in D).$$

Since now $\vec{z} = \vec{y}'$, the condition in (2.132) for a developable surface is satisfied.

(i.3) We may assume that the general cone generated by γ has its vertex in the origin; it has a parametric representation

$$\vec{x}(u^i) = \vec{0} + u^2 \vec{y}(u^1) \ ((u^1, u^2) \in D)$$

and the condition in (2.132) for a developable surface is satisfied.

(ii) Conversely, we assume that a ruled surface $RulS$ with a parametric representation (2.131) is a developable surface.

Then the condition in (2.132) holds by Theorem 2.8.15. We may assume that u^1 is the arc length along $\vec{y}$. Furthermore, we may assume by Remark 2.5.2 (a) that $\vec{y}' \perp \vec{z}$. Let γ be a curve on $RulS$ which is given by $u^2 = \varphi(u^1)$, that is, γ has a parametric representation $\vec{x}^*(u^1) = \vec{x}(u^1, \varphi(u^1))$. Denoting the derivatives with respect to u^1 by a dot, we obtain

$$\dot{\vec{x}}^* = \dot{\vec{y}} + \varphi\dot{\vec{z}} + \dot{\varphi}\vec{z}.$$

Since the unit vectors $\dot{\vec{y}}$, $\vec{z}$ and $\vec{N}$ are orthogonal to one another, we may express $\dot{\vec{z}}$ as a linear combination of them

$$\dot{\vec{z}} = a\dot{\vec{y}} + b\vec{z} + c\vec{N}, \text{ where } a = \dot{\vec{y}} \bullet \dot{\vec{z}},$$

$$b = \vec{z} \bullet \dot{\vec{z}} = \frac{1}{2}\frac{d}{du^1}\left(\vec{z}^2\right) = 0$$

and

$$c = \vec{N} \bullet \dot{\vec{z}} = \frac{1}{\sqrt{g}}\dot{\vec{z}} \bullet (\dot{\vec{y}} \times \vec{z}) = 0, \text{ by (2.132)}.$$

Therefore, it follows that

$$\dot{\vec{x}}^* = \dot{\vec{y}}(1 + \varphi(\dot{\vec{y}} \bullet \dot{\vec{z}})) + \dot{\varphi}\vec{z}.$$

The tangent of the curve γ is in the direction of $\vec{z}$, if

$$1 + \varphi(\dot{\vec{y}} \bullet \dot{\vec{z}}) = 0. \tag{2.133}$$

If $\dot{\vec{y}} \bullet \dot{\vec{z}} = 0$ for all u^1, then we have $\dot{\vec{z}} = \vec{0}$, hence $\vec{z}$ is a constant vector and the developable surface is a cylinder.

If $\dot{\vec{y}} \bullet \dot{\vec{z}} \neq 0$, then (2.133) implies

$$u^2 = \varphi(u^1) = -\frac{1}{\dot{\vec{y}} \bullet \dot{\vec{z}}}.$$

Consequently, there is at most one curve on $RulS$ with its tangents in the direction of $\vec{z}$, and one and only one curve for those parts of $RulS$ for which

$$\vec{0} \neq \dot{v}\vec{x}^* = \vec{z}\dot{\varphi}.$$

Parts of the surface with $\varphi = const$ have $\dot{\vec{x}}^* = \vec{c}$, where $\vec{c}$ is a constant vector. Then the straight lines in the directions of the vectors $\vec{z}$ have a common point; such parts of the surface are cones.

□

The next result shows that developable surfaces can be generated by lines of curvature on surfaces.

Theorem 2.8.17. *A curve γ on a surface S is a line of curvature if and only if the ruled surface generated by γ and the surface normal vectors along γ is a developable surface.*

Proof. Let S have a parametric representation $\vec{y}(u^i)$ $((u^1, u^2) \in D)$ and we may assume that the parameters are orthogonal. Furthermore, let γ be given by $\vec{y}^*(s) = \vec{y}(u^i(s))$ where s is the arc length along γ.

(i) First we assume that γ is a line of curvature on S.
We consider the ruled surface $RulS$ with a parametric representation

$$\vec{x}(s, t) = \vec{y}^*(s) + t\vec{N}^*(s),$$

where $\vec{N}^*(s) = \vec{N}(u^i(s))$. It follows that

$$\vec{x}_1 = \frac{\partial \vec{x}}{\partial s} = \frac{d\vec{y}^*}{ds} + t\frac{d\vec{N}^*}{ds} = \vec{y}_i \dot{u}^i + t\vec{N}_i \dot{u}^i \text{ and } \vec{x}_2 = \frac{\partial \vec{x}}{\partial t} = \vec{N}^* = \vec{N}.$$

Writing $\dot{\vec{y}}^* = d\vec{y}^*/ds$ and $\dot{\vec{N}}^* = d\vec{N}^*/ds$, we obtain by the Grassmann identity (1.5), the fact that the parameters on S are orthogonal, and the definition of the first and second fundamental coefficients

$$\dot{\vec{y}}^* \bullet (\vec{N}^* \times \dot{\vec{N}}^*) = \vec{y}_i \bullet (\vec{N} \times \vec{N}_k)\dot{u}^i\dot{u}^k$$
$$= \sqrt{g}\vec{y}_i \bullet \left((\vec{y}_1 \times \vec{y}_2) \times \vec{N}_k\right)\dot{u}^i\dot{u}^k$$
$$= -\sqrt{g}\vec{y}_i \bullet \left(\vec{N}_k \times (\vec{y}_1 \times \vec{y}_2)\right)\dot{u}^i\dot{u}^k$$
$$= -\sqrt{g}\vec{y}_i \bullet \left(\vec{y}_1(\vec{N}_k \bullet \vec{y}_2) - \vec{y}_2(\vec{N}_k \bullet \vec{y}_1)\right)\dot{u}^i\dot{u}^k$$
$$= -\sqrt{g}g_{1i}(-L_{2k})\dot{u}^i\dot{u}^k + \sqrt{g}g_{2i}(-L_{1k})\dot{u}^i\dot{u}^k$$
$$= \sqrt{g}\left(g_{1i}L_{2k}\dot{u}^i\dot{u}^k - g_{2i}L_{1k}\dot{u}^i\dot{u}^k\right).$$

The principal curvature λ satisfies

$$L_{1k}\dot{u}^k = \lambda g_{1i}\dot{u}^i \text{ and } L_{2k}\dot{u}^k = \lambda g_{2i}\dot{u}^i,$$

hence

$$g_{1i}L_{2k}\dot{u}^i\dot{u}^k = \lambda g_{2i}g_{1l}\dot{u}^i\dot{u}^l = L_{1k}g_{2i}\dot{u}^i\dot{u}^k.$$

Thus we have

$$\dot{\vec{y}}^* \bullet (\vec{N}^* \times \dot{\vec{N}}^*) = 0 \tag{2.134}$$

and $RulS$ is a developable surface by condition (2.132) in Theorem 2.8.15.

(b) Conversely, we assume that $RulS$ is a developable surface.
Then (2.134) is satisfied and we have

$$g_{1i}L_{2k}\dot{u}^i\dot{u}^k = L_{1k}g_{2i}\dot{u}^i\dot{u}^k$$

and consequently the principal directions are given by u^i.

$\square$

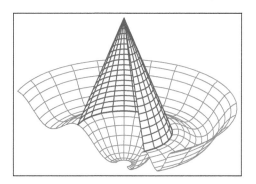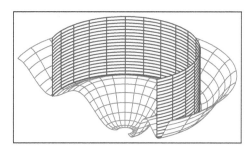

Figure 2.89 A circular cone (left) and a circular cylinder (right) generated by a line of curvature on a surface of rotation

Visualization 2.8.18. *Let RS be a surface of revolution, γ be a u^2–line on RS that corresponds to u_0^1. Then γ and the surface normal vectors along γ generate a circular cone, if $h'(u_0^1) \neq 0$, and a circular cylinder, if $h'(u_0^1) = 0$ (Figure 2.89).*

Proof. Let RS have a parametric representation

$$\vec{x}(u^i) = \{r(u^1)\cos u^2, r(u^1)\sin u^2, h(u^1)\} \ \left((u^1, u^2) \in D\right).$$

Then we have $g_{12} = L_{12}$, and consequently the parameter lines are lines of curvature by Theorem 2.8.6 (a). The surface normal vectors along the u^2–line corresponding to u_0^1 are given by

$$\vec{N}^*(u^2) = \vec{N}(u_0^1, u^2) = \frac{1}{\sqrt{(r'(u_0^1))^2 + (h'(u_0^1))^2}}\{-h'(u_0^1)\cos u^2, -h'(u_0^1)\sin u^2, r'(u_0^1)\}.$$

If $h'(u_0^1) \neq 0$, then the ruled surface with a parametric representation

$$\vec{x}^*(u^i) = \vec{x}(u_0^1, u^2) + u^1 \vec{N}(u_0^1, u^2) \ ((u^1, u^2) \in \mathbb{R} \times (0, 2\pi))$$

is a circular cone with its vertex in the point

$$P = \left(0, 0, h(u_0^1) + \frac{r'(u_0^1)r(u_0^1)}{h'(u_0^1)}\right),$$

since, for

$$u_1^1 = \frac{r(u_0^1)}{h'(u_0^1)} \cdot \sqrt{(r'(u_0^1))^2 + (h'(u_0^1))^2},$$

we have

$$\vec{x}^*(u_1^1, u_2) = \vec{x}(u_0^1, u^2) + u_1^1\vec{N}(u_0^1, u^2) = \overrightarrow{OP} \text{ for all } u^2 \in (0, 2\pi).$$

If $h'(u_0^1) = 0$, then we obtain

$$\vec{x}^*(u^1, u^2) = \vec{x}(u_0^1, u^2) + u^1\vec{e}^3,$$

that is, a circular cylinder with its axis along the x^3–axis. $\qquad\square$

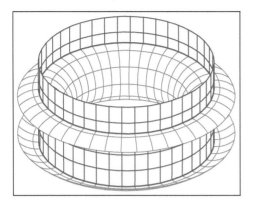 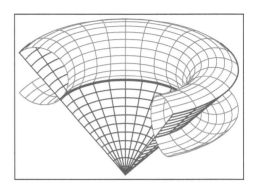

Figure 2.90 A circular cylinder and cone generated by a line of curvature on a torus

Visualization 2.8.19. *Let $r_0 > r_1 > 0$. We consider the torus with a parametric representation*

$$\vec{x}(u^i) = \{(r_0 + r_1 \cos u^1) \cos u^2, (r_0 + r_1 \cos u^1) \sin u^2, r_1 \sin u^1\} \ \left((u^1, u^2) \in (0, 2\pi)^2\right).$$

Then we have

$$r(u^1) = r_0 + r_1 \cos u^1, \ \ h(u^1) = r_1 \sin u^1,$$
$$r'(u^1) = -r_1 \sin u^1 \ and \ h'(u^1) = r_1 \cos u^1.$$

If $u_0^1 \neq \pi/2, 3\pi/2$, then $h'(u_0^1) \neq 0$, and the u_2–line that corresponds to u_0^1 and the surface normal vectors of the torus along that u_2–line generate a cone with vertex in the point

$$P = \left(0, 0, \frac{-r_1 \sin u_0^1 (r_0 + r_1 \cos u_0^1)}{r_1 \cos u_0^1} + r_1 \sin u_0^1\right) = (0, 0, -r_0 \tan u_0^1),$$

by Visualization 2.8.18 (right in Figure 2.90).
If $u_0^1 = \pi/2$, then $h'(u_0^1) = 0$ and we obtain the cylinder with a parametric representation (left in Figure 2.90)

$$\vec{x}^*(u^i) = \{r_0 \cos u^2, r_0 \sin u^2, r_1\} - u^1 \vec{e}^3 \ \left((u^1, u^2) \in \mathbb{R} \times 2\pi\right).$$

Visualization 2.8.20. *Let a_1 and a_2 be distinct positive constants. We consider the elliptic EC cone with a parametric representation*

$$\vec{x}(u^i) = \left\{\frac{1}{\sqrt{a_1}} u^1 \cos u^2, \frac{1}{\sqrt{a_2}} u^1 \sin u^2, u^1\right) \ \left((u^1, u^2) \in (0, \infty) \times (0, 2\pi)\right),$$

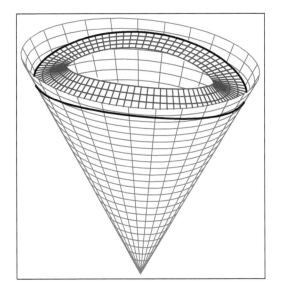

Figure 2.91 A tangent surface generated by a line of curvature on an elliptic cone

and the curve γ on EC given by

$$u^1(u^2) = \frac{k}{\sqrt{\dfrac{1}{a_1}\cos^2 u^2 + \dfrac{1}{a_2}\sin^2 u^2 + 1}},$$

where $k > 0$ is a constant.
*Then the ruled surface generated by γ and the surface normal vectors of EC along γ
is a tangent surface (Figure 2.91).*

Proof. We know from Visualization 2.8.10 that γ is a line of curvature on EC, hence
the ruled surface $RulS$ generated by γ and the surface normal vectors of EC along
γ is a developable surface by Theorem 2.8.17, hence a cone, a cylinder or a tangent
surface by Theorem 2.8.16.

(i) We are going to show that $RulS$ is neither a cone nor a cylinder.
    The surface normal vectors along γ are given by

$$\vec{N} = \vec{N}(u^2) = \frac{1}{\sqrt{a_1\cos^2 u^2 + a_1\cos^2 u^2 + 1}}\{-\sqrt{a_1}\cos u^2, -\sqrt{a_1}\sin u^2, 1\}.$$

We put $\bar{u}^1 = u^2$ and

$$\vec{y}(\bar{u}^1) = \vec{x}(u^1(u^2), u^2)$$

$$= \frac{k}{\sqrt{\dfrac{1}{a_1}\cos^2 \bar{u}^1 + \dfrac{1}{a_2}\sin^2 \bar{u}^1 + 1}}\left\{\frac{1}{\sqrt{a_1}}\cos \bar{u}^1, \frac{1}{\sqrt{a_2}}\sin \bar{u}^1, 1\right\}$$

and
$$\vec{z}(\bar{u}^1) = \vec{N}(\bar{u}^1).$$

Then $RulS$ has a parametric representation
$$\vec{x}(\bar{u}^i) = \vec{y}(\bar{u}^1) + \bar{u}^2 \vec{z}(\bar{u}^1).$$

(i.1) Since
$$\vec{z}(\bar{u}^1) \bullet \vec{e}^3 = \frac{1}{\sqrt{a_1 \cos^2 \bar{u}^1 + a_2 \sin^2 \bar{u}^1 + 1}},$$

the vectors $\vec{z}$ do not have a constant direction, hence $RulS$ cannot be a cylinder.

(ii.1) If $RulS$ were a cone, the straight lines with the parametric representations
$$\vec{x}(t) = \vec{y}(0) + t\vec{z}(0) \text{ and } \vec{x}^*(t^*) = \vec{y}(\pi/2) + t^*\vec{z}(\pi/2) \text{ for } t, t^* \in \mathbb{R}$$

would have a point of intersection P. Comparing the first two components of $\vec{x}(t)$ and $\vec{x}^*(t^*)$, we obtain
$$\frac{k}{\sqrt{1+a_1}} - t\frac{\sqrt{a_1}}{\sqrt{1+a_1}} = 0 \text{ and } \frac{k}{\sqrt{1+a_2}} - t^*\frac{\sqrt{a_2}}{\sqrt{1+a_2}} = 0,$$

hence
$$t = \frac{k}{\sqrt{a_1}} \text{ and } t^* = \frac{k}{\sqrt{a_2}}.$$

Thus we have
$$\overrightarrow{OS} = \vec{x}(k/\sqrt{a_1}) = \left\{0, 0, \frac{k\sqrt{a_1 + 1}}{\sqrt{a_1}}\right\}$$

and
$$\overrightarrow{OS} = \vec{x}^*(k/\sqrt{a_2}) = \left\{0, 0, \frac{k\sqrt{a_2 + 1}}{\sqrt{a_2}}\right\},$$

but $a_1 \neq a_2$ implies
$$\frac{k\sqrt{a_1 + 1}}{\sqrt{a_1}} \neq \frac{k\sqrt{a_2 + 1}}{\sqrt{a_2}},$$

and consequently $\vec{x}(k/\sqrt{a_1}) \neq \vec{x}^*(k/\sqrt{a_2})$. Thus $RulS$ cannot be a cone either.

□

2.9   TRIPLE ORTHOGONAL SYSTEMS

Triple orthogonal systems can be used to determine lines of curvature on certain surfaces, as we will see in Theorem 2.9.4.

Definition 2.9.1. We assume that the function $\vec{y} : D \subset \mathbb{R}^3 \to \mathbb{R}^3$ has continuous partial derivatives of order 2, that the vectors $\vec{y}_i = \partial\vec{y}/\partial u^i$ $(i = 1, 2, 3)$ are linearly independent and $\vec{y}_i \bullet \vec{y}_k = 0$ $(i \neq k)$. Then the surfaces given by $u^j = const$ form three families of mutually orthogonal surfaces, a so–called *triple orthogonal system of surfaces*.

Visualization 2.9.2 (Some triple orthogonal systems).
(a) Let

$$\vec{y}(u^1, u^2, u^3) = \{u^1 \cos u^2, u^1 \sin u^2, u^3\} \text{ for } (u^1, u^2, u^3) \in \mathbb{R} \times (0, 2\pi) \times \mathbb{R}.$$

If $u^1 = r > 0$ is constant, then we obtain a circular cylinder of radius r.
If $u^2 = \phi$ is constant, then we obtain a plane through the x^3–axis with an angle ϕ to the x^1–axis.
If $u^3 = h$ is constant, then we obtain a plane through $(0, 0, h)$, parallel to the $x^1 x^2$–plane.
Obviously $\vec{y}$ has continuous second-order partial derivatives,

$$\vec{y}_1 = \{\cos u^2, \sin u^2, 0\}, \quad \vec{y}_2 = \{-u^1 \sin u^2 0, u^1 \cos u^2, 0\},$$
$$\vec{y}_3 = \{0, 0, 1\} \text{ and } \vec{y}_i \bullet \vec{y}_k = 0 \text{ for } i \neq k.$$

Thus the surfaces form a triple orthogonal system.
(b) Let
$$\vec{y}(u^1, u^2, u^3) = \{u^1 \cos u^2 \cos u^3, u^1 \cos u^2 \sin u^3, u^1 \sin u^2\}$$

for $(u^1, u^2, u^3) \in (0, \infty) \times (-\pi/2, \pi/2) \times (0, 2\pi)$.
If $u^1 = r > 0$ is constant, then we obtain a sphere of radius r with its centre in the origin.
If $u^2 = \alpha \in (-\pi/2, \pi/2)$ is constant, then we obtain a circular cone with its vertex in the origin and meridians with an angle α to the $x^1 x^2$–plane.
If $u^3 = const$ is constant, then we obtain a plane orthogonal to the $x^1 x^2$–plane.
Obviously $\vec{y}$ has continuous second order partial derivatives

$$\vec{y}_1 = \{\cos u^2 \cos u^3, \cos u^2 \sin u^3, \sin u^2\},$$
$$\vec{y}_2 = \{-u^1 \sin u^2 \cos u^3, -u^1 \sin u^2 \sin u^3, u^1 \cos u^2\},$$
$$\vec{y}_3 = \{-u^1 \cos u^2 \sin u^3, u^1 \cos u^2 \cos u^3, 0\}$$
$$\text{and } \vec{y}_i \bullet \vec{y}_k = 0 \text{ for } i \neq k.$$

Thus the surfaces form a triple orthogonal system (Figure 2.92).

Visualization 2.9.3 (A triple system which is not orthogonal).
Let

$$\tilde{S} = \{(r, \phi, \theta) : r \in (0, \infty), \phi \in (0, 2\pi), \theta \in (-\pi, \pi)\}$$

and

$$\vec{y}(u^1, u^2, u^3) = \left\{u^3(1 + \cos u^1) \cos u^2, u^3(1 + \cos u^1) \sin u^2, u^3 \sin u^1\right\}$$

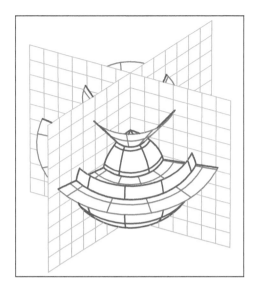

Figure 2.92 A triple orthogonal system of spheres, circular cones and planes

for $(u^1, u^2, u^3) \in \tilde{S}$.
If $u^3 = r > 0$ is constant, then $\vec{y}(u^1, u^2, r)$ for $(u^1, u^2) \in (-\pi, \pi) \times (0, 2\pi)$ is a parametric representation of a torus T_r.
If $u^2 = \phi \in (0, 2\pi)$ is constant, then

$$\vec{y}(u^1, \phi, u^3) = \left\{ u^3(1 + \cos u^1) \cos \phi, u^3(1 + \cos u^1) \sin \phi, u^3 \sin u^1 \right\}$$

for $(u^1, u^3) \in (-\pi, \pi) \times (0, \infty)$ is a parametric representation for a half plane H_ϕ.
If $u^1 = \theta \in (-\pi, \pi)$, then

$$\vec{y}(\theta, u^1, u^2) = \left\{ u^3(1 + \cos \theta) \cos u^2, u^3(1 + \cos \theta) \sin u^2, u^3 \sin \theta \right\}$$

for $(u^2, u^3) \in (0, 2\pi) \times (0, \infty)$ is a parametric representation of half a cone C_θ.
We consider the triple system $\mathcal{S} = \{T_r, H_\phi, C_\theta : (r, \phi, \theta) \in \tilde{S}\}$ and observe that for $u^1 \neq 0$

$$\frac{\partial \vec{y}}{\partial u^1} \bullet \frac{\partial \vec{y}}{\partial u^2} = \frac{\partial \vec{y}}{\partial u^2} \bullet \frac{\partial \vec{y}}{\partial u^3} = 0 \text{ and } \frac{\partial \vec{y}}{\partial u^1} \bullet \frac{\partial \vec{y}}{\partial u^3} = -u^3 \sin u^1 \neq 0.$$

Consequently the system $\mathcal{S}$ is not an orthogonal system (Figure 2.93).

The significance of triple orthogonal systems for lines of curvature lies in the next result.

Theorem 2.9.4 (Dupin, 1813). *The surfaces of a triple orthogonal system mutually intersect in lines of curvature.*

Proof. Partial differentiation of $\vec{y}_i \bullet \vec{y}_k = 0$ $(i \neq k)$ yields

$$\vec{y}_{12} \bullet \vec{y}_3 + \vec{y}_1 \bullet \vec{y}_{23} = \vec{y}_{23} \bullet \vec{y}_1 + \vec{y}_2 \bullet \vec{y}_{13} = \vec{y}_{21} \bullet \vec{y}_3 + \vec{y}_2 \bullet \vec{y}_{31} = 0,$$

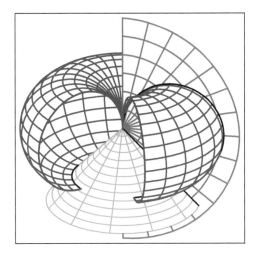

Figure 2.93 The triple system of Visualization 2.9.3

hence

$$\vec{y}_{12} \bullet \vec{y}_3 = \vec{y}_{31} \bullet \vec{y}_2 = \vec{y}_{23} \bullet \vec{y}_1 = 0.$$

Since the vectors $\vec{y}_1, \vec{y}_2$ and $\vec{y}_{12}$ are orthogonal to the vector $\vec{y}_3$, they are linearly dependent, that is, $\vec{y}_1 \bullet (\vec{y}_2 \times \vec{y}_{12}) = 0$. This means that we have

$$L_{12} = 0 \text{ and } g_{12} = \vec{y}_1 \bullet \vec{y}_2 = 0$$

for the surfaces given by $u^3 = const.$
Consequently, the parameter lines of these surfaces are lines of curvature by Theorem 2.8.6 (a).
The same holds true for the surfaces of the other two families. □

The next example shows that the converse of Dupin's theorem, Theorem 2.9.4, does not hold, in general.

Visualization 2.9.5. *We consider the families of surfaces consisting of spheres with their centres in the origin, elliptic – non-circular – cones with their vertices in the origin, and planes through the x^3–axis.*
The lines of intersection of the planes and spheres are meridians on the spheres, hence lines of curvature on the spheres, by Dupin's theorem, Theorem 2.9.4.
The lines of intersection of the planes and cones are the u^1–lines of the cones, hence lines of curvature on the cones, by Visualization 2.8.10.
The line of intersection of a cone, given by the parametric representation

$$\vec{x}(u^i) = \left\{ \frac{1}{\sqrt{a_1}} u^1 \cos u^2, \frac{1}{\sqrt{a_2}} u^1 \sin u^2, u^1 \right\} \ ((u^1, u^2) \in \mathbb{R} \times (0, 2\pi)),$$

where $a_1 \neq a_2$ are constants, and a sphere of radius $r > 0$, given by the equation

$$(x^1)^2 + (x^2)^2 + (x^3)^2 - r^2 = 0$$

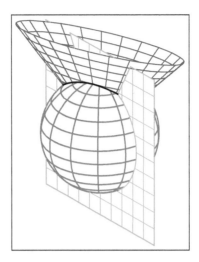

Figure 2.94 A line of curvature on an elliptic cone which is the intersection of the elliptic cone and a sphere

is determined by the equation

$$\frac{1}{a_1}(u^1)^2 \cos^2 u^2 + \frac{1}{a_2}(u^1)^2 \sin^2 u^2 + (u^1)^2 = r^2,$$

hence

$$u^1(u^2) = \frac{r}{\sqrt{\dfrac{1}{a_1}\cos^2 u^2 + \dfrac{1}{a_2}\sin^2 u^2 + 1}}$$

with respect to the parametric representation of the cone. Thus the line of intersection is a line of curvature on the cone by (2.127) in Visualization 2.8.10 (Figure 2.94).

Obviously, all points of a plane are umbilical points, and all points of any sphere are umbilical points by Example 2.5.25. Consequently, all curves on planes and spheres are lines of curvature.

Therefore, the surfaces of the families mutually intersect in their lines of curvature. But the surfaces do not form a triple orthogonal system, since the cones and planes do not intersect orthogonally, as we are going to see.

The surface normal vectors of a cone along its u^1–line corresponding to u_0^2 are given by

$$\vec{N} = \{-\sqrt{a_1}\cos u_0^2, -\sqrt{a_2}\sin u_0^2, 1\}\frac{1}{\sqrt{a_1 \cos^2 u_0^2 + a_2 \sin^2 u_0^2 + 1}},$$

and the surface normal vector of the corresponding plane of intersection is given by

$$\vec{N}^* = \{-\sqrt{a_1}\sin u_0^2, \sqrt{a_2}\cos u_0^2, 0\}\frac{1}{\sqrt{a_1 \sin^2 u_0^2 + a_2 \cos^2 u_0^2}}.$$

Therefore, we have

$$\vec{N} \bullet \vec{N}^{*} = \frac{(a_1 - a_2)\sin u_0^2 \sin u_0^2}{\sqrt{a_1 \cos^2 u_0^2 + a_2 \sin^2 u_0^2 + 1}\sqrt{a_1 \sin^2 u_0^2 + a_2 \cos^2 u_0^2}} \neq 0$$

for $u_0^2 \neq \pi/2, \pi, 3\pi/2$.

The next result and Dupin's theorem can be applied to determine the lines of curvature on ellipsoids and hyperboloids of one and two sheets.

Theorem 2.9.6. *Let $(a, b, c) \in \mathbb{R}^3$ be given with $0 < a^2 < b^2 < c^2$ and the functions $g_P : \mathbb{R} \setminus \{a^2, b^2, c^2\} \to \mathbb{R}$ be defined for every $P = (x^1, x^2, x^3) \in \mathbb{R}^3$ by*

$$g_P(\lambda) = \frac{(x^1)^2}{a^2 - \lambda} + \frac{(x^2)^2}{b^2 - \lambda} + \frac{(x^3)^2}{c^2 - \lambda} - 1.$$

Then the surfaces given by the equations

$$E_\lambda = \{P \in \mathbb{R}^3 : g_P(\lambda) = 0\} \text{ for } \lambda < a^2,$$
$$H_\lambda^1 = \{P \in \mathbb{R}^3 : g_P(\lambda) = 0\} \text{ for } a^2 < \lambda < b^2,$$

and

$$H_\lambda^2 = \{P \in \mathbb{R}^3 : g_P(\lambda) = 0\} \text{ for } b^2 < \lambda < c^2$$

form a triple orthogonal system of the families $\mathcal{E} = \{E_\lambda : \lambda < a^2\}$, $\mathcal{H}^1 = \{H_\lambda^1 : a^2 < \lambda < b^2\}$ and $\mathcal{H}^2 = \{H_\lambda^2 : b^2 < \lambda < c^2\}$ (Figure 2.95).

Proof. First, we observe that obviously E_λ is an ellipsoid for $\lambda < a^2$, H_λ^1 is a hyperboloid of one sheet for $a^2 < \lambda < b^2$, and H_λ^2 is a hyperboloid of two sheets for $b^2 < \lambda < c^2$.
Furthermore, for every fixed $P \in \mathbb{R}^3 \setminus \{(0,0,0)\}$, the function g_P is continuous on $\mathbb{R} \setminus \{a^2, b^2, c^2\}$, and we have

$$\lim_{\lambda \to a^2 \mp} g_P(\lambda) = \lim_{\lambda \to b^2 \mp} g_P(\lambda) = \lim_{\lambda \to a^2 \mp} g_P(\lambda) = \pm\infty$$

and

$$\lim_{\lambda \to \pm\infty} g_P(\lambda) = -1.$$

Therefore, there are at least one λ_1, one λ_2 and one λ_3 such that

$$\lambda_1 < a^2 < \lambda_2 < b^2 < \lambda_3 < c^2 \text{ and } g_P(\lambda_k) = 0 \ (k = 1, 2, 3).$$

On the other hand g_P has at most three zeros, since $g_P(\lambda) = 0$ is equivalent to a cubic equation. Thus, there is one and only one surface of each family through any given point $P \in \mathbb{R}^3 \setminus \{(0,0,0)\}$.

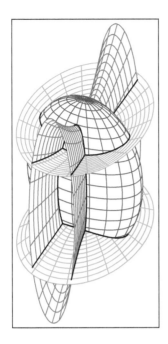

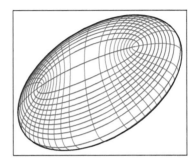

Figure 2.95 The triple orthogonal system of ellipsoids and hyperboloids of one and two sheets (left). Lines of curvature on an ellipsoid (right)

Furthermore, the surface normal vector at a point $P \in \mathbb{R}^3 \setminus \{(0,0,0)\}$ of a surface given by the equation $g_P(\lambda_k) = 0$ has the direction

$$\vec{d}_k = \frac{1}{2} \left\{ \frac{\partial}{\partial x^1}(g_P(\lambda_k)), \frac{\partial}{\partial x^2}(g_P(\lambda_k)), \frac{\partial}{\partial x^3}(g_P(\lambda_k)) \right\}$$

$$= \left\{ \frac{x^1}{a^2 - \lambda_k}, \frac{x^2}{b^2 - \lambda_k}, \frac{x^3}{c^2 - \lambda_k} \right\},$$

and we obtain for $i \neq k$

$$\vec{d}_i \bullet \vec{d}_k = \frac{(x^1)^2}{(a^2 - \lambda_i)(a^2 - \lambda_k)} + \frac{(x^2)^2}{(b^2 - \lambda_i)(b^2 - \lambda_k)} + \frac{(x^3)^2}{(c^2 - \lambda_i)(c^2 - \lambda_k)}$$

$$= \frac{1}{\lambda_k - \lambda_i} \left(\frac{(x^1)^2}{a^2 - \lambda_i} - \frac{(x^1)^2}{a^2 - \lambda_k} + \frac{(x^2)^2}{b^2 - \lambda_i} - \frac{(x^2)^2}{b^2 - \lambda_k} + \frac{(x^3)^2}{c^2 - \lambda_i} - \frac{(x^3)^2}{c^2 - \lambda_k} \right)$$

$$= \frac{1}{\lambda_k - \lambda_i} (g_P(\lambda_i) - g_P(\lambda_k)) = 0.$$

Thus we have shown that the families $\mathcal{E}$, $\mathcal{H}^1$ and $\mathcal{H}^2$ form a triple orthogonal system. □

Now we apply Dupin's Theorem and Theorem 2.9.6 to determine the lines of curvature on ellipsoids, and hyperboloids of one and two sheets.

Visualization 2.9.7. (The lines of curvature on ellipsoids and hyperboloids)
We define the function for every point $P = (x^1, x^2, x^3) \in \mathbb{R}^3 \setminus \{(0,0,0)\}$ by

$$\phi_P(\lambda) = (a^2 - \lambda)(b^2 - \lambda)(c^2 - \lambda)g_P(\lambda), \tag{2.135}$$

where g_P is the function defined in Theorem 2.9.6. We have

$$\begin{aligned}\phi_P(\lambda) = (x^1)^2(b^2 - \lambda)(c^2 - \lambda) + (x^2)^2(a^2 - \lambda)(c^2 - \lambda) \\ + (x^3)^2(a^2 - \lambda)(b^2 - \lambda) - (a^2 - \lambda)(b^2 - \lambda)(c^2 - \lambda)\end{aligned} \tag{2.136}$$

and $\phi_P(\lambda) = 0$ is a cubic equation with zeros λ_k, where

$$\lambda_1 < a^2 < \lambda_2 < b^2 < \lambda_3 < c^2.$$

Therefore, we may write

$$\phi_P(\lambda) = (\lambda - \lambda_1)(\lambda - \lambda_2)(\lambda - \lambda_3). \tag{2.137}$$

Choosing $\lambda = a^2$, we obtain from (2.136) and (2.137)

$$\phi_P(a^2) = (x^1)^2(b^2 - a^2)(c^2 - a^2) = (a^2 - \lambda_1)(a^2 - \lambda_2)(a^2 - \lambda_3),$$

hence

$$(x^1)^2 = \frac{(a^2 - \lambda_1)(a^2 - \lambda_2)(a^2 - \lambda_3)}{(a^2 - b^2)(a^2 - c^2)}. \tag{2.138}$$

Similarly, the choices $\lambda = b^2$ and $\lambda = c^2$ yield

$$(x^2)^2 = \frac{(b^2 - \lambda_1)(b^2 - \lambda_2)(b^2 - \lambda_3)}{(b^2 - a^2)(b^2 - c^2)} \text{ and } (x^3)^2 = \frac{(c^2 - \lambda_1)(c^2 - \lambda_2)(c^2 - \lambda_3)}{(c^2 - a^2)(c^2 - b^2)}. \tag{2.139}$$

If we choose one $\lambda_i = \mathrm{const}$, then the equations in (2.138) and (2.139) yield a parametrization of the surface given by $g_P(\lambda) = 0$ with respect to the other two parameters λ_j and λ_k. We put

$$\alpha = a^2 - \lambda_i, \ \beta = b^2 - \lambda_i, \ \gamma = c^2 - \lambda_i, \ u^1 = \lambda_j - \lambda_i \text{ and } u^2 = \lambda_k - \lambda_i.$$

Then the surface given by the equation

$$\frac{(x^1)^2}{\alpha} + \frac{(x^2)^2}{\beta} + \frac{(x^3)^2}{\gamma} = 1$$

has a parametric representation

$$\vec{x}(u^i) = \left\{ \pm\sqrt{\frac{\alpha(\alpha - u^1)(\alpha - u^2)}{(\alpha - \beta)(\alpha - \gamma)}}, \pm\sqrt{\frac{\beta(\beta - u^1)(\beta - u^2)}{(\beta - \alpha)(\beta - \gamma)}}, \ \pm\sqrt{\frac{\gamma(\gamma - u^1)(\gamma - u^2)}{(\gamma - \alpha)(\gamma - \beta)}} \right\}.$$

The lines of curvature of this surface are the parameter lines with respect to this parametric representation (right in Figure 2.95 and Figure 2.96).

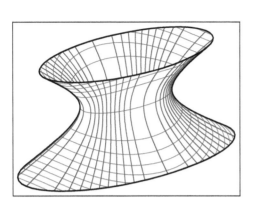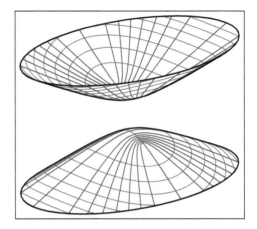

Figure 2.96 Lines of curvature on hyperboloids of one (left) and two sheets (right)

2.10 THE WEINGARTEN EQUATIONS

In this section, we prove the *Weingarten equations* which give a relation between the partial derivatives of the normal vector of a surface with a parametric representation $\vec{x}(u^i)$ and the vectors $\vec{x}_k$ ($k = 1, 2$). We also give a geometric interpretation of the Gaussian curvature.

Theorem 2.10.1 (The Weingarten equations).
Let S be a surface with a parametric representation $\vec{x}(u^i)$ ((u^1, u^2) $\in D$) and first and second fundamental coefficients g_{ik} and L_{ik} ($i, k = 1, 2$). We put

$$L_k^i = g^{ij} L_{jk} \text{ for } i, k = 1, 2. \tag{2.140}$$

Then the partial derivatives $\vec{N}_k$ of the surface normal vector of S satisfy the following identities

$$\vec{N}_k = -L_k^i \vec{x}_i \text{ for } k = 1, 2, \tag{2.141}$$

which are referred to as the Weingarten equations.

Proof. Since $\vec{N}^2 = 1$ implies $\vec{N} \bullet \vec{N}_k = 0$ for $k = 1, 2$, the vectors $\vec{N}_k$ are in the tangent plane of the surface, hence we may express them as linear combinations of the vectors $\vec{x}_1$ and $\vec{x}_2$

$$\vec{N}_k = B_k^m \vec{x}_m \text{ for } k = 1, 2. \tag{2.142}$$

It follows from (2.142) and the definition of the second fundamental coefficients

$$B_k^i = \delta_m^i B_k^m = g^{ij} g_{jm} B_k^m = g^{ij} \vec{x}_j \bullet \vec{x}_m B_k^m = g^{ij} \vec{x}_j \bullet \vec{N}_k$$
$$= -g^{ij} L_{jk} = -L_k^i \text{ for } i, k = 1, 2.$$

This implies (2.141). □

We observed in Example 2.4.5 that the second fundamental coefficients of a plane vanish identically. Applying the Weingarten equations, we can show that the converse implication also holds true.

Example **2.10.2.** *If the second fundamental coefficients of a surface S vanish identically, then S is a plane.*

Proof. Let $L_{ik} = 0$ for $i, k = 1, 2$. Then we have $L_i^k = L_{ij}g^{jk} = 0$ for $i, k = 1, 2$ by (2.140), and so by (2.141)

$$\vec{N}_k = -L_k^i \vec{x}_i = \vec{0} \text{ for } k = 1, 2,$$

hence $\vec{N} = \vec{c}$, where $\vec{c}$ is a constant vector. This shows that S is a plane. □

The Weingarten equations (2.141) reduce when the parameter lines of a surface are its lines of curvature.

Theorem 2.10.3 (The Rodrigues formulae).
The Weingarten equations (2.141) reduce to the Rodrigues formulae

$$\vec{N}_k = -\kappa_k \vec{x}_k \text{ for } k = 1, 2, \tag{2.143}$$

where κ_1 and κ_2 are the principal curvatures, if and only if the parameter lines of the surface are its lines of curvature.

Proof. Let S be a surface with a parametric representation $\vec{x}(u^i)$ $((u^1, u^2) \in D)$.

(i) First we show the necessity of the conditions in (2.143).
We assume that the parameter lines are the lines of curvature of S. Then it follows from Theorem 2.8.6 (a) that $g_{12} = L_{12} = 0$, hence

$$L_1^1 = g^{1j}L_{j1} = g^{11}L_{11} = \frac{L_{11}}{g_{11}}, \; L_2^1 = L_1^2 = g^{2j}L_{j1} = 0$$

and

$$L_2^2 = g^{2j}L_{j2} = g^{22}L_{22} = \frac{L_{22}}{g_{22}}.$$

Since the principal curvatures satisfy

$$\kappa_1 = \frac{L_{11}}{g_{11}} \text{ and } \kappa_2 = \frac{L_{22}}{g_{22}},$$

we obtain for the Weingarten equations (2.141)

$$\vec{N}_1 = -L_1^j \vec{x}_j = -\frac{L_{11}}{g_{11}}\vec{x}_1 = -\kappa_1 \vec{x}_1 \text{ and } \vec{N}_2 = -L_2^j \vec{x}_j = -\frac{L_{22}}{g_{22}}\vec{x}_2 = -\kappa_2 \vec{x}_2,$$
$$\tag{2.144}$$

that is, the Rodrigues formulae (2.143) are satisfied.

(ii) Now we show the sufficiency of the conditions in (2.143).
We assume that the identities in (2.144) hold. Then it follows that $L_1^2 = L_2^1 = 0$, hence

$$\vec{N}_1 \bullet \vec{x}_1 = -L_1^1 = L_{11}, \; \vec{N}_1 \bullet \vec{x}_2 = -L_1^2 = L_{12} = 0$$
$$\vec{N}_2 \bullet \vec{x}_1 = -L_2^1 = L_{21} = 0 \text{ and } \vec{N}_2 \bullet \vec{x}_2 = -L_2^2 = L_{22}$$

and

$$L_{12} = -g^{1j}L_{j2} = -g^{12}L_{22} = 0 \text{ and } L_{21} = -g^{2j}L_{j1} = -g^{12}L_{11} = 0.$$

If κ_1 and κ_2 are not both equal to zero, then we have $g^{12} = 0$ and so $g_{12} = 0$. If $\kappa_1 = \kappa_2 = 0$, then we may choose $g_{12} = 0$.
Thus we have in both cases $g_{12} = L_{12} = 0$ and consequently the parameter lines of the surface are its lines of curvature by Theorem 2.8.6 (a).

□

Now we establish the characterization of all surfaces that have only umbilical points.

Theorem 2.10.4. *If a surface S of class $C^r(D)$ for $r \geq 3$ has only umbilical points, then S is a sphere or a plane.*

Proof. Umbilical points satisfy by (2.73) in Remark 2.5.16 (b)

$$L_{ik}(u^j) = \lambda g_{ik}(u^j) \text{ for } i, k = 1, 2.$$

First $\lambda = 0$ implies $L_{ik} = 0$ for $i, k = 1, 2$ and so S is a plane by Example 2.10.2. If $\lambda \neq 0$, then every curve on S is a line of curvature, and we obtain by the Rodrigues formulae (2.143) in Theorem 2.10.3

$$\vec{N}_k = -\kappa_k \vec{x}_k \text{ for } k = 1, 2,$$

where κ_k are the principal curvatures, and so

$$\vec{N}_k = -\frac{L_{kk}}{g_{kk}}\vec{x}_k = -\lambda\vec{x}_k \text{ for } k = 1, 2. \tag{2.145}$$

It follows from (2.145) that

$$\vec{N}_{kj} = -\lambda_j\vec{x}_k - \lambda\vec{x}_{kj} \text{ and } \vec{N}_{jk} = -\lambda_k\vec{x}_j - \lambda\vec{x}_{jk}, \text{ where } \lambda_j = \partial\lambda/\partial u^j \ (j = 1, 2),$$

hence

$$\vec{0} = \vec{N}_{12} - \vec{N}_{21} = \lambda_1\vec{x}_2 - \lambda_2\vec{x}_1.$$

Since $\vec{x}_1$ and $\vec{x}_2$ are linearly independent, we must have $\lambda_1 = \lambda_2 = 0$, that is, $\lambda = const.$
Integrating (2.145), we obtain

$$\vec{N} = -\lambda\vec{x} + \vec{c} \text{ for some constant vector } \vec{c}.$$

Writing $\lambda = -1/r$ and $\vec{c} = -\vec{x}_0/r$, we have

$$r^2 = (r\vec{N}) \bullet (r\vec{N}) = (\vec{x} - \vec{x}_0)^2,$$

that is, S is a sphere of radius r and centre in X_0.

□

The next result gives a relation between the torsion of an asymptotic line and the Gaussian curvature.

Theorem 2.10.5 (Beltrami and Enneper, 1870).
Let S be a surface with a parametric representation $\vec{x}(u^i)$ $((u^1, u^2) \in D)$.
Then the torsion of an asymptotic line which is not a straight line is given by

$$|\tau| = \sqrt{-K}. \tag{2.146}$$

Proof. First, we observe that $K < 0$ along the asymptotic line by Remark 2.8.2 (b). Let the asymptotic line γ be given by a parametric representation $\vec{x}(s) = \vec{x}(u^i(s))$, where s is the arc length along γ, and $\vec{v}_k$ $(k = 1, 2, 3)$ denote the vectors of the trihedra along γ. Since γ has non–vanishing curvature, it follows from Theorem 2.8.7 that $\vec{N} = \pm\vec{v}_3$, hence $\dot{\vec{N}} = \pm\dot{\vec{v}}_3 = \mp\tau\vec{v}_2$ by Frenet's third formula (1.84). Using the Weingarten equations (2.141), we obtain

$$
\begin{aligned}
\tau &= \mp\dot{\vec{N}} \bullet \vec{v}_2 = \mp\dot{\vec{N}} \bullet (\vec{v}_3 \times \vec{v}_1) \\
&= \mp\dot{\vec{N}} \bullet (\vec{N} \times \dot{\vec{x}}) = \mp\dot{\vec{N}}_k \bullet (\vec{N} \times \dot{\vec{x}})\dot{u}^k \\
&= \pm L_k^i \vec{x}_i \bullet (\vec{N} \times \vec{x}_j)\dot{u}^k\dot{u}^j = \pm L_k^i \vec{N} \bullet (\vec{x}_j \times \vec{x}_i)\dot{u}^k\dot{u}^j \\
&= \pm\sqrt{g}\left(L_k^2\dot{u}^k\dot{u}^1 - L_k^1\dot{u}^k\dot{u}^2\right) = \pm\sqrt{g}\left(g^{2j}L_{jk}\dot{u}^k\dot{u}^1 - g^{1j}L_{jk}\dot{u}^k\dot{u}^2\right),
\end{aligned}
$$

that is,

$$\tau = \pm\sqrt{g}\left(\left(g^{12}L_{1k}\dot{u}^k + g^{22}L_{2k}\dot{u}^k\right)\dot{u}^1 - \left(g^{11}L_{1k}\dot{u}^k + g^{12}L_{2k}\dot{u}^k\right)\dot{u}^2\right). \tag{2.147}$$

Since the asymptotic line satisfies the differential equation (2.121) in Remark 2.8.2 (b)

$$L_{ik}\dot{u}^i\dot{u}^k = L_{11}(\dot{u}^1)^2 + 2L_{12}\dot{u}^1\dot{u}^2 + L_{22}(\dot{u}^2)^2 = 0,$$

we obtain

$$
\begin{aligned}
L_{11}^2(\dot{u}^1)^2 + 2L_{11}L_{12}\dot{u}^1\dot{u}^2 + L_{12}^2(\dot{u}^2)^2 &= \left(L_{11}\dot{u}^1 + L_{12}\dot{u}^2\right)^2 \\
&= L_{12}^2(\dot{u}^2)^2 - L_{11}L_{22}(\dot{u}^2)^2 = -L(\dot{u}^2)^2,
\end{aligned}
$$

hence

$$L_{1k}\dot{u}^k = \pm\dot{u}^2\sqrt{-L}, \text{ and similarly } L_{2k}\dot{u}^k = \mp\dot{u}^1\sqrt{-L}. \tag{2.148}$$

Substituting (2.148) in (2.147), we obtain

$$
\begin{aligned}
\tau &= \pm\sqrt{g}\left(-\frac{g_{12}}{g}\sqrt{-L}\dot{u}^1\dot{u}^2 - \frac{g_{11}}{g}\sqrt{-L}(\dot{u}^1)^2 - \frac{g_{22}}{g}\sqrt{-L}(\dot{u}^2)^2 - \frac{g_{12}}{g}\sqrt{-L}\dot{u}^1\dot{u}^2\right) \\
&= \pm\frac{\sqrt{-L}}{\sqrt{g}}(-g_{ik}\dot{u}^i\dot{u}^k) = \mp\sqrt{\frac{-L}{g}} = \mp\sqrt{-K}.
\end{aligned}
$$

This shows (2.146). □

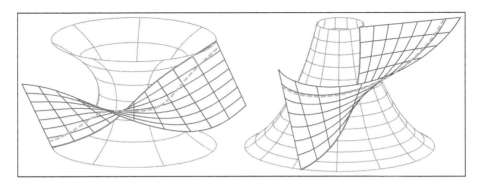

Figure 2.97 The torsion (dashed) along an asymptotic line on a catenoid (left) and a pseudo-sphere (right)

Visualization 2.10.6. *The Gaussian curvature of the catenoid K_c and the pseudo-sphere K_p are given by Visualization 2.5.19 as*

$$K_c = -\frac{1}{a^2 \cosh \frac{u^1}{a}} \quad and \quad K_p = -1.$$

Hence, the absolute values of the torsions τ_c and τ_p of the asymptotic lines on the catenoid and pseudo-sphere (Figure 2.97) are by (2.146) in the Beltrami–Enneper theorem, Theorem 2.10.5,

$$|\tau_c(u^1)| = \frac{1}{a^2 \cosh \frac{u^1}{a}} \quad and \quad |\tau_p| = 1.$$

Figure 2.98 shows a representation of the ± torsion along an asymptotic line on an explicit surface.

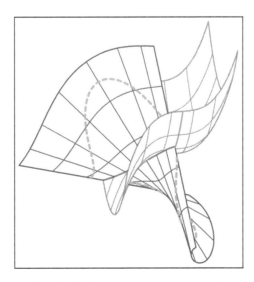

Figure 2.98 The ± torsion (dashed) along an asymptotic line on an explicit surface

Now we establish some useful relations between the derivatives of surface normal vectors and the Gaussian and mean curvature of surfaces.

Theorem 2.10.7. *Let S be a surface with a parametric representation $\vec{x}(u^i)$ $((u^1, u^2) \in D)$.*
Then the surface normal vectors of S satisfy the relation

$$\vec{N}_1 \times \vec{N}_2 = K\sqrt{g}\vec{N}; \qquad (2.149)$$

furthermore, we have

$$L_i^k L_{kj} - L_j^k L_{ki} = 0 \text{ for } i, j = 1, 2. \qquad (2.150)$$

If γ is a curve on S given by a parametric representation $\vec{x}(u^i(s))$, where s denotes the arc length along γ, then the surface normal vectors $\vec{N}$ along γ satisfy the relation

$$(\dot{\vec{N}})^2 = (2HL_{ik} - Kg_{ik})\, \dot{u}^i \dot{u}^k, \qquad (2.151)$$

where K and H are the Gaussian and mean curvature along γ.

Proof.

(i) First we prove (2.149).
    We apply the Weingarten equations (2.141) and obtain

$$\vec{N}_1 \times \vec{N}_2 = L_1^\ell L_2^j (\vec{x}_\ell \times \vec{x}_j) = \left(L_1^1 L_2^2 - L_1^2 L_2^1\right)\sqrt{g}\vec{N}$$

$$= \left(g^{1m}L_{m1}g^{2n}L_{n2} - g^{2i}L_{i1}g^{1k}L_{k2}\right)\sqrt{g}\vec{N}$$

$$= \Big[g^{11}L_{11}g^{21}L_{12} + g^{11}L_{11}g^{22}L_{22} + g^{12}L_{21}g^{21}L_{12} + g^{12}L_{21}g^{22}L_{22}$$

$$\qquad - \left(g^{21}L_{11}g^{11}L_{12} + g^{21}L_{11}g^{12}L_{22} + g^{22}L_{12}g^{11}L_{12}\right.$$

$$\qquad + \left. g^{22}L_{12}g^{12}L_{22}\right) \Big]\sqrt{g}\vec{N}$$

$$= \left(g^{11}g^{22}L_{11}L_{22} - \left(g^{12}\right)^2 L_{12}^2 - \left(g^{12}\right)^2 L_{11}L_{22} - g^{11}g^{22}L_{12}^2\right)\sqrt{g}\vec{N}$$

$$= \left(L_{11}L_{22}\left(g^{11}g^{22} - \left(g^{12}\right)^2\right) - L_{12}^2\left(g^{11}g^{22} - \left(g^{12}\right)^2\right)\right)\sqrt{g}\vec{N}$$

$$= \left(g^{11}g^{22} - \left(g^{12}\right)^2\right)\left(L_{11}L_{22} - L_{12}^2\right)\sqrt{g}\vec{N} = \frac{L}{g}\cdot\sqrt{g}\vec{N} = K\vec{N}\sqrt{g}.$$

Thus we have shown that (2.149) holds.

(ii) Now we prove (2.150).
    We have

$$L_i^k L_{kj} - L_j^k L_{ki} = g^{km}L_{mi}L_{kj} - g^{k\ell}L_{\ell j}L_{ki} = g^{km}L_{mi}L_{kj} - g^{m\ell}L_{\ell j}L_{mi}$$

$$= g^{km}L_{mi}L_{kj} - g^{km}L_{kj}L_{mi} = 0 \text{ for } i, j = 1, 2.$$

Thus we have shown (2.150).

(iii) Finally we prove (2.151).

Applying the Weingarten equations (2.141) again, we obtain

$$\vec{N}_1 \bullet \vec{N}_1 = L_1^{\ell} L_1^j g_{\ell j} = g^{\ell m} L_{m1} g^{jn} L_{n1} g_{\ell j} = \delta_j^m L_{m1} g^{jn} L_{n1} = L_{j1} g^{jn} L_{n1}$$
$$= L_{11} g^{1n} L_{n1} + L_{21} g^{2n} L_{n1} = L_{11} g^{1n} L_{ni} + L_{21} g^{2n} L_{n1} - L_{11} g^{2n} L_{n2}$$
$$= 2L_{11} H + g^{2n} (L_{12} L_{n1} - L_{11} L_{n2})$$
$$= 2L_{11} H + g^{21} (L_{12} L_{11} - L_{11} L_{12}) + g^{22} (L_{12} L_{21} - L_{11} L_{22})$$
$$= 2L_{11} H + g^{22} \left(L_{12}^2 - L_{11} L_{22} \right) = 2L_{11} H - g_{11} \frac{L}{g}$$
$$= 2L_{11} H - g_{11} K,$$

and similarly

$$\vec{N}_1 \bullet \vec{N}_2 = L_{j1} g^{jn} L_{n2} = L_{12} g^{2n} L_{n2} + L_{11} g^{1n} L_{n2}$$
$$= L_{12} g^{in} L_{ni} + g^{1n} (-L_{12} L_{n1} + L_{11} L_{n2})$$
$$= 2L_{12} H + g^{12} \left(L_{11} L_{22} - L_{12}^2 \right)$$
$$= 2L_{12} H - \frac{g_{12}}{g} L = 2L_{12} H - g_{12} K$$

and

$$\vec{N}_2 \bullet \vec{N}_2 = 2L_{22} H - g_{22} K.$$

Thus we have shown

$$\vec{N}_i \bullet \vec{N}_k = 2H L_{ik} - K g_{ik} \text{ for } i, k = 1, 2. \tag{2.152}$$

This implies

$$\dot{\vec{N}}^2 = \vec{N}_i \bullet \vec{N}_k \dot{u}^i \dot{u}^k = (2H L_{ik} - K g_{ik}) \dot{u}^i \dot{u}^k \text{ which is (2.151)}.$$

□

In view of (2.151), we define the *third fundamental coefficients*.

Definition 2.10.8 (Third fundamental coefficients).
Let S be a surface with a parametric representation $\vec{x}(u^i)$ $((u^1, u^2) \in D)$. Then the functions

$$c_{ik} = \vec{N}_i \bullet \vec{N}_k = 2H L_{ik} - K g_{ik} \ (i, k = 1, 2) \tag{2.153}$$

are called the *third fundamental coefficients of S*.

Visualization 2.10.9 (Parallel surfaces).
Let S be a surface with a parametric representation $\vec{x}(u^i)$ $((u^1, u^2) \in D)$ and $a \in \mathbb{R}$ be a constant. Then the surface S^* with a parametric representation

$$\vec{x}^*(u^i) = \vec{x}(u^i) + a\vec{N}(u^i),$$

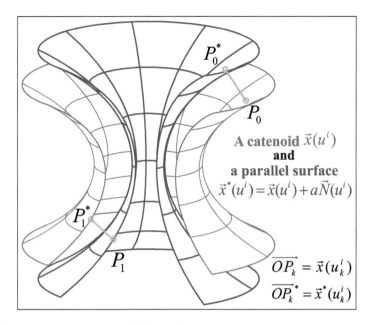

Figure 2.99 A catenoid and a parallel surface

where $\vec{N}$ denotes the surface normal vector of S, is called a parallel surface *(Figure 2.99)*.

If K and H are the Gaussian and mean curvature of S and $1 - 2aH + a^2K \neq 0$, then the Gaussian and mean curvature K^* and H^* of S^* are given by

$$K^* = \frac{K}{1 - 2aH + a^2K} \tag{2.154}$$

and

$$H^* = \frac{H - aK}{1 - 2aH + a^2K}. \tag{2.155}$$

Proof. We note that it follows from $\vec{x}_k^* = \vec{x}_k + a\vec{N}_k$ $(k = 1, 2)$ by (2.149) in Theorem 2.10.7 and the Weingarten equations (2.141) that

$$
\begin{aligned}
\vec{x}_1^* \times \vec{x}_2^* &= (\vec{x}_1 + a\vec{N}_1) \times (\vec{x}_2 + a\vec{N}_2) \\
&= \vec{x}_1 \times \vec{x}_2 + a(\vec{x}_1 \times \vec{N}_2 + \vec{N}_1 \times \vec{x}_2) + a^2(\vec{N}_1 \times \vec{N}_2) \\
&= \vec{x}_1 \times \vec{x}_2 - a\left(L_2^k \vec{x}_1 \times \vec{x}_k + L_1^k \vec{x}_k \times \vec{x}_2\right) + a^2 \sqrt{g} K \vec{N} \\
&= \vec{x}_1 \times \vec{x}_2 \left(1 + a^2 K - a\left(L_2^2 + L_1^1\right)\right) = (\vec{x}_1 \times \vec{x}_2)\left(1 + a^2 K - a g^{ik} L_{ik}\right) \\
&= (\vec{x}_1 \times \vec{x}_2)\left(1 + a^2 K - 2aH\right).
\end{aligned}
\tag{2.156}
$$

Therefore, we have for the surface normal vector $\vec{N}^*$ of the parallel surface S^*

$$\vec{N}^* = \vec{N}, \text{ if } 1 - 2aH + a^2K \neq 0. \tag{2.157}$$

We observe that (2.157) means that if S^* is a parallel surface of S, then S is also a parallel surface of S^*.

(i) First we prove (2.154).
It follows from (2.156) that

$$\sqrt{g^*} = \|\vec{x}_1^* \times \vec{x}_2^*\| = \sqrt{g}|1 - 2aH + a^2K|.$$

If $|a|$ is sufficiently small, then $1 - 2aH + a^2K > 0$, and we obtain for the Gaussian curvature K^* of the parallel surface S^* by (2.149)

$$K^* = \frac{1}{\sqrt{g^*}}\vec{N}^* \bullet \left(\vec{N}_1^* \times \vec{N}_2^*\right) = \frac{1}{\sqrt{g^*}}\vec{N} \bullet (\vec{N}_1 \times \vec{N}_2) = \frac{\sqrt{g}K}{\sqrt{g^*}} = \frac{K}{1 - 2aH + a^2K},$$

that is, (2.154) holds.

(ii) Now we prove (2.155).
Since S is obviously a parallel surface of S^*, we obtain by interchanging the roles of K and H with K^* and H^*, and replacing a by $-a$

$$K = \frac{K^*}{1 + 2aH^* + a^2K^*}.$$

This implies

$$2aH^*K = K^* - a^2K^*K - K = K^*(1 - a^2K) - K$$

$$= K\left(\frac{1 - a^2K}{1 - 2aH + a^2K} - 1\right)$$

$$= \frac{K}{1 - 2aH + a^2K}(1 - a^2K - 1 + 2aH - a^2K),$$

and we obtain for $K \neq 0$

$$H^* = \frac{H - aK}{1 - 2aH + a^2K},$$

that is, (2.155) holds.

□

Remark 2.10.10. *The identities in (2.154) and (2.155) may be used to find surfaces with given Gaussian or mean curvature.*

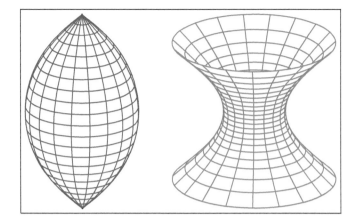

Figure 2.100   A hyperbolic spherical surface (left) and its parallel surface with constant mean curvature $H = -1/2$ (right)

Visualization 2.10.11 (Parallel surfaces of a surface of revolution).

Now we consider the surface of revolution RS, namely the elliptic spherical surface of Visualization 2.5.20, Case 3, with a parametric representation

$$r(u^1) = \lambda \cos\left(\frac{u^1}{c}\right) \text{ and } h(u^1) = \int \sqrt{1 - \frac{\lambda^2}{c^2} \sin^2\left(\frac{u^1}{c}\right)} \, du^1$$

for $u^1 \in I_1 \subset (-\pi/2, \pi/2)$, where λ and c are constants with $0 < \lambda < c$. Then RS has constant Gaussian curvature $K = 1/c^2$.

If we choose $a = c$, then it follows from (2.155) for the mean curvature H^* of the parallel surface RS^* for a of RS

$$H^* = \frac{H - \dfrac{1}{c}}{2(1 - cH)} = -\frac{1}{2c} \cdot \frac{1 - cH}{1 - cH} = -\frac{1}{2c},$$

that is, RS^* has constant mean curvature $H^* = -1/(2c)$ (Figure 2.100).

The same argument holds true if we choose RS as above but with (Figure 2.101)

$$\lambda > c \text{ and } u^1 \in \left(-c \sin^{-1} \frac{c}{\lambda}, c \sin^{-1} \frac{c}{\lambda}\right).$$

Now we consider the pseudo-sphere which has constant Gaussian curvature $K = -1$ by (2.80) in Visualization 2.5.19. If we choose $a = 1$, then it follows from (2.154) and (2.80) for the Gaussian curvature K^* of the parallel surface RS^* for a of RS (Figure 2.102)

$$K^* = -\frac{1}{1 - 2H - 1} = \frac{1}{2H} = \frac{\exp(-u^1)\sqrt{1 - \exp(-2u^1)}}{1 - 2\exp(-2u^1)}.$$

Now we give a geometric interpretation of the Gaussian curvature. The following definition is important.

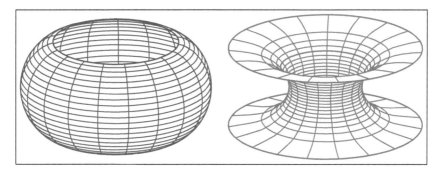

Figure 2.101 An elliptic spherical surface (left) and its parallel surface with constant mean curvature $H = -1/2$ (right)

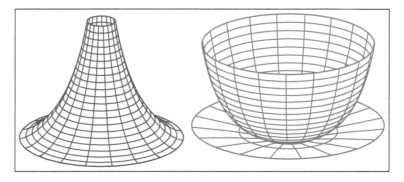

Figure 2.102 A pseudo-sphere (left) and its parallel surface (right) for $a = 1$

Definition 2.10.12. Let S be a surface with a parametric representation $\vec{x}(u^i)$ $((u^1, u^2) \in D)$. The map that assigns to every point P of S the surface normal vector at P is called the *spherical Gauss map* (Figures 2.103 and 2.104); the set of all image points on the sphere under the spherical Gauss map is called *spherical image of S*.

Remark 2.10.13. *The third fundamental coefficients (Definition 2.10.8) of a surface are the first fundamental coefficients of its spherical Gauss map.*

Let S be a surface with a parametric representation $\vec{x}(u^i)$ $((u^1, u^2) \in D)$, $P \in S$ be a point given by the parameters u_0^1 and u_0^2, and U be a neighbourhood of P, given by a subset D_U of D. The image of U under the spherical Gauss map defines a part U^* of the unit sphere. If A_U and A_{U^*} denote the surface areas of U and U^* – including their signs –, then it follows that

$$\frac{A_{U^*}}{A_U} = \frac{\iint_{D_u} \vec{N}(u^i) \bullet \left(\vec{N}_1(u^i) \times \vec{N}_2(u^i) \right) du^1 du^2}{\iint_{D_u} \vec{N}(u^i) \bullet (\vec{x}_1(u^i) \times \vec{x}_2(u^i)) du^1 du^2}. \tag{2.158}$$

The ratio in (2.158) is the bigger, the more the surface S is curved, since then the surface normal vectors cover a larger area of the unit sphere (Figure 2.105).

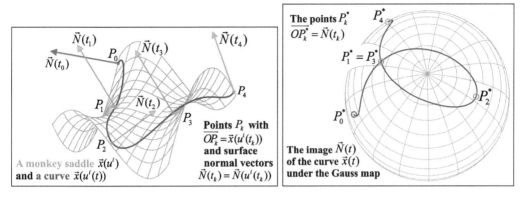

Figure 2.103 A curve on a monkey saddle and its image under the spherical Gauss map

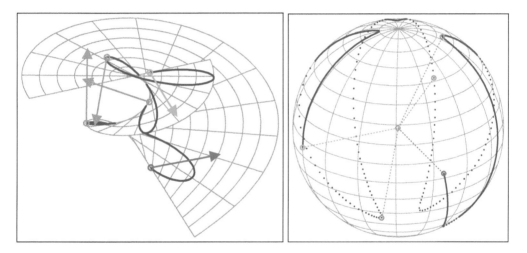

Figure 2.104 A curve on a conoid and its image under the spherical Gauss map

If the neighbourhood U of P shrinks to the point P, that is, if we take the limit $U \to \{P\}$ in (2.158), then we obtain using (2.149) in Theorem 2.10.7

$$\lim_{U \to \{P\}} \frac{A_{U^*}}{A_U} = \frac{\vec{N}(u_0^i) \bullet \left(\vec{N}_1(u_0^i) \times \vec{N}_2(u_0^i) \right)}{\vec{N}(u_0^i) \bullet (\vec{x}_1(u_0^i) \times \vec{x}_2(u_0^i))} = \frac{\vec{N}(u_0^i) \bullet \vec{N}(u_0^i)\sqrt{g(u^i)}K(u_0^i)}{\sqrt{g(u_0^i)}}$$

$$= K(u_0^i). \tag{2.159}$$

The value in (2.159) only depends on the point P on S and was originally introduced by Gauss as the measure of the curvature of a surface at a point.

The sign of K which determines the elliptic, parabolic or hyperbolic property of P, has the following geometric significance:

K is positive or negative depending on whether the Gauss map preserves or reverses the orientation (Figure 2.106).

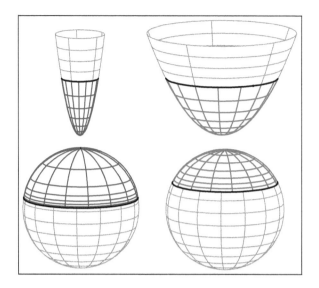

Figure 2.105 The spherical images of the same domain of two different paraboloids

Visualization 2.10.14. *(a) We verify (2.159) without the use of the above results in the case of a torus with a parametric representation (Visualization 2.3.8)*

$$\vec{x}(u^i) = \{r(u^1)\cos u^2, r(u^1)\sin u^2, h(u^1)\} \quad \left((u^1, u^2) \in D = (0, 2\pi)^2\right),$$

where

$$r(u^1) = (r_0 + r_1)\cos u^1 \text{ and } h(u^1) = r_1 \sin u^1,$$

and r_0 and r_1 are constants with $0 < r_1 < r_0$. Furthermore, we choose a point P_0 on the torus given by the parameters $(u_0^1, u_0^2) \in D$ and a neighbourhood U of P_0 defined by the rectangle $R = [u_0^1 - w^1, u_0^1 + w^1] \times [u_0^2 - w^2, u_0^2 + w^2]$, where w^1 and w^2 are positive reals and $R \subset D$ (Figure 2.107). Then the surface area of U is, since $g(u^i) = r_1^2(r_0 + r_1\cos u^1)^2$ by (2.30),

$$A_U = \iint_U \sqrt{g(u^i)}\, du^1\, du^2 = \int_{u_0^1 - w^1}^{u_0^1 + w^1} \int_{u_0^2 - w^2}^{u_0^2 + w^2} r_1(r_0 + r_1\cos u^1)\, du^1\, du^2$$

$$= 2r_1 w^2 \int_{u_0^1 - w^1}^{u_0^1 + w^1} (r_0 + r_1\cos u^1)\, du^1$$

$$= 4r_0 r_1 w^1 w^2 + 2r_1^2 w^2 \left(\sin\left(u_0^1 + w^1\right) - \sin\left(u_0^1 + w^1\right)\right)$$

$$= 4r_0 r_1 w^1 w^2 + 4r_1^2 w^2 \sin w^1 \cos u_0^1$$

$$= 4r_1 w^2 \left(r_0 w^1 + r_1 \sin w^1 \cos u_0^1\right).$$

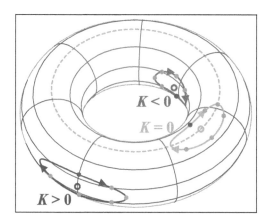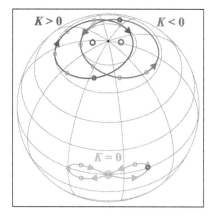

Figure 2.106 Curves and their orientations in the neighbourhoods of points with $K > 0$, $K = 0$ and $K < 0$ and their images under the Gauss map

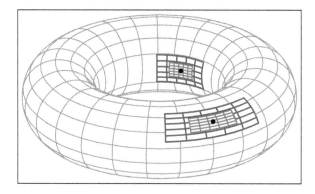

Figure 2.107 Neighbourhoods of an elliptic and hyperbolic point on a torus

Since $\sqrt{g(u^1)} = r_1(r_0 + r_1 \cos u^1)$, we have $K(u^1) = \cos u^1/\sqrt{g(u^1)}$ by (2.118) in Visualization 2.6.4 (d), and it follows for the surface area of the image U^* under the Gauss map by (2.149) in Theorem 2.10.7

$$A_{U^*} = \iint_{U^*} \vec{N}(u^1) \bullet (\vec{N}_1(u^1) \times \vec{N}_2(u^2))\, du^1\, du^2$$

$$= \iint_{U^*} \sqrt{g(u^1)} K(u^i)\, du^1\, du^2 = 2w^2 \int_{u_0^1 - w^1}^{u_0^1 + w^1} \cos u^1\, du^1$$

$$= 2w^2(\sin(u_0^1 + w^1) - \sin(u_0^1 - w^1)) = 4w^2 \sin w^1 \cos u_0^1.$$

Thus it follows that

$$\frac{A_{U^*}}{A_U} = \frac{\sin w^1 \cos u_0^1}{r_1(r_0 w^1 + r_1 \sin w^1 \cos u_0^1)} = \frac{\cos u_0^1}{r_1(r_0 \dfrac{w^1}{\sin w^1} + r_1 \cos u_0^1)}$$

$$\longrightarrow \frac{\cos u_0^1}{r_1(r_0 + r_1 \cos u_0^1)} = K(u_0^1) \; for \; (w^1, w^2) \to (0, 0).$$

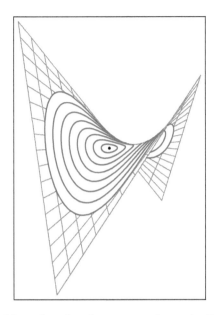

Figure 2.108 Circular neighbourhoods of zero on a hyperbolic paraboloid

(b) Now we verify (2.159) without the use of the above results in the case of the hyperbolic paraboloid with a parametric representation

$$\vec{x}(u^i) = \{u^1, u^2, u^1 u^2\} \ \left((u^1, u^2) \in D = \mathbb{R}^2\right).$$

We choose the point P_0 on the hyperbolic paraboloid given by the parameters $(u_0^1, u_0^2) = (0,0) \in D$ and a neighbourhood $U_r \ (r > 0)$ of P_0, given by the circle line in the parameter plane with radius r and centre in $(0,0)$ (Figure 2.108). It follows from

$$\vec{x}_1(u^i) = \{1, 0, u^2\}, \ \vec{x}_2(u^i) = \{0, 1, u^1\}, \ \vec{x}_{12} = \{0, 0, 1\},$$
$$\vec{x}_{11}(u^i) = \vec{x}_{22}(u^i) = \vec{0}, \ \vec{n}(u^i) = \vec{x}_1(u^i) \times \vec{x}_2(u^i) = \{-u^2, -u^1, 1\},$$
$$g_{11}(u^i) = 1 + (u^2)^2, \ g_{12}(u^i) = u^1 u^2, \ g_{22}(u^i) = 1 + (u^1)^2,$$
$$g(u^i) = 1 + (u^1)^2 + (u^2)^2, \ \vec{N}(u^i) = \vec{n}(u^i) \frac{1}{\sqrt{g(u^i)}},$$
$$L_{kk}(u^i) = \vec{N}(u^i) \bullet \vec{x}_{kk}(u^i) = 0 \ (k = 1, 2) \ and \ L_{12}(u^i) = \vec{N}(u^i) \bullet \vec{x}_{12}(u^i) = \frac{1}{\sqrt{g(u^i)}}$$

that the Gaussian curvature of the hyperbolic paraboloid is given by

$$K(u^i) = \frac{L(u^i)}{g(u^i)} = -\frac{L_{12}^2(u^i)}{g(u^i)} = -\frac{1}{(1 + (u^1)^2 + (u^2)^2)^2},$$

hence $K(0,0) = -1$. We put

$$u^1 = u^1(\rho, \phi) = \rho \cos \phi \ and \ u^2 = u^2(\rho, \phi) = \rho \sin \phi \ for \ (\rho, \phi) \in (0, r) \times (0, 2\pi)$$

and obtain for the surface area of U_r

$$A_{U_r} = \iint\limits_{U_r} \sqrt{g(u^i)}\, du^1 du^2 = \int\limits_0^{2\pi} \int\limits_0^r \rho \sqrt{1+\rho^2}\, d\rho d\phi$$

$$= \frac{2\pi}{3} \left(\sqrt{1+\rho^2} \right)^3 \Big|_{\rho=0}^r = \frac{2\pi}{3} \left(\left(\sqrt{1+r^2} \right)^3 - 1 \right).$$

Furthermore, we have

$$\vec{N}_1(u^i) = \{0, -1, 0\} \frac{1}{\sqrt{g(u^i)}} + \alpha_1(u^i)\vec{n}(u^i)$$

and

$$\vec{N}_2(u^i) = \{-1, 0, 0\} \frac{1}{\sqrt{g(u^i)}} + \alpha_2(u^i)\vec{n}(u^i),$$

where

$$\alpha_k(u^i) = \frac{\partial}{\partial u^k} \left(\frac{1}{\sqrt{g(u^i)}} \right) \quad for \ k = 1, 2,$$

and it follows that

$$\vec{N}(u^i) \bullet \left(\vec{N}_1(u^i) \times \vec{N}_2(u^i) \right) = \vec{N}(u^i) \bullet \left[\left(\{0, -1, 0\} \frac{1}{\sqrt{g(u^i)}} + \alpha_1(u^i)\vec{n}(u^i) \right) \right.$$

$$\left. \times \left(\{-1, 0, 0\} \frac{1}{\sqrt{g(u^i)}} + \alpha_2(u^i)\vec{n}(u^i) \right) \right]$$

$$= \vec{N}(u^i) \frac{1}{g(u^i)} \bullet (\{0, 1, 0\} \times \{1, 0, 0\}) = -\frac{1}{\left(\sqrt{g(u^i)} \right)^3},$$

and

$$A_{U_r^*} = \iint\limits_{u_r} \vec{N}(u^i) \bullet \left(\vec{N}_1(u^i) \times \vec{N}_2(u^i) \right) du^1 du^2 = -\int\limits_0^{2\pi} \int\limits_0^r \frac{\rho}{\left(\sqrt{1+\rho^2} \right)^3}\, d\rho d\phi$$

$$= \frac{2\pi}{\sqrt{1+\rho^2}} \Big|_{\rho=0}^r = \frac{2\pi \left(1 - \sqrt{1+r^2} \right)}{\sqrt{1+r^2}}.$$

Therefore, we obtain

$$\frac{A_{U_r^*}}{A_{U_r}} = -3 \frac{\sqrt{1+r^2} - 1}{\sqrt{1+r^2} \left(\left(\sqrt{1+r^2} \right)^3 - 1 \right)} = -3 \frac{\sqrt{1+r^2} - 1}{(1+r^2)^2 - \sqrt{1+r^2}}$$

$$= -3\frac{1 + \dfrac{r^2}{2} + o(r^3) - 1}{1 + 2r^2 + o(r^3) - 1 - \dfrac{r^2}{2} + o(r^3)} = -3\frac{\dfrac{r^2}{2} + o(r^3)}{\dfrac{3r^2}{2} + o(r^3)} = -1 + o(1) \ (r \to 0),$$

hence

$$\lim_{r \to 0} \frac{A_{U_r^*}}{A_{U_r}} = -1 = K(0, 0).$$

CHAPTER 3

The Intrinsic Geometry of Surfaces

In this chapter, we study the *intrinsic geometry of surfaces*. By this, we mean the geometric properties of surfaces that depend on measurements on the surface itself, that is, on the first fundamental coefficients and their derivatives only.

The most important topics and subjects in this chapter are

- *the Christoffel symbols of first and second kind and the geodesic curvature* in Section 3.1

- *Liouville's theorem*, Theorem 3.1.12 in Section 3.1

- *geodesic lines on surfaces with orthogonal parameters and on surfaces of revolution* in Sections 3.3 and 3.4

- *the minimum property of geodesic lines*, Theorem 3.5.2 in Section 3.5

- *orthogonal and geodesic parameters* in Section 3.6

- *geodesic parallel and geodesic polar coordinates* in Section 3.6

- *Levi–Cività parallelism* in Section 3.7

- *the derivation formulae by Gauss, and Mainardi and Codazzi*, Theorem 3.8.1 in Section 3.8

- *Theorema egregium*, Theorem 3.8.3 in Section 3.8

- *conformal, isometric and area preserving maps of surfaces* in Section 3.9

- *the Gauss–Bonnet theorem*, Theorem 3.10.1 in Section 3.10

- *minimal surfaces and the Weierstrass formulae*, (3.158) in Section 3.11.

DOI: 10.1201/9781003370567-3

3.1   THE CHRISTOFFEL SYMBOLS

In Section 2.4, we introduced the geodesic and normal curvature κ_g and κ_n of a curve γ on a surface. This was done by splitting the vector $\ddot{\vec{x}}$ of curvature of γ into two components

$$\ddot{\vec{x}}(s) = \kappa_g(s)\vec{t}(s) + \kappa_n(s)\vec{N}(s), \tag{3.1}$$

where $\vec{N}(s)$ denotes the normal vector of the surface at s, and

$$\vec{t}(s) = \vec{N}(s) \times \dot{\vec{x}}(s) \tag{3.2}$$

is a vector in the tangent plane of the surface at s. In order to find κ_n and κ_g, we introduced the functions Γ_{ik}^r and the second fundamental coefficients L_{ik} by (Section 2.4)

$$\vec{x}_{ik} = \Gamma_{ik}^r \vec{x}_r + L_{ik}\vec{N} \quad (i, k = 1, 2). \tag{3.3}$$

The remainder of Chapter 2 was mainly devoted to the study of the normal curvature and various concepts arising in this context.

Now we are going to take a closer look at the geodesic curvature of curves on a surface. The first step will be to give some geometric meaning to the coefficient functions Γ_{ik}^r that have only been formally introduced.

We start with a result for the geodesic curvature of a curve which is similar to Meusnier's theorem, Theorem 2.5.7, for the normal curvature.

Theorem 3.1.1. *The geodesic curvature of a curve on a surface is equal to its curvature multiplied by the Cosine of the angle between the binormal vector of the curve and the normal vector of the surface, that is, the Cosine of the angle between the osculating plane of the curve and the tangent plane of the surface (Figure 3.1).*

Proof. Let γ be a curve on a surface given by a parametric representation $\vec{x}(s) = \vec{x}(u^i(s))$ $(s \in I)$, where s denotes the arc length along γ, and let $\vec{v}_k$ be the vectors of the trihedra of γ. Then we have

$$\kappa_g = \ddot{\vec{x}} \bullet \vec{t} = \kappa \vec{v}_2 \bullet (\vec{N} \times \vec{v}_1) = \kappa \vec{N} \bullet (\vec{v}_1 \times \vec{v}_2) = \kappa \vec{N} \bullet \vec{v}_3 = \kappa \cos \alpha.$$

$\square$

Definition 3.1.2. Let S be a surface with parametric representation $\vec{x}(u^j)$ $((u^1, u^2) \in D \subset \mathbb{R}^2)$ and first fundamental coefficients g_{jk}. We recall that the inverse of the matrix (g_{lr}) is denoted by (g^{lr}).

(a)   The functions $[ikl] : D \to \mathbb{R}$ with

$$[ikl] = \vec{x}_{ik} \bullet \vec{x}_l \ (i, k, l = 1, 2)$$

are called the *first Christoffel symbols of the surface S*.

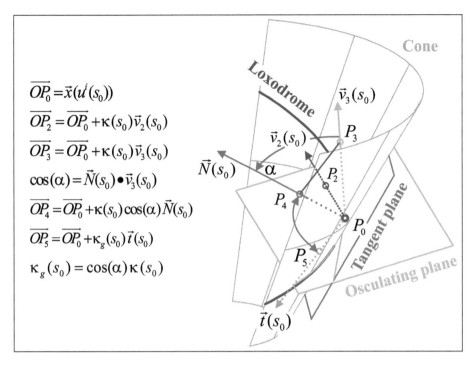

$$\overrightarrow{OP_0} = \vec{x}(u^i(s_0))$$

$$\overrightarrow{OP_2} = \overrightarrow{OP_0} + \kappa(s_0)\vec{v}_2(s_0)$$

$$\overrightarrow{OP_3} = \overrightarrow{OP_0} + \kappa(s_0)\vec{v}_3(s_0)$$

$$\cos(\alpha) = \vec{N}(s_0) \bullet \vec{v}_3(s_0)$$

$$\overrightarrow{OP_4} = \overrightarrow{OP_0} + \kappa(s_0)\cos(\alpha)\vec{N}(s_0)$$

$$\overrightarrow{OP_5} = \overrightarrow{OP_0} + \kappa_g(s_0)\vec{t}(s_0)$$

$$\kappa_g(s_0) = \cos(\alpha)\kappa(s_0)$$

Figure 3.1 The relation between the curvature and geodesic curvature of a curve on a surface (Theorem 3.1.1)

(b) The functions $\left\{ {r \atop ik} \right\} : D \to \mathbb{R}$ with

$$\left\{ {r \atop ik} \right\} = g^{lr}[ikl] \ (i, k, r = 1, 2)$$

are called the *second Christoffel symbols of the surface S*.

Example **3.1.3** (**The Christoffel symbols for explicit surfaces**).
Let $f : D \subset \mathbb{R}^2 \to \mathbb{R}$ *be a function of class* $C^r(D)$ *for* $r \geq 2$ *and ES be the explicit surface with a parametric representation*

$$\vec{x}(u^i) = \{u^1, u^2, f(u^1, u^2)\} \ \left((u^1, u^2) \in D\right).$$

Then the first and second Christoffel symbols of ES are given by

$$[ijk] = [ijk](u^1, u^2) = \vec{x}_{ij}(u^1, u^2) \bullet \vec{x}_k(u^1, u^2) = f_{ij}(u^i, u^2)f_k(u^1, u^2) \tag{3.4}$$

for $i, j, k = 1, 2,$ *and*

$$\left\{ {i \atop jk} \right\} = \frac{f_i f_{jk}}{1 + f_1^2 + f_2^2} \ for \ i, j, k = 1, 2. \tag{3.5}$$

Proof.

(i) First we prove (3.4).

We have for the first Christoffel symbols by Definition 3.1.2 (a) and Example 2.3.4

$$[ijk] = [ijk](u^1, u^2) = \vec{x}_{ij}(u^1, u^2) \bullet \vec{x}_k(u^1, u^2) = f_{ij}(u^i, u^2)f_k(u^1, u^2)$$

for $i, j, k = 1, 2$, that is, (3.4) holds.

(ii) Now we prove (3.5).

It follows from Example 2.3.4 that

$$g^{11} = \frac{g_{22}}{g} = \frac{1 + f_2^2}{1 + f_1^2 + f_2^2}, \quad g^{12} = -\frac{g_{12}}{g} = -\frac{f_1 f_2}{1 + f_1^2 + f_2^2}$$

and

$$g^{22} = \frac{g_{11}}{g} = \frac{1 + f_1^2}{1 + f_1^2 + f_2^2}.$$

Now we obtain by Definition 3.1.2 (b), omitting the arguments u^1 and u^2, for the second Christoffel symbols

$$\left\{ \begin{matrix} 1 \\ jk \end{matrix} \right\} = g^{m1}[jkm] = \frac{1}{1 + f_1^2 + f_2^2}\left((1 + f^2)^2 f_{jk}f_1 - f_1 f_2 f_{jk} f_2 \right) = \frac{f_1 f_{jk}}{1 + f_1^2 + f_2^2},$$

and similarly

$$\left\{ \begin{matrix} 2 \\ jk \end{matrix} \right\} = \frac{f_2 f_{jk}}{1 + f_1^2 + f_2^2} \quad \text{for } j, k = 1, 2.$$

Thus we have shown that (3.5) holds.

□

It is useful to establish some important formulae for the first and second Christoffel symbols.

Lemma 3.1.4. *The following formulae hold for the first and second Christoffel symbols.*

$$\text{(a)} \quad [ikl] = [kil] \qquad\qquad (i, k, l = 1, 2)$$

$$\text{(b)} \quad \left\{ \begin{matrix} l \\ ik \end{matrix} \right\} = \left\{ \begin{matrix} l \\ ki \end{matrix} \right\} \qquad\qquad (i, k, l = 1, 2)$$

$$\text{(c)} \quad [ikl] = \frac{1}{2}\left(\frac{\partial g_{il}}{\partial u^k} - \frac{\partial g_{ik}}{\partial u^l} + \frac{\partial g_{kl}}{\partial u^i} \right) \quad (i, k, l = 1, 2)$$

$$\text{(d)} \quad \frac{\partial g_{ik}}{\partial u^l} = [ilk] + [kli] \qquad\qquad (i, k, l = 1, 2)$$

With respect to new parameters u^{*k} *($k = 1, 2$) given by*

$$u^j = u^j(u^{*k}) = u^j(u^{*1}, u^{*2}) \quad (j = 1, 2),$$

we have

(e) $\quad g^{*kl} = g^{jm} \dfrac{\partial u^{*k}}{\partial u^j} \dfrac{\partial u^{*l}}{\partial u^m}$ $\hspace{3cm} (k, l = 1, 2)$

(f) $\quad [ikl]^* = \left([\alpha\beta\gamma] \dfrac{\partial u^\alpha}{\partial u^{*i}} \dfrac{\partial u^\beta}{\partial u^{*k}} + g_{\alpha\gamma} \dfrac{\partial^2 u^\alpha}{\partial u^{*i} \partial u^{*k}} \right) \dfrac{\partial u^\gamma}{\partial u^{*l}}$ $\quad (i, k, l = 1, 2)$

(g) $\quad \left\{ {r \atop ik} \right\}^* = \left(\left\{ {\gamma \atop \alpha\beta} \right\} \dfrac{\partial u^\alpha}{\partial u^{*i}} \dfrac{\partial u^\beta}{\partial u^{*k}} + \dfrac{\partial^2 u^\gamma}{\partial u^{*i} \partial u^{*k}} \right) \dfrac{\partial u^{*r}}{\partial u^\gamma}$ $\quad (i, k, r = 1, 2)$.

Proof.

(a,b) Parts (a) and (b) are immediate consequences of the definitions of the first and second Christoffel symbols in Definition 3.1.2 and the fact that $\vec{x}_{ik} = \vec{x}_{ki}$ $(i, k = 1, 2)$.

(c) We obtain from $g_{il} = \vec{x}_i \bullet \vec{x}_l$

$$\frac{\partial g_{il}}{\partial u^k} = \vec{x}_{ik} \bullet \vec{x}_l + \vec{x}_{lk} \bullet \vec{x}_i = [ilk] + [lki]$$

and similarly

$$-\frac{\partial g_{ik}}{\partial u^l} = -\vec{x}_{il} \bullet \vec{x}_k - \vec{x}_{kl} \bullet \vec{x}_i = -([ilk] + [kli])$$

and

$$\frac{\partial g_{kl}}{\partial u^i} = \vec{x}_{ki} \bullet \vec{x}_l + \vec{x}_{li} \bullet \vec{x}_k = [ilk] + [kli],$$

and so

$$\frac{\partial g_{il}}{\partial u^k} - \frac{\partial g_{ik}}{\partial u^l} + \frac{\partial g_{kl}}{\partial u^i} = 2[ikl]$$

by Part (a).

(d) Part (d) is the third identity in Part (c) of the proof.

(e) We obtain from the formula (2.18) of transformation for the first fundamental coefficients by interchanging u and u^*

$$g_{mr} = g^*_{nt} \frac{\partial u^{*n}}{\partial u^m} \frac{\partial u^{*t}}{\partial u^r} \quad (m, r = 1, 2)$$

and

$$\begin{aligned}
g^{*kl} &= g^{*ks} \delta^{*l}_s = g^{*ks} \frac{\partial u^{*l}}{\partial u^j} \frac{\partial u^j}{\partial u^{*s}} = g^{*ks} \frac{\partial u^{*l}}{\partial u^j} \frac{\partial u^r}{\partial u^{*s}} \delta^j_r \\
&= g^{*ks} \frac{\partial u^{*l}}{\partial u^j} \frac{\partial u^r}{\partial u^{*s}} g_{mr} g^{mj} = g^{*ks} \frac{\partial u^{*l}}{\partial u^j} \frac{\partial u^r}{\partial u^{*s}} g^*_{nt} \frac{\partial u^{*n}}{\partial u^m} \frac{\partial u^{*t}}{\partial u^r} g^{mj} \\
&= g^{*ks} g^*_{nt} \frac{\partial u^{*t}}{\partial u^r} \frac{\partial u^r}{\partial u^{*s}} \frac{\partial u^{*l}}{\partial u^j} \frac{\partial u^{*n}}{\partial u^m} g^{mj} = g^{mj} g^{*ks} g^*_{nt} \delta^{*t}_s \frac{\partial u^{*l}}{\partial u^j} \frac{\partial u^{*n}}{\partial u^m}
\end{aligned}$$

$$= g^{mj} g^{*ks} g^*_{ns} \frac{\partial u^{*l}}{\partial u^j} \frac{\partial u^{*n}}{\partial u^m} = g^{mj} \delta^{*k}_n \frac{\partial u^{*l}}{\partial u^j} \frac{\partial u^{*n}}{\partial u^m}$$

$$= g^{mj} \frac{\partial u^{*l}}{\partial u^j} \frac{\partial u^{*k}}{\partial u^m} = g^{jm} \frac{\partial u^{*k}}{\partial u^j} \frac{\partial u^{*l}}{\partial u^m} \quad (k, l = 1, 2).$$

(f) From

$$\vec{x}^*_i = \vec{x}_\alpha \frac{\partial u^\alpha}{\partial u^{*i}} \quad (i = 1, 2)$$

and

$$\vec{x}^*_{ik} = \frac{\partial}{\partial u^{*k}} \left(\vec{x}_\alpha \frac{\partial u^\alpha}{\partial u^{*i}} \right) = \frac{\partial}{\partial u^{*k}} (\vec{x}_\alpha) \frac{\partial u^\alpha}{\partial u^{*i}} + \vec{x}_\alpha \frac{\partial}{\partial u^{*k}} \left(\frac{\partial u^\alpha}{\partial u^{*i}} \right)$$

$$= \vec{x}_{\alpha\beta} \frac{\partial u^\alpha}{\partial u^{*i}} \frac{\partial u^\beta}{\partial u^{*k}} + \vec{x}_\alpha \frac{\partial^2 u^\alpha}{\partial u^{*i} \partial u^{*k}} \quad (i, k = 1, 2),$$

we obtain

$$[ikl]^* = \vec{x}^*_{ik} \bullet \vec{x}^*_l = \left(\vec{x}_{\alpha\beta} \frac{\partial u^\alpha}{\partial u^{*i}} \frac{\partial u^\beta}{\partial u^{*k}} + \vec{x}_\alpha \frac{\partial^2 u^\alpha}{\partial u^{*i} \partial u^{*k}} \right) \bullet \vec{x}_\gamma \frac{\partial u^\gamma}{\partial u^{*l}}$$

$$= \left([\alpha\beta\gamma] \frac{\partial u^\alpha}{\partial u^{*i}} \frac{\partial u^\beta}{\partial u^{*k}} + g_{\alpha\gamma} \frac{\partial^2 u^\alpha}{\partial u^{*i} \partial u^{*k}} \right) \frac{\partial u^\gamma}{\partial u^{*l}} \quad (i, k, l = 1, 2).$$

(g) Finally, we obtain from Parts (e) and (f)

$$\left\{ \begin{matrix} r \\ ik \end{matrix} \right\}^* = g^{*rl} [ikl]^*$$

$$= g^{\phi\epsilon} \frac{\partial u^{*r}}{\partial u^\phi} \frac{\partial u^{*l}}{\partial u^\epsilon} \left([\alpha\beta\gamma] \frac{\partial u^\alpha}{\partial u^{*i}} \frac{\partial u^\beta}{\partial u^{*k}} + g_{\alpha\gamma} \frac{\partial^2 u^\alpha}{\partial u^{*i} \partial u^{*k}} \right) \frac{\partial u^\gamma}{\partial u^{*l}}$$

$$= \frac{\partial u^\gamma}{\partial u^{*l}} \frac{\partial u^{*l}}{\partial u^\epsilon} g^{\phi\epsilon} \left([\alpha\beta\gamma] \frac{\partial u^\alpha}{\partial u^{*i}} \frac{\partial u^\beta}{\partial u^{*k}} + g_{\alpha\gamma} \frac{\partial^2 u^\alpha}{\partial u^{*i} \partial u^{*k}} \right) \frac{\partial u^{*r}}{\partial u^\phi}$$

$$= \delta^\gamma_\epsilon g^{\phi\epsilon} \left([\alpha\beta\gamma] \frac{\partial u^\alpha}{\partial u^{*i}} \frac{\partial u^\beta}{\partial u^{*k}} + g_{\alpha\gamma} \frac{\partial^2 u^\alpha}{\partial u^{*i} \partial u^{*k}} \right) \frac{\partial u^{*r}}{\partial u^\phi}$$

$$= \left(g^{\phi\gamma} [\alpha\beta\gamma] \frac{\partial u^\alpha}{\partial u^{*i}} \frac{\partial u^\beta}{\partial u^{*k}} + g^{\phi\gamma} g_{\alpha\gamma} \frac{\partial^2 u^\alpha}{\partial u^{*i} \partial u^{*k}} \right) \frac{\partial u^{*r}}{\partial u^\phi}$$

$$= \left(\left\{ \begin{matrix} \phi \\ \alpha\beta \end{matrix} \right\} \frac{\partial u^\alpha}{\partial u^{*i}} \frac{\partial u^\beta}{\partial u^{*k}} + \frac{\partial^2 u^\phi}{\partial u^{*i} \partial u^{*k}} \right) \frac{\partial u^{*r}}{\partial u^\phi}$$

$$= \left(\left\{ \begin{matrix} \gamma \\ \alpha\beta \end{matrix} \right\} \frac{\partial u^\alpha}{\partial u^{*i}} \frac{\partial u^\beta}{\partial u^{*k}} + \frac{\partial^2 u^\gamma}{\partial u^{*i} \partial u^{*k}} \right) \frac{\partial u^{*r}}{\partial u^\gamma} \quad (r, i, k = 1, 2),$$

where we changed the index of summation ϕ to γ in the last step.

□

We obtain as an immediate consequence of Lemma 3.1.4

Corollary 3.1.5. *We obtain for $i \neq k$ (no summation)*

$$[iii] = \frac{1}{2}\frac{\partial g_{ii}}{\partial u^i}, \quad [iki] = [kii] = \frac{1}{2}\frac{\partial g_{ii}}{\partial u^k},$$

$$[iik] = \frac{1}{2}\left(\frac{\partial g_{ik}}{\partial u^i} - \frac{\partial g_{ii}}{\partial u^k} + \frac{\partial g_{ik}}{\partial u^i}\right) = \frac{\partial g_{12}}{\partial u^i} - \frac{1}{2}\frac{\partial g_{ii}}{\partial u^k},$$

$$\begin{Bmatrix} i \\ ii \end{Bmatrix} = g^{ri}[iir] = g^{ii}[iii] + g^{ik}[iik]$$

$$= \frac{1}{g}\left(g_{kk}\frac{1}{2}\frac{\partial g_{ii}}{\partial u^i} - g_{12}\left(\frac{\partial g_{12}}{\partial u^i} - \frac{1}{2}\frac{\partial g_{ii}}{\partial u^k}\right)\right)$$

$$= \frac{1}{2g}\left(g_{kk}\frac{\partial g_{ii}}{\partial u^i} + g_{12}\left(\frac{\partial g_{ii}}{\partial u^k} - 2\frac{\partial g_{12}}{\partial u^i}\right)\right),$$

$$\begin{Bmatrix} k \\ ii \end{Bmatrix} = g^{lk}[iil] = g^{kk}[iik] + g^{ik}[iii]$$

$$= \frac{1}{g}\left(g_{ii}\left(\frac{\partial g_{12}}{\partial u^i} - \frac{1}{2}\frac{\partial g_{ii}}{\partial u^k}\right) - g_{12}\frac{\partial g_{ii}}{\partial u^i}\right)$$

$$= \frac{1}{2g}\left(g_{ii}\left(2\frac{\partial g_{12}}{\partial u^i} - \frac{\partial g_{ii}}{\partial u^k}\right) - g_{12}\frac{\partial g_{ii}}{\partial u^i}\right)$$

and

$$\begin{Bmatrix} k \\ ik \end{Bmatrix} = \begin{Bmatrix} k \\ ki \end{Bmatrix} = g^{rk}[ikr] = g^{kk}[ikk] + g^{ik}[iki]$$

$$= \frac{1}{g}\left(g_{ii}\frac{1}{2}\frac{\partial g_{kk}}{\partial u^i} - g_{12}\frac{1}{2}\frac{\partial g_{ii}}{\partial u^k}\right)$$

$$= \frac{1}{2g}\left(g_{ii}\frac{\partial g_{kk}}{\partial u^i} - g_{12}\frac{\partial g_{ii}}{\partial u^k}\right).$$

For orthogonal parameters (Definition 2.3.1 (c)), where $g_{12} = 0$ and $g = g_{11}g_{22}$, we have for $i \neq k$

$$[iii] = \frac{1}{2}\frac{\partial g_{ii}}{\partial u^i}, \quad [iik] = -\frac{1}{2}\frac{\partial g_{ii}}{\partial u^k}, \quad [iki] = [kii] = \frac{1}{2}\frac{\partial g_{ii}}{\partial u^k}, \tag{3.6}$$

$$\begin{Bmatrix} i \\ ii \end{Bmatrix} = \frac{1}{2g_{ii}}\frac{\partial g_{ii}}{\partial u^i} = \frac{\partial(\log\sqrt{g_{ii}})}{\partial u^i}, \quad \begin{Bmatrix} k \\ ii \end{Bmatrix} = -\frac{1}{2g_{kk}}\frac{\partial g_{ii}}{\partial u^k} \tag{3.7}$$

and

$$\begin{Bmatrix} k \\ ik \end{Bmatrix} = \begin{Bmatrix} k \\ ki \end{Bmatrix} = \frac{1}{2g_{kk}}\frac{\partial g_{kk}}{\partial u^i} = \frac{\partial(\log\sqrt{g_{kk}})}{\partial u^i}. \tag{3.8}$$

If the parameters are orthogonal and the first fundamental coefficients satisfy

$$g_{kk}(u^i) = g_{kk}(u^1) \text{ for } k = 1, 2, \tag{3.9}$$

then it easily follows from (3.7) and (3.8) that

$$\left\{ \begin{matrix} 1 \\ 11 \end{matrix} \right\} = \frac{1}{2g_{11}} \frac{dg_{11}}{du^1}, \ \left\{ \begin{matrix} 1 \\ 22 \end{matrix} \right\} = -\frac{1}{2g_{11}} \frac{dg_{22}}{du^1} = 0, \ \left\{ \begin{matrix} 1 \\ 12 \end{matrix} \right\} = \left\{ \begin{matrix} 1 \\ 21 \end{matrix} \right\} = 0,$$

$$\left\{ \begin{matrix} 2 \\ 11 \end{matrix} \right\} = \left\{ \begin{matrix} 2 \\ 22 \end{matrix} \right\} = 0 \ and \ \left\{ \begin{matrix} 2 \\ 12 \end{matrix} \right\} = \left\{ \begin{matrix} 2 \\ 21 \end{matrix} \right\} = \frac{1}{2g_{22}} \frac{dg_{22}}{du^1}. \tag{3.10}$$

Remark 3.1.6. *By Definition 3.1.2 and Lemma 3.1.4, the Christoffel symbols only depend on the first fundamental coefficients and their derivatives. Quantities in the theory of surfaces that depend on the first fundamental coefficients only are called quantities of the internal or intrinsic geometry.*

Example **3.1.7 (The Christoffel symbols for surfaces of revolution).**
Let RS be a surface of revolution with a parametric representation

$$\vec{x}(u^i) = \{r(u^1)\cos u^2, r(u^1)\sin u^2, h(u^1)\}.$$

Then the parameters are orthogonal, hence $g_{12} = 0$. We also have $g_{ii}(u^1, u^2) = g_{ii}(u^1)$ $(i = 1, 2)$ and consequently

$$\frac{\partial g_{ii}}{\partial u^2} = 0.$$

It follows from (3.6) that

$$[111] = \frac{1}{2}\frac{\partial g_{11}}{\partial u^1} = r'(u^1)r''(u^1) + h'(u^1)h''(u^1), \ [112] = -\frac{1}{2}\frac{\partial g_{11}}{\partial u^2} = 0,$$

$$[121] = [211] = \frac{1}{2}\frac{\partial g_{11}}{\partial u^2} = 0, \ [222] = \frac{1}{2}\frac{\partial g_{22}}{\partial u^2} = 0$$

and

$$[221] = -\frac{1}{2}\cdot\frac{\partial g_{22}}{\partial u^1} = -r(u^1)r'(u^1).$$

Furthermore, we have from (3.7)

$$\left\{ \begin{matrix} 1 \\ 11 \end{matrix} \right\} = \frac{1}{2g_{11}}\frac{\partial g_{11}}{\partial u^1} = \frac{r''(u^1)r'(u^1) + h''(u^1)h'(u^1)}{(r'(u^1))^2 + (h'(u^1))^2},$$

$$\left\{ \begin{matrix} 1 \\ 12 \end{matrix} \right\} = \left\{ \begin{matrix} 1 \\ 21 \end{matrix} \right\} = \frac{1}{2g_{11}}\frac{\partial g_{11}}{\partial u^2} = 0,$$

$$\left\{ \begin{matrix} 1 \\ 22 \end{matrix} \right\} = -\frac{1}{2g_{11}}\frac{\partial g_{22}}{\partial u^1} = -\frac{r'(u^1)r(u^1)}{(r'(u^1))^2 + (h'(u^1))^2},$$

$$\left\{ \begin{matrix} 2 \\ 11 \end{matrix} \right\} = -\frac{1}{2g_{22}}\frac{\partial g_{11}}{\partial u^2} = 0,$$

$$\left\{ \begin{matrix} 2 \\ 12 \end{matrix} \right\} = \left\{ \begin{matrix} 2 \\ 21 \end{matrix} \right\} = \frac{1}{2g_{22}}\frac{\partial g_{22}}{\partial u^1} = \frac{r'(u^1)}{r(u^1)}$$

and

$$\left\{ \begin{matrix} 2 \\ 22 \end{matrix} \right\} = \frac{1}{2g_{22}}\frac{\partial g_{22}}{\partial u^2} = 0.$$

In particular, it follows for the pseudo-sphere, where $g_{11} = 0$ and $r(u^1) = e^{-u^1}$, that

$$\left\{ \begin{matrix} 1 \\ 11 \end{matrix} \right\} = \left\{ \begin{matrix} 1 \\ 12 \end{matrix} \right\} = \left\{ \begin{matrix} 1 \\ 21 \end{matrix} \right\} = 0, \quad \left\{ \begin{matrix} 1 \\ 22 \end{matrix} \right\} = e^{-2u^1}, \tag{3.11}$$

$$\left\{ \begin{matrix} 2 \\ 11 \end{matrix} \right\} = \left\{ \begin{matrix} 2 \\ 22 \end{matrix} \right\} = 0 \text{ and } \left\{ \begin{matrix} 2 \\ 12 \end{matrix} \right\} = \left\{ \begin{matrix} 2 \\ 21 \end{matrix} \right\} = -1. \tag{3.12}$$

Theorem 3.1.8. *Let S be a surface with a parametric representation $\vec{x}(u^i)$ and γ a curve on S given by a parametric representation $\vec{x}(u^i(s))$, where s is the arc length along γ. We put*

$$\epsilon_{mi} = \vec{N} \bullet (\vec{x}_m \times \vec{x}_i) \text{ for } m, i = 1, 2,$$

that is,

$$\epsilon_{11} = \epsilon_{22} \text{ and } \epsilon_{12} = -\epsilon_{21} = \|\vec{x}_1 \times \vec{x}_2\| = g.$$

Then the geodesic curvature of γ is given by

$$\kappa_g = \epsilon_{mi}\dot{u}^m \left(\ddot{u}^i + \left\{ \begin{matrix} i \\ jk \end{matrix} \right\} \dot{u}^j \dot{u}^k \right). \tag{3.13}$$

Thus the geodesic curvature of a curve is a quantity of the internal geometry.

Proof. We know from the second identity in (2.46) of Remark 2.4.3 (c) that

$$\kappa_g = \left(\ddot{u}^r + \Gamma^r_{jk}\dot{u}^j\dot{u}^k \right) \vec{x}_r \bullet \vec{t} \text{ for } r = 1, 2, \tag{3.14}$$

where $\vec{t} = \vec{N} \times \dot{\vec{x}}$. Now

$$\vec{x}_r \bullet \vec{t} = \vec{x}_r \bullet (\vec{N} \times \vec{x}_l)\dot{u}^l = \vec{N} \bullet (\vec{x}_l \times \vec{x}_r)\dot{u}^l = \epsilon_{lr}\dot{u}^l$$

for $r = 1, 2$ implies

$$\kappa_g = \epsilon_{lr}\dot{u}^l \left(\ddot{u}^r + \Gamma^r_{jk}\dot{u}^j\dot{u}^k \right).$$

Furthermore, it follows from (3.3) that

$$\vec{x}_{jk} = \Gamma^i_{jk}\vec{x}_i + L_{jk}\vec{N} \text{ for } j, k = 1, 2,$$

so

$$\vec{x}_{jk} \bullet \vec{x}_l = [jkl] = \Gamma^i_{jk}\vec{x}_i \bullet \vec{x}_l = \Gamma^i_{jk}g_{il} \text{ for } j, k, l = 1, 2,$$

$$\Gamma^r_{jk} = \delta^r_i\Gamma^i_{jk} = g^{mr}\left(g_{im}\Gamma^i_{jk} \right) = g^{mr}[jkm] = \left\{ \begin{matrix} r \\ jk \end{matrix} \right\} \text{ for } r, j, k = 1, 2 \tag{3.15}$$

and consequently

$$\kappa_g = \epsilon_{mi}\dot{u}^m \left(\ddot{u}^i + \left\{ \begin{matrix} i \\ jk \end{matrix} \right\} \dot{u}^j \dot{u}^k \right).$$

$\square$

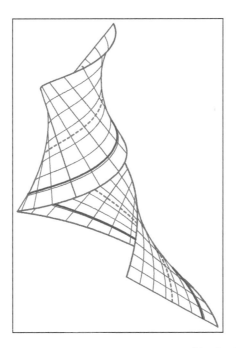

Figure 3.2 Representation of $-0.9\kappa_g$ along a loxodrome (dashed) on the pseudo-sphere

We use the formula in (3.13) of Theorem 3.1.8 to find the geodesic curvature of a curve on a surface of revolution. We may use the same principle as in the case of the normal curvature in (2.58) in Section 2.5 to represent the geodesic curvature along a curve.

If S is a surface with a parametric representation $\vec{x}(u^i)$ $((u^1, u^2) \in D \subset \mathbb{R}^2)$ and γ is a curve on S with a parametric representation $\vec{x}(t) = \vec{x}(u^i(t))$ $(t \in I)$ and geodesic curvature $\kappa_g(t) = \kappa_g(u^i(t))$, then we may represent $\kappa_g(t)$ by the curve γ_g, given by a parametric representation

$$\vec{x}^*(t) = \vec{x}(t) + \kappa_g(t)\vec{t}(t) \ (t \in I),$$

where $\vec{t} = \vec{N} \times \vec{v}_1$. Writing $u^{*1} = t$, we see that γ_g is a curve on the ruled surface RS that has a parametric representation

$$\vec{x}^*(u^{*i}) = \vec{y}(u^{*1}) + u^{*2}\vec{z}(u^{*1}) \ \left((u^{*1}, u^{*2}) \in I \times \mathbb{R}\right), \tag{3.16}$$

where $\vec{y}(u^{*1}) = \vec{x}(u^i(u^{*1}))$ and $\vec{z}(u^{*1}) = \vec{t}(u^i(u^{*1}))$, and γ_g considered as a curve on RS is given by putting $u^{*2} = \kappa_g(u^i(u^{*1}))$ in (3.16) (Figures 3.2–3.5).

Visualization 3.1.9. (Geodesic curvature of certain curves on surfaces of revolution) *Let RS be a surface of revolution with a parametric representation*

$$\vec{x}(u^i) = \{r(u^1)\cos u^2, r(u^1)\sin u^2, h(u^1)\} \ \left((u^1, u^2) \in D = I_1 \times I_2 \subset \mathbb{R} \times (0, 2\pi)\right),$$

where we assume, as always, that $r(u^1) > 0$ and $|r'(u^1)| + |h^1(u^1)| > 0$ on I_1. Furthermore, let γ be a curve on RS given by a parametric representation $\vec{x}(u^i(s))$ $(s \in I)$,

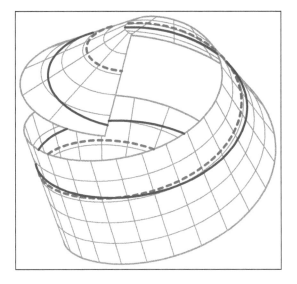

Figure 3.3 Representation of $-3\kappa_g$ along a loxodrome (dashed) on the unit sphere

where I is such that $u^1(I) \subset I_1$ and s denotes the arc length along γ. Then the geodesic curvature κ_g along γ is given by (3.13) and the formulae for the second Christoffel symbols in Example 3.1.7

$$\kappa_g(s) = \kappa_g(u^1(s), u^2(s))$$
$$= \epsilon_{mi}\left(u^1(s), u^2(s)\right) \dot{u}^m(s)\left(\ddot{u}^i(s) + \begin{Bmatrix} i \\ jk \end{Bmatrix}\left(u^1(s), u^2(s)\right)\dot{u}^j(s)\dot{u}^k(s)\right)$$
$$= \sqrt{g(u^1(s))}\left[\dot{u}^1(s)\left(\ddot{u}^2(s) + 2\frac{r'(u^1(s))}{r(u^1(s))}\dot{u}^1(s)\dot{u}^2(s)\right)\right.$$
$$-\dot{u}^2(s)\left(\ddot{u}^1(s) + \frac{r''(u^1(s))r'(u^1(s)) + h''(u^1(s))h'(u^1(s))}{(r'(u^1(s)))^2 + (h'(u^1(s)))^2}(\dot{u}^1(s))^2\right.$$
$$\left.\left.-\frac{r'(u^1(s))r(u^1(s))}{(r'(u^1(s)))^2 + (h'(u^1(s)))^2}(\dot{u}^2(s))^2\right)\right].$$

If we assume that u^1 is the arc length along the curve that generates the surface of revolution RS, then we have $(r'(u^1(s))^2 + (h'(u^1(s))^2 = 1$ and consequently

$$r''(u^1(s))r'(u^1(s)) + h''(u^1(s))h'(u^1(s)) = 0,$$

hence

$$\kappa_g(s) = r(u^1(s))\left[\dot{u}^1(s)\left(\ddot{u}^2(s) + 2\frac{r'(u^1(s))}{r(u^1(s))}\dot{u}^1(s)\dot{u}^2(s)\right)\right.$$
$$\left.- \dot{u}^2(s)\left(\ddot{u}^1(s) - r'(u^1(s))r(u^1(s))(\dot{u}^2(s))^2\right)\right]. \tag{3.17}$$

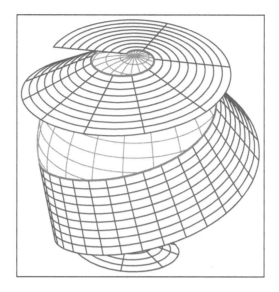

Figure 3.4 The ruled surface generated by a loxodrome on the unit sphere and the vectors $\vec{t}(s)$

As a special case, we consider a loxodrome λ_β on RS, where $\beta \in (0, \pi/2)$ is the constant angle between the loxodrome λ_β and the meridians of RS. Then λ_β is given by (2.32) in Visualization 2.3.11

$$u^2(u^1) = \tan \beta \int \sqrt{\frac{g_{11}(u^1)}{g_{22}(u^1)}}\, du^1 = \tan \beta \int \frac{1}{r(u^1)}\, du^1.$$

Writing $u^1(t) = t$ and $c = \tan \beta$, we obtain a parametric representation $\vec{x}^*(t) = \vec{x}(t, u^2(t))$ for λ_β, so

$$\left\| \frac{d\vec{x}^*(t)}{dt} \right\|^2 = g_{11}(t) + g_{22}(t) \left(\frac{du^2(t)}{dt} \right)^2 = 1 + g_{22}(t) \frac{c^2}{g_{22}(t)} = 1 + c^2,$$

hence the arc length along λ_β is given by $s(t) = \sqrt{1 + c^2}\, t$.
Writing $c_1 = 1/\sqrt{1 + c^2}$, we obtain a parametric representation of λ_β with respect to its arc length s

$$\vec{x}(s) = \vec{x}^* \left(c_1 s, u^2(u^1((c_1 s))) \right) = \vec{x}^*(u^{*1}(s), u^{*2}(s)),$$

and it follows that

$$u^{*1}(s) = u^1(t(s)) = u^1(c_1 s), \ \dot{u}^{*1}(s) = \frac{du^1}{dt} \frac{dt}{ds} = c_1, \ \ddot{u}^{*1}(s) = 0,$$

$$u^{*2}(s) = u^2(u^1(t(s))), \ \dot{u}^{*2}(s) = \frac{du^2}{du^1} \dot{u}^{*1}(s) = \frac{cc_1}{r(u^{*1}(s))}$$

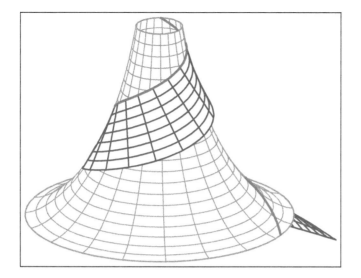

Figure 3.5 The ruled surface generated by a loxodrome on a pseudo-sphere and the vectors $\vec{t}(s)$

and

$$\ddot{u}^{*2}(s) = \frac{d^2u^2}{d(u^1)^2}\left(\dot{u}^{*1}\right)^2 + \frac{du^2}{du^1}\ddot{u}^{*1} = -cc_1^2\frac{r'(u^{*1}(s))}{r^2(u^{*1}(s))}.$$

Thus we obtain for the geodesic curvature κ_g^* along λ_β from (3.17)

$$\kappa_g^*(s) = r(u^{*1}(s))\left[\dot{u}^{*1}(s)\left(-\frac{cr'(u^{*1}(s))}{r^2(u^{*1}(s))}\left(\dot{u}^{*1}(s)\right)^2\right.\right.$$
$$\left.+2\frac{r'(u^{*1}(s))}{r(u^*(s))}\dot{u}^{*1}(s)\frac{c}{r(u^{*1}(s))}\dot{u}^{*1}(s)\right)$$
$$\left.+\frac{c}{r(u^{*1}(s))}\dot{u}^{*1}(s)r'(u^{*1}(s))r(u^{*1}(s))\left(\dot{u}^{*1}(s)\right)^2\right]$$

$$= c\left(\dot{u}^{*1}(s)\right)\left(\frac{r'(u^{*1}(s))}{r(u^{*1}(s))}+r(u^{*1}(s))\right) = cc_1^3\frac{r'(u^{*1}(s))}{r(u^{*1}(s))}\left(1+r^2(u^{*1}(s))\right).$$

In particular, we obtain for the unit sphere where $r(u^1) = \cos u^1$

$$\kappa_g^*(s) = -cc_1^3\tan\left(c_1 s\right)\left(1+\cos^2\left(c_1 s\right)\right),$$

and for the pseudo-sphere with

$$r(u^1) = \exp(-u^1) \text{ and } h(u^1) = \int\sqrt{1-\exp\left(-2u^1\right)}\,du^1$$

$$\kappa_g^*(s) = -cc_1^3\left(1+\exp\left(-2c_1 s\right)\right).$$

Finally, we have for the geodesic curvature κ_g along the loxodrome λ_β with respect to the parameter t (Figures 3.2 and 3.3)

$$\kappa_g(t) = -cc_1^3 \tan t \left(1 + \cos^2 t\right) \ and \ \kappa_g(t) = -cc_1^3 \left(1 + \exp\left(-2t\right)\right).$$

Now we establish formulae for the geodesic curvature of the parameter lines of surfaces.

Lemma 3.1.10. *The geodesic curvature κ_{gm} of a u^m–line on a surface S with a parametric representation $\vec{x}(u^i)$ is given by*

$$\kappa_{gm} = (-1)^{m+1} \frac{\sqrt{g}}{(g_{mm})^{3/2}} \left\{ \begin{matrix} k \\ mm \end{matrix} \right\} \ for \ k \neq m. \tag{3.18}$$

In particular, if the parameters are orthogonal and s_k denotes the arc length along the u^k–lines then

$$\kappa_{gm} = (-1)^m \frac{d}{ds_k} \left(\log \sqrt{g_{mm}}\right) \ for \ k \neq m. \tag{3.19}$$

Proof. Let s_1 denote the arc length along a u^1-line of S. Since $u^2 = const$ along the u^1–lines, the formula in (3.13) for the geodesic curvature κ_{g_1} of the u^1–lines reduces to

$$\kappa_{g_1} = \epsilon_{12} \dot{u}^1 \left\{ \begin{matrix} 2 \\ 11 \end{matrix} \right\} (\dot{u}^1)^2 = \sqrt{g} \left\{ \begin{matrix} 2 \\ 11 \end{matrix} \right\} (\dot{u}^1)^3.$$

As s_1 is the arc length along the u^1–line, it follows that $g_{11}(\dot{u}^1)^2 = 1$, hence

$$\kappa_{g_1} = \frac{\sqrt{g}}{(g_{11})^{3/2}} \left\{ \begin{matrix} 2 \\ 11 \end{matrix} \right\}.$$

Similarly, we obtain for the geodesic curvature κ_{g_2} of a u^2–line

$$\kappa_{g_2} = -\frac{\sqrt{g}}{(g_{22})^{3/2}} \left\{ \begin{matrix} 1 \\ 22 \end{matrix} \right\}.$$

Thus we have proved (3.18).

If the parameters are orthogonal then it follows from (3.7) and (3.8) that

$$\kappa_{g_1} = \frac{\sqrt{g_{22}}}{g_{11}} \left(-\frac{1}{2g_{22}} \frac{\partial g_{11}}{\partial u^2}\right) = -\frac{1}{2\sqrt{g_{22}}} \frac{1}{g_{11}} \frac{\partial g_{11}}{\partial u^2} \tag{3.20}$$

and

$$\kappa_{g_2} = \frac{\sqrt{g_{11}}}{g_{22}} \frac{1}{2g_{11}} \frac{\partial g_{22}}{\partial u^1} = \frac{1}{2\sqrt{g_{11}}} \frac{1}{g_{22}} \frac{\partial g_{22}}{\partial u^1}. \tag{3.21}$$

Finally, we obtain

$$\frac{d\left(\log \sqrt{g_{11}}\right)}{ds_2} = \frac{1}{2g_{11}} \frac{\partial g_{11}}{\partial u^2} \frac{du^2}{ds_2} = \frac{1}{2\sqrt{g_{22}}} \frac{1}{g_{11}} \frac{\partial g_{11}}{\partial u^2}$$

and

$$\frac{d\left(\log \sqrt{g_{22}}\right)}{ds_1} = \frac{1}{2g_{22}} \frac{\partial g_{22}}{\partial u^1} \frac{du^1}{ds_1} = \frac{1}{2\sqrt{g_{11}}} \frac{1}{g_{22}} \frac{\partial g_{22}}{\partial u^1}.$$

Thus we have shown (3.19). □

Visualization 3.1.11. (Geodesic curvature of parameter lines on explicit surfaces) *We apply (3.18) in Lemma 3.1.10 to compute the geodesic curvature along the parameter lines of an explicit surface with a parametric representation*

$$\vec{x}(u^i) = \{u^1, u^2, f(u^1, u^2)\} \left((u^1, u^2) \in D \right).$$

First we consider the u^1–line corresponding to u_0^2. We obtain for the geodesic curvature κ_{g_1} along this u^1–line by (3.18) and (3.5) in Example 3.1.3

$$\kappa_{g_1}(t) = \kappa_{g_1}(t, u_0) = \frac{\sqrt{g(t, u_0^2)}}{(g_{11}(t, u_0^2))^{3/2}} \left\{ \begin{matrix} 2 \\ 11 \end{matrix} \right\}$$

$$= \frac{\sqrt{1 + (f_1(t, u_0^2))^2 + (f_2(t, u_0^2))^2}}{(1 + (f_1(t, u_0^2))^2)^{3/2}} \cdot \frac{f_2(t, u_0^2) f_{11}(t, u_0^2)}{1 + (f_1(t, u_0^2))^2 + (f_2(t, u_0^2))^2}$$

$$= \frac{f_2(t, u_0^2) f_{11}(t, u_0^2)}{(1 + (f_1(t, u_0^2))^2)^{3/2} \sqrt{1 + (f_1(t, u_0^2))^2 + (f_2(t, u_0^2))^2}}.$$

Similarly, we obtain for the geodesic curvature κ_{g_2} along the u^2–line corresponding to u_0^1

$$\kappa_{g_2}(t) = \kappa_{g_2}(u_0^1, t) = -\frac{f_1(u_0^1, t) f_{22}(u_0^1, t)}{(1 + (f_2(u_0^1, t))^2)^{3/2} \sqrt{1 + (f_1(u_0^1, t))^2 + (f_2(u_0^2, t))^2}}.$$

As a special case, we consider the hyperbolic paraboloid as an explicit surface given by $f(u^1, u^2) = (u^1)^2 - (u^2)^2$ $((u^1, u^2) \in \mathbb{R}^2)$. Then we have

$$f_1(u^i) = 2u^1, \quad f_2(u^i) = -2u^2, \quad f_{11}(u^i) = -f_{22}(u^i) = 2$$

and the geodesic curvature along the u^1–line corresponding to u_0^2 and the u^2–line corresponding to u_0^1 are given by (Figure 3.6)

$$\kappa_{g_1}(t) = -\frac{4u_0^2}{(1 + 4t^2)^{3/3} \sqrt{1 + 4(t^2 + (u_0^2))}}$$

and

$$\kappa_{g_2}(t) = \frac{4u_0^1}{(1 + 4t^2)^{3/3} \sqrt{1 + 4(t^2 + (u_1^2))}}.$$

The next result expresses the geodesic curvature of a curve on a surface with orthogonal parameters in terms of the geodesic curvature of the parameter lines and the angle between the curve and the u^1–lines; it is the analogue for the geodesic curvature of Euler's theorem, Theorem 2.5.31, for the normal curvature.

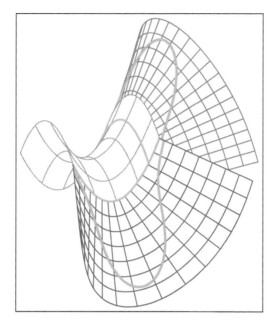

Figure 3.6 Geodesic curvature along the parameter lines of a hyperbolic paraboloid

Theorem 3.1.12 (Liouville).
Let S be a surface with a parametric representation $\vec{x}(u^i)$ and orthogonal parameters, γ be a curve on S with a parametric representation $\vec{x}(u^i(s))$, where s is the arc length along γ, and $\Theta(s)$ be the angle between γ and the u^1–lines of S. Furthermore, let κ_{g_i} denote the geodesic curvature of the u^i–lines ($i = 1, 2$).
Then we have

$$\kappa_g = \frac{d\Theta}{ds} + \kappa_{g_1} \cos\Theta + \kappa_{g_2} \sin\Theta.$$

Proof. Since s is the arc length along γ and $g_{12} = 0$, we have

$$\cos\Theta = \frac{\dot{\vec{x}} \bullet \vec{x}_1}{\|\vec{x}_1\|} = \sqrt{g_{11}}\dot{u}^1 \text{ and } \sin\Theta = \frac{\dot{\vec{x}} \bullet \vec{x}_2}{\|\vec{x}_2\|} = \sqrt{g_{22}}\dot{u}^2,$$

hence

$$\dot{u}^1 = \frac{\cos\Theta}{\sqrt{g_{11}}} \text{ and } \dot{u}^2 = \frac{\sin\Theta}{\sqrt{g_{22}}}.$$

We put

$$A = \sqrt{g}\left(\dot{u}^1\ddot{u}^2 - \dot{u}^2\ddot{u}^1\right) = \sqrt{g}\left(\frac{\cos\Theta}{\sqrt{g_{11}}}\frac{d}{ds}\left(\frac{\sin\Theta}{\sqrt{g_{22}}}\right) - \frac{\sin\Theta}{\sqrt{g_{22}}}\frac{d}{ds}\left(\frac{\cos\Theta}{\sqrt{g_{11}}}\right)\right)$$

$$= \sqrt{g}\left[\frac{\cos^2\Theta}{\sqrt{g_{11}g_{22}}}\frac{d\Theta}{ds} + \frac{\sin^2\Theta}{\sqrt{g_{11}g_{22}}}\frac{d\Theta}{ds} + \dot{u}^1\sin\Theta\frac{d}{ds}\left(\frac{1}{\sqrt{g_{22}}}\right) - \dot{u}^2\cos\Theta\frac{d}{ds}\left(\frac{1}{\sqrt{g_{11}}}\right)\right]$$

$$= \frac{d\Theta}{ds} + \sqrt{g}\left(\dot{u}^1\sin\Theta\frac{d}{ds}\left(\frac{1}{\sqrt{g_{22}}}\right) - \dot{u}^2\cos\Theta\frac{d}{ds}\left(\frac{1}{\sqrt{g_{11}}}\right)\right).$$

Now we obtain by (3.7) and (3.8) in Corollary 3.1.5

$$\frac{d}{ds}\left(\frac{1}{\sqrt{g_{22}}}\right) = -\frac{1}{2(g_{22})^{3/2}}\frac{\partial g_{22}}{\partial u^j}\dot{u}^j = -\frac{1}{\sqrt{g_{22}}}\begin{Bmatrix}2\\12\end{Bmatrix}\dot{u}^1 - \frac{1}{\sqrt{g_{22}}}\begin{Bmatrix}2\\22\end{Bmatrix}\dot{u}^2,$$

and similarly

$$\frac{d}{ds}\left(\frac{1}{\sqrt{g_{11}}}\right) = -\frac{1}{\sqrt{g_{11}}}\begin{Bmatrix}1\\11\end{Bmatrix}\dot{u}^1 - \frac{1}{\sqrt{g_{11}}}\begin{Bmatrix}1\\21\end{Bmatrix}\dot{u}^2.$$

Thus we have

$$A = \frac{d\Theta}{ds} - \sqrt{g}\left[\dot{u}^1\left(\begin{Bmatrix}2\\12\end{Bmatrix}\dot{u}^1\dot{u}^2 + \begin{Bmatrix}2\\22\end{Bmatrix}(\dot{u}^2)^2\right) - \dot{u}^2\left(\begin{Bmatrix}1\\11\end{Bmatrix}(\dot{u}^1)^2 + \begin{Bmatrix}1\\21\end{Bmatrix}\dot{u}^1\dot{u}^2\right)\right].$$

Finally, we obtain

$$\kappa_g = A + \sqrt{g}\left(\dot{u}^1\begin{Bmatrix}2\\ik\end{Bmatrix}\dot{u}^i\dot{u}^k - \dot{u}^2\begin{Bmatrix}1\\ik\end{Bmatrix}\dot{u}^i\dot{u}^k\right) =$$

$$= \frac{d\Theta}{ds} + \sqrt{g}\left(\dot{u}^1\left(\begin{Bmatrix}2\\12\end{Bmatrix}\dot{u}^1\dot{u}^2 + \begin{Bmatrix}2\\11\end{Bmatrix}(\dot{u}^1)^2\right)\right)$$

$$+ \sqrt{g}\left(\dot{u}^2\left(-\begin{Bmatrix}1\\12\end{Bmatrix}\dot{u}^1\dot{u}^2 + \begin{Bmatrix}1\\22\end{Bmatrix}(\dot{u}^2)^2\right)\right),$$

and by Corollary 3.1.5, (3.20) and (3.21)

$$\sqrt{g}\begin{Bmatrix}2\\12\end{Bmatrix}(\dot{u}^1)^2\dot{u}^2 = \cos^2\Theta\sin\Theta\frac{1}{\sqrt{g_{11}}}\begin{Bmatrix}2\\12\end{Bmatrix}$$

$$= \cos^2\Theta\sin\Theta\frac{1}{\sqrt{g_{11}}}\frac{1}{2g_{22}}\frac{\partial g_{22}}{\partial u^1} = \kappa_{g_2}\cos^2\Theta\sin\Theta,$$

$$\sqrt{g}\begin{Bmatrix}2\\11\end{Bmatrix}(\dot{u}^1)^3 = \cos^3\Theta\frac{\sqrt{g}}{g_{11}^{3/2}}\begin{Bmatrix}2\\11\end{Bmatrix} = \kappa_{g_1}\cos^3\Theta,$$

$$-\sqrt{g}\begin{Bmatrix}1\\12\end{Bmatrix}\dot{u}^1)(\dot{u}^2)^2 = -\sin^2\Theta\cos\Theta\frac{1}{\sqrt{g_{11}}}\begin{Bmatrix}2\\12\end{Bmatrix}$$

$$= -\sin^2\Theta\cos\Theta\frac{1}{\sqrt{g_{22}}}\frac{1}{2g_{11}}\frac{\partial g_{11}}{\partial u^2} = \kappa_{g_1}\sin^2\Theta\cos\Theta$$

and

$$-\sqrt{g}\begin{Bmatrix}1\\22\end{Bmatrix}(\dot{u}^2)^3 = \sin^3\Theta\frac{\sqrt{g}}{g_{22}^{3/2}}\begin{Bmatrix}1\\22\end{Bmatrix} = \kappa_{g_2}\sin^3\Theta,$$

whence

$$\kappa_g = \frac{d\Theta}{ds} + \kappa_{g_1}\cos\Theta\left(\cos^2\Theta + \sin^2\Theta\right) + \kappa_{g_2}\sin\Theta\left(\cos^2\Theta + \sin^2\Theta\right)$$

$$= \frac{d\Theta}{ds} + \kappa_{g_1}\cos\Theta + \kappa_{g_2}\sin\Theta.$$

□

We close this section with a few examples.

Example **3.1.13** (**Geodesic curvature of parallels on spheres**).
(a) For the sphere with a parametric representation

$$\vec{x}(u^i) = r\{\cos u^1 \cos u^2, \cos u^1 \sin u^2, \sin u^1\} \ \left((u^1, u^2) \in (-\pi/2, \pi/2) \times (0, 2\pi)\right),$$

we obviously have

$$\vec{N}(u^i) = -\vec{x}(u^i) \cdot \frac{1}{r}.$$

The curvature of the u^2–line corresponding to $u^1 = \alpha$ is

$$\kappa = \frac{1}{r \cos \alpha},$$

and the binormal vector of this u^2–line is given by $\vec{v}_3 = \vec{e}^3$.
Therefore, by Theorem 3.1.12, the geodesic curvature of the u^2–line is

$$\kappa_g = -\frac{1}{r \cos \alpha} \cdot \sin \alpha = -\frac{1}{r} \cdot \tan \alpha.$$

(b) **A curve of constant geodesic curvature on a sphere is a circle line.**
Since

$$\kappa_n = \vec{N} \bullet \ddot{\vec{x}} = -\frac{1}{r}(\vec{x} \bullet \ddot{\vec{x}}) = \frac{1}{r},$$

$\kappa_g = const$ implies $\kappa = const$. Furthermore, we have

$$\dot{\vec{N}} = -\frac{1}{r}\dot{\vec{x}} = -\frac{1}{r}\vec{v}_1,$$

hence

$$0 = \dot{k}_g = \kappa \left(\vec{N} \bullet \frac{d\vec{v}_3}{ds}\right) = -\kappa\tau(\vec{N} \bullet \vec{v}_2) = \frac{\tau}{r}.$$

Thus $\tau = 0$ and the curve is in a plane, that is, a circle.

Example **3.1.14** (**Geodesic curvature of helices on circular cylinders**).
Helices on circular cylinder with a parametric representation

$$\vec{x}(u^i) = \{r \cos u^2, r \sin u^2, u^1\} \ \left((u^1, u^2) \in \mathbb{R} \times (0, 2\pi)\right),$$

where $r > 0$ is a constant, have identically vanishing geodesic curvature.

Proof. Since the first fundamental coefficients are

$$g_{11} = 1, \ g_{12} = 0, \ g_{22} = r^2, \ g = r^2,$$

we obtain the first and second Christoffel symbols by Example 3.1.7

$$[ijk] = 0 \text{ and } \left\{ \begin{matrix} i \\ jk \end{matrix} \right\} = 0 \ (i, j, k = 1, 2).$$

Furthermore, it follows that

$$\epsilon_{11} = \epsilon_{22} = 0 \text{ and } \epsilon_{12} = -\epsilon_{21} = \sqrt{g} = r.$$

We compute the geodesic curvature of a helix with a parametric representation

$$\vec{x}(u^i(s)) = \{r\cos\omega s, r\sin\omega s, h\omega s\},$$

where $\omega = 1/\sqrt{r^2 + h^2}$, by applying identity (3.13). From

$$u^1(s) = h\omega s, \ u^2(s) = \omega s, \ \dot{u}^1(s) = h\omega, \ \dot{u}^2(s) = \omega,$$

$$\ddot{u}^1(s) = 0 \text{ and } \ddot{u}^2(s) = 0,$$

we conclude

$$\kappa_g = \epsilon_{12}\dot{u}^1\left(\ddot{u}^2 + \left\{{2 \atop jk}\right\}\dot{u}^j\dot{u}^k\right) + \epsilon_{21}\dot{u}^2\left(\ddot{u}^1 + \left\{{1 \atop jk}\right\}\dot{u}^j\dot{u}^k\right)$$
$$= rh\omega(0+0) - r\omega(0+0) = 0.$$

Since

$$\vec{v}_3 = \omega\{h\sin\omega s, -h\cos\omega s, r\} \text{ and } \vec{N} = -\{\cos u^2, \sin u^2, 0\},$$

we obtain from Theorem 3.1.12 for the geodesic curvature of the helix

$$\kappa_g = \kappa(\vec{N} \bullet \vec{v}_3) = 0.$$

□

3.2 GEODESIC LINES

In the plane, curves of vanishing curvature, that is, straight lines, play an eminent role, because they are the curves of minimum length joining two points. It will turn out in Theorem 3.5.2 that, generally on surfaces, curves with vanishing geodesic curvature have the same minimum property.

Definition 3.2.1. A curve on a surface with identically vanishing geodesic curvature is called a *geodesic line*.

There is an analogy in the definitions of asymptotic and geodesic lines, namely that along asymptotic lines the normal curvature vanishes identically, whereas along geodesic lines the geodesic curvature vanishes identically.

The next result will give characterizations of geodesic lines.

Theorem 3.2.2. *Geodesic lines on a surface with a parametric representation $\vec{x}(u^i)$ $((u^1, u^2) \in D)$ are characterized by either of the following conditions*

$$\ddot{u}^i + \left\{{i \atop jk}\right\}\dot{u}^j u^k = 0 \ (i = 1, 2) \tag{3.22}$$

or

$$\ddot{\vec{x}} \bullet (\vec{N} \times \dot{\vec{x}}) = 0. \tag{3.23}$$

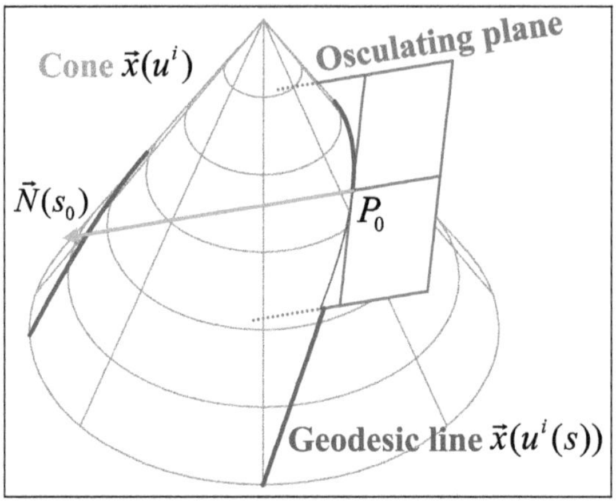

Figure 3.7 Geodesic line on a cone and surface normal vector in the osculating plane at a point

Proof. The condition in (3.23) is an immediate consequence of (3.1) and (3.3). By Remark 2.4.3, we have

$$\ddot{\vec{x}} = \left(\ddot{u}^r + \left\{ \begin{matrix} r \\ ik \end{matrix} \right\} \dot{u}^i \dot{u}^k \right) \vec{x}_r + (L_{ik}\dot{u}^i\dot{u}^k)\vec{N}.$$

Now (3.1) implies

$$\kappa_g \vec{t} = \left(\ddot{u}^r + \left\{ \begin{matrix} r \\ ik \end{matrix} \right\} \dot{u}^i \dot{u}^k \right) \vec{x}_r.$$

Since the vectors $\vec{x}_1$ and $\vec{x}_2$ are linearly independent, it follows that $\kappa_g = 0$ if and only if the conditions in (3.22) hold. □

We obtain the following geometric characterization of geodesic lines as an immediate consequence of equation (3.23) in Theorem 3.2.2.

Remark 3.2.3. *If $\ddot{\vec{x}} \neq \vec{0}$ for a curve on a surface with a parametric representation $\vec{x}(u^i)$ ($(u^1, u^2) \in D$), then the geodesic lines are characterized by the fact that their osculating planes contain the normal vectors of the surface (Figure 3.7).*

Whereas asymptotic lines do not exist in the neighbourhood of elliptic points of a surface, that is, at points with $L > 0$, by Theorem 2.8.2 (b), geodesic lines exist at every point of any surface in some neighbourhood of the point, as the next result will show.

Theorem 3.2.4. *Let P be any point of a surface S, and $\vec{d}$ be a unit vector in the tangent plane of S at P. Then there is one and only one geodesic line through P in the direction given by $\vec{d}$, and the geodesic line exists in some neighbourhood of the point P.*

Proof. Let P have the parameters u_0^k $(k = 1, 2)$ and the direction $\vec{d}$ at P be given by $\vec{d} = \xi^k \vec{x}_k$ where $g_{ik}(u_0^j)\xi^i\xi^k = 1$.

A geodesic line through P has to be a solution of the system (3.22) of differential equations with the initial conditions

$$u^i(0) = u_0^i \quad \text{and} \quad \dot{u}^i(0) = \xi^i \quad (i = 1, 2). \tag{3.24}$$

It follows from the existence and uniqueness theorem for systems of ordinary differential equations that there is one and only one solution $(u^1(s), u^2(s))$ of (3.22) that satisfies the initial conditions in (3.24).

To show that this solution is a geodesic line we have to prove that s is the arc length along the curve given by $u^k(s)$ $(k = 1, 2)$. Then the assertion follows from (3.22). We consider the function f defined by $f(s) = g_{ik}(u^j(s))\dot{u}^i\dot{u}^k$, and shall show $\dot{f}(s) = 0$ for all s. Then we have $f(0) = g_{ik}(u_0^j)\xi^i\xi^k = 1$ by the initial conditions, and this implies $f(s) = 1$ for all s, which means that s is the arc length along the curve. Now we obtain

$$\dot{f}(s) = \frac{\partial g_{ik}}{\partial u^j}\dot{u}^j\dot{u}^i\dot{u}^k + g_{ik}\ddot{u}^i\dot{u}^k + g_{ik}\dot{u}^i\ddot{u}^k.$$

Lemma 3.1.4 (d) yields

$$\frac{\partial g_{ik}}{\partial u^j} = [ijk] + [kji] = g_{kr}\left\{{r \atop ij}\right\} + g_{ir}\left\{{r \atop kj}\right\},$$

hence

$$\dot{f}(s) = g_{kr}\left\{{r \atop ij}\right\}\dot{u}^j\dot{u}^i\dot{u}^k + g_{ir}\left\{{r \atop kj}\right\}\dot{u}^j\dot{u}^i\dot{u}^k + g_{ik}\ddot{u}^i\dot{u}^k + g_{ik}\dot{u}^i\ddot{u}^k$$

$$= g_{kr}\left(\ddot{u}^r + \left\{{r \atop ij}\right\}\dot{u}^j\dot{u}^i\right)\dot{u}^k + g_{ir}\left(\ddot{u}^r + \left\{{r \atop kj}\right\}\dot{u}^j\dot{u}^k\right)\dot{u}^i.$$

Since the functions $u^i(s)$ $(i = 1, 2)$ are solutions of the system (3.22), the terms in the parentheses vanish, and so $\dot{f}(s) = 0$. □

It is useful to give the differential equations for geodesic lines with respect to an arbitrary parameter t.

Theorem 3.2.5. *Let S be a surface with a parametric representation $\vec{x}(u^i)$ $((u^1, u^2) \in D)$.*

(a) If a geodesic line on S is given by $((u^1(t), u^2(t))$, where t is an arbitrary parameter, then we have

$$\frac{d^2u^i}{dt^2} + \left\{{i \atop jk}\right\}\frac{du^j}{dt}\frac{du^k}{dt} = \lambda\frac{du^i}{dt} \quad (i = 1, 2) \text{ where } \lambda = -\frac{d^2t}{ds^2}\left(\frac{ds}{dt}\right)^2. \tag{3.25}$$

(b) A curve on S given by $(u^1(t), u^2(t))$, where t is an arbitrary parameter, is a geodesic line if and only if

$$\frac{d^2u^1}{dt^2}\frac{du^2}{dt} - \frac{d^2u^2}{dt^2}\frac{du^1}{dt} + \left(\frac{du^2}{dt}\left\{{1 \atop ik}\right\} - \frac{du^1}{dt}\left\{{2 \atop ik}\right\}\right)\frac{du^i}{dt}\frac{du^k}{dt} = 0. \tag{3.26}$$

Proof. Let $\bar{u}^i(s) = u^i(t(s))$ $(i = 1, 2)$. Then we have

$$\dot{\bar{u}}^i(s) = \frac{du^i}{dt}\frac{dt}{ds} \text{ and } \ddot{\bar{u}}^i(s) = \frac{d^2u^i}{dt^2}\left(\frac{dt}{ds}\right)^2 + \frac{du^i}{dt}\frac{d^2t}{ds^2} \text{ for } i = 1, 2.$$

(a) If γ_g is a geodesic line given by $(\bar{u}^1(s), \bar{u}^2(s))$, then it follows from (3.22) in Theorem 3.2.2 that

$$0 = \ddot{\bar{u}}^i + \left\{ \begin{matrix} i \\ jk \end{matrix} \right\} \dot{\bar{u}}^j \dot{\bar{u}}^k$$

$$= \frac{d^2u^i}{dt^2}\left(\frac{dt}{ds}\right)^2 + \frac{du^i}{dt}\frac{d^2t}{ds^2} + \left\{ \begin{matrix} i \\ jk \end{matrix} \right\} \frac{du^j}{dt}\frac{du^k}{dt}\left(\frac{dt}{ds}\right)^2 \text{ for } i = 1, 2,$$

hence

$$\frac{d^2u^i}{dt^2} + \left\{ \begin{matrix} i \\ jk \end{matrix} \right\} \frac{du^j}{dt}\frac{du^k}{dt} = -\frac{du^i}{dt}\frac{d^2t}{ds^2}\frac{1}{\left(\dfrac{dt}{ds}\right)^2} = -\frac{du^i}{dt}\frac{d^2t}{ds^2}\left(\frac{ds}{dt}\right)^2$$

$$= \lambda\frac{du^i}{dt} \text{ for } i = 1, 2,$$

that is, (3.25) holds.

(b) Now we prove Part (b).

(b.i) First we show the necessity of (3.26) for a curve to be a geodesic line. If γ_g is a geodesic line given by $(u^1(t), u^2(t))$, where t is an arbitrary parameter, then it follows from Part (a) that

$$\frac{d^2u^1}{dt^2} + \left\{ \begin{matrix} 1 \\ jk \end{matrix} \right\} \frac{du^j}{dt}\frac{du^k}{dt} = \lambda\frac{du^1}{dt} \text{ and } \frac{d^2u^2}{dt^2} + \left\{ \begin{matrix} 2 \\ jk \end{matrix} \right\} \frac{du^j}{dt}\frac{du^k}{dt} = \lambda\frac{du^2}{dt}. \quad (3.27)$$

We multiply the first equation in (3.27) by du^2/dt and the second equation in (3.27) by du^1/dt, and obtain the difference

$$\frac{d^2u^1}{dt^2}\frac{du^1}{dt} - \frac{d^1u^1}{dt^2}\frac{du^2}{dt} + \left(\left\{ \begin{matrix} 1 \\ jk \end{matrix} \right\}\frac{du^2}{dt} - \left\{ \begin{matrix} 2 \\ jk \end{matrix} \right\}\frac{du^1}{dt}\right)\frac{du^j}{dt}\frac{du^k}{dt} = 0,$$

that is, (3.26) holds.

(b.ii) Finally, we show the sufficiency of (3.26) for a curve to be a geodesic line. We assume that (3.26) holds for a curve γ given by $(u^1(t), u^2(t))$. Writing $u^i(t) = \bar{u}^i(s(t))$ $(i = 1, 2)$, we obtain

$$\frac{du^i}{dt} = \dot{\bar{u}}^i\frac{ds}{dt}, \quad \frac{d^2u^i}{dt^2} = \ddot{\bar{u}}^i\left(\frac{ds}{dt}\right)^2 + \dot{\bar{u}}^i\frac{d^2s}{dt^2} \text{ for } i = 1, 2,$$

The first step is (3.26) and the second step is (3.13) for the geodesic curvature κ_g,

$$0 = \left(\ddot{\bar{u}}^1\left(\frac{ds}{dt}\right)^2 + \dot{\bar{u}}^1\frac{d^2s}{dt^2}\right)\dot{\bar{u}}^2\frac{ds}{dt} - \left(\ddot{\bar{u}}^2\left(\frac{ds}{dt}\right)^2 + \dot{\bar{u}}^2\frac{d^2s}{dt^2}\right)\dot{\bar{u}}^1\frac{ds}{dt}$$

$$+ \left(\left\{ {1 \atop jk} \right\} \dot{u}^2 - \left\{ {2 \atop jk} \right\} \dot{u}^1 \right) \dot{u}^j \dot{u}^k \left(\frac{ds}{dt} \right)^3$$

$$= \ddot{u}^1 \dot{u}^2 \left(\frac{ds}{dt} \right)^3 + \dot{u}^1 \dot{u}^2 \frac{d^2 s}{dt^2} \frac{ds}{dt} + \left\{ {1 \atop jk} \right\} \dot{u}^2 \dot{u}^j \dot{u}^k \left(\frac{ds}{dt} \right)^3$$

$$- \ddot{u}^2 \dot{u}^1 \left(\frac{ds}{dt} \right)^3 - \dot{u}^2 \dot{u}^1 \frac{d^2 s}{dt^2} \frac{ds}{dt} - \left\{ {2 \atop jk} \right\} \dot{u}^1 \dot{u}^j \dot{u}^k \left(\frac{ds}{dt} \right)^3$$

$$= \left(\frac{ds}{dt} \right)^3 \left(\dot{u}^2 \left(\ddot{u}^1 + \left\{ {1 \atop jk} \right\} \dot{u}^j \dot{u}^k \right) - \dot{u}^1 \left(\ddot{u}^2 + \left\{ {2 \atop jk} \right\} \dot{u}^j \dot{u}^k \right) \right)$$

$$= \left(\frac{ds}{dt} \right)^3 \left(-\frac{\kappa_g}{g} \right).$$

Since $ds/dt \neq 0$ this implies $\kappa_g = 0$, and so γ is a geodesic line.

□

Now we give a few examples.

First we determine the geodesic lines on a circular cylinder. We already know from Example 3.1.14 that helices are geodesic lines on circular cylinders. Now we show that if a curve on a circular cylinder is a geodesic line, then it is a helix or a parameter line.

Example **3.2.6 (Geodesic lines on circular cylinders).**
We consider the circular cylinder with a parametric representation

$$\vec{x}(u^i) = \{ r \cos u^2, r \sin u^2, h u^1 \} \quad \left((u^1, u^2) \in \mathbb{R} \times (0, 2\pi) \right),$$

where r and h are positive constants.
Then the geodesic line through the point with $(u^1(0), u^2(0)) = (u_0^1, u_0^2)$ at an angle $\Theta_0 \in (-\pi/2, \pi/2)$ to the u^2–line through $u^1(0)$ is given by

$$u^1(s) = \sin \Theta_0 \cdot s + u_0^1 \text{ and } u^2(s) = \frac{\cos \Theta_0}{r} \cdot s + u_0^2 \text{ for } s \in \mathbb{R}. \tag{3.28}$$

Proof. Since $r(u^1) = r$ and $h(u^1) = u^1$, it follows from (2.24) in Example 2.3.5 that

$$g_{11}(u^i) = g_{11}(u^1) = 1, \ g_{12}(u^i) = 0 \text{ and } g_{22}(u^i) = g_{22}(u^1) = \frac{1}{r}$$

and we obtain from the formulae in Example (3.1.7)

$$\left\{ {i \atop jk} \right\} = 0 \text{ for all } i, j, k = 1, 2.$$

So the differential equations (3.22) in Theorem 3.2.2 for the geodesic lines on the circular cylinder reduce to

$$\ddot{u}^i = 0 \text{ for } i = 1, 2$$

with the solutions

$$u^i = c_i s + d_i \ (i = 1, 2) \text{ for all } s \in \mathbb{R},$$

where c_i and d_i $(i = 1, 2)$ are constants.
The initial conditions $u^i(0) = u_0^i$ yield $d_i = u_0^i$ for $i = 1, 2$. We also have

$$\dot{u}^1(0) = \frac{\sin \Theta_0}{\sqrt{g_{11}(u^i(0))}} = \sin \Theta_0 = c_1$$

and

$$\dot{u}^2(0) = \frac{\cos \Theta_0}{\sqrt{g_{22}(u^i(0))}} = \frac{\cos \Theta_0}{r} = c_2.$$

□

Now we determine the geodesic lines on the *hyperbolic plane* or *Poincaré half–plane*, which is a model of the *hyperbolic non–Euclidean geometry*.

Definition 3.2.7. The *hyperbolic plane* or *Poincaré half–plane* is the set

$$H = \{(x, y) \in \mathbb{R}^2 : y > 0\}$$

with the first fundamental form

$$(ds)^2 = \frac{1}{y^2} \cdot \left((dx)^2 + (dy)^2 \right).$$

Example 3.2.8 (Geodesic lines on the Poincaré half–plane).
The geodesic lines on the Poincaré half–plane are segments of Euclidean circle lines or straight lines that meet the real axis $y = 0$ at right angles.

Proof. We put $u^1 = y$ and $u^2 = x$. Since

$$g_{11}(u^i) = g_{22}(u^i) = \frac{1}{(u^1)^2} \text{ and } g_{12}(u^i) = 0,$$

we obtain from (3.10) in Corollary 3.1.5

$$\left\{ \begin{matrix} 1 \\ 11 \end{matrix} \right\} = \frac{1}{2g_{11}} \cdot \frac{\partial g_{11}}{\partial u^1} = \frac{1}{2g_{11}} \cdot \frac{\partial g_{22}}{\partial u^1} = -\left\{ \begin{matrix} 1 \\ 22 \end{matrix} \right\} = -\frac{1}{u^1},$$

$$\left\{ \begin{matrix} 1 \\ 12 \end{matrix} \right\} = \left\{ \begin{matrix} 1 \\ 21 \end{matrix} \right\} = \left\{ \begin{matrix} 2 \\ 11 \end{matrix} \right\} = \left\{ \begin{matrix} 2 \\ 22 \end{matrix} \right\} = 0$$

and

$$\left\{ \begin{matrix} 2 \\ 12 \end{matrix} \right\} = \left\{ \begin{matrix} 2 \\ 21 \end{matrix} \right\} = \frac{1}{2g_{22}} \cdot \frac{\partial g_{22}}{\partial u^1} = -\frac{1}{u^1}.$$

Consequently the differential equations (3.22) in Theorem 3.2.2 for the geodesic lines on H reduce to

$$\ddot{u}^1 = \frac{1}{u^1} \cdot \left((\dot{u}^1)^2 - (\dot{u}^2)^2 \right) \text{ and } \ddot{u}^2 = \frac{2\dot{u}^1 \dot{u}^2}{u^1}. \tag{3.29}$$

(i) If $\dot{u}^2 = 0$, then $u^2 = const$ and we obtain a straight line segment that meets the x–axis at a right angle.

(ii) If $\dot{u}^2 \neq 0$, then the second equation in (3.29) yields

$$\left(\frac{\dot{u}^2}{(u^1)^2}\right)^{\boldsymbol{\cdot}} = \frac{\ddot{u}^2(u^1)^2 - 2\dot{u}^2\dot{u}^1 u^1}{(u^1)^4} = \frac{1}{(u^1)^2} \cdot \left(\ddot{u}^2 - \frac{2\dot{u}^1\dot{u}^2}{u^1}\right) = 0,$$

hence

$$\dot{u}^2 = c \cdot u^1 \neq 0 \text{ for some constant } c \neq 0.$$

Since the arc length is the parameter of the geodesic lines, we have that

$$g_{ik}u^i u^k = \frac{1}{(u^1)^2}\left((\dot{u}^1)^2 + (\dot{u}^2)^2\right) = 1, \text{ that is, } (\dot{u}^1)^2 = (u^1)^2 - (\dot{u}^2)^2$$

and it follows that

$$\left(\frac{\dot{u}^1}{\dot{u}^2}\right)^2 = \frac{(u^1)^2 - (\dot{u}^2)^2}{(\dot{u}^2)^2} = \frac{1}{c^2(u^1)^2} - 1.$$

This implies

$$\left(\frac{du^1}{du^2}\right)^2 = \frac{1 - c^2(u^1)^2}{c^2(u^1)^2} \text{ for } u^1 \leq \frac{1}{|c|}.$$

Since the improper integral

$$\int\limits_{t=0}^{1/|c|} \frac{t\, dt}{\sqrt{1 - c^2 t^2}} \text{ exists,}$$

we obtain for all $u^1 \leq 1/|c|$

$$u^2 = \pm|c| \int\limits_{t}^{u^1} \frac{t\, dt}{\sqrt{1 - c^2 t^2}} = \mp\frac{1}{|c|}\left(\sqrt{1 - c^2 t^2}\right)\Big|_{t=0}^{u^1} = \mp\frac{1}{|c|}\left(\sqrt{1 - c^2(u^1)^2} - 1\right).$$

This yields

$$\left(|c|u^2 \mp 1\right)^2 = c^2 u^2 \mp 2|c|u^2 + 1 = 1 - c^2(u^1)^2$$

and so

$$\left(u^2 \mp \frac{1}{|c|}\right)^2 + (u^1) = \frac{1}{|c|} \text{ for } u^1 \leq \frac{1}{|c|}.$$

This is the equation of a semi–circle line in H, with the radius $r = 1/|c|$ and centre in the point $(x, y) = (\pm 1/|c|, 0)$; such a semi–circle line meets the x–axis at a right angle.

$\square$

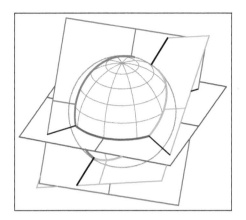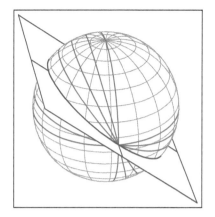

Figure 3.8 Principal circles (left) and geodesic lines (right) on a sphere

Visualization 3.2.9 (Geodesic lines on planes and spheres).
(a) The geodesic lines in a plane are exactly the straight lines.
(b) The geodesic lines on a sphere are exactly the principal circle lines, *that is, the intersections of the sphere with planes through the centre of the sphere (left in Figure 3.8).*

Proof.

(a) If $\vec{x}(s)$ is a curve in a plane, then $\dot{\vec{x}} \perp \vec{N}$, hence $\vec{a} = \dot{\vec{x}} \times \vec{N} \neq \vec{0}$. Furthermore, since $\vec{a} \perp \vec{N}$, $\vec{a} \perp \dot{\vec{x}}$ and $\ddot{\vec{x}} \perp \dot{\vec{x}}, \vec{N}$, identity (3.23) implies $\dddot{\vec{x}} = \vec{0}$. Thus $\vec{x}$ is a straight line.

Conversely, any straight line satisfies $\ddot{\vec{x}} = 0$, hence identity (3.23).

(b) Now we prove Part (b).

(b.i) First we show that any geodesic line on a sphere is a principal circle line. If $\vec{x}(s)$ is a geodesic line on a sphere of radius r, then $\vec{x}(s)$ is a circle line by Example 3.1.13 (b). Furthermore, $\kappa_g = 0$ implies $\kappa_n^2 = \kappa^2$. On the sphere, we have

$$\vec{x} \bullet \vec{x} = r, \text{ hence } \vec{x} \bullet \dot{\vec{x}} = 0 \text{ and } \ddot{\vec{x}} \bullet \vec{x} + \dot{\vec{x}} \bullet \dot{\vec{x}} = 0.$$

It follows that

$$\ddot{\vec{x}} \bullet \vec{x} = -1, \text{ and consequently } \kappa_n = \vec{N} \bullet \ddot{\vec{x}} = -\frac{1}{r}(\vec{x} \bullet \ddot{\vec{x}}) = \frac{1}{r},$$

and the circle line is a principal circle line.

(b.ii) Now we show that any principal circle line is a geodesic line on the sphere. If $\vec{x}(s)$ is a principal circle line, then we have $\kappa = 1/r$ for its curvature and therefore

$$\kappa_g^2 = \kappa^2 - \kappa_n^2 = \frac{1}{r^2} - \frac{1}{r^2} = 0.$$

Thus $\vec{x}(s)$ is a geodesic line (right in Figure 3.8).

□

Visualization 3.2.10 (Geodesic lines on a circular cone).
We consider the circular cone with a parametric representation

$$\vec{x}(u^i) = \left\{ \frac{1}{\sqrt{a}} \cdot u^1 \cos u^2, \frac{1}{\sqrt{a}} \cdot u^1 \sin u^2, u^1 \right\} \quad \left((u^1, u^2) \in D = (0, \infty) \times (0, 2\pi) \right)$$

for given $a > 0$.
We put $\alpha = \sqrt{a/(a+1)}$.

(a) *Then the geodesic lines are the meridians and the curves with*

$$\left\{ \begin{array}{ll} u^1(s) & = \sqrt{\alpha^2(s+s_0)^2 + \beta^2} \\ u^2(s) & = \dfrac{k}{\alpha\beta} \cdot \tan^{-1}\left(\dfrac{\alpha}{\beta}(s+s_0) \right) + d \end{array} \right\} \quad \textit{for } s \in \mathbb{R}, \qquad (3.30)$$

where k and d are real constants and $\beta^2 = k^2/a$ (Figure 3.9).

(b) *Let $(u_0^1, u_0^2) \in D$ be given and Θ_0 denote the angle between the parallel through u_0^1 and the geodesic line through the point with position vector $\vec{x}_0 = \vec{x}(u_0^i)$.*
Then the constants k, s_0, β and d in (3.30) are given by

$$k = \sqrt{a} \cdot u_0^1 \cos \Theta_0, \ \ s_0 = \frac{u_0^1 \sin \Theta_0}{\alpha}, \ \ \beta = u_0^1 \cos \Theta_0$$

and

$$d = u_0^2 - \sqrt{a+1} \cdot \tan^{-1}\left(\tan \tilde{\Theta}_0 \right)$$

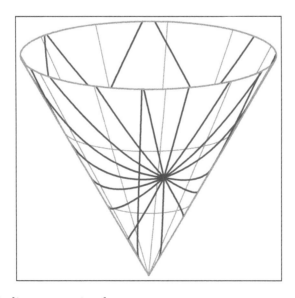

Figure 3.9 Geodesic lines on a circular cone

with

$$\tilde{\Theta}_0 = \begin{cases} \Theta_0 & \left(\Theta_0 \in \left(\dfrac{-\pi}{2}, \dfrac{\pi}{2}\right)\right) \\ \Theta_0 - \pi & \left(\Theta_0 \in \left(\dfrac{\pi}{2}, \dfrac{3\pi}{2}\right)\right) \end{cases}.$$

Proof.

(a) We obtain from Example 3.1.7

$$\begin{Bmatrix} 1 \\ 11 \end{Bmatrix} = \begin{Bmatrix} 1 \\ 12 \end{Bmatrix} = \begin{Bmatrix} 1 \\ 21 \end{Bmatrix} = 0, \quad \begin{Bmatrix} 1 \\ 22 \end{Bmatrix} = -\frac{u^1}{a+1},$$

$$\begin{Bmatrix} 2 \\ 11 \end{Bmatrix} = \begin{Bmatrix} 2 \\ 22 \end{Bmatrix} = 0 \text{ and } \begin{Bmatrix} 2 \\ 12 \end{Bmatrix} = \begin{Bmatrix} 2 \\ 21 \end{Bmatrix} = \frac{1}{u^1}$$

and the differential equations (3.22) in Theorem 3.2.2 reduce to

$$\ddot{u}^1 - \frac{u^1}{a+1}(\dot{u}^2)^2 = 0 \tag{3.31}$$

and

$$\ddot{u}^2 + \frac{2}{u^1} \cdot \dot{u}^1 \dot{u}^2 = 0. \tag{3.32}$$

First it follows from (3.32) that

$$\ddot{u}^2 (u^1)^2 + 2\dot{u}^1 \dot{u}^2 u^1 = \frac{d}{ds}\left(\dot{u}^2 \cdot (u^1)^2\right) = 0,$$

whence

$$\dot{u}^2 = \frac{k}{(u^1(s))^2}, \tag{3.33}$$

where $k \in \mathbb{R}$ is a constant.

(a.i) First we assume that $k = 0$.
Then $\dot{u}^2 = 0$, and since s is the arc length along the geodesic line, it follows that

$$(\dot{u}^1)^2 = \frac{1}{g_{11}} = \frac{a}{a+1} = \alpha^2$$

and so

$$u^1 = \alpha s + \tilde{d} \text{ for all } s > -\frac{\tilde{d}}{\alpha},$$

where $\tilde{d} \in \mathbb{R}$ is a constant.
Thus we have obtained the meridians, which are geodesic lines by (3.18) in Lemma 3.1.10, since $\begin{Bmatrix} 2 \\ 11 \end{Bmatrix} = 0$.

(a.ii) Now we assume that $k \neq 0$.
Then we multiply (3.31) by $\dot{u}^1$, substitute (3.33), and obtain

$$\ddot{u}^1 \dot{u}^1 - \frac{1}{a+1} \cdot \frac{k^2}{(u^1)^3} \cdot \dot{u}^1 = \frac{1}{2} \cdot \frac{d}{ds}\left((\dot{u}^1)^2 + \frac{k^2}{a+1}\frac{1}{(u^1)^2}\right) = 0,$$

hence

$$(\dot{u}^1)^2 = \tilde{d} - \frac{k^2}{a+1} \cdot \frac{1}{(u^1)^2},$$

where $\tilde{d} > 0$ is a constant. This implies

$$\frac{\dot{u}^1 u^1}{\sqrt{\tilde{d}(u^1)^2 - \dfrac{k^2}{a+1}}} = \pm 1 \text{ for } u^1 > \frac{|k|}{\sqrt{\tilde{d}(a+1)}}.$$

Integration and solving for u^1 yields

$$u^1(s) = \sqrt{\tilde{d}(s+s_0)^2 + \frac{k^2}{\tilde{d}(a+1)}} \text{ for all } s,$$

where $s_0 \in \mathbb{R}$ is a constant.
Since s is the arc length along the geodesic line, we obtain

$$1 = g_{11}(\dot{u}^1)^2 + g_{22}(\dot{u}^2)^2 = \frac{a+1}{a}\left(\tilde{d} - \frac{k^2}{a+1} \cdot \frac{1}{(u^1)^2}\right) + \frac{k^2}{a} \cdot \frac{1}{(u^1)^2},$$

that is,

$$\tilde{d} = \frac{a}{a+1} = \alpha^2.$$

This yields the first identity in (3.30), where $\beta^2 = k^2/2$. Substituting this in (3.33) and integrating, we finally obtain the second identity in (3.30).

Thus we have proved Part (a).

(b) It follows from

$$\dot{u}^2(0) = \frac{\cos \Theta_0}{\sqrt{g_{22}(u_0^1)}}$$

that

$$k = \sqrt{a} \cdot u_0^1 \cos \Theta_0 \text{ and } \beta^2 = (u_0^1)^2 \cdot \cos^2 \Theta_0.$$

Furthermore,

$$\dot{u}^1(0) = \frac{\sin \Theta_0}{\sqrt{g_{11}(u_0^1)}} \text{ implies } s_0 = \frac{u_0^1 \cdot \sin \Theta_0}{\alpha}.$$

Since $\tan^{-1}$ is an odd function, we may choose $\beta = u_0^1 \cdot \cos \Theta_0$, and

$$u_0^2 = \frac{k}{\alpha\beta} \cdot \tan^{-1}\left(\frac{\alpha}{\beta} \cdot s_0\right) + d$$

finally yields

$$d = u_0^2 - \sqrt{a+1} \cdot \tan^{-1}(\tan \tilde{\Theta}_0)$$

with

$$\tilde{\Theta}_0 = \begin{cases} \Theta_0 & \left(\Theta_0 \in \left(\dfrac{-\pi}{2}, \dfrac{\pi}{2}\right)\right) \\ \Theta_0 - \pi & \left(\Theta_0 \in \left(\dfrac{\pi}{2}, \dfrac{3\pi}{2}\right)\right). \end{cases}$$

□

Visualization 3.2.11 (Geodesic lines on a general cone).
Let S be the sphere with radius $r > 0$ and its centre in the origin, and γ be a curve on S with a parametric representation $\vec{y}(s^*)$ ($s^* \in I \subset \mathbb{R}$), where s^* is the arc length along γ. By GC, we denote the general cone generated by the straight lines that join the origin and the points on γ. We put $u^2 = s^*$, and obtain the following parametric representation for the general cone GC

$$\vec{x}(u^i) = u^1 \vec{y}(u^2) \quad \left((u^1, u^2) \in (0, \infty) \times I\right).$$

(a) Then the u^1–lines are geodesic lines on GC.
(b) The geodesic line through $u^1(0) = \underset{0}{u^1} > 0$, $u^2(0) = \underset{0}{u^2} > 0$ at an angle $\Theta_0 \neq \pi/2, 3\pi/2$ to u^2–line through $\underset{0}{u^1}$ is given by (Figure 3.10)

$$\left\{ \begin{aligned} u^1(s) &= \frac{1}{r}\sqrt{(s + r\underset{0}{u^1}\sin\Theta_0)^2 + (r\underset{0}{u^1}\cos\Theta_0)^2} \\ &\text{and} \\ u^2(s) &= r\tan^{-1}\left(\frac{s + r\underset{0}{u^1}\sin\Theta_0}{r\underset{0}{u^1}\cos\Theta_0}\right) + \underset{0}{u^2} - r\tan^{-1}(\tan\Theta_0) \end{aligned} \right\} \text{ for } s > 0. \quad (3.34)$$

Proof. It follows that

$$\vec{x}_1 = \vec{y}, \ \vec{x}_2 = u^1\vec{y}', \ \vec{x}_{11} = \vec{0}, \ \vec{x}_{12} = \vec{y}' \text{ and } \vec{x}_{22} = u^1\vec{y}''.$$

Since $\vec{y}^2 = r^2$ and $(\vec{y}')^2 = 1$, we have

$$\vec{y}' \bullet \vec{y} = \vec{y}'' \bullet \vec{y}' = 0 \text{ and } \vec{y}'' \bullet \vec{y} = -1.$$

Thus we obtain

$$g_{11} = r^2, \ g_{12} = 0, \ g_{22} = (u^1)^2 \text{ and } g = (u^1)^2,$$

and so the parameters are orthogonal and the first fundamental coefficients satisfy the condition in (3.9) in Corollary 3.1.5. Therefore, we have by (3.10)

$$\left\{ \begin{matrix} 1 \\ 11 \end{matrix} \right\} = \frac{1}{2g_{11}}\frac{\partial g_{11}}{\partial u^1} = 0, \ \left\{ \begin{matrix} 1 \\ 22 \end{matrix} \right\} = -\frac{1}{2g_{11}}\frac{\partial g_{22}}{\partial u^1} = -\frac{u^1}{r^2},$$

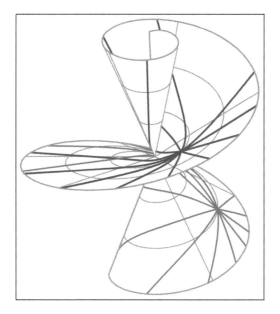

Figure 3.10 Geodesic lines on a cone generated by a loxodrome on a sphere

$$\left\{\begin{matrix} 1 \\ 12 \end{matrix}\right\} = \left\{\begin{matrix} 1 \\ 21 \end{matrix}\right\} = 0, \quad \left\{\begin{matrix} 2 \\ 11 \end{matrix}\right\} = \left\{\begin{matrix} 2 \\ 22 \end{matrix}\right\} = 0$$

and

$$\left\{\begin{matrix} 2 \\ 12 \end{matrix}\right\} = \left\{\begin{matrix} 2 \\ 21 \end{matrix}\right\} = \frac{1}{2g_{22}} \frac{\partial g_{22}}{\partial u^1} = \frac{1}{u^1}.$$

The system (3.22) of differential equations for the geodesic lines on the general cone reduces to

$$\ddot{u}^1 - \frac{u^1}{r^2}(\dot{u}^2)^2 = 0 \tag{3.35}$$

and

$$\ddot{u}^2 + \frac{2}{u^1}\dot{u}^1\dot{u}^2 = 0. \tag{3.36}$$

We multiply equation (3.36) by $(u^1)^2$ and obtain

$$\ddot{u}^2(u^1)^2 + 2u^1\dot{u}^1\dot{u}^2 = \frac{d}{ds}\left(\dot{u}^2(u^1)^2\right) = 0,$$

hence

$$\dot{u}^2(s) = \frac{k}{(u^1(s))^2}, \tag{3.37}$$

where $k \in \mathbb{R}$ is a constant.

(a) First we assume $k = 0$.

Since s is the arc length along the geodesic line, we have

$$(\dot{u}^1)^2 = \frac{1}{g_{11}} = \frac{1}{r^2}$$

and the geodesic lines are given by

$$u^1(s) = \frac{1}{r}(s + s_0) \ (s > -s_0)$$

for some constant s_0, and $u^2(s) = c$, where $c \in I$ is a constant; they are the u^1–lines corresponding to $c \in I$.

(b) Now let $k \neq 0$.
We multiply equation (3.35) by $\dot{u}^1$, substitute (3.37), and obtain

$$\ddot{u}^1\dot{u}^1 - \frac{k^2\dot{u}^1}{r^2(u^1)^3} = \frac{1}{2}\frac{d}{ds}\left((\dot{u}^1)^2 + \frac{k^2}{r^2(u^1)^2}\right) = 0.$$

This implies

$$(\dot{u}^1)^2 + \frac{k^2}{r^2(u^1)^2} = c^2,$$

where $c \neq 0$ is a constant,

$$\frac{\dot{u}^1 u^1}{\sqrt{c^2(u^1)^2 - \frac{k^2}{r^2}}} = \pm 1 \ \left(u^1 > \left|\frac{k}{cr}\right|\right)$$

and

$$(u^1)^2 - \frac{k^2}{r^2c^2} = c^2(s + s_0)^2,$$

where $s_0 \in \mathbb{R}$ is a constant. Therefore, we have

$$u^1 = \sqrt{c^2(s + s_0)^2 + \frac{k^2}{r^2c^2}} \ \text{ for all } s.$$

Now (3.37) implies

$$\dot{u}^2 = \frac{k}{c^2(s + s_0)^2 + \frac{k^2}{r^2c^2}} = \frac{r^2c^2}{k\left(\frac{r^2c^4(s+s_0)^2}{k^2} + 1\right)}$$

and

$$u^2(s) = r\tan^{-1}\left(\frac{rc^2(s + s_0)}{k}\right) + d,$$

where $d \in \mathbb{R}$ is a constant. Since s is the arc length along the geodesic line, we must have

$$1 = g_{11}(\dot{u}^1)^2 + g_{22}(\dot{u}^2)^2 = r^2\left(c^2 - \frac{k^2}{(u^1)^2}\right) + (u^1)^2\frac{k^2}{(u^1)^4} = r^2c^2,$$

hence $|c| = 1/r$. Thus the geodesic lines are given by

$$\left\{ \begin{array}{l} u^1(s) = \dfrac{1}{r}\sqrt{(s+s_0)^2 + k^2 r^2} \\ \text{and} \\ u^2(s) = r\tan^{-1}\left(\frac{s+s_0}{kr}\right) + d \end{array} \right\} \quad (s \in \mathbb{R}), \qquad (3.38)$$

where $k \neq 0$, s_0 and d are constants.
For the initial conditions

$$u^1(0) = u_0^1 > 0, \ u^2(0) = u_0^2 > 0$$

and the angle Θ_0 between the geodesic line and the u^2–line through u_0^1, we obtain

$$\dot{u}^2(0) = \frac{\cos\Theta_0}{\sqrt{g_{22}(u_0^2)}} = \frac{\cos\Theta_0}{u_0^1} \text{ and } k = \dot{u}^2(0)(u^1(0))^2 = u_0^1\cos\Theta_0,$$

and then

$$\dot{u}^1(0) = \frac{\sin\Theta_0}{\sqrt{g_{11}(u_0^1)}} \text{ implies } s_0 = u_0^1 r\sin\Theta_0.$$

Finally, we have

$$d = u_0^2 - r\tan^{-1}\left(\frac{s_0}{kr}\right) = u_0^2 - r\tan^{-1}(\tan\Theta_0)$$

$$= \left\{ \begin{array}{ll} u_0^2 - r\Theta_0 & (\Theta_0 \in (-\pi/2, \pi/2)) \\ u_0^2 - r(\Theta_0 + \pi) & (\Theta_0 \in (\pi/2, 3\pi/2)). \end{array} \right.$$

Thus the geodesic lines for $\Theta_0 \neq \pi/2, 3\pi/2$ are given by (3.34) (Figure 3.10).

$\square$

Remark 3.2.12. *We can obtain the geodesic lines on the circular cone in Visualization 3.2.10 as a special case of the geodesic lines on a general cone as in Visualization 3.2.11.*
Putting

$$\vec{y}(u^2) = \left\{ \frac{1}{\sqrt{a}}\cdot\cos u^2, \frac{1}{\sqrt{a}}\cdot\sin u^2, 1 \right\},$$

we obtain $\vec{x}(u^i) = u^1\vec{y}(u^2)$. *We also have*

$$\|\vec{y}(u^2)\| = \sqrt{\frac{1}{a}+1} = \sqrt{\frac{a+1}{a}} = r$$

and

$$s(t) = \int_0^t \|\vec{y}'(\tau)\| \, dt = \int_0^t \left\|\left\{ -\frac{1}{\sqrt{a}}\sin\tau, \frac{1}{\sqrt{a}}\cos\tau, 0 \right\}\right\| \, d\tau = \frac{1}{\sqrt{a}}\int_0^t d\tau = \frac{t}{\sqrt{a}},$$

that is, $t = \sqrt{a}s$. We put $u^1 = u^{*1}$ and $u^2 = \sqrt{a}u^{*2}$, and obtain

$$\vec{x}^*(u^{*i}) = u^{*1}\vec{y}(u^{*2})$$

for the corresponding representation as a general cone.
We obtain for the geodesic lines by Visualization 3.2.11 by the first identity in (3.34)

$$u^1(s) = u^{*1}(s) = \frac{1}{r}\sqrt{(s + ru_0^{*1}\sin\Theta_0)^2 + (u_0^{*1}\cos\Theta_0)^2}$$

$$= \sqrt{(as + u_0^1)^2 + (u_0^{*1}\cos\Theta_0)^2} = \sqrt{\alpha^2(s + s_0)^2 + \beta^2},$$

where $\alpha s_0 = u_0^1\sin\Theta_0$ and $\beta = u_0^1\cos\Theta_0$. This is the first identity in (3.30).
The second identity in (3.34) yields for the geodesic line

$$u^2(s) = \sqrt{a}u^{*2}(s) = \sqrt{a}\left(r\tan^{-1}\left(\frac{s + ru_0^{*1}\sin\Theta_0}{ru_0^{*1}\cos\Theta_0}\right) + u_0^{*2} - r\tan^{-1}(\tan\Theta_0)\right)$$

$$= \sqrt{a+1}\tan^{-1}\left(\frac{\frac{s}{r} + u_0^1\sin\Theta_0}{u_0^1\cos\Theta_0}\right) + u_0^2 - \sqrt{a+1}\tan^{-1}(\tan\Theta_0)$$

$$= \frac{k}{\alpha\beta} \cdot \tan^{-1}\left(\frac{\alpha}{\beta}(s + s_0)\right) + d,$$

where

$$\frac{k}{\alpha\beta} = \frac{\sqrt{a}u_0^1\cos\Theta_0}{\alpha u_0^1\cos\Theta_0} = \sqrt{a+1}, \quad \frac{\frac{s}{r} + u_0^1\sin\Theta_0}{u_0^1\cos\Theta_0} = \frac{\alpha(s + s_0)}{\beta}$$

and $d = u_0^2 - \sqrt{a+1}\tan^{-1}(\tan\tilde{\Theta}_0)$.
Thus we obtained the second identity in (3.30).

Visualization 3.2.13 (Geodesic lines on a pseudo-sphere).

Let PS be the pseudo-sphere of Visualization 2.5.19 (c), and $(u_0^1, u_0^2) \in (0,\infty) \times (0, 2\pi)$ be given. Then the u^1–line corresponding to u_0^2 is a geodesic line. Furthermore, if $\Theta_0 \in [0, 2\pi) \setminus \{\pi/2, 3\pi/2\}$, then the geodesic line through (u_0^1, u_0^2) with an angle of Θ_0 to the u^2–line through (u_0^1, u_0^2) is given by (Figure 3.11)

$$\left.\begin{cases} u^1(s) = -\log\left(|\delta|\cosh(s + s_0)\right) \\ and \\ u^2(s) = \frac{1}{\delta}\tanh(s + s_0) + c_1 \\ \qquad for\ s < \log\left(\frac{1 + \sqrt{1 - \delta^2}}{|\delta|}\right) - s_0, \end{cases}\right\} \qquad (3.39)$$

where

$$\left.\begin{cases} \delta = e^{-u_0^1}\cos\Theta_0, \ c_1 = u_0^2 + e^{u_0^1}\tan\Theta_0 \\ and \\ s_0 = \log\left(\sqrt{\frac{1 - \sin\Theta_0}{1 + \sin\Theta_0}}\right). \end{cases}\right\} \qquad (3.40)$$

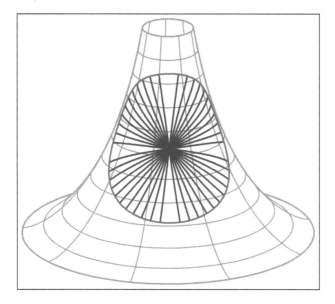

Figure 3.11 Geodesic lines on a pseudo-sphere

Proof. By (3.11) and (3.12), the differential equations (3.22) for the geodesic lines on PS reduce to

$$\ddot{u}^1 + e^{-2u^1}\left(\dot{u}^2\right)^2 = 0 \tag{3.41}$$

and

$$\ddot{u}^2 - 2\dot{u}^1\dot{u}^2 = 0. \tag{3.42}$$

If $\dot{u}^2 = 0$, then we obtain the u^1–line corresponding to $u^2 = u_0^2$, that is, for a suitable orientation of the arc length s,

$$u^1(s) = s + u_0^1 \text{ and } u^2(s) = u_0^2 \text{ for } s > -u_0^1.$$

Now we assume $\dot{u}^2 \neq 0$. Then (3.42) yields

$$\frac{\ddot{u}^2}{\dot{u}^2} = 2\dot{u}^1, \text{ that is, } \dot{u}^2 = \delta e^{2u^1} \tag{3.43}$$

for some constant $\delta \neq 0$. Substituting this in (3.41), we obtain

$$\ddot{u}^1 + \delta^2 e^{2u^1} = 0, \text{ that is, } \frac{d}{ds}\left((\dot{u}^1)^2 + \delta^2 e^{2u^1}\right) = 0,$$

hence

$$\left(\dot{u}^1\right)^2 = d^2 - \delta^2 e^{2u^1} \text{ for some constant } d \neq 0 \text{ and } u^1 \leq \log\left|\frac{d}{\delta}\right|. \tag{3.44}$$

Since s is the arc length along the geodesic line, we have by the second identity in (3.43) and (3.44)

$$g_{11}(u^1(s))\left(\dot{u}^1(s)\right)^2 + g_{22}(u^1(s))\left(\dot{u}^2(s)\right)^2 = d^2 - \delta^2 e^{2u^1(s)} + e^{-2u^1(s)}\delta^2 e^{4u^1(s)} = d^2 = 1,$$

that is, $d = \pm 1$. Now (3.44) yields

$$\dot{u}^1 = \pm\sqrt{1 - \delta^2 e^{2u^1}} \text{ for } 0 < u^1 \le -\log|\delta|, \tag{3.45}$$

that is, $|\delta| < 1$, and

$$I(u^1) = \int \frac{du^1}{\sqrt{1 - \delta^2 e^{2u^1}}} = \pm(s + s_0)$$

for some constant s_0. To evaluate the integral $I(u^1)$, we substitute $z = 1/|\delta|e^{-u^1}$, so $du^1/dz = -1/z$, observe that $z > 1$, since $u^1 < -\log|\delta|$, and obtain

$$I(u^1) = -\int \frac{dz}{z\sqrt{1 - z^{-2}}} = -\int \frac{dz}{\sqrt{z^2 - 1}} = -\log\left(z + \sqrt{z^2 - 1}\right).$$

Writing $\beta = \mp(s + s_0)$, we obtain $z^2 - 1 = (e^\beta - z)^2$, that is, $2ze^\beta = 1 + e^{2\beta}$, and so $z = \cosh\beta$. Since $\cosh\beta = \cosh(-\beta)$, this implies

$$u^1(s) = -\log(|\delta|\cosh(s + s_0)).$$

We note that $\cosh(s + s_0) \ge 1$ for all s implies $u^1(s) \le -\log|\delta|$ for all s. Furthermore, $u^1(s) > 0$ implies $\cosh(s + s_0) < 1/|\delta|$, that is,

$$s < \log\left(\frac{1}{|\delta|} + \sqrt{\frac{1}{\delta^2} - 1}\right) - s_0 = \log\left(\frac{1 + \sqrt{1 - \delta^2}}{|\delta|}\right) - s_0.$$

Now it follows from the second identity in (3.43) that

$$\dot{u}^2 = \frac{1}{\delta\cosh^2(s + s_0)},$$

and so

$$u^2(s) = \frac{1}{\delta}\tanh(s + s_0) + c_1$$

for some constant $c_1 \in \mathbb{R}$. Therefore, we have shown that the general solution of the differential equations (3.41) and (3.42) is given by (3.39).

Now we determine the constants δ, s_0 and c_1 such that $u^i(0) = u_0^i$ for $i = 1, 2$ and Θ_0 is the angle between the geodesic line and the u^2–line through (u_0^1, u_0^2). Let $\vec{y}(s)$ be a parametric representation of the geodesic line. Then it follows from

$$\frac{\dot{\vec{y}}(0) \bullet \vec{x}_1(u_0^1)}{\|\vec{x}_1(u_0^1)\|} = \sqrt{g_{11}(u_0^1)}\frac{du^1(0)}{ds} = -\tanh s_0 = \sin\Theta_0$$

that

$$s_0 = \tanh^{-1}(-\sin\Theta_0) = \frac{1}{2}\log\left(\frac{1 - \sin\Theta_0}{1 + \sin\Theta_0}\right),$$

which is the third identity in (3.40). Furthermore,

$$\frac{\dot{\vec{y}}(0) \bullet \vec{x}_2(u_0^1)}{\|\vec{x}_2(u_0^1)\|} = \sqrt{g_{22}(u_0^1)}\frac{du^2(0)}{ds} = e^{-u_0^1}\delta e^{2u_0^1} = \cos\Theta_0$$

implies $\delta = e^{-u_0^1} \cos \Theta_0$, which is the first identity in (3.40). Finally,

$$u_0^2 = \frac{1}{\delta} \tanh s_0 + c_1$$

yields

$$c_1 = u_0^2 + \frac{e^{u_0^1} \sin \Theta_0}{\cos \Theta_0} = u_0^2 + e^{u_0^1} \tan \Theta_0,$$

which is the second identity in (3.40) (Figure 3.11).   □

3.3   GEODESIC LINES ON SURFACES WITH ORTHOGONAL PARAMETERS

In general, it is difficult to find geodesic lines on surfaces by solving the system of differential equations (3.22). The solution can, however, explicitly be given by an integral, if the first fundamental coefficients g_{ik} of the surface satisfy the following conditions

$$g_{12}(u^k) = 0 \text{ and } g_{11}(u^k) = g_{11}(u^1) \text{ and } g_{22}(u^k) = g_{22}(u^1). \tag{3.46}$$

We already know from (3.8) in Corollary 3.1.5 that then the second Christoffel symbols satisfy

$$\left\{\begin{matrix} 1 \\ 11 \end{matrix}\right\} = \frac{1}{2g_{11}} \frac{dg_{11}}{du^1}, \quad \left\{\begin{matrix} 1 \\ 12 \end{matrix}\right\} = \left\{\begin{matrix} 1 \\ 21 \end{matrix}\right\} = 0, \quad \left\{\begin{matrix} 1 \\ 22 \end{matrix}\right\} = -\frac{1}{2g_{11}} \frac{dg_{22}}{du^1},$$

$$\left\{\begin{matrix} 2 \\ 11 \end{matrix}\right\} = 0, \quad \left\{\begin{matrix} 2 \\ 12 \end{matrix}\right\} = \left\{\begin{matrix} 2 \\ 21 \end{matrix}\right\} = \frac{1}{2g_{22}} \frac{dg_{22}}{du^1}, \quad \left\{\begin{matrix} 2 \\ 22 \end{matrix}\right\} = 0. \tag{3.47}$$

For instance, the first fundamental coefficients of surfaces of revolution satisfy the conditions in (3.46).

Theorem 3.3.1 (Clairaut).
Let S be a surface with first fundamental coefficients that satisfy the conditions in (3.46), γ be a geodesic line on S, and Θ be the angle between γ and the u^2–lines of S. Then we have (Figure 3.12)

$$\sqrt{g_{22}} \cos \Theta = c = const. \tag{3.48}$$

Proof. Let the geodesic line γ be given by $u^k(s)$ $(k = 1, 2)$. Then it follows from (3.22) and (3.47) that

$$\ddot{u}^2 + \left\{\begin{matrix} 2 \\ ik \end{matrix}\right\} \dot{u}^i \dot{u}^k = \ddot{u}^2 + \frac{1}{g_{22}} \frac{dg_{22}}{du^1} \dot{u}^1 \dot{u}^2 = 0, \tag{3.49}$$

and so

$$\frac{d}{ds} \left(g_{22} \dot{u}^2 \right) = g_{22} \ddot{u}^2 + \frac{dg_{22}}{du^1} \dot{u}^1 \dot{u}^2 = 0 \text{ implies } g_{22} \dot{u}^2 = c \tag{3.50}$$

for some constant $c \in \mathbb{R}$.

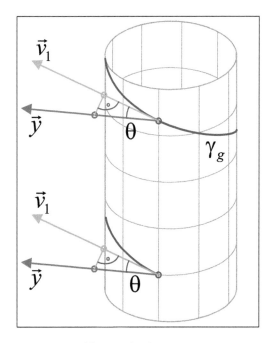
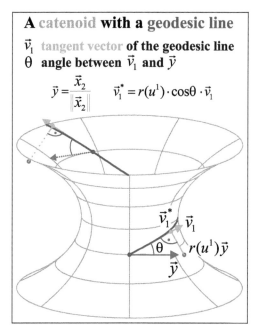

Figure 3.12   Clairaut's theorem

Furthermore, we have for the angle Θ between the geodesic line with a parametric representation $\vec{x}_\gamma$ and the u^2–lines of S

$$\cos\Theta = \frac{\dot{\vec{x}}_\gamma \bullet \vec{x}_2}{\|\vec{x}_2\|} = \frac{g_{22}\dot{u}^2}{\sqrt{g_{22}}} = \frac{c}{\sqrt{g_{22}}}, \tag{3.51}$$

since $\vec{x}_1 \bullet \vec{x}_2 = 0$.
Now (3.48) is an immediate consequence. $\qquad\square$

The following result gives the explicit solution of the differential equations (3.22) for surfaces with first fundamental coefficients satisfying (3.46).

Theorem 3.3.2. *Let S be a surface with a parametric representation $\vec{x}(u^i)$ $((u^1, u^2) \in D)$ and first fundamental coefficients that satisfy the conditions in (3.46). Then we have:*
(a) The u^1–lines are geodesic lines.
(b) The u^2–lines are geodesic lines if and only if

$$\frac{dg_{22}}{du^1} = 0. \tag{3.52}$$

(c) If $c \in \mathbb{R}$ and u_0^1 are chosen such that $g_{22}(u_0^1) > c^2$, then a curve γ with a parametric representation $\vec{x}_c(u^1) = \vec{x}_c(u^1, u^2(u^1))$ is a geodesic line in some neighbourhood of u_0^1 if and only if

$$u^2(u^1) - u_0^2 = \pm c \int_{u_0^1}^{u^1} \frac{\sqrt{g_{11}(u)}}{\sqrt{g_{22}(u)}\sqrt{g_{22}(u) - c^2}} \, du. \tag{3.53}$$

Proof. By (3.22) and (3.47), the geodesic lines on S are the solutions of (3.50) and

$$\ddot{u}^1 + \frac{1}{2g_{11}}\frac{dg_{11}}{du^1}(\dot{u}^1)^2 - \frac{1}{2g_{11}}\frac{dg_{22}}{du^1}(\dot{u}^2)^2 = 0. \tag{3.54}$$

(a) For the u^1–lines, we have $\dot{u}^2 = \ddot{u}^2 = 0$ and so (3.49) is satisfied. Furthermore, $g_{11}(\dot{u}^1)^2 = 1$ implies

$$0 = \frac{d}{ds}\left(g_{11}(\dot{u}^1)^2\right) = 2\ddot{u}^1\dot{u}^1 g_{11} + \frac{dg_{11}}{du^1}(\dot{u}^1)^3 = 2\dot{u}^1 g_{11}\left(\ddot{u}^1 + \frac{1}{g_{11}}\frac{dg_{11}}{du^1}(\dot{u}^1)^2\right),$$

and so (3.54) is also satisfied.
Thus the u^1–lines are geodesic lines.

(b) For the u^2–lines, we have $\dot{u}^1 = \ddot{u}^1 = 0$ and the differential equations (3.54) and (3.49) reduce to

$$\frac{1}{2g_{11}}\frac{dg_{22}}{du^1}(\dot{u}^2)^2 = 0 \text{ and } \ddot{u}^2 = 0.$$

These differential equations are satisfied if and only if condition (3.52) holds.

(c) Now we prove Part (c).

(c.i) First we show the necessity of (3.53).
We assume that a geodesic line γ_g is given by the functions $u^k(s)$ $(k = 1, 2)$. Then these functions satisfy the differential equations (3.54) and (3.49). As in (3.50), (3.49) implies $g_{22}\dot{u}^2 = c$ for some constant $c \in \mathbb{R}$. Since s is the arc length along γ_g, we have

$$1 = g_{ik}\dot{u}^i\dot{u}^k = g_{11}(\dot{u}^1)^2 + g_{22}(\dot{u}^2)^2 = g_{11}(\dot{u}^1)^2 + \frac{c^2}{g_{22}},$$

and so

$$\dot{u}^1 = \pm\sqrt{\left(1 - \frac{c^2}{g_{22}}\right)\frac{1}{g_{11}}} = \pm\sqrt{\frac{g_{22} - c^2}{g_{11}g_{22}}} \text{ for } g_{22} \geq c^2. \tag{3.55}$$

Let $u_0^1 > 0$ be chosen such that $g_{22}(u_0^1) > c^2$. Then we have by (3.55) and the fact that $g_{22}\dot{u}^2 = c$

$$\frac{du^2}{du^1} = \frac{\dot{u}^2}{\dot{u}^1} = \pm\frac{c}{g_{22}}\sqrt{\frac{g_{11}g_{22}}{g_{22} - c^2}} = \pm\frac{c}{\sqrt{g_{22}}}\sqrt{\frac{g_{11}}{g_{22} - c^2}}.$$

This implies (3.53).

(c.ii) Finally, we show the sufficiency of (3.53).
We assume that the curve γ_g is given by $u^t = t$ and $u^2(t)$ as in (3.53). Then we have

$$\frac{du^1}{dt} = 1, \quad \frac{d^2u^1}{dt^2} = 0, \quad \frac{du^2}{dt} = \pm c\frac{\sqrt{g_{11}(t)}}{\sqrt{g_{22}(t)}\sqrt{g_{22}(t) - c^2}},$$

and we obtain by (3.47)

$$\frac{d^2u^2}{dt^2} = \pm \frac{c}{2} \frac{1}{\sqrt{g_{22}(t)}\sqrt{g_{22}(t)-c^2}} \frac{1}{\sqrt{g_{11}(t)}} \cdot \frac{dg_{11}(t)}{dt}$$

$$\mp \frac{c}{2} \frac{\sqrt{g_{11}(t)}}{\left(\sqrt{g_{22}(t)}\sqrt{g_{22}(t)-c^2}\right)^3} \left((g_{22}-c^2)\frac{dg_{22}(t)}{dt} + g_{22}\frac{dg_{22}(t)}{dt}\right)$$

$$= \pm c \frac{\sqrt{g_{11}(t)}}{\sqrt{g_{22}(t)}\sqrt{g_{22}(t)-c^2}} \left[\frac{1}{2g_{11}(t)} \frac{dg_{11}(t)}{dt}\right.$$

$$\left. - \frac{1}{2g_{22}(t)(g_{22}(t)-c^2)} \left((g_{22}(t)-c^2)\frac{dg_{22}(t)}{dt} + g_{22}(t)\frac{g_{22}(t)}{dt}\right)\right]$$

$$= \frac{du^2}{dt} \left(\left\{\begin{matrix}1\\11\end{matrix}\right\} - \frac{1}{g_{22}(t)}\frac{dg_{22}(t)}{dt} - \frac{c^2}{2g_{22}(t)(g_{22}-c^2)}\frac{dg_{22}(t)}{dt}\right)$$

$$= \frac{du^2}{dt} \left(\left\{\begin{matrix}1\\11\end{matrix}\right\} - 2\left\{\begin{matrix}2\\12\end{matrix}\right\} - \frac{1}{2g_{11}(t)}\frac{dg_{22}(t)}{dt}\left(\frac{du^2}{dt}\right)^2\right)$$

$$= \frac{du^2}{dt} \left(\left\{\begin{matrix}1\\11\end{matrix}\right\} + \left\{\begin{matrix}2\\11\end{matrix}\right\}\left(\frac{du^2}{dt}\right)^2 - 2\left\{\begin{matrix}2\\12\end{matrix}\right\}\right),$$

hence (3.26) is satisfied, and so γ_g is a geodesic line by Theorem 3.2.5.

□

We close this section with an example.

Example 3.3.3 (Geodesic lines on general cylinders and cones).
(a) Geodesic lines on a general cylinder
Let γ be a curve with a parametric representation $\vec{y}(u^1)$ ($u^1 \in I_1$) in the x^1x^2–plane. Then a general cylinder GCyl is generated by the straight lines through the points of γ orthogonal to the x^1x^2–plane. A parametric representation of GCyl is given by

$$\vec{x}(u^i) = \vec{y}(u^1) + u^2\vec{e}^3 \ \left((u^1, u^2) \in I_1 \times \mathbb{R}\right).$$

It follows from $\vec{x}_1(u^i) = \vec{y}'(u^1)$ and $\vec{x}_2(u^i) = \vec{e}^3$ that

$$g_{11}(u^i) = \left(\vec{y}'(u^1)\right)^2, \ g_{12}(u^i) = 0 \ and \ g_{22}(u^i) = 1,$$

and the first fundamental coefficients of GCyl satisfy the conditions in (3.46). By Theorem 3.3.2 (a) and (b), the parameter lines are geodesic lines on GCyl. Furthermore, by Theorem 3.3.2 (c), the geodesic line through the point with parameters (u_0^1, u_0^2) at an angle if $\Theta_0 \neq 0, \pi/2, \pi, 3/2\pi$ to the u^2–line through u_0^1 is given by

$$u^2(u^1) = u_0^2 + \frac{\cos\Theta_0}{|\sin\Theta_0|} \int_{u_0^1}^{u^1} \|\vec{y}'(u)\| \, du.$$

If we denote the arc length along the curve γ between u_0^1 and u^1 by $s(u^1)$, then the geodesic line is given by

$$u^2(u^1) = u_0^2 + \frac{\cos \Theta_0}{|\sin \Theta_0|} \cdot s(u^1).$$

(b) Geodesic lines on a general cone
The first fundamental coefficients of the general cone GC in Visualization 3.2.11 satisfy the conditions in (3.46). By Theorem 3.3.2, the geodesic line that starts at the point given by (u_0^1, u_0^2) at an angle $\Theta_0 \in (0, \pi/2)$ to the u^2–line through u_0^1 is given by

$$u^2(u^1) = u_0^2 + c \int_{u_0^1}^{u^1} \frac{1}{u\sqrt{u^2 - c^2}} \, du, \tag{3.56}$$

where $c = u_0^1 \cos \Theta_0$. To evaluate the integral in (3.56), we substitute $t = c/u$, hence $dt/du = -c/u^2$, and obtain

$$c \int_{u_0^1}^{u^1} \frac{1}{u\sqrt{u^2 - c^2}} \, du = -\int_{\frac{c}{u_0^1}}^{\frac{c}{u^1}} \frac{1}{\sqrt{1 - t^2}} \, dt$$

$$= \cos^{-1}\left(\frac{c}{u^1}\right) - \cos^{-1}\left(\frac{c}{u_0^1}\right)$$

$$= \cos^{-1}\left(\frac{u_0^1 \cos \Theta_0}{u^1}\right) - \cos^{-1} \cos \Theta_0.$$

Therefore, the geodesic line on the general cone CG is given by

$$u^2(u^1) = u_0^2 + \cos^{-1}\left(\frac{u_0^1 \cos \Theta_0}{u^1}\right) - \Theta_0 \quad (u^1 > u_0^1 \cos \Theta_0).$$

3.4 GEODESIC LINES ON SURFACES OF REVOLUTION

We already observed that the first fundamental coefficients of surfaces of rotation satisfy the conditions in (3.46). Therefore, we can apply the results of Section 3.3 to find the geodesic lines on surfaces of revolution; it turns out that they have one of three characteristic shapes.

Throughout this section, we assume that RS is a surface of rotation with a parametric representation

$$\vec{x}(u^i) = \{r(u^1)\cos u^1, r(u^1)\sin u^1, h(u^1)\} \quad \left((u^1, u^2) \in I_1 \times I_2 \subset \mathbb{R} \times (0, 2\pi)\right)$$

and we make the usual assumptions $r(u^1) > 0$ and $|r(u^1)| + |h'(u^1)| > 0$ on I_1. We recall that the first fundamental coefficients of RS are

$$g_{11}(u^i) = (r'(u^1))^2 + (h'(u^1))^2, \ g_{12}(u^i) = 0 \text{ and } g_{22}(u^i) = (r(u^1))^2. \tag{3.57}$$

We assume that a geodesic line γ with a parametric representation $\vec{x}_c(u^1, u^2(u^1))$ starts at a point P_0 with parameters u_0^1 and u_0^2 at an angle $\Theta_0 \in [0, \pi/2]$ to the parallel that corresponds to $u^1 = u_0^1$. We obtain from Theorem 3.3.2:

If $\Theta_0 = \pi/2$, then we obtain the meridian through P_0 as a geodesic line.

If $\Theta_0 = 0$, then the parallel through P_0 is a geodesic line if and only if

$$\frac{dg_{22}}{du^1}(u_0^1) = 2r'(u_0^1)r(u_0^1) = 0.$$

Since $r(u_0^1) \neq 0$, this is the case if and only if $r'(u_0^1) = 0$.

If $\Theta_0 \in (0, \pi/2)$, then we have

$$0 < c = \sqrt{g_{22}(u_0^1)} \cos \Theta_0 = r(u_0^1) \cos \Theta_0 < r(u_0^1),$$

and consequently γ exists in neighbourhood of u_0^1. It follows from (3.53) of Theorem 3.3.2 (c)

$$u^2(u^1) = u_0^2 + c \int\limits_{u_0^1}^{u^1} \frac{\sqrt{(r'(u))^2 + (h'(u))^2}}{r(u)\sqrt{(r(u))^2 - c^2}}. \tag{3.58}$$

The characteristic shape of the geodesic line depends on whether the integral in (3.58) exists for all $u^1 > u_0^1$ or has a removable or non–removable singularity.

First characteristic shape

If $r(u^1) > c$ for all $u^1 \in I_1$ with $u^1 > u_0$, then the integral on the right hand side of (3.58) exits for all such values of u^1.

Visualization 3.4.1 (Geodesic line on a circular cylinder and a catenoid).
(a) We consider the circular cylinder with radius $r > 0$ and a parametric representation given by $r(u^1) = r$ and $h(u^1) = u^1$. Let $(u_0^1, u_0^2) \in I_1 \times (0, 2\pi)$ be an arbitrary point.
If $\Theta_0 = \pi/2$, then we obtain the meridian through (u_0^1, u_0^2) as a geodesic line.
If $\Theta_0 = 0$, then $r'(u_0^1) = 0$ implies that the parallel through (u_0^1, u_0^2) is a geodesic line.
If $\Theta_0 \in (0, \pi/2)$, then we have $c = r(u_0^1) \cos \Theta_0 < r(u_0^1) = r = r(u^1)$ for all $u^1 \in I_1$, hence the geodesic line takes the first characteristic shape. Observing $c = r \cos \Theta_0$, we obtain from (3.58) for the geodesic line

$$u^2(u^1) = u_0^2 + c \int\limits_{u_0^1}^{u^1} \frac{du^1}{r\sqrt{r^2 - c^2}} = u_0^2 + \frac{c}{r\sqrt{r^2 - c^2}} \cdot \left(u^1 - u_0^1\right)$$

$$= u_0^2 + \frac{r \cos \Theta_0}{r\sqrt{r^2 - r^2 \cos^2 \Theta_0}} \left(u^1 - u_0^1\right) = u_0^2 + \frac{\cos \Theta_0}{r \sin \Theta_0} \left(u^1 - u_0^1\right)$$

$$= u_0^2 + \frac{\cot \Theta_0}{r} \left(u^1 - u_0^1\right)$$

for all $u^1 \in I_1$.

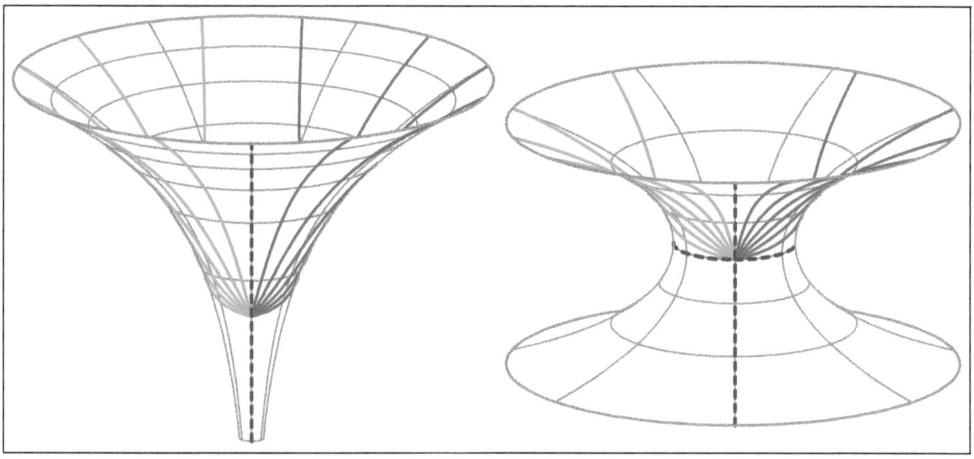

Figure 3.13 Geodesic lines of the first characteristic shape

(b) We consider the catenoid given by $r(u^1) = a\cosh(u^1/a)$ and $h(u^1) = u^1$. Let (u_0^1, u_0^2) be an arbitrary point with $u_0^1 \geq 0$. Since $r'(u^1) = \sinh(u^1/a) = 0$ if and only if $u^1 = 0$, the parallel through (u_0^1, u_0^2) is a geodesic line if and only if $u_0 = 0$. If $\Theta_0 \in (0, \pi/2)$ then $c = r(u_0)\cos\Theta_0 = a\cosh(u_0^1/a) < r(u^1) = a\cosh(u^1/a)$ for all $u^1 > u_0^1$, since $\cosh$ is monotone increasing on $(0, \infty)$. Hence the geodesic line takes the first characteristic shape. We obtain from (3.58) for the geodesic line (Figure 3.13)

$$u^2(u^1) = u_0^2 + c\int_{u_0^1}^{u^1} \frac{\cosh\left(\dfrac{u}{a}\right)}{a\cosh\left(\dfrac{u}{a}\right)\sqrt{a^2\cosh^2\left(\dfrac{u}{a}\right) - c^2}}\, du$$

$$= u_0^2 + \frac{c}{a^2}\int_{u_0^1}^{u^1} \frac{du}{\sqrt{\cosh^2\left(\dfrac{u}{a}\right) - \dfrac{c^2}{a^2}}}.$$

The common properties of the second and third characteristic shapes

Now let $r(u^1) = c$ for some $u^1 > u_0^1$. We put

$$\hat{u}^1 = \inf\{u^1 > u_0^1 : r(u^1) = c\}, \tag{3.59}$$

and observe that $r(\hat{u}^1) = c$, since r is continuous. We study the geodesic line as $u^1 \to \hat{u}^1$, that is, the improper integral

$$I(\hat{u}^1; c) = \int_{u_0}^{\hat{u}^1} \frac{\sqrt{(r'(u))^2 + (h'(u))^2}}{r(u)\sqrt{(r(u))^2 - c^2}}. \tag{3.60}$$

We assume $r \in C^{m+1}(I_1)$ and $k = \min\{m \in \mathbb{N} : r^{(m)}(\hat{u}^1) \neq 0\}$. It follows from Taylor's formula that

$$r(u^1) = c + \frac{r^{(k)}(\hat{u}^1)}{k!}(u^1 - \hat{u}^1)^k + o\left((u^1 - \hat{u}^1)^k\right) \quad (u^1 \to \hat{u}^1),$$

hence

$$(r(u^1))^2 - c^2 = 2c\frac{r^{(k)}(\hat{u}^1)}{k!}(u^1 - \hat{u}^1)^k + o\left((u^1 - \hat{u}^1)^k\right) \quad (u^1 \to \hat{u}^1). \qquad (3.61)$$

Thus, by (3.61), the improper integral converges if and only if $k = 1$. If Θ denotes the angle between the parallel through u^1 and the tangent vector of the geodesic line, then we obviously have

$$\frac{\dfrac{d\vec{x}_c}{du^1}}{\left\|\dfrac{d\vec{x}_c}{du^1}\right\|} = \sin\Theta \cdot \frac{\vec{x}_1}{\|\vec{x}_1\|} + \cos\Theta \cdot \frac{\vec{x}_2}{\|\vec{x}_2\|}$$

$$= \sqrt{1 - \frac{c^2}{(r(u^1))^2}} \cdot \frac{\vec{x}_1}{\|\vec{x}_1\|} + \frac{c}{r(u^1)} \cdot \frac{\vec{x}_2}{\|\vec{x}_2\|},$$

and consequently

$$\frac{\dfrac{d\vec{x}_c(u^1, u^2(u^1))}{du^1}}{\left\|\dfrac{d\vec{x}_c(u^1, u^2(u^1))}{du^1}\right\|} \to \frac{\vec{x}_2}{\|\vec{x}_2\|} \quad (u^1 \to \hat{u}^1). \qquad (3.62)$$

The second characteristic shape

We assume $k = 1$, that is, $r'(\hat{u}^1) \neq 0$. Then the improper integral in (3.60) converges, that is, the limit

$$\hat{u}^2 = \lim_{u^1 \to \hat{u}^1} u^2(u^1)$$

exists and the geodesic line is tangent to the parallel that corresponds to $\hat{u}^1$, which is not a geodesic line, since $r'(u^1) \neq 0$ (Figures 3.14 and 3.15).

If a geodesic line takes the second characteristic shape for some initial values of u_0^1, u_0^2 and Θ_0, then it will also take the second characteristic shape in some open interval I_Θ that contains Θ_0, when u_0^1 and u_0^2 remain fixed. Then the value of $\hat{u}^1$ depends on $\Theta \in I_\Theta$, say $\hat{u}^1(\Theta)$, where $r(\hat{u}^1(\Theta)) = c(\Theta) = r(u_0^1)\cos\Theta$. We assume that the function r has an inverse ϕ on an interval $I_r \supset \{\hat{u}^1(\Theta) : \Theta \in I_\Theta\}$. Then we consider the *bounding line* for the geodesic lines for Θ varying in I_Θ (Figure 3.16), given by $\hat{u}^1(\Theta) = \phi(c(\Theta))$ and

$$\hat{u}^2(\Theta) = u_0^2 + c(\Theta) \lim_{u^1 \to \hat{u}_1(\Theta)} \left(\int_{u_0^1}^{u^1} \frac{\sqrt{(r'(u))^2 + (h'(u))^2}}{r(u)\sqrt{r^2(u) - c^2(\Theta)}} \, du \right) \quad (\Theta \in I_\Theta).$$

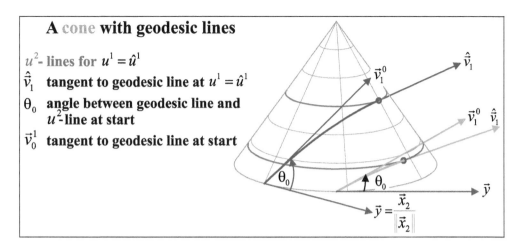

Figure 3.14 Geodesic lines of the second characteristic shape with tangent vectors

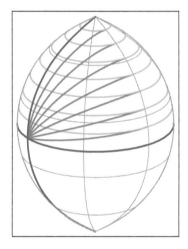

Figure 3.15 Geodesic lines of the second characteristic shape

Visualization 3.4.2 (Geodesic lines on a circular cone).
We consider the circular cone with a parametric representation

$$\vec{x}(u^i) = \left\{ (1 - u^1)\frac{1}{\sqrt{a}}\cos u^2, (1 - u^1)\frac{1}{\sqrt{a}}\sin u^2, u^1 \right\}$$

for $(u^1, u^2) \in (0,1) \times (0, 2\pi)$, where $a > 0$ is a constant (Figure 3.14).
This Visualization is similar to Visualization 3.2.11.
Let $(u_0^1, u_0^2) \in (0,1) \times (0, 2\pi)$ be given.
If $\Theta_0 = 0$, then the meridian through u_0^2 is a geodesic line.
If $\Theta_0 \neq 0$, then $r'(u_0^1) = -1/\sqrt{a} \neq 0$ and the parallel through u_0^1 is not a geodesic line.

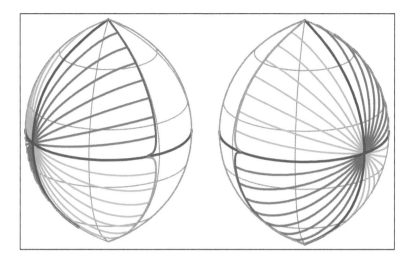

Figure 3.16 Geodesic lines of the second characteristic shape and the bounding lines for $u^1 = \hat{u}^1$

If $\Theta_0 \in (0, \pi/2)$, then we have

$$c = \cos\Theta_0 \sqrt{g_{22}(u_0^1)} = \cos\Theta_0 \frac{(1 - u_0^1)}{\sqrt{a}} < \frac{(1 - u_0^1)}{\sqrt{a}},$$

and

$$r'(\hat{u}^1) \neq 0 \text{ for } \hat{u}^1 = 1 - \cos\Theta_0(1 - u_0^1).$$

Putting $\alpha := \sqrt{(a + 1)/a}$, we obtain

$$u^2(u^1) = u_0^2 + c \int_{u_0^1}^{u^1} \frac{\sqrt{\dfrac{1}{a} + 1}}{\dfrac{1 - u}{\sqrt{a}} \sqrt{\dfrac{(1 - u)^2}{a} - c^2}}\, du$$

$$= u_0^2 + ac\alpha \int_{u_0^1}^{u^1} \frac{du}{(1 - u)\sqrt{(1 - u)^2 - ac^2}}$$

$$= u_0^2 - ac\alpha \int_{1 - u_0^1}^{1 - u^1} \frac{dt}{t\sqrt{t^2 - ac^2}}.$$

To evaluate the integral, we make the substitution $y = \sqrt{a}c/t$ and obtain

$$\int_{1 - u_0^1}^{1 - u^1} \frac{dt}{t\sqrt{t^2 - ac^2}} = \int_{1 - u_0^1}^{1 - u^1} \frac{dt}{t^2\sqrt{1 - \dfrac{ac^2}{t^2}}} = -\frac{1}{c\sqrt{a}} \int_{c\sqrt{a}/(1 - u_0^1)}^{c\sqrt{a}/(1 - u^1)} \frac{dy}{\sqrt{1 - y^2}}$$

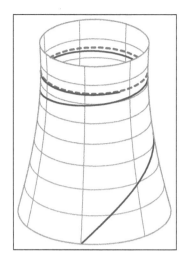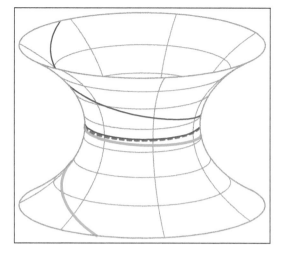

Figure 3.17 Geodesic lines of the third characteristic shape and their asymptotes

$$= \frac{1}{c\sqrt{a}} \left(\cos^{-1} \left(\frac{c\sqrt{a}}{1 - u^1} \right) - \cos^{-1} \left(\frac{c\sqrt{a}}{1 - u_0^1} \right) \right).$$

Since $c\sqrt{a} = (1 - u_0^1) \cos \Theta_0$ *and* $\Theta_0 \in (0, \pi/2)$, *we finally obtain*

$$u^2(u^1) = u_0^2 - \sqrt{a+1} \left(\cos^{-1} \left(\cos \Theta_0 \frac{1 - u_0^1}{1 - u^1} \right) - \Theta_0 \right)$$

for the geodesic line.

The third characteristic shape

We assume $k \geq 2$, that is, in particular $r'(\hat{u}^1) = 0$. Then the improper integral in (3.60) diverges and $u^2(u^1) \to \infty$ as $u^1 \to \hat{u}^1$. It follows from (3.62) that $\vec{x}_c$ approaches the u^2–line through $\hat{u}^1$ which now is a geodesic line since $r'(\hat{u}^1) = 0$. The geodesic line cannot be tangent to this u^2–line because of the uniqueness of geodesic lines stated in Theorem 3.2.4; it asymptotically approaches the u^2–line (Figure 3.17).

Visualization 3.4.3 (Geodesic lines on a hyperboloid of one sheet).
We consider the hyperboloid of one sheet given by parametric representation

$$\vec{x}(u^i) = \left\{ \frac{1}{\sqrt{a}} \cosh u^1 \cos u^2, \frac{1}{\sqrt{a}} \cosh u^1 \sin u^2, \sqrt{-\frac{1}{b}} \sinh u^1 \right\}$$

for $((u^1, u^2) \in D = \mathbb{R} \times (0, 2\pi))$, *where* $a > 0$ *and* $b < 0$ *are constants.*
Let $(u_0^1, u_0^2) \in D$ *be given. If we choose*

$$\Theta_0 = \cos^{-1} \left(\frac{1}{\cosh u_0^1} \right), \tag{3.63}$$

then it follows that

$$r(u^1) = \frac{1}{\sqrt{a}} \cosh u^1$$

implies

$$c = r(u_0^1) \cos \Theta_0 = \frac{1}{\sqrt{a}} \cosh u_0^1 \cdot \frac{1}{\cosh u_0^1}$$

and $r(u^1) > 0$ for all $u^1 \neq 0$, and we obtain for $\hat{u}^1 = 0$

$$r(\hat{u}^1) = c \text{ and } r'(\hat{u}^1) = \frac{1}{\sqrt{a}} \sinh \hat{u}^1 = 0.$$

Therefore, the geodesic line takes the third characteristic shape.
Putting $\alpha = 1 - a/b > 1$ (since $b < 0$) and observing $c = 1/\sqrt{a}$, we obtain for $u^1 < 0$

$$u^2(u^1) = u_0^2 + c\sqrt{a} \int \frac{\sqrt{\alpha \cosh^2 u - 1}}{\cosh u \sqrt{\cosh^2 u - c^2 a}} \, du$$

$$= u_0^2 - c\sqrt{a} \int \frac{\sqrt{\alpha \cosh^2 u - 1}}{\cosh u \sinh u} \, du.$$

We substitute

$$x = \sqrt{\alpha \cosh^2 u - 1}, \quad \frac{dx}{du} = \frac{\alpha \sinh u \cosh u}{\sqrt{\alpha \cosh^2 u - 1}},$$

$$\cosh u = \sqrt{\frac{x^2 + 1}{\alpha}} \text{ and } \sinh u = -\sqrt{\frac{x^2 + 1 - \alpha}{\alpha}}.$$

Then the integral becomes

$$I = \int \frac{\sqrt{\alpha \cosh^2 u - 1}}{\cosh u \sinh u} \, du = \alpha \int \frac{x^2}{(x^2 + 1)(x^2 + 1 - \alpha)} \, dx$$

$$= (\alpha - 1) \int \frac{dx}{(x^2 + 1 - \alpha)} + \int \frac{dx}{x^2 + 1}$$

$$= \frac{\sqrt{\alpha - 1}}{2} \left(\int \frac{dx}{x - \sqrt{\alpha - 1}} - \int \frac{dx}{x + \sqrt{\alpha - 1}} \right) + \tan^{-1} x$$

$$= \frac{\sqrt{\alpha - 1}}{2} \log \left(\frac{x - \sqrt{\alpha - 1}}{x + \sqrt{\alpha - 1}} \right) + \tan^{-1} x + k,$$

where k is a constant of integration. We put

$$k(u_0^1) = \frac{\sqrt{\alpha - 1}}{2} \log \left(\frac{\sqrt{\alpha \cosh^2 u_0^1 - 1} - \sqrt{\alpha - 1}}{\sqrt{\alpha \cosh^2 u_0^1 - 1} + \sqrt{\alpha - 1}} \right) + \tan^{-1} \left(\sqrt{\alpha \cosh^2 u_0^1 - 1} \right)$$

and finally obtain for the geodesic line (Figure 3.17)

$$u^2(u^1) = u_0^2 - c\sqrt{a} \left(\frac{\sqrt{\alpha - 1}}{2} \log \left(\frac{\sqrt{\alpha \cosh^2 u^1 - 1} - \sqrt{\alpha - 1}}{\sqrt{\alpha \cosh^2 u^1 - 1} + \sqrt{\alpha - 1}} \right) \right.$$

$$\left. + \tan^{-1} \left(\sqrt{\alpha \cosh^2 u^1 - 1} \right) - k(u_0^1) \right) \text{ for } u^1 < 0.$$

3.5 THE MINIMUM PROPERTY OF GEODESIC LINES

A curve of minimal arc length between two distinct points in the plane is a section of a straight line. We need the next lemma to be able to prove that geodesic lines on surfaces have the same minimal property.

Lemma 3.5.1 (Fundamental lemma of calculus of variation).
Let D be a domain in n–dimensional space $\mathbb{E}^n$, f be continuous on D and

$$\int_D fg = 0 \text{ for every } g \in C^\infty(D) \text{ with } g|_{\partial D} \equiv 0,$$

where ∂D denotes the boundary of the domain D. Then $f \equiv 0$ on D.

Proof. We assume $f(x_0) \neq 0$ for some $\vec{x}_0 \in D$. It follows from the continuity of f that there exist a point with the position vector $\vec{x}_0$ in the interior of D, and a neighbourhood $N_\rho(\vec{x}_0) = \{\vec{x} : \|vx - \vec{x}_0\| < \rho\}$ of $\vec{x}_0$ in D such that $f(\vec{x}) \neq 0$ for all $\vec{x} \in N_\rho(\vec{x}_0)$. We may assume $f(vx) > 0$ on $N_\rho(\vec{x}_0)$ and define the function $g : \mathbb{E}^n \to \mathbb{R}$ by

$$g(\vec{x}) = \begin{cases} \exp\left(-\dfrac{1}{\rho^2 - \|\vec{x} - \vec{x}_0\|^2}\right) & (x \in N_\rho(\vec{x}_0)) \\ 0 & (x \notin N_\rho(\vec{x}_0)). \end{cases}$$

Then we have $g \in C^\infty(D)$, $g|_{\partial D} \equiv 0$ for the restriction of g to the boundary of D, and $\int_D fg > 0$. $\qquad\square$

Theorem 3.5.2. *If γ is a curve of minimum length between two points of a surface, then γ is a section of a geodesic line.*

Proof. Let S be a surface, and $(u^i(t) = (u^1(t), u^2(t)))$ $(t \in I = [t_0, t_1])$ be a parametric representation of a curve γ on S of minimal arc length between the distinct points P_0 and P_1 of S with position vectors $\vec{x}(t_0)$ and $\vec{x}(t_1)$. Furthermore, let $\mathcal{C}_\tau = \{\gamma_\varepsilon : |\varepsilon| < \tau\}$ for $\tau > 0$ be a family of curves $\gamma_\varepsilon \in S$ with the parametric representations

$$u^i(t; \varepsilon), \text{ with } u^i(t; 0) = u^i(t), \ u^i(t_0; \varepsilon) = u^i(t_0) \text{ and } u^i(t_1; \varepsilon) = u^i(t_1) \text{ for } i = 1, 2$$

that join the points P_0 and P_1 and contains the curve of minimal arc length for $\varepsilon = 0$. Then the lengths $L(\varepsilon)$ of the curves $\gamma_\varepsilon \in \mathcal{C}$ must satisfy

$$L'(0) = 0.$$

We write $u^i = u^i(t; \varepsilon)$ $(i = 1, 2)$, for short, and

$$w(t; \varepsilon) = \sqrt{g_{ik}(u^j)u^{i'} u^{k'}},$$

and obtain

$$L(\varepsilon) = \int_{t_0}^{t_1} w(t; \varepsilon)\, dt. \tag{3.64}$$

Partial differentiation of the integrand in (3.64) with respect to ε yields

$$\frac{\partial w}{\partial \varepsilon} = \frac{1}{2w} \cdot \left(\frac{\partial g_{ik}}{\partial u^j} \cdot \frac{\partial u^j}{\partial \varepsilon} u^{i\prime} u^{k\prime} + g_{ik} \cdot \frac{\partial^2 u^i}{\partial t \partial \varepsilon} u^{k\prime} + g_{ik} \cdot \frac{\partial^2 u^k}{\partial t \partial \varepsilon} u^{i\prime} \right)$$

$$= \frac{1}{2w} \cdot \left(\frac{\partial g_{ik}}{\partial u^j} \cdot \frac{\partial u^j}{\partial \varepsilon} u^{i\prime} u^{k\prime} + 2 \cdot g_{ij} \cdot \frac{\partial^2 u^j}{\partial t \partial \varepsilon} u^{i\prime} \right),$$

since $g_{ik} = g_{ki}$. If

$$u^i(t; \varepsilon) = u^i(t) + \varepsilon v^i(t), \text{ where } v^i(t_0) = v^i(t_1) = 0 \text{ for } i = 1, 2,$$

and t is the arc length along the curve γ_0, then we have

$$\sqrt{w(t; 0)} = 1 \text{ and } \frac{\partial u^i}{\partial \varepsilon} = v^i \text{ for } i = 1, 2.$$

This implies

$$L'(0) = \frac{1}{2} \int_{t_0}^{t_1} \left(\frac{\partial g_{ik}}{\partial u^j} \cdot \dot{u}^i \dot{u}^k v^j + 2 g_{ij} \dot{u}^i \dot{v}^j \right) \, ds.$$

Partial integration of the second term of the integrand yields, since $v^i(t_0) = v^i(t_1) = 0$,

$$L'(0) = \frac{1}{2} \int_{t_0}^{t_1} \left(\frac{\partial g_{ik}}{\partial u^j} \cdot \dot{u}^i \dot{u}^k v^j - 2 \frac{d(g_{ij} \dot{u}^i)}{ds} v^j \right) \, ds = 0.$$

Since $L'(0) = 0$ for all functions v^j, it follow from Lemma 3.5.1 that

$$\frac{1}{2} \cdot \frac{\partial g_{ik}}{\partial u^j} \cdot \dot{u}^i \dot{u}^k - \frac{d(g_{ij} \dot{u}^i)}{ds} = 0 \text{ for } j = 1, 2. \tag{3.65}$$

Also

$$\frac{d(g_{ij} \dot{u}^i)}{ds} = \frac{\partial g_{ij}}{\partial u^k} \dot{u}^k \dot{u}^i + g_{rj} \ddot{u}^r$$

and so (3.65) implies

$$\frac{1}{2} \cdot \frac{\partial g_{ik}}{\partial u^j} \dot{u}^k \dot{u}^i - \frac{\partial g_{ij}}{\partial u^k} \dot{u}^k \dot{u}^i - g_{rj} \ddot{u}^r = 0 \text{ for } j = 1, 2.$$

If we write

$$\frac{\partial g_{ij}}{\partial u^k} \dot{u}^k \dot{u}^i = \frac{1}{2} \left(\frac{\partial g_{ij}}{\partial u^k} \dot{u}^k \dot{u}^i + \frac{\partial g_{kj}}{\partial u^i} \dot{u}^k \dot{u}^i \right),$$

then it follows from Lemma 3.1.4 (c) that

$$[ikj] \dot{u}^i \dot{u}^k + g_{rj} \ddot{u}^r = 0 \text{ for } j = 1, 2,$$

and so, by the definition of the second Christoffel symbols in Definition 3.1.2 (b),

$$g^{lj}\left(g_{rj}\ddot{u}^r + [ikj]\dot{u}^i\dot{u}^k\right) = \delta_r^l\ddot{u}^r + \begin{Bmatrix} l \\ ik \end{Bmatrix}\dot{u}^i\dot{u}^k$$

$$= \ddot{u}^l + \begin{Bmatrix} l \\ ik \end{Bmatrix}\dot{u}^i\dot{u}^k = 0 \text{ for } l = 1, 2.$$

Thus the curve γ satisfies the differential equations (3.22) in Theorem 3.2.2 for geodesic lines. $\qquad\square$

Remark 3.5.3. *Obviously the converse of Theorem 3.5.2 is not true in general. We shall, however, show later that if a section γ of a geodesic line is a member of a so-called geodesic field, then γ is a curve of minimum length between the initial and the final points of γ. To achieve this, we need to consider transformations of parameters.*

Let S be a given surface with a parametric representation $\vec{x}(u^i)$ $((u^1, u^2) \in D)$. We draw the u^1- and u^2-lines in the u^i-parameter plane with a Cartesian coordinate system, and introduce new parameters $\bar{u}^i$ $(i = 1, 2)$ as follows:

Given two families of curves in the u^i-plane, we choose new parameters $\bar{u}^i$ $(i = 1, 2)$ such that the curves of one family become the $\bar{u}^1-$lines and the other ones of the other family become the $\bar{u}^2-$lines. Based on the families of curves

$$C_1(\bar{u}_2) : \left\{ \begin{array}{l} u^1 = u_1^1(t; \bar{u}^2) \\ u^2 = u_1^2(t; \bar{u}^2) \end{array} \right\} \quad \text{and} \quad C_2(\bar{u}_1) : \left\{ \begin{array}{l} u^1 = u_2^1(t; \bar{u}^1) \\ u^2 = u_2^2(t; \bar{u}^1) \end{array} \right\}$$

we search for a one-to-one correspondence $(u^1, u^2) \leftrightarrow (\bar{u}^1, \bar{u}^2)$. It is our aim to determine $u^1 = u^1(\bar{u}^1, \bar{u}^2)$ and $u^2 = u^2(\bar{u}^1, \bar{u}^2)$.

Example **3.5.4 (Straight lines in the u^i-plane).**
We consider the families of curves

$$C_1(\bar{u}_2) : \left\{ \begin{array}{l} u^1 = t \\ u^2 = c_1 t + \bar{u}^2 \end{array} \right\} \quad \text{and} \quad C_2(\bar{u}_1) : \left\{ \begin{array}{l} u^1 = t \\ u^2 = c_2 t + \bar{u}^1 \end{array} \right\},$$

where c_1 and c_2 are constants. We obtain

$$\left\{ \begin{array}{l} u^2 = c_1 u^1 + \bar{u}^2 \\ u^2 = c_2 u^1 + \bar{u}^1 \end{array} \right\}, \quad \text{hence} \quad \left\{ \begin{array}{l} \bar{u}^1 = -c_2 u^1 + u^2 \\ \bar{u}^2 = -c_1 u^1 + u^2 \end{array} \right\}.$$

If $c_1 \neq c_2$, then this yields

$$u^1 = u^1(\bar{u}^i) = \frac{\bar{u}^1 - \bar{u}^2}{c_1 - c_2} \text{ and } u^2 = u^2(\bar{u}^i) = \frac{c_1\bar{u}^1 - c_2\bar{u}^2}{c_1 - c_2}$$

with the Jacobian

$$\left(\frac{\partial u^i}{\partial \bar{u}^k} \right) = \begin{vmatrix} \dfrac{\partial u^1}{\partial \bar{u}^1} & \dfrac{\partial u^2}{\partial \bar{u}^1} \\ \dfrac{\partial u^1}{\partial \bar{u}^2} & \dfrac{\partial u^2}{\partial \bar{u}^2} \end{vmatrix} = \frac{1}{(c_1 - c_2)^2} \cdot \begin{vmatrix} 1 & c_1 \\ -1 & -c_1 \end{vmatrix} = \frac{1}{c_1 - c_2}.$$

Now we study the technique of computing parameter transformations.

(i) For given (u^1, u^2), we determine $(\bar{u}^1, \bar{u}^2)$.

   (i.1) We search $\bar{u}^2$ and t_1 such that

$$\left\{ \begin{array}{l} u^1 = u_1^1(t_1; \bar{u}^2) \\ u^2 = u_1^2(t_1; \bar{u}^2) \end{array} \right\}.$$

   We eliminate t_1 and compute

$$\bar{u}^2 = \bar{u}^2(u^1, u^2).$$

   (i.2) We search $\bar{u}^1$ and t_2 such that

$$\left\{ \begin{array}{l} u^1 = u_2^1(t_2; \bar{u}^1) \\ u^2 = u_2^2(t_2; \bar{u}^1) \end{array} \right\}.$$

   We eliminate t_2 and compute

$$\bar{u}^1 = \bar{u}^1(u^1, u^2).$$

(ii) For given $(\bar{u}^1, \bar{u}^2)$, we determine (u^1, u^2).
   We eliminate t_1 and t_2 from

$$\left\{ \begin{array}{l} u_1^1(t_1; \bar{u}^2) = u_2^1(t_2; \bar{u}^1) \\ u_1^2(t_1; \bar{u}^2) = u_2^2(t_2; \bar{u}^1) \end{array} \right\}$$

and obtain

$$t_1 = t_1(\bar{u}^1, \bar{u}^2) \text{ and } t_1 = t_1(\bar{u}^1, \bar{u}^2).$$

There are two ways to compute (u^1, u^2):

(ii.1)

$$\left\{ \begin{array}{l} u^1 = u_1^1(t_1(\bar{u}^1, \bar{u}^2); \bar{u}^2) = u^1(\bar{u}^1, \bar{u}^2) \\ u^2 = u_1^2(t_1(\bar{u}^1, \bar{u}^2); \bar{u}^2) = u^2(\bar{u}^1, \bar{u}^2) \end{array} \right\}$$

   or

(ii.2)

$$\left\{ \begin{array}{l} u^1 = u_2^1(t_2(\bar{u}^1, \bar{u}^2); \bar{u}^1) = u^1(\bar{u}^1, \bar{u}^2) \\ u^2 = u_2^2(t_2(\bar{u}^1, \bar{u}^2); \bar{u}^1) = u^2(\bar{u}^1, \bar{u}^2) \end{array} \right\}.$$

Example **3.5.5. (A surface of revolution with loxodromes as parameter lines)** *We consider a surface RS of revolution with a parametric representation*

$$\vec{x}(u^i) = \{r(u^1) \cos u^2, r(u^1) \sin u^2, h(u^1)\}, \quad \left((u^1, u^2) \in D = I \times (0, 2\pi)\right),$$

and introduce new parameters $\bar{u}^i$ for $i = 1, 2$ such that the $\bar{u}^i$–lines are the loxodromes at a constant angle β_i to the u^2–lines of RS. We assume that $\beta_2 \neq \beta_1 + k\pi$ $(k \in \mathbb{Z})$. First we put

$$c_k = |\cot \beta_k| \text{ for } k = 1, 2 \text{ and } \varphi(t) = \int \frac{\sqrt{g_{11}(t)}}{\sqrt{g_{22}(t)}} \, dt,$$

and consider the families of curves

$$\mathcal{C}_1(\bar{u}^2) : \left\{ \begin{array}{l} u^1 = u_1^1(t) = t \\ u^2 = u_1^2(t; \bar{u}^2) = c_1 \cdot \varphi(t) + \bar{u}^2 \end{array} \right\}$$

and

$$\mathcal{C}_2(\bar{u}^1) : \left\{ \begin{array}{l} u^1 = u_2^1(t) = t \\ u^2 = u_2^2(t; \bar{u}^1) = c_2 \cdot \varphi(t) + \bar{u}^1 \end{array} \right\}.$$

We obtain

$$\bar{u}^1 - \bar{u}^2 = (c_1 - c_2) \cdot \varphi(u^1).$$

Since $\varphi'(u^1) \neq 0$ for all u^1, the inverse function φ^{-1} exists, hence

$$u^1 = \varphi^{-1} \left(\frac{\bar{u}^1 - \bar{u}^2}{c_1 - c_2} \right).$$

Furthermore, we have $(c_1 - c_2)u^2 = c_1\bar{u}^1 - c_2\bar{u}^2$, and so, because of $c_2 \neq c_1$,

$$u^2 = \frac{c_1\bar{u}^1 - c_2\bar{u}^2}{c_1 - c_2}.$$

Finally, we obtain for the Jacobian

$$\frac{\partial(u^i)}{\partial(\bar{u}^k)} = \begin{vmatrix} \dfrac{\partial u^1}{\bar{u}^1} & \dfrac{\partial u^2}{\bar{u}^1} \\ \dfrac{\partial u^1}{\bar{u}^2} & \dfrac{\partial u^2}{\bar{u}^2} \end{vmatrix} = \frac{1}{\varphi'(u^1)} \cdot \begin{vmatrix} \dfrac{1}{c_1 - c_2} & \dfrac{c_1}{c_1 - c_2} \\ -\dfrac{1}{c_1 - c_2} & -\dfrac{c_2}{c_1 - c_2} \end{vmatrix}$$

$$= \frac{1}{\varphi'(u^1)} \cdot \frac{c_1 - c_2}{(c_1 - c_2)^2} = \frac{1}{c_1 - c_2} \cdot \frac{1}{\varphi'(u^1)}.$$

In particular, we obtain for the sphere with

$$r(u^1) = r \cos u^1 \text{ and } h(u^1) = r \sin u^1 \text{ for } u^1 \in \left(-\frac{\pi}{2}, \frac{\pi}{2} \right)$$

where $r > 0$ is a constant, that

$$\varphi(u^1) = \log \left(\tan \left(\frac{u^1}{2} + \frac{\pi}{4} \right) \right),$$

whence

$$u^1(\bar{u}^1, \bar{u}^2) = 2 \cdot \tan^{-1} \left(\exp \frac{\bar{u}^1 - \bar{u}^2}{c_1 - c_2} \right) - \frac{\pi}{2} \text{ and } u^2(\bar{u}^1, \bar{u}^2) = \frac{c_1\bar{u}^1 - c_2\bar{u}^2}{c_1 - c_2}.$$

It follows from

$$\sin 2x = 2 \cdot \sin x \cos x = \frac{2\tan x}{1 + \tan^2 x} \ \ and \ \ \cos 2x = \cos^2 x - \sin^2 x = \frac{1 - \tan^2 x}{1 + \tan^2 x}$$

that

$$\sin u^1 = -\cos\left(2 \cdot \tan^{-1}\left(\exp\frac{\bar{u}^1 - \bar{u}^2}{c_1 - c_2}\right)\right) = -\frac{1 - \exp\left(2 \cdot \frac{\bar{u}^1 - \bar{u}^2}{c_1 - c_2}\right)}{1 + \exp\left(2 \cdot \frac{\bar{u}^1 - \bar{u}^2}{c_1 - c_2}\right)} = \tanh\frac{\bar{u}^1 - \bar{u}^2}{c_1 - c_2}$$

and

$$\cos u^1 = \sin\left(2 \cdot \tan^{-1}\left(\exp\frac{\bar{u}^1 - \bar{u}^2}{c_1 - c_2}\right)\right) = \frac{2 \cdot \exp\left(\frac{\bar{u}^1 - \bar{u}^2}{c_1 - c_2}\right)}{1 + \exp\left(2 \cdot \frac{\bar{u}^1 - \bar{u}^2}{c_1 - c_2}\right)} = \frac{1}{\cosh\frac{\bar{u}^1 - \bar{u}^2}{c_1 - c_2}}$$

and we obtain the following parametric representation of the sphere with respect to the new parameters $\bar{u}^1$ and $\bar{u}^2$ (right in Figure 2.6)

$$\vec{x}(\bar{u}^i) = \frac{r}{\cosh\frac{\bar{u}^1 - \bar{u}^2}{c_1 - c_2}} \cdot \left\{\cos\frac{c_1\bar{u}^1 - c_2\bar{u}^2}{c_1 - c_2}, \sin\frac{c_1\bar{u}^1 - c_2\bar{u}^2}{c_1 - c_2}, \sinh\frac{\bar{u}^1 - \bar{u}^2}{c_1 - c_2}\right\}.$$

This is the parametric representation of a sphere in Visualization 2.1.7 (c).

3.6 ORTHOGONAL AND GEODESIC PARAMETERS

It is often convenient to have *orthogonal* and *geodesic* parameters. Therefore, we study in this section how to construct them.

First we deal with the construction of orthogonal parameters.

If $g_{12} \neq 0$ for the parameters u^i of a surface, then we introduce new parameters u^{*1} and u^{*2} by

$$u^1 = u^{*1} \ \text{and} \ u^2 = u^2(u^{*1}, u^{*2})$$

such that

$$g_{12}^* = 0$$

for the first fundamental coefficients of the surface with respect to the new parameters u^{*k}. Since

$$\frac{\partial u^1}{\partial u^{*1}} = 1 \ \text{and} \ \frac{\partial u^1}{\partial u^{*2}} = 0,$$

the transformation formula

$$g_{12}^* = g_{ik}\frac{\partial u^i}{\partial u^{*1}}\frac{\partial u^k}{\partial u^{*2}}$$

reduces to

$$g_{12}^* = g_{12}\frac{\partial u^2}{\partial u^{*1}} + g_{22}\frac{\partial u^2}{\partial u^{*1}}\frac{\partial u^2}{\partial u^{*2}} = \left(g_{12} + \frac{\partial u^2}{\partial u^{*1}}g_{22}\right)\frac{\partial u^2}{\partial u^{*2}}. \tag{3.66}$$

For admissible parameter transformations, we must have

$$\frac{\partial(u^i)}{\partial(u^{*k})} = \begin{vmatrix} \dfrac{\partial u^1}{\partial u^{*1}} & \dfrac{\partial u^2}{\partial u^{*1}} \\[2mm] \dfrac{\partial u^1}{\partial u^{*2}} & \dfrac{\partial u^2}{\partial u^{*2}} \end{vmatrix} = \begin{vmatrix} 1 & \dfrac{\partial u^2}{\partial u^{*1}} \\[2mm] 0 & \dfrac{\partial u^2}{\partial u^{*2}} \end{vmatrix} = \frac{\partial u^2}{\partial u^{*2}} \neq 0.$$

Thus the differential equation (3.66) yields

$$g_{12}(u^{*1}, u^2(u^{*1}, u^{*2})) + \frac{\partial u^2}{\partial u^{*1}} g_{22}(u^{*1}, u^2(u^{*1}, u^{*2})) = 0. \tag{3.67}$$

If we put $x = u^{*1}$, $y = u^{*2}$ and $z = u^2$, then (3.67) becomes

$$g_{12}(x, z(x, y)) + g_{22}(x, z(x, y)) \frac{\partial z}{\partial x} = 0.$$

Example **3.6.1.** *Let S be a screw surface with a parametric representation*

$$\vec{x}(u^i) = \{u^1 \cos u^2, u^1 \sin u^2, hu^2 + \xi(u^1)\} \; \left((u^1, u^2) \in I \times (0, 2\pi)\right),$$

where $h \in \mathbb{R}$ is a constant and $\xi \in C^3(I)$. We may assume $h \neq 0$ for otherwise S is a surface of revolution with orthogonal parameters. Since

$$\vec{x}_1 = \{\cos u^2, \sin u^2, \xi'(u^1)\}, \; \vec{x}_2 = \{-u^1 \sin u^2, u^1 \cos u^2, h\},$$

$$g_{11} = 1 + \left(\xi'(u^1)\right)^2, \; g_{12} = h\xi'(u^1) \neq 0 \text{ for } h \neq 0 \text{ and } \xi'(u^1) \neq 0$$

and

$$g_{22} = (u^1)^2 + h^2,$$

we have to solve the differential equation

$$h\xi'(u^{*1}) + \left((u^{*1})^2 + h^2\right) \frac{\partial u^2}{\partial u^{*1}} = 0.$$

From

$$\frac{\partial u^2}{\partial u^{*1}} = -h \frac{\xi'(u^{*1})}{(u^{*1})^2 + h^2},$$

we obtain

$$u^2 = -h \int \frac{\xi'(u^{*1})}{(u^{*1})^2 + h^2} \, du^{*1} + u^{*2}$$

say, and with

$$H(u^{*1}) = h \int \frac{\xi'(u^{*1})}{(u^{*1})^2 + h^2} \, du^{*1}$$

the following transformation formulae

$$u^1 = u^{*1} \ and \ u^2 = -H(u^{*1}) + u^{*2}.$$

Thus a parametric representation of the screw surface S with respect to the new orthogonal parameters u^{*1} and u^{*2} is given by

$$\vec{x}^{*}(u^{*i}) = \Big\{ u^{*1} \cos(u^{*2} - H(u^{*1})), u^{*1} \sin(u^{*2} - H(u^{*1})),$$

$$h \cdot (u^{*2} - H(u^{*1})) + \xi(u^{*1}) \Big\}.$$

Now we introduce the concept of *geodesic parameters*.

Definition 3.6.2. Orthogonal parameters u^{*i} $(i = 1, 2)$ for a surface such that the u^{*1}–lines are geodesic lines are called *geodesic parameters*.

The following result holds.

Theorem 3.6.3. *The u^k–lines of a surface are geodesic lines if and only if*

$$\left\{ {i \atop kk} \right\} = 0 \quad (i \neq k). \tag{3.68}$$

Proof. (i) First we assume that a u^k–line is a geodesic line. Then $u^i = const$ for $(i \neq k)$, hence $\dot{u}^i = \ddot{u}^i = 0$ and so

$$0 = \ddot{u}^i + \left\{ {i \atop jk} \right\} \dot{u}^j \dot{u}^k = \left\{ {i \atop kk} \right\} (\dot{u}^k)^2$$

implies $\left\{ {i \atop kk} \right\} = 0$ for $i \neq k$.

(ii) Conversely we assume $\left\{ {i \atop kk} \right\} = 0$ for $k \neq i$. Then $u^i = const$ clearly satisfies the differential equation for geodesic lines.

□

Theorem 3.6.4. *(a) If $g_{12} = 0$, then a u^k–line is a geodesic line if and only if g_{kk} depends on u^k only.*
(b) If a surface is given with respect to geodesic parameters, then we may assume

$$g_{11} = 1 \ and \ g_{12} = 0. \tag{3.69}$$

(c) If a surface is given with respect to geodesic parameters, then the u^2–lines are geodesically parallel; this means that the distances between two u^2–lines measured along the u^1–lines are constant.

Proof.

(a) If $g_{12} = 0$, then we have by (3.10) in Corollary 3.1.5

$$\left\{ \begin{matrix} i \\ kk \end{matrix} \right\} = -\frac{1}{2g_{ii}} \cdot \frac{\partial g_{kk}}{\partial u^i} \text{ for } (k \neq i)$$

and the statement follows by Theorem 3.6.3.

(b) We assume that $\bar{u}^i$ $(i = 1, 2)$ are geodesic parameters. Then their orthogonality implies

$$\bar{g}_{12} = 0. \tag{3.70}$$

Since the $\bar{u}^1$–lines are geodesic lines, we obtain by Part (a)

$$\bar{g}_{11}(\bar{u}^i) = \bar{g}_{11}(\bar{u}^1). \tag{3.71}$$

We introduce new parameters u^i by putting

$$u^1 = \int \sqrt{\bar{g}_{11}(\bar{u}^1)} \, d\bar{u}^1 \text{ and } u^2 = \bar{u}^2.$$

Since

$$\frac{\partial u^1}{\partial \bar{u}^1} = \sqrt{\bar{g}_{11}(\bar{u}^1)}, \ \frac{\partial u^1}{\partial \bar{u}^2} = \frac{\partial u^2}{\partial \bar{u}^1} = 0 \text{ and } \frac{\partial u^2}{\partial \bar{u}^2} = 1,$$

we get

$$\frac{\partial(u^i)}{\partial(\bar{u}^k)} = \begin{vmatrix} \sqrt{\bar{g}_{11}(\bar{u}^1)} & 0 \\ 0 & 1 \end{vmatrix} = \sqrt{\bar{g}_{11}(\bar{u}^1)} \neq 0,$$

and consequently the parameter transformation is admissible.

Furthermore, it follows from (3.70) and the transformation formulae for the first fundamental coefficients in (2.18) of Remark 2.3.3 (a), since $\partial \bar{u}^2 / \partial u^1 = 0$ and $\partial \bar{u}^1 / \partial u^2 = 0$, that

$$g_{11} = \bar{g}_{ik} \frac{\partial \bar{u}^i}{\partial u^1} \cdot \frac{\partial \bar{u}^k}{\partial u^1} = \bar{g}_{11} \left(\frac{\partial \bar{u}^1}{\partial u^1} \right)^2 + \bar{g}_{22} \left(\frac{\partial \bar{u}^2}{\partial u^1} \right)^2 = \bar{g}_{11} \cdot \frac{1}{\bar{g}_{11}} = 1,$$

$$g_{12} = \bar{g}_{ik} \frac{\partial \bar{u}^i}{\partial u^1} \cdot \frac{\partial \bar{u}^k}{\partial u^2} = \bar{g}_{11} \frac{\partial \bar{u}^1}{\partial u^1} \cdot \frac{\partial \bar{u}^1}{\partial u^2} + \bar{g}_{22} \frac{\partial \bar{u}^2}{\partial u^1} \cdot \frac{\partial \bar{u}^2}{\partial u^2} = 0$$

and

$$g_{22} = \bar{g}_{ik} \frac{\partial \bar{u}^i}{\partial u^2} \cdot \frac{\partial \bar{u}^k}{\partial u^2} = \bar{g}_{11} \left(\frac{\partial \bar{u}^1}{\partial u^2} \right)^2 + \bar{g}_{22} \left(\frac{\partial \bar{u}^2}{\partial u^2} \right)^2 = \bar{g}_{22}.$$

(c) We may assume by Part (b) that $g_{11} = 1$. We obtain for the arc length of the sections of the u^1–lines between the u^2–lines with $u^1 = c_1$ and $u^1 = c_2$, since $\dot{u}^1 = 1$ and $\dot{u}^2 = 0$,

$$\int_{c_1}^{c_2} \sqrt{g_{ik} \dot{u}^i \dot{u}^k} \, ds = \int_{c_1}^{c_2} \sqrt{g_{11}} \, ds = \int_{c_1}^{c_2} ds = c_2 - c_1.$$

$\square$

Definition 3.6.5. A family $\mathcal{F}$ of geodesic lines on a surface S is called a *geodesic field for a region* $S' \subset S$ if there is one and only one line of $\mathcal{F}$ through every point of S'.

Now we are able to prove the converse result of Theorem 3.5.2.

Theorem 3.6.6. *If a part of a geodesic line is a member of a geodesic field $\mathcal{F}$ for a region G, then it is a curve of minimum length joining its endpoints which is possible inside G.*

Proof. We choose geodesic parameters for the surface. Then the first fundamental form ((2.15) in Definition 2.3.1 (b)) is given by

$$(ds)^2 = (du^1)^2 + g_{22}(u^j)(du^2)^2. \tag{3.72}$$

Let γ, given by $u^2 = u^2(u^1)$ $(u^1 \in [a, b])$, be a piecewise smooth curve that joins the endpoints of the given geodesic arc. Then we obtain for the arc length of γ

$$\left| \int_a^b \sqrt{1 + g_{22} \left(u^{2'} \right)^2} \, du^1 \right| \geq \left| \int_a^b du^1 \right| = |b - a|.$$

Now let γ be a given curve which is not represented in the form $u^2 = u^2(u^1)$, which does not necessarily intersect a line $u^1 = const$ only once. Then we divide γ into subarcs which can uniquely be represented by functions $u^2(u^1)$. Now the statement follows as in the first part of the proof. □

Visualization 3.6.7. *(a) The parameters of a surface of revolution with a parametric representation*

$$\vec{x}(u^i) = \{r(u^1)\cos u^2, r(u^1)\sin u^2, h(u^1)\}$$

are geodesic parameters (Figure 3.18).
(b) On a surface with first fundamental coefficients

$$g_{11} = g_{11}(u^1), \ g_{12} = 0 \ and \ g_{22} = g_{22}(u^1),$$

we introduce geodesic parameters u^{*i} $(i = 1, 2)$ as follows:
Let the u^{*1}–lines be the geodesic lines that start on the u^2–line through u_0^1 at an angle $\Theta_0 \in (0, \pi/2)$ with the u^2–line, and let the u^{*2}–lines be the orthogonal trajectories *of the geodesic lines (Figure 3.19).*
*   This means, the u^{*2}–lines intersect the geodesic lines at right angles. We put $c = \sqrt{g_{22}(u_0^1)} \cos \Theta_0$ and obtain by Theorem 3.3.2*

$$\begin{cases} u_1^1(t) = t \\ u_1^2(t) = c \displaystyle\int_{u_0^1}^{t} \frac{\sqrt{g_{11}(\tau)}}{\sqrt{g_{22}(\tau)}\sqrt{g_{22}(\tau) - c^2}} \, d\tau + u_0^2 \end{cases}$$

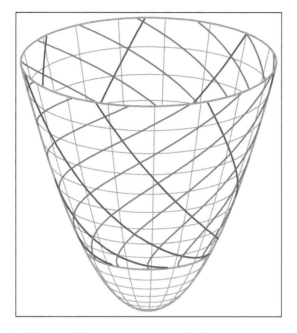

Figure 3.18 Parameter lines with respect to geodesic parameters on a paraboloid of revolution

for the geodesic lines in a suitable interval I. Their orthogonal trajectories given by $u_2^1(t) = t$ and $u_2^2(t)$ must satisfy the condition

$$g_{11} + g_{22}\frac{du_1^2}{dt}\frac{du_2^2}{dt} = 0.$$

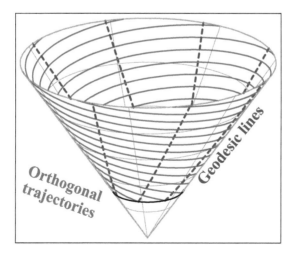

Figure 3.19 A family of geodesic lines (dashed) on a cone and their orthogonal trajectories

This yields

$$\begin{cases} u_2^1(t) = t \\ u_2^2 = -\dfrac{1}{c} \int\limits_{u_0^1}^{t} \dfrac{\sqrt{g_{11}(\tau)}}{\sqrt{g_{22}(\tau)}} \sqrt{g_{22}(\tau) - c^2} \, d\tau + u_0^2. \end{cases}$$

We obtain for the new parameters u^{*i}

$$u^{*1} - u^{*2} = \varphi(u^1) = \dfrac{1}{c} \int\limits_{u_0^1}^{u^1} \dfrac{\sqrt{g_{11}(u)g_{22}(u)}}{\sqrt{g_{22}(u) - c^2}} \, du.$$

Since $\varphi'(u^1) \neq 0$, *this can be solved for* u^1 *and thus yields a transformation formula*

$$u^1 = u^1(u^{*1}, u^{*2}).$$

Finally, from

$$u^2(u^1) = c \int\limits_{u_0^1}^{u^1} \dfrac{\sqrt{g_{11}(u)}}{\sqrt{g_{22}(u)}\sqrt{g_{22}(u) - c^2}} \, du + u^{*2},$$

we obtain a transformation formula

$$u^2 = u^2(u^{*1}, u^{*2}).$$

In particular, if we consider the circular cone with a parametric representation

$$\vec{x}(u^i) = \left\{ \dfrac{u^1}{\sqrt{a}} \cos u^2, \dfrac{u^1}{\sqrt{a}} \sin u^2, u^1 \right\} \left((u^1, u^2) \in (-\infty, 0) \times (0, 2\pi) \right),$$

where $a > 0$ *is a constant, and put*

$$\psi(u^{*i}) = \sqrt{\left(\dfrac{u^{*1} - u^{*2}}{\sqrt{a+1}} - \tan \Theta_0 \right)^2 + 1},$$

then the transformation formulae are

$$\begin{cases} u^1(u^{*k}) = -c\sqrt{a}\psi(u^{*k}) \\ \quad and \\ u^2(u^{*k}) = \sqrt{a+1} \left(\cos^{-1}\left(-\dfrac{1}{\psi(u^{*k})} \right) + \Theta_0 - \pi \right) + u^{*2} \end{cases}$$

for

$$0 \leq u^{*1} - u^{*2} < \sqrt{a+1} \tan \Theta_0$$

and

$$0 \leq u^{*2} < \sqrt{a+1} \cdot \Theta_0 + 2\pi.$$

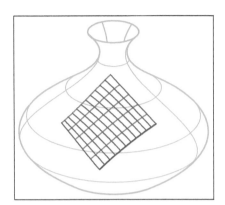
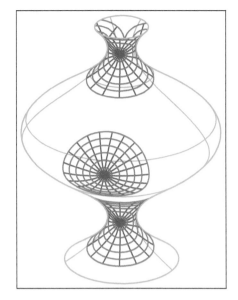

Figure 3.20 Geodesic parallel and polar coordinates on surfaces of revolution

Now we introduce *geodesic parallel coordinates* and *geodesic polar coordinates*, which are generalizations of Cartesian and polar coordinates of the plane (Figure 3.20).

First we construct geodesic parallel coordinates as follows: Let S be a surface with a parametric representation $\vec{x}(u^i)$ $((u^1, u^2) \in D)$, and $\vec{x}(s) = \vec{x}(u^i(s))$ $(s \in I)$ be a curve γ on S, where s is the arc length along γ. At each s along γ, let $\vec{y}_s(t)$ be a parametric representation of the geodesic line through s orthogonal to γ. We assume that $\vec{y}_s(t)$ is given by $v_s^i(t)$ with $v_s^i(0) = u^i(s)$ $(i = 1, 2)$. Given a fixed value of t, a so-called *parallel curve* γ_t^* of γ is generated by the points with position vectors $\vec{y}_s(t)$ $(s \in I)$. The points of γ_t^* have a constant distance from γ measured along the geodesic line $\vec{y}_s(t)$.

Choosing the new parameters $\bar{u}^j$ $(j = 1, 2)$ such that the $\bar{u}^1$ lines are the parallel curves and the $\bar{u}^2$–lines are the geodesic lines, we obtain a system of geodesic parallel coordinates. The transformation formulae are

$$u^i = v_{\bar{u}^1}^i(\bar{u}^2) \ (i = 1, 2).$$

Visualization 3.6.8 (Parallel curves of loxodromes on a circular cone).
We consider the circular cone with a parametric representation

$$\vec{x}(u^i) = \left\{ \frac{1}{\sqrt{a}} \cdot u^1 \cos u^2, \frac{1}{\sqrt{a}} \cdot u^1 \sin u^2, u^1 \right\} \ \left((u^1, u^2) \in (0, \infty) \times (-\pi, \pi) \right)$$

for given $a > 0$ and the loxodrome γ through the point with $(u_0^1, u_0^2) = (1, 0)$ with a constant angle $\beta \in (-\pi/2, \pi/2) \setminus \{0\}$ to the parallels of the cone; also let $I_1 = [b_1, b_2] \subset (0, \infty)$ denote the parameter interval of γ.
If we introduce new parameters $\bar{u}^i$ $(i = 1, 2)$ such that the $\bar{u}^1$–lines are the parallel curves of γ and the u^2–lines are the geodesic lines, then we obtain for the

transformation formulae, putting

$$\alpha = \sqrt{\frac{a}{a+1}}, \ c_1 = \alpha \cdot \sin^2 \beta \ \text{and} \ c_2 = \cot \beta \cdot \sqrt{a+1},$$

$$\left\{ \begin{array}{ll} u^1 & = u^1(\bar{u}^i) \\ & = \sqrt{(c_1\bar{u}^1 + 1)^2 - 2\bar{u}^2\alpha \sin \beta \cdot (c_1\bar{u}^1 + 1) + \alpha^2(\bar{u}^2)^2} \\ u^2 & = u^2(\bar{u}^i) \\ & = \sqrt{a+1} \cdot \tan^{-1}\left(\dfrac{\alpha\bar{u}^2}{(c_1\bar{u}^1 + 1)\cos \beta} - \tan \beta\right) \\ & \quad + c_2 \cdot \log\left(c_1\bar{u}^1 + 1\right) + \sqrt{a+1}\beta \end{array} \right\} \tag{3.73}$$

for $(\bar{u}^1, \bar{u}^2) \in I_s \times I_t$, *where*

$$I_s = \left[\frac{b_1 - 1}{c_1}, \frac{b_2 - 1}{c_1}\right] \ \text{and} \ I_t \subset \left(\frac{b_1}{\alpha} \cdot \sin \beta, \infty\right)$$

(Figure 3.21).

Proof. Since $r'(u^1) = 1/\sqrt{a}$ and $h'(u^1) = 1$, we obtain for the loxodrome γ by (2.32) in Visualization 2.3.11 (interchanging the roles of u^1 and u^2)

$$u^2 = \cot \beta \int_1^{u^1} \frac{\sqrt{(r'(u))^2 + (h'(u))^2}}{r(u)} \, du = \cot \beta \sqrt{a+1} \int_1^{u^1} \frac{du}{u}$$
$$= \cot \beta \sqrt{a+1} \cdot \log u^1 \ \text{for} \ u^1 \in I_1.$$

The arc length along γ is given by

$$s(u^1) = \int_1^{u^1} \sqrt{g_{11}(u) + g_{22}(u)\cot^2 \beta \cdot \frac{g_{11}(u)}{g_{22}(u)}} \, du = (1 + \cot^2 \beta) \int_1^{u^1} \sqrt{g_{11}(u)} \, du$$
$$= (1 + \cot^2 \beta)\sqrt{\frac{a+1}{a}}(u^1 - 1),$$

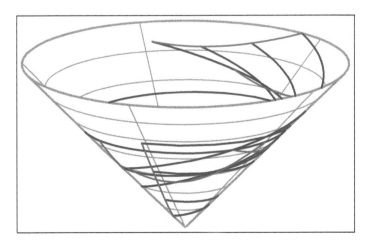

Figure 3.21 Parallel curves of loxodromes on a circular cone

and so

$$u^1(s) = \sqrt{\frac{a}{a+1} \cdot \frac{1}{1+\cot^2 \beta}} \cdot s + 1 = c_1 s + 1 \text{ and } u^2(s) = c_2 \log (c_1 s + 1)$$

for

$$s \in I_s = \left[\frac{b_1 - 1}{c_1}, \frac{b_2 - 1}{c_1} \right].$$

The angle $\Theta(s)$ between the geodesic line $(v_s^i(t)$ and the parallel corresponding to $u^i(s)$ satisfies

$$\Theta(s) = \beta - \frac{\pi}{2} \in (-\pi, 0) \setminus \{-\pi/2\}.$$

We obtain by Visualization 3.2.10 (b)

$$\left\{ \begin{array}{ll} v_s^1(t) & = \sqrt{(\alpha(t + t_0(s)))^2 + \delta^2} \\ v_s^2(t) & = \dfrac{k}{\alpha\delta} \tan^{-1} \left(\dfrac{\alpha}{\delta} \cdot (t + t_0(s)) \right) + d \end{array} \right\},$$

where

$$\alpha = \sqrt{\frac{a}{a+1}}, \ t_0(s) = -\frac{u^1(s) \cdot \sin \beta}{\alpha}, \ \delta = u^1(s) \cdot \cos \beta,$$
$$k = \sqrt{a} \cdot u^1(s) \cdot \cos \beta \text{ and } d = u^2(s) + \sqrt{a+1} \cdot \beta,$$

hence

$$\left\{ \begin{array}{ll} v_s^1(t) & = \sqrt{\alpha^2 t^2 - 2t\alpha \sin \beta \cdot u^1(s) + (u^1(s))^2} \\ v_s^2(t) & = \sqrt{a+1} \cdot \tan^{-1} \left(\dfrac{\alpha t}{u^1(s) \cdot \cos \beta} - \tan \beta \right) + u^2(s) + \sqrt{a+1} \cdot \beta \end{array} \right\}.$$

It follows from

$$\alpha^2 t^2 - 2t\alpha \sin \beta \cdot u^1(s) + (u^1(s))^2 = \left(\alpha t - \sin \beta u^1(s) \right)^2 + (u^1(s))^2(1 - \sin^2 \beta)$$
$$\geq (u^1(s))^2 \cos^2 \beta$$

that $v_s^1(t) \geq u^1 \cdot \cos \beta$. We choose

$$t > \frac{u^1(s) \cdot \sin \beta}{\alpha}$$

since $\sin \beta < 0$,

$$t > -\frac{b_1}{\alpha} \cdot \sin \beta,$$

an interval

$$I_t \subset \left(-\frac{b_1}{\alpha} \cdot \sin \beta, \infty \right),$$

and obtain the parallel curve of the loxodrome for each fixed $t \in I_t$ from

$$
\begin{cases}
u_t^1(s) &= \sqrt{(c_1 s + 1)^2 - 2t\alpha \sin\beta \cdot (c_1 s + 1) + \alpha^2 t^2} \\
u_t^2(s) &= \sqrt{a+1}\, \tan^{-1}\left(\dfrac{\alpha t}{(c_1 s + 1)\cos\beta} - \tan\beta \right) \\
&\quad + c_2 \cdot \log\left(c_1 s + 1\right) + \sqrt{a+1} \cdot \beta
\end{cases}
$$

for $s \in I_s$.

If we introduce new parameters $\bar{u}^i$ such that the $\bar{u}^1$–lines are the parallel curves of the loxodrome, and the $\bar{u}^2$–lines are the geodesic lines, then we obtain the transformation formulae in (3.73). ☐

Visualization 3.6.9. (Geodesic parallel coordinates on a circular cylinder)
We consider the circular cylinder with a parametric representation

$$
\vec{x}(u^i) = \{r\cos u^2, r\sin u^2, u^1\} \quad \left((u^1, u^2) \in \mathbb{R} \times (0, 2\pi) \right). \tag{3.74}
$$

Let γ be the geodesic line through the point $X(0,0)$ at an angle Θ_0 to the u^2–line through $u_0^1 = 0$. Then the transformation formulae for the geodesic parallel coordinates are

$$
u^1 = \bar{u}^1 \sin\Theta_0 + \bar{u}^2 \cos\Theta_0 \text{ and } u^2 = \frac{1}{r} \cdot \left(\bar{u}^1 \cos\Theta_0 - \bar{u}^2 \sin\Theta_0 \right) \tag{3.75}
$$

(Figure 3.22).

Proof. By (3.28) in Example 3.2.6, the geodesic line γ is given by

$$
u^1(s) = \sin\Theta_0 \cdot s \text{ and } u^2(s) = \frac{\cos\Theta_0}{r} \cdot s \text{ for all } s \in \mathbb{R}. \tag{3.76}
$$

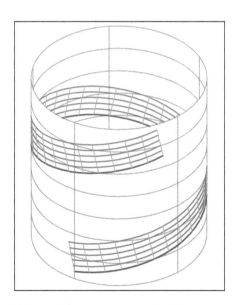

Figure 3.22 Geodesic parallel coordinates on a cylinder

Since $g_{22}(u^i) = r$, it follows from (3.48) in Clairaut's theorem, Theorem 3.3.1, that the angle between γ and the parallels of the cylinder is constant. Therefore, at each s, we have $\Theta = \Theta_0 + \pi/2$ for the angle between the geodesic line γ_s^* with $(v_s^i(t))$ and the parallel through $u^1(s)$. Thus γ_s^* is given by

$$v_s^1(t) = \cos\Theta_0 \cdot t + u^1(s) \text{ and } v_s^2(t) = -\frac{\sin\Theta_0}{r} \cdot t + u^2(s) \text{ for all } t \in \mathbb{R}.$$

We introduce new parameters $\bar{u}^i$ ($i = 1, 2$) such that the $\bar{u}^1$–lines are the parallel curves of the geodesic line γ and the $\bar{u}^2$–lines are the geodesic lines γ_s^*. Then the transformation formulae are given by (3.75). □

Visualization 3.6.10 (Geodesic parallel coordinates on a circular cone).
We consider the circular cone of Visualization 3.6.8. By Visualization 3.2.10 (b), the geodesic line γ with the initial conditions $u_0^1 > 0$, $u_0^2 = 0$ and $\Theta_0 \in (0, \pi/2)$ is given by

$$\left\{ \begin{array}{ll} u^1(s) & = \sqrt{a^2 s^2 + 2u_0^1 a \sin\Theta_0 \cdot s + (u_0^1)^2} \\ u^2(s) & = \sqrt{a+1}\left(\tan^{-1}\left(\frac{as}{u_0^1 \cos\Theta_0} + \tan\Theta_0\right) - \Theta_0\right) \end{array} \right\}, \qquad (3.77)$$

where $\alpha = \sqrt{a/(a+1)}$. In a sufficiently small neighbourhood of (u_0^1, u_0^2), the angle $\Theta(s)$ between γ and the parallel through $u^1(s)$ satisfies

$$\Theta(s) \in \left(0, \frac{\pi}{2}\right),$$

and the angle $\Theta_v(s)$ of the geodesic line γ_s^* given by $(v_s^i(t))$ satisfies $\Theta_v(s) = \Theta(s) + \pi/2$.

Applying Visualization 3.2.10 (b) again, we find that γ_s^* is given by

$$\left\{ \begin{array}{ll} v_s^1(t) & = \sqrt{\left(\alpha(t + t_0(s))\right)^2 + \beta^2} \\ v_s^2(t) & = \frac{k}{\alpha\beta} \cdot \tan^{-1}\left(\frac{\alpha}{\beta} \cdot (t + t_0(s))\right) + d \end{array} \right\},$$

where

$$t_0(s) = \frac{u^1(s)\sin\Theta_v(s)}{\alpha}, \ \beta = u^1(s)\cos\Theta_v(s),$$
$$d = u^2(s) - \sqrt{a+1} \cdot \tilde{\Theta}_v(s) = u^2(s) - \sqrt{a+1}(\Theta_v(s) - \pi)$$

and

$$k = \sqrt{a} \cdot u^1(s)\cos\Theta_v(s).$$

Since $\Theta(s) \in (0, \pi/2)$ it follows from (3.48) in Clairaut's theorem, Theorem 3.3.1 that

$$\sin\Theta_v(s) = \cos\Theta(s) = \cos\Theta_0 \cdot \frac{\sqrt{g_{22}(u_0^1)}}{\sqrt{g_{22}(u^1(s))}} = \cos\Theta_0 \cdot \frac{u_0^1}{u_1(s)},$$

and so

$$t_0(s) = \frac{u^1(s) \cdot \cos \Theta(s)}{\alpha} = \frac{u_0^1 \cdot \cos \Theta_0}{\alpha}$$

and

$$v_s^1(t) = \sqrt{\alpha^2 t^2 + 2u_0^1 \alpha \cos \Theta_0 \cdot t + (u^1(s))^2}.$$

Furthermore, we obtain

$$\frac{k}{\alpha\beta} = \frac{\sqrt{a} \cdot u_1(s) \cos \Theta_v(s)}{\alpha \cdot u_1(s) \cos \Theta_v(s)} = \sqrt{a+1},$$

$$\frac{\alpha}{\beta}(t + t_0(s)) = \frac{1}{u^1(s) \cdot \cos \Theta_v(s)} \left(\alpha t + u^1(s) \cdot \cos \Theta(s) \right)$$

$$= -\frac{1}{u^1(s) \cdot \sin \Theta(s)} \left(\alpha t + u_0^1 \cdot \cos \Theta_0 \right)$$

$$= -\frac{1}{u^1(s)\sqrt{1 - \cos^2 \Theta(s)}} \left(\alpha t + u_0^1 \cdot \cos \Theta_0 \right)$$

$$= -\frac{1}{\sqrt{(u_1(s))^2 - (u_0^1)^2 \cos^2 \Theta_0}} \left(\alpha t + u_0^1 \cdot \cos \Theta_0 \right)$$

and

$$\tilde{\Theta}_v(s) = \Theta_s - \frac{\pi}{2} = \cos^{-1} \frac{u_0^1 \cdot \cos \Theta_0}{u^1(s)} - \frac{\pi}{2},$$

hence

$$v_s^2(t) = u^2(s) - \sqrt{a+1} \left[\tan^{-1} \left(\frac{\alpha t + u_0^1 \cdot \cos \Theta_0}{\sqrt{(u_1(s))^2 - (u_0^1)^2 \cos^2 \Theta_0}} \right) \right.$$

$$\left. + \cos^{-1} \left(\frac{u_0 \cdot \cos \Theta_0}{u^1(s)} \right) - \frac{\pi}{2} \right]$$

(Figure 3.23).

Now we construct geodesic polar coordinates as follows: Let S be a surface with a parametric representation $\vec{x}(u^i)$ $((u^1, u^2) \in D \subset \mathbb{R})$ and $P \in S$ be a given point with position vector $\vec{x}(u_0^i)$. Furthermore, let γ_Θ with a parametric representation $\vec{y}_\Theta(s)$ denote the geodesic line through the point P at an angle Θ to the u^2–line through P. We assume that s is the arc length along γ_Θ and $\vec{y}_\Theta(s)$ is given by $v_\Theta^i(s)$ with $v_\Theta^i(0) = u_0^i$ $(i = 1, 2)$. For each fixed value of s, the vectors $\vec{y}_\Theta(s)$ $(\Theta \in [0, 2\pi])$ generate a circle line γ_s^* of constant distance s from the point P measured along the geodesic line γ_Θ.

Choosing the new parameters $\bar{u}^j$ $(j = 1, 2)$ such that the $\bar{u}^1$–lines are the geodesic lines γ_Θ $(\Theta \in [0, 2\pi])$ and the $\bar{u}^2$–lines are the circle lines γ_s^*, we obtain a system of geodesic polar coordinates. The transformation formulae are

$$u^i = v_{\bar{u}^2}^i(\bar{u}^1) \ (i = 1, 2). \tag{3.78}$$

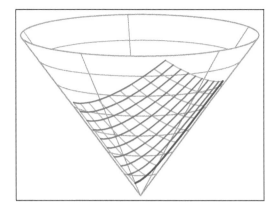

Figure 3.23 Geodesic parallel coordinates on a circular cone

Visualization 3.6.11 (Geodesic polar coordinates on a circular cylinder).
Again we consider the circular cylinder of Visualization 3.6.9 with a parametric representation (3.74). It follows from the representation of the geodesic line γ through the point $X(0,0)$ at an angle Θ_0 to the u^2–line through $u_0^1 = 0$ and the transformation formulae (3.78) for the corresponding geodesic polar coordinates that

$$u^1(\bar{u}^i) = v_{\bar{u}^2}^1(\bar{u}^1) = \bar{u}^1 \sin \bar{u}^2 \text{ and } u^2(\bar{u}^i) = v_{\bar{u}^2}^2(\bar{u}^1) = \frac{\bar{u}^1}{r} \cos \bar{u}^2.$$

Hence the parametric representation of the circular cylinder with respect to the new parameters $\bar{u}^i$ $(i = 1, 2)$ in a suitable neighbourhood of the point $X(0,0)$ is given by

$$\bar{\bar{x}}(\bar{u}^i) = \left\{ r \cos \left(\frac{\bar{u}^1}{r} \cdot \cos \bar{u}^2 \right), r \sin \left(\frac{\bar{u}^1}{r} \cdot \cos \bar{u}^2 \right), \bar{u}^1 \cdot \sin \bar{u}^2 \right\}$$

(Figure 3.24).

Visualization 3.6.12 (Geodesic polar coordinates on a circular cone).
Again we consider the circular cone of Visualization 3.6.10. It follows from the representation (3.77) of the geodesic line γ through the point $X(u_0^1, 0)$ at an angle Θ_0 to the u^2–line through $u_0^1 > 0$ and the transformation formulae (3.78) that the parameters $\bar{u}^i$ $(i = 1, 2)$ for the geodesic polar coordinates satisfy

$$\left\{ \begin{aligned} u^1(\bar{u}^i) \quad &= v_{\bar{u}^2}^1(\bar{u}^i) = \sqrt{\alpha(\bar{u}^1)^2 + 2u_0^1\alpha\bar{u}^1 \cdot \sin \bar{u}^2 + (u_0^1)^2} \\ u^2(\bar{u}^i) \quad &= v_{\bar{u}^2}^2(\bar{u}^i) = \sqrt{a+1} \left(\arctan \left(\frac{\alpha\bar{u}^1}{u_0^1 \cdot \cos \bar{u}^1} + \tan \bar{u}^2 \right) \right) \\ &\qquad\qquad - \arctan \left(\tan \widetilde{\bar{u}^2} \right) \end{aligned} \right\},$$

where $\alpha = \sqrt{a/(a+1)}$, as before (Figure 3.25).

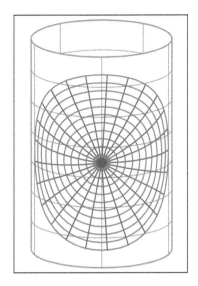

Figure 3.24 Geodesic polar coordinates on a cylinder

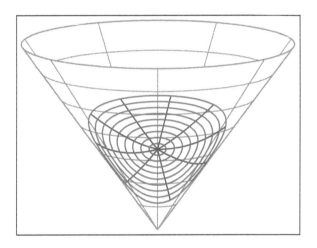

Figure 3.25 Geodesic polar coordinates on a circular cone

3.7 LEVI-CIVITÀ PARALLELISM

In Section 3.6, we studied the construction of *geodesically parallel curves* of a given curve.

Now we consider the problem of a sensible definition of the parallel translation of special vectors, the so–called *surface vectors*, which are introduced as follows.

Definition 3.7.1. Let S be a surface with a parametric representation $\vec{x}(u^i)$ $((u^1, u^2) \in D)$, P be a point of S and $T(P)$ denote the tangent plane to S at P. Then a vector $\vec{z}$ at P in $T(P)$ is called a *surface vector of S at P*. We write

$$\vec{z} = (\xi^1, \xi^2),$$

where ξ^i $(i = 1, 2)$ are the components of the vectors $\vec{x}_i$ at P.

If the surface vectors $\vec{z}(s)$ of a plane are moved parallel along a straight line, that is, along a geodesic line, then the angle between the vector and the straight line, that is, the angle between $\vec{z}(s)$ and the tangent to the geodesic line, remains constant, that is, $\dot{\vec{z}}(s) = 0$.

The following generalization to arbitrary surfaces is due to *Levi–Cività*.

Definition 3.7.2. Let S be a surface and γ be a curve on S with a parametric representation $\vec{x}(u^i(s))$. The surface vectors $\vec{z}(s)$ are said to be *geodesically parallel along* γ, if the components of $\dot{\vec{z}}(s)$ in the direction of the tangents to γ at s are equal to 0, that is,

$$\vec{x}_i(u^i(s)) \bullet \dot{\vec{z}}(s) = 0 \text{ for } i = 1, 2. \tag{3.79}$$

Remark 3.7.3. *We omit the parameters and write* $\vec{z} = \vec{x}_k \xi^k$. *Then the condition in (3.79) yields*

$$0 = \vec{x}_i \bullet \frac{d}{ds}(\vec{x}_k \xi^k) = \vec{x}_i \bullet \vec{x}_k \dot{\xi}^k + \vec{x}_i \bullet x_{kj} \dot{u}^j \xi^k = g_{ik} \dot{\xi}^k + [kij] \dot{u}^i \xi^k.$$

Multiplying by g^{ir} and summing with respect to i we obtain

$$0 = g^{ir}\left(g_{ik}\dot{\xi}^k + [kij]\dot{u}^i\xi^k\right) = \delta_{rk}\dot{\xi}^k + g^{ir}[kij]\dot{u}^i\xi^k = \dot{\xi}^r + \begin{Bmatrix} r \\ kj \end{Bmatrix}\dot{u}^j\xi^k \text{ for } r = 1, 2. \tag{3.80}$$

We write

$$\frac{\delta\xi^i}{\delta s} = \frac{d\xi^i}{ds} + \begin{Bmatrix} i \\ kj \end{Bmatrix}\dot{u}^j\xi^k \text{ for } i = 1, 2. \tag{3.81}$$

If t is an arbitrary parameter of the curve, the (3.80) becomes

$$\frac{\delta\xi^i}{dt} = \frac{d\xi^i}{dt} + \begin{Bmatrix} i \\ kj \end{Bmatrix}\frac{du^i}{dt}\xi^k \text{ for } i = 1, 2. \tag{3.82}$$

Apart from the parameters $\xi^i(s)$ and $u^i(s)$ $(i = 1, 2)$ of the vector field $\vec{z}$ the curve γ, only the second Christoffel symbols $\begin{Bmatrix} i \\ kj \end{Bmatrix}$ appear in the formula in (3.80); hence the geodesic parallelism is a quantity of the intrinsic geometry of a surface.

We consider the plane with Cartesian parameters. Then $\begin{Bmatrix} i \\ kj \end{Bmatrix} = 0$ for all i, j and k, and (3.82) reduces to

$$\frac{d\xi^i}{dt} = 0 \text{ for } i = 1, 2,$$

that is, the components of the surface vectors are constant.

The concept of geodesic parallelism is also independent of the parametric representation of the surface, since the vector $\dot{\vec{z}}$, which is orthogonal to the surface, is defined independently of the parameters of the surface.

The next result is an immediate consequence of the existence and uniqueness theorem for the system (3.80) of differential equations.

Theorem 3.7.4. *If γ is a curve on a surface S and $\vec{z}_0$ is a surface vector at the point $P_0 \in S$, then there exists a unique field of geodesically parallel surface vectors $\vec{z}(s)$ along γ such that $\vec{z}(0) = \vec{z}_0$.*

We refer to a curve on a surface given by $u^i(s)$ $(i = 1, 2)$ as *straightest curve*, if its tangent vectors form a geodesically parallel vector field along itself. Then it follows from (3.80) that

$$\ddot{u}^r + \left\{ \begin{matrix} r \\ jk \end{matrix} \right\} \dot{u}^j \dot{u}^k = 0,$$

and so the next result holds.

Theorem 3.7.5. *The geodesic lines of a surface are also its straightest lines.*

Theorem 3.7.6. *Let γ given by $u^i(s)$ $(i = 1, 2)$ be a curve on a surface, and $\vec{v}(s)$ and $\vec{w}(s)$ given by $v^i(s)$ and $w^i(s)$ $(i = 1, 2)$ be two geodesically parallel vector fields along γ. Then the angle between the vectors $\vec{v}(s)$ and $\vec{w}(s)$ along γ is constant and the lengths of all vectors $\vec{v}(s)$ are equal.*

Proof. We have

$$\frac{d}{ds} \left(g_{ik}(u^j(s)) v^i w^k \right) = \frac{\partial g_{ik}}{\partial u^j} \dot{u}^j v^i w^k + g_{ik} \dot{v}^i w^k + g_{ik} v^i \dot{w}^k$$

and, since the vector fields are parallel, it follows from (3.80) that

$$\dot{v}^i = - \left\{ \begin{matrix} i \\ mr \end{matrix} \right\} v^m \dot{u}^r \text{ and } \dot{w}^k = - \left\{ \begin{matrix} k \\ lr \end{matrix} \right\} w^l \dot{u}^r,$$

hence

$$\frac{d}{ds} \left(g_{ik}(u^j(s)) v^i w^k \right) = \frac{\partial g_{ik}}{\partial u^j} \dot{u}^j v^i w^k - g_{ik} \left(\left\{ \begin{matrix} i \\ mr \end{matrix} \right\} \dot{u}^r v^m w^k + \left\{ \begin{matrix} k \\ lr \end{matrix} \right\} \dot{u}^r v^i w^l \right).$$

Changing the indices of summation i, k and j to m, l and r in the first term on the right hand side of the last identity, k to l in the second term, and i to m in the last term we obtain

$$\frac{d}{ds} \left(g_{ik}(u^j(s)) v^i w^k \right) = \left(\frac{\partial g_{ml}}{\partial u^r} - g_{il} \left\{ \begin{matrix} i \\ mr \end{matrix} \right\} - g_{mk} \left\{ \begin{matrix} k \\ lr \end{matrix} \right\} \right) \dot{u}^r v^m w^l.$$

Now it follows from the definition of the second Christoffel symbols, Definition 3.1.2, that

$$g_{il} \left\{ \begin{matrix} i \\ mr \end{matrix} \right\} = g_{il} g^{ji} [mrj] = \delta_l^j [mrj] = [mrl]$$

and

$$g_{mk} \left\{ \begin{matrix} k \\ lr \end{matrix} \right\} = g_{mk} g^{jk} [lrj] = \delta_m^j [mrj] = [lrm],$$

hence by Lemma 3.1.4 (d)

$$\frac{d}{ds}\left(g_{ik}(u^j(s))v^i w^k\right) = \left(\frac{\partial g_{ml}}{\partial u^r} - [mrl] - [lrm]\right)\dot{u}^r v^m w^l = 0.$$

From this, we conclude that the lengths and angles are constant. □

Visualization 3.7.7. (Parallel movement along a curve on a surface of revolution) *We consider the parallel movement along an arbitrary curve γ on any surface of revolution RS with a parametric representation $\vec{x}(u^i)$ $((u^1, u^2) \in D)$. Let P_0 be a point of RS with position vector $\vec{x}(u_0^i)$. We assume that the surface vector $\vec{z}(t)$ with $\vec{z}(0) = \vec{x}(u_0^i)$ has an angle $\Theta_0 \in [0, \pi/2]$ at P_0 to the parallel with $u^1 = u_0^1$.*
We obtain for the second Christoffel symbols of RS from Example 3.1.7

$$\left\{\begin{matrix}1\\11\end{matrix}\right\} = \frac{1}{2g_{11}}\cdot\frac{dg_{11}}{du^1}, \quad \left\{\begin{matrix}1\\22\end{matrix}\right\} = -\frac{1}{2g_{11}}\cdot\frac{dg_{22}}{du^1},$$

$$\left\{\begin{matrix}2\\12\end{matrix}\right\} = \left\{\begin{matrix}2\\21\end{matrix}\right\} = \frac{1}{2g_{22}}\cdot\frac{dg_{22}}{du^1}$$

and

$$\left\{\begin{matrix}i\\jk\end{matrix}\right\} = 0 \text{ for all other values } i, j \text{ and } k.$$

Hence the differential equations in (3.82) are

$$\frac{d\xi^1}{dt} + \frac{1}{2g_{11}}\cdot\frac{dg_{11}}{du^1}\xi^1\frac{du^1}{dt} - \frac{1}{2g_{11}}\cdot\frac{dg_{22}}{du^1}\xi^2\frac{du^2}{dt} = 0 \qquad (3.83)$$

and

$$\frac{d\xi^2}{dt} + \frac{1}{2g_{22}}\cdot\frac{dg_{22}}{du^1}\xi^1\frac{du^2}{dt} + \frac{1}{2g_{22}}\cdot\frac{dg_{22}}{du^1}\xi^2\frac{du^1}{dt} = 0. \qquad (3.84)$$

We may assume by Theorem 3.7.6 that the surface vector has constant length 1 when moved along γ, that is,

$$g_{11}(\xi^1)^2 + g_{22}(\xi^2)^2 = 1,$$

whence

$$\xi^2 = \pm\frac{1}{g_{22}}\cdot\sqrt{1 - g_{11}(\xi^1)^2}. \qquad (3.85)$$

Since the surface vector $\vec{z}(0)$ at $t = 0$ satisfies

$$\vec{z}(0) = \xi^1(0)\vec{x}_1(u^i(0)) + \xi^2(0)\vec{x}_2(u^i(0)) = \frac{\sin\Theta_0}{g_{11}(u^i(0))}\cdot\vec{x}_1(u^i(0)) + \frac{\cos\Theta_0}{g_{22}(u^i(0))}\cdot\vec{x}_2(u^i(0)),$$

that is,

$$\xi^1(0) = \frac{\sin \Theta_0}{g_{11}(u^i(0))} \text{ and } \xi^2(0) = \frac{\cos \Theta_0}{g_{22}(u^i(0))},$$

we choose the upper sign in (3.85), that is,

$$\xi^2 = \frac{1}{g_{22}} \cdot \sqrt{1 - g_{11}(\xi^1)^2}. \tag{3.86}$$

Substituting (3.86) in (3.83), we obtain an ordinary first order differential equation for ξ^1 the solution of which and ξ^2 from (3.86) describe the parallel movement of the surface vector along γ.

In particular we consider the parallel movement along a loxodrome γ on RS given by

$$u^1(t) = t \text{ and } u^2(t) = |\cot \beta| \int \sqrt{\frac{g_{11}(t)}{g_{22}(t)}} \, dt,$$

where $\beta \in (-\pi/2, \pi/2) \setminus \{0\}$ is the constant angle between γ and the parallels of RS.

Now we have to solve the differential equation

$$\frac{d\xi^1}{dt} = -\frac{1}{2g_{11}} \cdot \frac{dg_{11}}{du^1} \xi^1 + \frac{1}{2\sqrt{g}} \cdot \frac{dg_{22}}{du^1} \cdot |\cot \beta| \xi^2 \tag{3.87}$$

with

$$g = g_{11}g_{22} \text{ and } \xi^2 = \frac{\sqrt{1 - g_{11}(\xi^1)^2}}{\sqrt{g_{22}}}.$$

If RS is a sphere of radius $r > 0$, we obtain

$$g_{11} = r^2 \text{ and } g_{22} = r^2 \cos^2 u^1,$$

and (3.87) yields

$$\frac{d\xi^1}{dt} = -\frac{|\cot \beta| \cdot \sin t}{r \cos t} \cdot \sqrt{1 - r^2(\xi^1)^2}.$$

This implies

$$\int \frac{r}{\sqrt{1 - r^2(\xi^1)^2}} \, d\xi^1 = \sin^{-1}(r\xi^1) = -|\cot \beta| \int \frac{\sin t}{\cos t} \, dt$$

$$= |\cot \beta| \cdot \log |\cos t| + c,$$

where c is a constant of integration, and so

$$\xi^1(t) = \frac{1}{r} \cdot \sin\left(|\cot \beta| \cdot \log |\cos t| + c\right).$$

Since

$$\xi^1(0) = \frac{\sin \Theta_0}{\sqrt{g_{11}}} = \frac{\sin \Theta_0}{r},$$

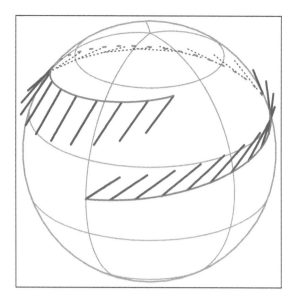

Figure 3.26 Parallel movement along a loxodrome on a sphere

we obtain (Figure 3.26)

$$\xi^1(t) = \frac{1}{r} \cdot \sin\left(|\cot\beta| \cdot \log|\cos t| + \Theta_0\right)$$

$$for\ t < \cos^{-1}\left(\exp\left(\frac{-\Theta_0 - \frac{\pi}{2}}{|\cot\beta|}\right)\right)$$

and

$$\xi^2(t) = \frac{1}{r\cos t} \cdot \cos\left(|\cot\beta| \cdot \log|\cos t| + \Theta_0\right).$$

Visualization 3.7.8. (Parallel movement along a geodesic line on a surface of revolution) *Now we consider parallel movements on surfaces of revolution RS with a parametric representation $\vec{x}(u^i)$ $((u^1, u^2) \in D)$ along geodesic lines γ which are parameterized with respect to their arc lengths, that is, $\vec{x}_\gamma(u^i(s))$. Again we assume that the surface vector $\vec{z}(t)$ at $t = 0$ at the point P_0 corresponding to the parameter point $(u_0^1, u_0^2) = (u^1(0), u^2(0))$ has an angle $\Theta_0 \in [0, \pi/2]$ with the parallel through u_0^1. The angle β between γ and the surface vector $\vec{z}(t)$ of constant length 1 is constant under the movement along γ. We obtain*

$$\xi^1(s) = \frac{c \cdot \sin\beta}{\sqrt{g_{11}(u^1(s)) g_{22}(u^1(s))}} + \cos\beta \cdot \dot{u}^1(s) \tag{3.88}$$

and

$$\xi^2(s) = \frac{1}{g_{22}(u^1(s))}\left(c \cdot \cos\beta - \sin\beta\sqrt{g_{22}(u^1(s)) - c^2}\right), \tag{3.89}$$

where $c = \sqrt{g_{22}(u_0^1)} \cdot \cos\Theta_0$.

In particular, we obtain for the circular cone of Visualization 3.2.10 and the parallel movement along the geodesic line γ with the initial conditions $u^1(0) = u_0^1 > 0$, $u^2(0) = 0$ and $\Theta_0 \in (0, \pi/2)$

$$\xi^1(s) = \alpha \cdot \frac{\alpha s + u_0^1 \cdot \sin(\Theta_0 + \beta)}{\sqrt{(\alpha s + u_0^1 \cdot \sin\Theta_0)^2 + (u_0^1 \cdot \cos\Theta_0)^2}} \tag{3.90}$$

and

$$\xi^2(s) = \frac{\sqrt{a} \cdot (-\alpha s \cdot \sin\beta + u_1^0 \cdot \cos(\Theta_0 + \beta))}{(\alpha s + u_0^1 \cdot \sin\Theta_0)^2 + (u_0^1 \cdot \cos\Theta_0)^2} \ for \ s \geq 0, \tag{3.91}$$

where $\alpha = \sqrt{a/(a+1)}$ (left in Figure 3.27).

Proof. We put $\delta = \cos\beta$ and have to solve the equations

$$g_{ik}\xi^i\frac{du^k}{ds} = g_{11}\xi^1\dot{u}^1 + g_{22}\xi^2\dot{u}^2 = \delta \text{ for } |\delta| \leq 1 \tag{3.92}$$

and

$$g_{ik}\xi^i\xi^k = g_{11}(\xi^1)^2 + g_{22}(\xi^2)^2 = 1. \tag{3.93}$$

It follows from (3.48) in Clairaut's theorem, Theorem 3.3.1 that

$$g_{22} \cdot \dot{u}^2 = g_{22}(u^1(s)) \cdot \dot{u}^2(s) = c = const.$$

and so by (3.92) and (3.93)

$$\left(\xi^2 \cdot c - \delta\right)^2 = \left(\xi^2 g_{22}\dot{u}^2 - \left(\xi^1 g_{11}\dot{u}^1 + \xi^2 g_{22}\dot{u}^2\right)\right)^2 = g_{11}^2(\xi^1)^2(\dot{u}^1)^2$$
$$= g_{11}\left(1 - g_{22}(\xi^2)^2\right)(\dot{u}^1)^2,$$

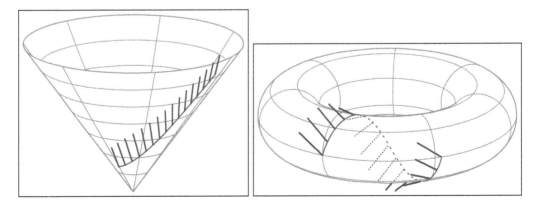

Figure 3.27 Parallel movement along a geodesic line on a cone and a torus

hence, since s is the arc length along γ

$$
\begin{aligned}
0 &= (\xi^2)^2 \left(c^2 + g_{11}g_{22}(\dot{u}^1)^2\right) - 2\xi^2 c\delta + \delta^2 - g_{11}(\dot{u}^1)^2 \\
&= g_{22}(\xi^2)^2 \left(g_{22}(\dot{u}^2)^2 + g_{11}(\dot{u}^1)^2\right) - 2\xi^2 c\delta + \delta^2 - g_{11}(\dot{u}^1)^2 \\
&= g_{22}(\xi^2)^2 - 2\xi^2 c\delta + \delta^2 - g_{11}(\dot{u}^1)^2.
\end{aligned}
$$

This quadratic equation has the solutions

$$
\begin{aligned}
\xi^2_{1,2} &= \frac{1}{g_{22}} \left(c\delta \pm \sqrt{c^2\delta^2 - g_{22}\left(\delta^2 - g_{11}(\dot{u}^1)^2\right)}\right) \\
&= \frac{1}{g_{22}} \left(c\delta \pm \sqrt{c^2\delta^2 - g_{22}\delta^2 + g_{22} - c^2}\right) = \frac{1}{g_{22}} \left(c\delta \pm \sqrt{(g_{22} - c^2)(1 - \delta^2)}\right) \\
&= \frac{1}{g_{22}} \left(c \cdot \cos\beta \pm |\sin\beta| \sqrt{g_{22} - c^2}\right).
\end{aligned}
$$

If $\beta \in [0, \pi]$, then it follows from $c = \sqrt{g_{22}(u_0^1)} \cdot \cos\Theta_0$ that

$$
\xi^2_{1,2}(0) = \frac{1}{\sqrt{g_{22}(u_0^1)}} (\cos\Theta_0 \cdot \cos\beta \pm \sin\Theta_0 \cdot \sin\beta) = \frac{\cos(\Theta_0 \mp \beta)}{\sqrt{g_{22}(u_0^1)}}.
$$

Since on the other hand

$$
\xi^2 = \frac{\cos(\Theta_0 + \beta)}{\sqrt{g_{22}(u_0^1)}},
$$

we obtain (3.89).

Furthermore, since $\sqrt{g_{22} - c^2} = \sqrt{g_{11}g_{22}} \cdot |\dot{u}^1|$, it follows from (3.89) by (3.92)

$$
\begin{aligned}
\xi^1 &= \frac{\delta - c\xi^2}{g_{11}\dot{u}^1} = \frac{\delta g_{22} - c^2 \cos\beta + c \cdot \sin\beta \cdot \sqrt{g_{11}g_{22}} \cdot |\dot{u}^1|}{g_{11}g_{22}\dot{u}^1} \\
&= \mathrm{sgn}(\dot{u}^1) \cdot \frac{c \cdot \sin\beta}{\sqrt{g_{11}g_{22}}} + \cos\beta \cdot \dot{u}^1.
\end{aligned}
$$

Finally, the identity in (3.88) follows, since

$$
\xi^1(0) = \frac{\sin(\Theta_0 + \beta)}{\sqrt{g_{11}g_{22}}}.
$$

We obtain for the geodesic line on the circular cone from Visualization 3.2.10

$$
u^1(s) = \sqrt{\left(\alpha s + u_0^1 \cdot \sin\Theta_0\right)^2 + \left(u_0^1 \cdot \cos\Theta_0\right)^2}
$$

and

$$
\dot{u}^1(s) = \frac{\alpha \left(\alpha s + u_0^1 \cdot \sin\Theta_0\right)}{u^1(s)}.
$$

Since

$$\sqrt{g_{22} - c^2} = \sqrt{g_{11}g_{22}} \cdot |\dot{u}^1| = \sqrt{\frac{a+1}{a} \cdot \frac{(u^1)^2}{a}} \cdot \frac{\alpha \left(\alpha s + u_0^1 \cdot \sin \Theta_0 \right)}{u^1(s)}$$

$$= \frac{1}{\sqrt{a}} \left(\alpha s + u_0^1 \cdot \sin \Theta_0 \right),$$

(3.89) yields (3.91) and (3.88) yields (3.90).

$\square$

3.8   THEOREMA EGREGIUM

One of the most important results of the theory of surfaces states that the Gaussian curvature K of a surface is a quantity of its intrinsic geometry. This result is *Gauss's theorema egregium*. To be able to prove this we have to express the Gaussian curvature K in terms of the first fundamental coefficients and their derivatives. Since $K = L/g$, we have to establish a relation between the first and second fundamental coefficients.

We need the following result.

Theorem 3.8.1. (Derivation formulae by Gauss, and Mainardi and Codazzi) *Let S be a surface with a parametric representation $\vec{x}(u^j)$. We denote the partial differentiation with respect to the parameter u^j by $_{;j}$, and put for $i, k, j, m = 1, 2$*

$$R_{ikj}^m = \left\{ \begin{matrix} m \\ ik \end{matrix} \right\}_{;j} - \left\{ \begin{matrix} m \\ ij \end{matrix} \right\}_{;k} + \left\{ \begin{matrix} r \\ ik \end{matrix} \right\} \left\{ \begin{matrix} m \\ rj \end{matrix} \right\} - \left\{ \begin{matrix} r \\ ij \end{matrix} \right\} \left\{ \begin{matrix} m \\ rk \end{matrix} \right\} \tag{3.94}$$

$$G_{ikj}^m = R_{ikj}^m - \left(L_{ik}L_j^m - L_{ij}L_k^m \right) \tag{3.95}$$

and

$$C_{ikj} = L_{ik;j} - L_{ij;k} + \left\{ \begin{matrix} r \\ ik \end{matrix} \right\} L_{rj} - \left\{ \begin{matrix} r \\ ij \end{matrix} \right\} L_{rk}. \tag{3.96}$$

Then we have

$$G_{ikj}^m = 0 \quad (m, i, k, j = 1, 2) \quad \text{(Gauss)} \tag{3.97}$$

and

$$C_{ikj} = 0 \qquad (i, k, j = 1, 2) \quad \text{(Mainardi and Codazzi)}. \tag{3.98}$$

Proof. We know from (3.15) in the proof of Theorem 3.1.8 that

$$\vec{x}_{ik} = \left\{ \begin{matrix} r \\ ik \end{matrix} \right\} \vec{x}_r + L_{ik}\vec{N} \text{ for } i, k = 1, 2. \tag{3.99}$$

Those formulae are referred to as the *Gauss derivation formulae*. Now it follows from (3.99) and the Weingarten equations (2.141) in Theorem 2.10.1 that

$$
\vec{x}_{ikj} = \frac{\partial}{\partial u^j}\left(\left\{\begin{matrix} r \\ ik \end{matrix}\right\}\vec{x}_r + L_{ik}\vec{N}\right) = \left\{\begin{matrix} r \\ ik \end{matrix}\right\}_{;j}\vec{x}_r + \left\{\begin{matrix} r \\ ik \end{matrix}\right\}\vec{x}_{rj} + L_{ik;j}\vec{N} + L_{ik}\vec{N}_j
$$

$$
= \left\{\begin{matrix} r \\ ik \end{matrix}\right\}_{;j}\vec{x}_r + \left\{\begin{matrix} r \\ ik \end{matrix}\right\}\left(\left\{\begin{matrix} m \\ rj \end{matrix}\right\}\vec{x}_m + L_{rj}\vec{N}\right) + L_{ik;j}\vec{N} - L_{ik}L_j^r\vec{x}_r
$$

$$
= \left(\left\{\begin{matrix} m \\ ik \end{matrix}\right\}_{;j} + \left\{\begin{matrix} r \\ ik \end{matrix}\right\}\left\{\begin{matrix} m \\ rj \end{matrix}\right\} - L_{ik}L_j^m\right)\vec{x}_m + \vec{N}\left(L_{ik;j} + \left\{\begin{matrix} r \\ ik \end{matrix}\right\}L_{rj}\right).
$$

Interchanging the roles of j and k, we obtain

$$
\vec{x}_{ijk} = \left(\left\{\begin{matrix} m \\ ij \end{matrix}\right\}_{;k} + \left\{\begin{matrix} r \\ ij \end{matrix}\right\}\left\{\begin{matrix} m \\ rk \end{matrix}\right\} - L_{ij}L_k^m\right)\vec{x}_m + \vec{N}\left(L_{ij;k} + \left\{\begin{matrix} r \\ ij \end{matrix}\right\}L_{rk}\right),
$$

hence by (3.94), (3.95) and (3.96)

$$
\vec{0} = \vec{x}_{ikj} - \vec{x}_{ijk}
$$

$$
= \left(\left\{\begin{matrix} m \\ ik \end{matrix}\right\}_{;j} - \left\{\begin{matrix} m \\ ij \end{matrix}\right\}_{;k} + \left\{\begin{matrix} r \\ ik \end{matrix}\right\}\left\{\begin{matrix} m \\ rj \end{matrix}\right\} - \left\{\begin{matrix} r \\ ij \end{matrix}\right\}\left\{\begin{matrix} m \\ rk \end{matrix}\right\} - L_{ik}L_j^m + L_{ij}L_k^m\right)\vec{x}_m
$$

$$
+ \left(L_{ik;j} - L_{ij;k} + \left\{\begin{matrix} r \\ ik \end{matrix}\right\}L_{rj} - \left\{\begin{matrix} r \\ ij \end{matrix}\right\}L_{rk}\right)\vec{N}
$$

$$
= \left(R_{ikj}^m - L_{ik}L_j^m + L_{ij}L_k^m\right)\vec{x}_m + C_{ikj}\vec{N}
$$

$$
= G_{ikj}^m\vec{x}_m - C_{ijk}\vec{N} \text{ for } i, k, j = 1, 2.
$$

Since the vectors $\vec{x}_1$, $\vec{x}_2$ and $\vec{N}$ are linearly independent, the identities in (3.97) and (3.98) follow. □

Remark 3.8.2. *The values R_{ikj}^m form the so-called* Riemann tensor of curvature; *they are quantities of the intrinsic geometry of a surface by (3.94). We define*

$$
R_{ijkl} = g_{ir}R_{jkl}^r \text{ for } i, j, k, l = 1, 2. \tag{3.100}
$$

Theorem 3.8.3 (Theorema egregium; Gauss).
The Gaussian curvature K is a quantity of the intrinsic geometry of a surface; it can be expressed by the first fundamental coefficients and their derivatives up to the second order. The following holds

$$
K = \frac{g_{2r}R_{112}^r}{g} = \frac{R_{1212}}{g}. \tag{3.101}
$$

Proof. It follows from the identities in (3.94) and (3.97) and the definition of L_k^r in (2.140) that

$$
R_{ikj}^r = G_{ikj}^r + (L_{ik}L_j^r - L_{ij}L_k^r) = L_{ik}L_j^r - L_{ij}L_k^r = L_{ik}g^{rm}L_{mj} - L_{ij}g^{rm}L_{mk}
$$

$$
= g^{rm}\left(L_{ik}L_{mj} - L_{ij}L_{mk}\right).
$$

We obtain for $i = k = 1$ and $j = m = 2$

$$g_{2r} R^r_{112} = L_{11} L_{22} - L^2_{12} = L. \tag{3.102}$$

By the definition of the Riemann tensor of curvature in (3.94), the values on the lefthand side of (3.102) depend on the first fundamental coefficients and their derivatives up to the second order only, hence are quantities of the intrinsic geometry of the surface.

Finally, (3.102) and (3.100) yield

$$K = \frac{L}{g} = \frac{g_{2r} R^r_{112}}{g} = \frac{R_{1212}}{g}.$$

$\square$

Example **3.8.4 (The Gaussian curvature of Poincaré's half–plane).**
Poincaré's half–plane of Definition (3.2.7) has constant Gaussian curvature $K = -1$.

Proof. We saw in the proof of Example 3.2.8 that

$$\left\{ \begin{matrix} 1 \\ 11 \end{matrix} \right\} = \left\{ \begin{matrix} 2 \\ 12 \end{matrix} \right\} = \left\{ \begin{matrix} 2 \\ 21 \end{matrix} \right\} = -\frac{1}{u^1} = -\left\{ \begin{matrix} 1 \\ 22 \end{matrix} \right\} \text{ and } \left\{ \begin{matrix} i \\ jk \end{matrix} \right\} = 0 \text{ otherwise.}$$

Since $g_{11} = g_{22} = 1/u^2$ and $g_{12} = 0$, we obtain by the definition of the Riemannian tensor of curvature in (3.94)

$$R^2_{112} = \left\{ \begin{matrix} 2 \\ 11 \end{matrix} \right\}_{;2} - \left\{ \begin{matrix} 2 \\ 12 \end{matrix} \right\}_{;1} + \left\{ \begin{matrix} r \\ 11 \end{matrix} \right\} \left\{ \begin{matrix} 2 \\ r2 \end{matrix} \right\} - \left\{ \begin{matrix} r \\ 12 \end{matrix} \right\} \left\{ \begin{matrix} 2 \\ r1 \end{matrix} \right\}$$

$$= 0 - \frac{1}{(u^1)^2} + \left\{ \begin{matrix} 1 \\ 11 \end{matrix} \right\} \left\{ \begin{matrix} 2 \\ 12 \end{matrix} \right\} + 0 - 0 - \left\{ \begin{matrix} 2 \\ 12 \end{matrix} \right\} \left\{ \begin{matrix} 2 \\ 21 \end{matrix} \right\}$$

$$= -\frac{1}{(u^1)^2} + \left(-\frac{1}{u^1} \right) \cdot \left(-\frac{1}{u^1} \right) - \left(-\frac{1}{u^1} \right) \cdot \left(-\frac{1}{u^1} \right) = -\frac{1}{(u^1)^2},$$

and so by (3.101)

$$K = \frac{g_{2r} R^r_{112}}{g} = \frac{g_{22} R^2_{112}}{g} = \frac{R^2_{112}}{g_{11}} = -\frac{1}{(u^1)^2} \cdot (u^1)^2 = -1.$$

$\square$

We obtain the following result as a consequence of (3.101).

Lemma 3.8.5. *The Gaussian curvature K of a surface with orthogonal parameters u^k $(k = 1, 2)$ is given by*

$$K = -\frac{1}{2\sqrt{g}} \left(\frac{\partial}{\partial u^2} \left(\frac{1}{\sqrt{g}} \frac{\partial g_{11}}{\partial u^2} \right) + \frac{\partial}{\partial u^1} \left(\frac{1}{\sqrt{g}} \frac{\partial g_{22}}{\partial u^1} \right) \right). \tag{3.103}$$

If the surface has geodesic parameters then

$$\frac{\partial^2 (\sqrt{g_{22}})}{\partial (u^1)^2} + K \sqrt{g_{22}} = 0. \tag{3.104}$$

Proof. Since $g_{12} = 0$, we have by (3.101) and (3.94)

$$K = \frac{g_{22}}{g} R_{112}^2 = \frac{1}{g_{11}} \left(\left\{ \begin{matrix} 2 \\ 11 \end{matrix} \right\}_{;2} - \left\{ \begin{matrix} 2 \\ 12 \end{matrix} \right\}_{;1} + \left\{ \begin{matrix} r \\ 11 \end{matrix} \right\} \left\{ \begin{matrix} 2 \\ r2 \end{matrix} \right\} - \left\{ \begin{matrix} r \\ 12 \end{matrix} \right\} \left\{ \begin{matrix} 2 \\ r1 \end{matrix} \right\} \right).$$

(i) First we show the identity in (3.103).

It follows from (3.7) and (3.8) in Corollary 3.1.5 that

$$\left\{ \begin{matrix} 1 \\ 11 \end{matrix} \right\} = \frac{1}{2g_{11}} \frac{\partial g_{11}}{\partial u^1} = \frac{1}{2g_{11}} g_{11\,;1}, \quad \left\{ \begin{matrix} 2 \\ 11 \end{matrix} \right\} = -\frac{1}{2g_{22}} g_{11\,;2},$$

$$\left\{ \begin{matrix} 1 \\ 12 \end{matrix} \right\} = \frac{1}{2g_{11}} g_{11\,;2}, \quad \left\{ \begin{matrix} 2 \\ 12 \end{matrix} \right\} = \frac{1}{2g_{22}} g_{22\,;1} \text{ and } \left\{ \begin{matrix} 2 \\ 22 \end{matrix} \right\} = \frac{1}{2g_{22}} g_{22\,;2},$$

hence

$$\left\{ \begin{matrix} 2 \\ 11 \end{matrix} \right\}_{;2} = \frac{g_{22\,;2} g_{11\,;2}}{2(g_{22})^2} - \frac{g_{11\,;22}}{2g_{22}}, \quad \left\{ \begin{matrix} 2 \\ 12 \end{matrix} \right\}_{;1} = \frac{g_{22\,;11}}{2g_{22}} - \frac{(g_{22\,;1})^2}{2(g_{22})^2},$$

$$\left\{ \begin{matrix} 1 \\ 11 \end{matrix} \right\} \left\{ \begin{matrix} 2 \\ 12 \end{matrix} \right\} = \frac{1}{4g_{11}g_{22}} g_{11\,;1} g_{22\,;1}, \quad \left\{ \begin{matrix} 2 \\ 11 \end{matrix} \right\} \left\{ \begin{matrix} 2 \\ 22 \end{matrix} \right\} = -\frac{1}{4(g_{22})^2} g_{11\,;2} g_{22\,;2},$$

$$-\left\{ \begin{matrix} 1 \\ 12 \end{matrix} \right\} \left\{ \begin{matrix} 2 \\ 11 \end{matrix} \right\} = \frac{1}{4g_{11}g_{22}} (g_{11\,;2})^2, \quad -\left\{ \begin{matrix} 2 \\ 12 \end{matrix} \right\} \left\{ \begin{matrix} 2 \\ 21 \end{matrix} \right\} = -\frac{(g_{22\,;1})^2}{4(g_{22})^2},$$

and so by the identity at the beginning of the proof

$$K = \frac{1}{g_{11}} \left(\frac{g_{11\,;2} g_{22\,;2}}{2(g_{22})^2} - \frac{g_{11\,;22}}{2g_{22}} - \frac{g_{22\,;11}}{2g_{22}} + \frac{(g_{22\,;1})^2}{2(g_{22})^2} + \right.$$

$$\left. + \frac{g_{11\,;1} g_{22\,;1}}{4g_{11}g_{22}} - \frac{g_{11\,;2} g_{22\,;2}}{4(g_{22})^2} + \frac{(g_{11\,;2})^2}{4g_{11}g_{22}} - \frac{(g_{22\,;1})^2}{4(g_{22})^2} \right). \quad (3.105)$$

On the other hand, we have

$$\left(\frac{g_{11\,;2}}{\sqrt{g}} \right)_{;2} = \frac{g_{11\,;22}}{\sqrt{g_{11}g_{22}}} - \frac{g_{11\,;2}}{2(g_{11}g_{22})^{3/2}} (g_{11\,;2} g_{22} + g_{11} g_{22\,;2}),$$

$$\left(\frac{g_{22\,;1}}{\sqrt{g}} \right)_{;1} = \frac{g_{22\,;11}}{\sqrt{g_{11}g_{22}}} - \frac{g_{22\,;1}}{2(g_{11}g_{22})^{3/2}} (g_{11\,;1} g_{22} + g_{11} g_{22\,;1})$$

and

$$-\frac{1}{2\sqrt{g}} \left(\left(\frac{g_{11\,;2}}{\sqrt{g}} \right)_{;2} + \left(\frac{g_{22\,;1}}{\sqrt{g}} \right)_{;1} \right) = \frac{1}{g_{11}} \left(-\frac{g_{11\,;22}}{2g_{22}} + \frac{(g_{11\,;2})^2}{4g_{11}g_{22}} + \frac{g_{11\,;2} g_{22\,;2}}{4(g_{22})^2} - \right.$$

$$\left. - \frac{g_{22\,;11}}{2g_{22}} + \frac{g_{22\,;1} g_{11\,;1}}{4g_{22}g_{11}} + \frac{(g_{22\,;1})^2}{4(g_{22})^2} \right). \quad (3.106)$$

Comparing (3.105) and (3.106), we obtain (3.103).

(ii) Now we show the identity in (3.104).

For geodesic parameters, we have $g_{12} = 0$ and $g_{11} = 1$, and so $g_{11\,;i} = g_{11\,;ik} = 0$ for $i, k = 1, 2$. We obtain from (3.103)

$$-\sqrt{g_{22}}K = \frac{1}{2}\left(\frac{1}{\sqrt{g_{22}}}g_{22\,;1}\right)_{;1} = \frac{g_{22\,;11}}{2\sqrt{g_{22}}} - \frac{(g_{22\,;1})^2}{4(g_{22})^{3/2}}.$$

On the other hand we have

$$(\sqrt{g_{22}})_{;11} = \left(\frac{g_{22\,;1}}{2\sqrt{g_{22}}}\right)_{;1},$$

hence $K\sqrt{g_{22}} + (\sqrt{g_{22}})_{;11} = 0$, which is (3.104).

□

3.9   MAPS BETWEEN SURFACES

In this section, we study certain maps between surfaces, namely *conformal, isometric* and *area preserving* maps. Some of the results presented here will be used later in the proof of the *Gauss–Bonnet theorem*, Theorem 3.10.1, and the study of *minimal surfaces* in Section 3.11.

The most important application of maps between surfaces is in cartography.

Definition 3.9.1. Let S and S^* be surfaces with admissible parametric representations $\vec{x}(u^i)$ and $\vec{\bar{x}}(\bar{u}^i)$. Then a map from S onto S^* is said to be *admissible* if it satisfies the following conditions:

(i)   the map is one-to-one;

(ii)   it is given by

$$\bar{u}^k = h^k(u^i) \text{ with } h^k \in C^r(D) \text{ for } k = 1, 2, \tag{3.107}$$

where $r \geq 1$ is chosen according to need;

(iii)   the corresponding Jacobian has no zeros.

Remark 3.9.2. *(a) Since we will always assume that maps between surfaces are admissible, we will omit the word* admissible.
(b) We may introduce new parameters u^{*i} $(i = 1, 2)$ on the surface S^* by putting

$$\bar{u}^k = h^k(u^{*i}) \text{ for } k = 1, 2 \tag{3.108}$$

with the functions h^k from (3.107). Then the map from S to S^* has the simple form

$$u^{*k} = u^k \text{ for } k = 1, 2.$$

So the values of the parameters of each image point are equal to those of corresponding pre–image point. In such case, we say that S and S^* have the same parameter systems.

Example **3.9.3 (Map of a circular cylinder onto a part of the plane).**
We consider the circular cylinder $S = Cyl$ with a parametric representation

$$\vec{x}(u^i) = \{r\cos u^2, r\sin u^2, u^1\} \ \left((u^1, u^2) \in D = \mathbb{R} \times (0, 2\pi)\right),$$

and the strip S^* of the plane with a parametric representation

$$\vec{\bar{x}}(\bar{u}^i) = \{\bar{u}^1, \bar{u}^2, 0\} \ \left((\bar{u}^1, \bar{u}^2) \in \tilde{D} = \mathbb{R} \times (0, 2\pi)\right).$$

Furthermore, let the functions h^k be defined on D by

$$\bar{u}^1 = h^1(u^i) = u^1 \ and \ \bar{u}^2 = h^2(u^i) = ru^2.$$

The corresponding map from S to S^* obviously is one-to-one with $h^k \in C^\infty(D)$ for $k = 1, 2$ and Jacobian satisfying

$$\frac{\partial(\bar{u}^i)}{\partial(u^k)} = \begin{vmatrix} 1 & 0 \\ 0 & r \end{vmatrix} = r \neq 0 \ for \ all \ (u^1, u^2) \in D.$$

Thus the map is admissible.
We introduce new parameters u^{*i} $(i = 1, 2)$ on S^* by

$$\bar{u}^1 = u^{*1} \ and \ \bar{u}^2 = ru^{*2}.$$

Then

$$\vec{x}^*(u^{*i}) = \{u^{*1}, ru^{*2}, 0\} \ for \ all \ (u^{*i}, u^{*2}) \in D^* = D$$

is a new parametric representation for S^*.
Finally, the part of the helix on S with

$$u^1(t) = ct \ and \ u^2(t) = t \ (t \in (0, \alpha) \subset (0, 2\pi)),$$

(where $c \in \mathbb{R}$ is a given constant) between the points $P_0 = (1, 0, 0)$ and $P_\alpha = (r\cos\alpha, r\sin\alpha, c\alpha\}$ is mapped onto the straight line segment on S^* between the points $P_0^* = (0, 0, 0)$ and $P_\alpha^* = (c\alpha, r\alpha, 0)$.

First we consider *conformal* or *angle preserving maps*.

Definition 3.9.4. An admissible map of a surface S onto a surface S^* is said to be *conformal* or *angle preserving*, if for each pair of sections of curves on S that intersect at a point $P \in S$ the angle of intersection of the image curves at $P^* \in S^*$ is the same as the angle of intersection of the original curves at P.

The next result is a simple characterization of conformal maps.

Theorem 3.9.5. *A map of a surface S onto a surface S^* is conformal if and only if, with respect to the same parameter systems on S and S^*, the first fundamental coefficients g_{ik} of S and g_{ik}^* of S^* are proportional at corresponding points, that is, if*

$$g_{ik}^* = \eta(u^j)g_{ik} \ (\eta > 0; \ i, k = 1, 2). \tag{3.109}$$

Proof.

(i) First we show the sufficiency of the condition in (3.109).
We assume that the condition in (3.109) is satisfied. Let γ and $\bar{\gamma}$ be curves on S given by $u^i(s)$ and $\bar{u}^i(s)$ $(i = 1, 2)$ and γ^* and $\bar{\gamma}^*$ given by $u^{*i}(s)$ and $\bar{u}^{*i}(s)$ $(i = 1, 2)$ their images on S^*. Then the angles α and α^* between the curves at their points of intersection on S and S^* are given by

$$\cos\alpha = \frac{g_{ik}\dot{u}^i\dot{\bar{u}}^k}{\sqrt{g_{ik}\dot{u}^i\dot{u}^k} \cdot \sqrt{g_{ik}\dot{\bar{u}}^i\dot{\bar{u}}^k}} \text{ and } \cos\alpha^* = \frac{g^*_{ik}\dot{u}^{*i}\dot{\bar{u}}^{*k}}{\sqrt{g^*_{ik}\dot{u}^{*i}\dot{u}^{*k}} \cdot \sqrt{g^*_{ik}\dot{\bar{u}}^{*i}\dot{\bar{u}}^{*k}}}.$$

It follows from (3.109) that $\cos\alpha = \cos\alpha^*$.

(ii) Now we show the necessity of the condition in (3.109).
We assume that a conformal map from S onto S^* is given and that both surfaces have the same parameter system, that is, the map is given by

$$u^{*k} = u^k \ (k = 1, 2).$$

Let P be an arbitrary point on a segment γ_1 of a curve on S; we assume that $\{\xi^1_{(1)}, \xi^2_{(1)}\}$ is the direction of the tangent to γ_1 at P. Then the image point P^* of P on S^* has the same parameter values as P on S, the image γ^*_1 on S^* of γ_1 on S is given by the same parametric representation as γ_1 on S, and, by the conformity of the map, the direction of the tangent of γ^*_1 at P^* is also given by $\{\xi^1_{(1)}, \xi^2_{(1)}\}$.
Let γ_2 be a segment of another curve on S, which intersects γ at a point P, with $\{\xi^1_{(2)}, \xi^2_{(2)}\}$ as the direction of its tangent at P. The angle α between γ_1 and γ_2 at P is given by

$$\cos\alpha = \frac{g_{ik}\xi^i_{(1)}\xi^k_{(2)}}{\sqrt{g_{ik}\xi^i_{(1)}\xi^k_{(1)}} \cdot \sqrt{g_{ik}\xi^i_{(2)}\xi^k_{(2)}}}.$$

If we write $\vec{t}_{(1)}$ and $\vec{t}_{(2)}$ for the tangent vectors to γ_1 and γ_2 at P, then we obtain

$$\cos\alpha = \frac{\vec{t}_{(1)} \bullet \vec{t}_{(2)}}{\|\vec{t}_{(1)}\| \cdot \|\vec{t}_{(2)}\|}$$

and

$$\sin^2\alpha = \frac{\|\vec{t}_{(1)}\|^2 \cdot \|\vec{t}_{(1)}\|^2 - \left(\vec{t}_{(1)} \bullet \vec{t}_{(2)}\right)^2}{\|\vec{t}_{(1)}\|^2 \cdot \|\vec{t}_{(1)}\|^2} = \frac{\left(\vec{t}_{(1)} \times \vec{t}_{(2)}\right)^2}{\|\vec{t}_{(1)}\|^2 \cdot \|\vec{t}_{(1)}\|^2}$$

$$= \frac{g\left(\xi^1_{(1)}\xi^2_{(2)} - \xi^2_{(1)}\xi^1_{(2)}\right)^2}{g_{ik}\xi^i_{(1)}\xi^k_{(1)} \cdot g_{ik}\xi^i_{(2)}\xi^k_{(2)}},$$

hence

$$\sin\alpha = \sqrt{g} \cdot \frac{\left|\xi^1_{(1)}\xi^2_{(2)} - \xi^2_{(1)}\xi^1_{(2)}\right|}{\sqrt{g_{ik}\xi^i_{(1)}\xi^k_{(1)}} \cdot \sqrt{g_{ik}\xi^i_{(2)}\xi^k_{(2)}}}. \tag{3.110}$$

Analogously, we obtain for the angle α^* between the image curves γ_1^* and γ_2^* at the image point P^* of P

$$\sin \alpha^* = \sqrt{g^*} \cdot \frac{\left| \xi^1_{(1)} \xi^2_{(2)} - \xi^2_{(1)} \xi^1_{(2)} \right|}{\sqrt{g^*_{ik} \xi^i_{(1)} \xi^k_{(1)}} \cdot \sqrt{g^*_{ik} \xi^i_{(2)} \xi^k_{(2)}}}. \tag{3.111}$$

Since $\alpha = \alpha^*$, (3.110) and (3.111) imply

$$\frac{\sqrt{g}}{\sqrt{g^*}} = \frac{\sqrt{g^*_{ik} \xi^i_{(1)} \xi^k_{(1)}} \cdot \sqrt{g^*_{ik} \xi^i_{(2)} \xi^k_{(2)}}}{\sqrt{g_{ik} \xi^i_{(1)} \xi^k_{(1)}} \cdot \sqrt{g_{ik} \xi^i_{(2)} \xi^k_{(2)}}}.$$

We choose γ_2 as the u^1–line. Then γ_2^* is the u^{*1}–line and we have

$$\frac{\sqrt{g}}{\sqrt{g^*}} = \frac{\sqrt{g^*_{ik} \xi^i_{(1)} \xi^k_{(1)}}}{\sqrt{g_{ik} \xi^i_{(1)} \xi^k_{(1)}}} \cdot \frac{\sqrt{g^*_{11}}}{\sqrt{g_{11}}}$$

or

$$\frac{\sqrt{g}}{\sqrt{g^*}} \cdot \frac{\sqrt{g_{11}}}{\sqrt{g^*_{11}}} = \frac{\sqrt{g^*_{ik} \xi^i_{(1)} \xi^k_{(1)}}}{\sqrt{g_{ik} \xi^i_{(1)} \xi^k_{(1)}}} = \sqrt{\eta(u^1, u^2)}, \tag{3.112}$$

where η only depends on P, but not on the considered direction of the tangents. Since (3.112) has to hold for all directions $\{\xi^1_{(1)}, \xi^2_{(1)}\}$, (3.109) follows.

$\square$

Example 3.9.6. *The map of Example 3.9.3 is conformal.*

Proof. It follows from Example 3.9.3 that

$$\vec{x}_1(u^i) = \{0, 0, 1\}, \ \vec{x}_2(u^i) = r\{- \sin u^2, \cos u^2, 0\}, \ g_{11} = 1, \ g_{12} = 0, \ g_{22} = r^2,$$

$$\vec{x}_1^*(u^{*i}) = \{1, 0, 0\}, \ \vec{x}_2^*(u^{*i}) = \{0, r, 0\}, \ g^*_{11} = 1, \ g^*_{12} = 0 \text{ and } g^*_{22} = r^2.$$

Thus the identities in (3.109) are satisfied with $\eta(u^i) = 1 > 0$, and consequently the map is conformal by Theorem 3.9.5. $\square$

Visualization 3.9.7 (The Riemann sphere and stereographic projection).
The Riemann sphere and stereographic projection play an important role in complex analysis concerning the extended complex plane $\overline{\mathbb{C}}$.
We embed the complex plane $\mathbb{C}$ in three-dimensional space $\mathbb{R}^3$ in which we introduce a Cartesian coordinate system with coordinates ξ, η and ζ in such a way that the ξ– and η–axes coincide with the x– and y–axes, respectively, of $\mathbb{C}$. The sphere in $\mathbb{R}^3$

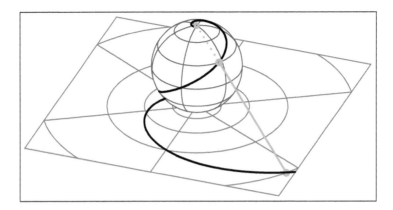

Figure 3.28 The stereographic projection

with radius $1/2$ and its centre in the point $(0, 0, 1/2)$ is referred to as the Riemann sphere; *it is given by*

$$S = S_{1/2}\left(\left(0, 0, \frac{1}{2}\right)\right) = \left\{(\xi, \eta, \zeta) \in I\!\!R^3 : \xi^2 + \eta^2 + \left(\zeta - \frac{1}{2}\right)^2 = \frac{1}{4}\right\}.$$

The point $N = (0, 0, 1) \in S$ is referred to as the north pole. *The stereographic projection from the complex plane to the Riemann sphere is defined as follows: Given a point $z \in \mathbb{C}$, we determine the straight line through the points z and N which intersects the Riemann sphere in one and only one point (other than the north pole) which we denote by $P(z)$ (Figure 3.28). In this way, we obtain a one–to–one correspondence between the points of the complex plane and the points of the Riemann sphere minus the north pole. We write ∞ for the element which is formally assigned to the north pole and obtain the* extended complex plane $\tilde{\mathbb{C}} = \mathbb{C} \cup \{\infty\}$.
(a) The stereographic projection is conformal.
(b) Loxodromes on the Riemann sphere correspond to logarithmic spirals in the xy–plane.

Proof. (a) Let S be the part of the Riemann sphere given by the parametric representation

$$\vec{x}(u^i) = \frac{1}{2} \cdot \left\{\cos u^1 \cos u^2, \cos u^1 \sin u^2, 1 + \sin u^1\right\}$$

for $(u^1, u^2) \in D = (-\pi/2, \pi/2) \times (0, \infty)$.
If $(\xi, \eta, \zeta) \in S$ is given and (x^*, y^*) is the corresponding point under the stereographic projection in complex plane, then

$$\xi = \frac{1}{2} \cdot \cos u^1 \cos u^2, \ \eta = \frac{1}{2} \cdot \cos u^1 \sin u^2 \text{ and } \zeta = \frac{1}{2}(1 + \sin u^1)$$

and we obtain (Figure 3.29)

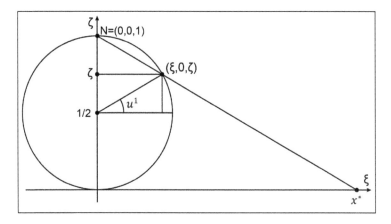

Figure 3.29 The principle of the stereographic projection

$$x^* = \frac{\xi}{1 - \zeta} = \frac{\cos u^1 \cos u^2}{1 - \sin u^1} \text{ and } y^* = \frac{\eta}{1 - \zeta} = \frac{\cos u^1 \sin u^2}{1 - \sin u^1}.$$

Let S^* denote the image of S under the stereographic projection. Then S^* has a parametric representation

$$\vec{x}^*(u^{*i}) = \frac{\cos u^{*1}}{1 - \sin u^{*1}} \cdot \left\{ \cos u^{*2}, \sin u^{*2}, 0 \right\} \text{ for } (u^{*1}, u^{*2}) \in D.$$

We obtain

$$\vec{x}_1 = \frac{1}{2} \cdot \left\{ -\sin u^1 \cos u^2, -\sin u^1 \sin u^2, \cos u^1 \right\},$$

$$\vec{x}_2 = \frac{1}{2} \cdot \left\{ -\cos u^1 \sin u^2, \cos u^1 \cos u^2, 0 \right\},$$

$$g_{11} = \frac{1}{4}, \ g_{12} = 0, \ g_{22} = \frac{1}{4} \cdot \cos^2 u^1,$$

$$\vec{x}_1^* = \frac{-\sin u^{*1}(1 - \sin u^{*1}) + \cos^2 u^{*1}}{(1 - \sin u^{*1})^2} \cdot \{\cos u^{*2}, \sin u^{*2}, 0\}$$

$$= \frac{1}{1 - \sin u^{*1}} \cdot \{\cos u^{*2}, \sin u^{*2}, 0\},$$

$$\vec{x}_2^* = \frac{\cos u^{*1}}{1 - \sin u^{*1}} \cdot \{-\sin u^{*2}, \cos u^{*2}, 0\},$$

$$\overset{*}{g}_{11} = \frac{1}{(1 - \sin u^{*1})^2}, \ \overset{*}{g}_{12} = 0 \text{ and } \overset{*}{g}_{22} = \frac{\cos^2 u^{*2}}{(1 - \sin u^{*1})^2}.$$

We write $u^{*i} = u^i$ for $i = 1, 2$. Then identities in (3.109) are satisfied with

$$\eta(u^i) = \frac{4}{(1 - \sin u^{*1})^2} > 0 \text{ on } D$$

and so the stereographic projection is conformal by Theorem 3.9.5.

(b) Any loxodrome γ on the Riemann sphere S intersects the meridians of S at a constant angle. Obviously the images γ^* of the meridians under the stereographic projection are the rays in the complex plane S^* that originate in the origin. Since the stereographic projection is conformal by Part (a), γ^* intersects those rays at a constant angle, and consequently is a logarithmic spiral by Visualization 2.3.12 (a). Figure 3.30 shows the Riemann sphere, a loxodrome on it and the surface generated by moving a constant distance along the rays of the stereographic projection of the points of the loxodrome.

□

One of the most popular maps from the sphere into the plane in cartography is the *Mercator projection*.

Example 3.9.8 (Mercator Projection). *Let a sphere S be given by a parametric representation*

$$\vec{x}(u^i) = \{r\cos u^2 \cos u^1, r\cos u^2 \sin u^1, r\sin u^2\}\ \left((u^1, u^2) \in (0, 2\pi) \times \left(\frac{\pi}{2}, \frac{\pi}{2}\right)\right).$$
(3.113)

(We note that u^1 and u^2 are interchanged in the usual parametric representation of a sphere.)
We use the transformations

$$x_1^* = u^1 \text{ and } x_2^* = \log\left(\tan\left(\frac{u^2}{2} + \frac{\pi}{4}\right)\right) \text{ for } |u^2| < \frac{\pi}{2}$$
(3.114)

and introduce x_1^* and x_2^* as Cartesian coordinates in the plane S^*. Then the Mercator projection of the sphere S into the plane S^* is given by (3.114).
(a) The Mercator projection is conformal.
(b) The images of the meridians and parallels, that is, the u^2–lines and u^1–lines on S are straight lines in the plane S^*.
(c) The pre–images of arbitrary straight lines in the plane S^* are loxodromes on S.

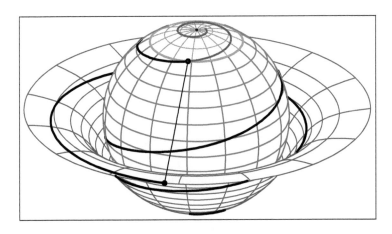

Figure 3.30 Intermediate image of the stereographic projection

Proof. We choose $u^{*i} = u^i$ for the parameters in the plane S^*, and obtain the following parametric representation for S^*

$$\vec{x}^*(u^{*i}) = \left\{ u^{*1}, \log\left(\tan\left(\frac{u^{*2}}{2} + \frac{\pi}{4}\right)\right) \right\}.$$

(a)   The first fundamental coefficients of S are

$$g_{11}(u^i) = r^2 \cdot \cos^2 u^2, \ \ g_{12}(u^i) = 0 \text{ and } g_{22}(u^i) = r^2.$$

Furthermore,

$$\vec{x}_1^*(u^{*i}) = \{1, 0\}$$

and

$$\vec{x}_2^*(u^{*i}) = \left\{ 0, \frac{1}{2} \cdot \frac{1}{\tan\left(\dfrac{u^{*2}}{2} + \dfrac{\pi}{4}\right)} \cdot \frac{1}{\cos^2\left(\dfrac{u^{*2}}{2} + \dfrac{\pi}{4}\right)} \right\}$$

$$= \left\{ 0, \frac{1}{2 \cdot \sin\left(\dfrac{u^{*2}}{2} + \dfrac{\pi}{4}\right) \cdot \cos\left(\dfrac{u^{*2}}{2} + \dfrac{\pi}{4}\right)} \right\}$$

$$= \left\{ 0, \frac{1}{\sin\left(u^{*2} + \dfrac{\pi}{2}\right)} \right\} = \left\{ 0, \frac{1}{\cos u^{*2}} \right\},$$

yield the first fundamental coefficients g_{ik}^* for S^*

$$g_{11}^*(u^{*i}) = 1, \ \ g_{12}^*(u^{*i}) = 0 \text{ and } g_{22}^*(u^{*i}) = \frac{1}{\cos^2 u^{*2}}.$$

Since $u^i = u^{*i}$ for $i = 1, 2$, we obtain

$$g_{ik}^*(u^{*j}) = \eta(u^j) g_{ik}(u^j) \text{ for } i, k = 1, 2 \text{ where } \eta(u^j) = \frac{1}{r^2 \cdot \cos^2 u^2},$$

hence the Mercator projection is conformal by (3.109) in Theorem 3.9.5.

(b)   It is an immediate consequence of the transformation formulae in (3.114) that the images of the meridians and parallels are the straight lines with

$$x_1^* = u_0^1, \text{ and } x_2^* = \log\left(\tan\left(\frac{u_0^2}{2} + \frac{\pi}{2}\right)\right).$$

(c) An arbitrary straight line γ in the plane E is given by the equation

$$ax_1^* + bx_2^* + c = 0.$$

It follows from the transformation formulae in (3.114) that the inverse image of γ under the Mercator projection is given by the equation

$$au^1 + b\log\left(\tan\left(\frac{u^2}{2} + \frac{\pi}{4}\right)\right) + c = 0;$$

thus the inverse image of γ under the Mercator projection is a loxodrome on the sphere.

□

Parameters of a surface for which the first fundamental coefficients satisfy the condition in (3.109) are of special interest.

Definition 3.9.9. Let u^1 and u^2 be Cartesian parameters in a plane Pl and T be a map of a surface S into the plane Pl. Then we can introduce u^1 and u^2 as parameters for the surface S. If, in particular, the map T is conformal, then, by (3.109) in Theorem 3.9.5, the first fundamental form of S has the form

$$(ds)^2 = \eta(u^i)\left((du^1)^2 + (du^2)^2\right)^2. \tag{3.115}$$

Admissible parameters u^1 and u^2 for which the first fundamental form satisfies (3.115) are called *isothermal*.

Example **3.9.10.** *The parameters of Poincaré's half–plane of Definition 3.2.7 are isothermal.*

Visualization 3.9.11. *Let S be a surface of revolution given by a parametric representation*

$$\vec{x}(u^i) = \{r(u^1)\cos u^2, r(u^1)\sin u^2, h(u^1)\}\ \left((u^1, u^2) \in D = I \times (0, 2\pi)\right)$$

with $r(t) > 0$ and $f(t) = (r'(t))^2 + (h'(t))^2 > 0$ for all $t \in I$. We put

$$\phi(t) = \int_{t_0}^{t} \frac{\sqrt{f(\tau)}}{r(\tau)}\,d\tau \text{ for } t_0, t \in I.$$

Then $\phi'(t) > 0$ on I, and so the inverse ψ of ϕ exists on $\phi(I)$ and

$$\psi'(\phi(t)) = \frac{1}{\phi'(t)} \text{ for all } t \in I.$$

We introduce new parameters u^{*i} $(i = 1, 2)$ by

$$u^1 = u^1(u^{*i}) = \psi\left(\frac{u^{*1} - u^{*2}}{\sqrt{2}}\right) \text{ and } u^2 = \frac{u^{*1} + u^{*2}}{\sqrt{2}}.$$

Then we have

$$\frac{\partial u^1}{\partial u^{*1}} = \frac{1}{\sqrt{2}} \cdot \frac{1}{\phi'(u^1(u^{*i}))} = -\frac{\partial u^1}{\partial u^{*2}} \text{ and } \frac{\partial u^2}{\partial u^{*1}} = \frac{\partial u^2}{\partial u^{*2}} = \frac{1}{\sqrt{2}}.$$

It is easy to see that the transformation is admissible and the transformation formulae (2.18) in Remark 2.3.3 (a) for the first fundamental coefficients

$$g_{jk}^* = g_{lm} \frac{\partial u^l}{\partial u^{*j}} \frac{\partial u^m}{\partial u^{*k}} \text{ for } j, k = 1, 2$$

yield

$$g_{11}^* = g_{11} \left(\frac{\partial u^1}{\partial u^{*1}}\right)^2 + g_{22} \left(\frac{\partial u^2}{\partial u^{*1}}\right)^2 = g_{11} \left(\frac{\partial u^1}{\partial u^{*2}}\right)^2 + g_{22} \left(\frac{\partial u^2}{\partial u^{*2}}\right)^2$$

$$= g_{22}^* = \frac{1}{2} \cdot \frac{f(u^1(u^{*1}))}{(\phi'(u^1(u^{*1})))^2} + \frac{1}{2} \cdot r^2(u^1(u^{*1}))$$

$$= \frac{f(u^1(u^{*1}))}{2} \cdot \frac{r^2(u^1(u^{*1}))}{f(u^1(u^{*1}))} + \frac{1}{2} \cdot r^2(u^1(u^{*1})) = r^2(u^1(u^{*1}))$$

and

$$g_{12}^* = g_{11} \frac{\partial u^1}{\partial u^{*1}} \frac{\partial u^1}{\partial u^{*2}} + g_{22} \frac{\partial u^2}{\partial u^{*1}} \frac{\partial u^2}{\partial u^{*2}}$$

$$= -g_{11} \left(\frac{\partial u^1}{\partial u^{*1}}\right)^2 + g_{22} \left(\frac{\partial u^2}{\partial u^{*2}}\right)^2 = 0.$$

Thus u^{*1} and u^{*2} are isothermal parameters for S (Figure 3.31).

The following important general result holds; it is also needed in the proof of the Gauss–Bonnet theorem, Theorem 3.10.1.

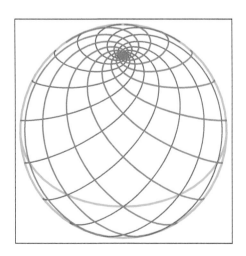
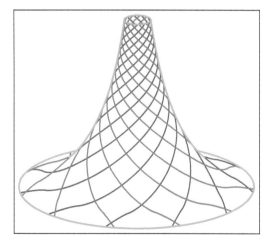

Figure 3.31 Isothermal parameters on a sphere and a pseudo-sphere

Theorem 3.9.12. *Any simply connected part S of a surface of class $r \geq 3$ can be mapped conformally into the plane.*

Proof. Let u^{*k} $(k = 1, 2)$ denote the parameters of S. We show that we can introduce isothermal parameters u^k $(k = 1, 2)$ on S by proving the existence of an admissible parameter transformation

$$u^{\alpha} = u^{\alpha}(u^{*1}, u^{*2}) \ (\alpha = 1, 2) \tag{3.116}$$

such that the first fundamental coefficients $g_{\alpha\beta}$ and g_{ik}^* satisfy the relation in (3.109). Since the parameter transformation is admissible, the inverse transformation also exists. It follows that the transformation (3.116) must satisfy

$$\eta \delta_{\mu\nu} = g_{\alpha\beta}^* \frac{\partial u^{*\alpha}}{\partial u^{\mu}} \frac{\partial u^{*\beta}}{\partial u^{\nu}} \ \text{ and } \ \frac{\delta^{\mu\nu}}{\eta} = g^{*\alpha\beta} \frac{\partial u^{\mu}}{\partial u^{*\alpha}} \frac{\partial u^{\nu}}{\partial u^{*\beta}} \ \ (\eta > 0; \mu, \nu = 1, 2).$$

If $\mu = 1$ and $\nu = 2$, then this yields

$$g^{*\alpha\beta} \frac{\partial u^1}{\partial u^{*\alpha}} \frac{\partial u^2}{\partial u^{*\beta}} = 0, \tag{3.117}$$

and if $\mu = \nu = 1$ or $\mu = \nu = 2$, then we obtain

$$\frac{1}{\eta} = g^{*\alpha\beta} \frac{\partial u^1}{\partial u^{*\alpha}} \frac{\partial u^1}{\partial u^{*\beta}} \ \text{ or } \ \frac{1}{\eta} = g^{*\alpha\beta} \frac{\partial u^2}{\partial u^{*\alpha}} \frac{\partial u^2}{\partial u^{*\beta}},$$

whence

$$g^{*\alpha\beta} \frac{\partial u^1}{\partial u^{*\alpha}} \frac{\partial u^1}{\partial u^{*\beta}} = g^{*\sigma\tau} \frac{\partial u^2}{\partial u^{*\sigma}} \frac{\partial u^2}{\partial u^{*\tau}}. \tag{3.118}$$

We put $\epsilon^{\mu\nu} = \epsilon_{\alpha\beta} g^{\mu\alpha} g^{\nu\beta}$ $(\mu, \nu = 1, 2)$ where $\epsilon_{11} = \epsilon_{22} = 0$ and $\epsilon_{12} = -\epsilon_{21} = \sqrt{g}$. Then it is easy to see that $\epsilon^{11} = \epsilon^{22} = 0$, $\epsilon^{12} = -\epsilon^{21} = 1/\sqrt{g}$, $\epsilon^{\alpha\gamma} \epsilon_{\beta\gamma} = \delta_{\beta}^{\alpha}$ and $\epsilon^{\alpha\gamma} \epsilon_{\gamma\beta} = -\delta_{\beta}^{\alpha}$ $(\alpha, \beta = 1, 2)$. Now we show:

(A) If the functions μ^{α} $(\alpha = 1, 2)$ in (3.116) satisfy

$$\frac{\partial u^1}{\partial u^{*\alpha}} = g_{\alpha\kappa}^* \epsilon^{*\kappa\lambda} \frac{\partial u^2}{\partial u^{*\lambda}} \ \text{ for } \alpha = 1, 2, \tag{3.119}$$

then they satisfy (3.117) and (3.118).
First $\epsilon^{\mu\nu} = \epsilon_{\alpha\beta} g^{\mu\alpha} g^{\nu\beta}$ for $\mu, \nu = 1, 2$ implies

$$g_{\mu\tau} g_{\nu\sigma} \epsilon^{\mu\nu} = \epsilon_{\alpha\beta} g_{\mu\tau} g^{\mu\alpha} g_{\nu\sigma} g^{\nu\beta} = \epsilon_{\alpha\beta} \delta_{\tau}^{\alpha} \delta_{\sigma}^{\beta} = \epsilon_{\tau\sigma} \ \text{ for } \tau, \sigma = 1, 2. \tag{3.120}$$

Since $\epsilon^{\alpha\gamma} \epsilon_{\beta\gamma} = \delta_{\beta}^{\alpha}$ for $\alpha, \beta = 1, 2$, we have from (3.120)

$$\begin{aligned} \epsilon^{\alpha\gamma} \epsilon_{\beta\gamma} g^{\beta\kappa} &= \delta_{\beta}^{\alpha} g^{\beta\kappa} = g^{\alpha\kappa} = \epsilon^{\alpha\gamma} g_{\mu\beta} g_{\nu\gamma} \epsilon^{\mu\nu} g^{\beta\kappa} \\ &= \epsilon^{\alpha\gamma} \epsilon^{\mu\nu} g_{\mu\beta} g^{\beta\kappa} g_{\nu\gamma} = \epsilon^{\alpha\gamma} \epsilon^{\mu\nu} \delta_{\mu}^{\kappa} g_{\nu\gamma} \\ &= \epsilon^{\alpha\gamma} \epsilon^{\kappa\nu} g_{\nu\gamma} \ \text{ for } \alpha, \kappa = 1, 2, \end{aligned} \tag{3.121}$$

and so by (3.119)

$$g^{*\alpha\beta} \frac{\partial u^1}{\partial u^{*\alpha}} \frac{\partial u^2}{\partial u^{*\beta}} = g^{*\alpha\beta} g^*_{\alpha\kappa} \epsilon^{*\kappa\lambda} \frac{\partial u^2}{\partial u^{*\lambda}} \frac{\partial u^2}{\partial u^{*\beta}} = \delta^{*\beta}_\kappa \epsilon^{*\kappa\lambda} \frac{\partial u^2}{\partial u^{*\lambda}} \frac{\partial u^2}{\partial u^{*\beta}}$$

$$= \epsilon^{*\beta\lambda} \frac{\partial u^2}{\partial u^{*\lambda}} \frac{\partial u^2}{\partial u^{*\beta}} = \epsilon^{*12} \frac{\partial u^2}{\partial u^{*2}} \frac{\partial u^2}{\partial u^{*1}} - \epsilon^{*21} \frac{\partial u^2}{\partial u^{*1}} \frac{\partial u^2}{\partial u^{*2}} = 0.$$

Hence (3.117) holds. Furthermore, we have by (3.119) and (3.121)

$$g^{*\alpha\beta} \frac{\partial u^1}{\partial u^{*\alpha}} \frac{\partial u^1}{\partial u^{*\beta}} = g^{*\alpha\beta} g^*_{\alpha\kappa} \epsilon^{*\kappa\lambda} g^*_{\beta\delta} \epsilon^{*\delta\mu} \frac{\partial u^2}{\partial u^{*\lambda}} \frac{\partial u^2}{\partial u^{*\mu}} = \delta^{*\beta}_\kappa \epsilon^{*\kappa\lambda} g^*_{\beta\delta} \epsilon^{*\delta\mu} \frac{\partial u^2}{\partial u^{*\lambda}} \frac{\partial u^2}{\partial u^{*\mu}}$$

$$= \epsilon^{*\beta\lambda} g^*_{\beta\delta} \epsilon^{*\delta\mu} \frac{\partial u^2}{\partial u^{*\lambda}} \frac{\partial u^2}{\partial u^{*\mu}} = g^{*\lambda\mu} \frac{\partial u^2}{\partial u^{*\lambda}} \frac{\partial u^2}{\partial u^{*\mu}}.$$

So (3.118) is also satisfied. This completes the proof of the statement in (A).

The conditions of integrability for (3.119) are

$$\frac{\partial^2 u^1}{\partial u^{*1} \partial u^{*2}} = \frac{\partial^2 u^1}{\partial u^{*2} \partial u^{*1}} \text{ or equivalently } \sqrt{g^*} \epsilon^{*\alpha\beta} \frac{\partial^2 u^1}{\partial u^{*\alpha} \partial u^{*\beta}} = 0.$$

Now (3.119), $\sqrt{g^*} \epsilon^{*\alpha\beta} = \pm 1$ or $\sqrt{g^*} \epsilon^{*\alpha\beta} = 0$ and (3.121) together imply

$$\sqrt{g^*} \epsilon^{*\alpha\beta} \frac{\partial^2 u^1}{\partial u^{*\alpha} \partial u^{*\beta}} = \sqrt{g^*} \epsilon^{*\alpha\beta} \frac{\partial}{\partial u^{*\beta}} \left(\frac{\partial u^1}{\partial u^{*\alpha}} \right) = \sqrt{g^*} \epsilon^{*\alpha\beta} \frac{\partial}{\partial u^{*\beta}} \left(g^*_{\alpha\kappa} \epsilon^{*\kappa\lambda} \frac{\partial u^2}{\partial u^{*\lambda}} \right)$$

$$= \frac{\partial}{\partial u^{*\beta}} \left(\sqrt{g^*} \epsilon^{*\alpha\beta} g^*_{\alpha\kappa} \epsilon^{*\kappa\lambda} \frac{\partial u^2}{\partial u^{*\lambda}} \right) = \frac{\partial}{\partial u^{*\beta}} \left(\sqrt{g^*} g^{*\beta\lambda} \frac{\partial u^2}{\partial u^{*\lambda}} \right) = 0.$$

$$(3.122)$$

This is a linear partial differential equation with continuously differentiable coefficients. It has a non–constant solution in any sufficiently small part S_1 of S, as can be shown by the method of successive approximations.

A similar differential equation follows when we solve (3.119) for $\partial u^2/\partial u^{*\lambda}$.

From (3.120) and (3.119), we obtain

$$\frac{\partial u^2}{\partial u^{*\sigma}} = \delta^{*\beta}_\sigma \frac{\partial u^2}{\partial u^{*\beta}} = \epsilon^{*\beta\tau} \epsilon^*_{\sigma\tau} \frac{\partial u^2}{\partial u^{*\beta}} = \epsilon^{*\beta\tau} g^*_{\mu\sigma} g^*_{\nu\tau} \epsilon^{*\mu\nu} \frac{\partial u^2}{\partial u^{*\beta}}$$

$$= -g^*_{\mu\sigma} \epsilon^{*\mu\nu} g^*_{\nu\tau} \epsilon^{*\tau\beta} \frac{\partial u^2}{\partial u^{*\beta}} = -g^*_{\mu\sigma} \epsilon^{*\mu\nu} \frac{\partial u^1}{\partial u^{*\nu}},$$

that is,

$$\frac{\partial u^2}{\partial u^{*\sigma}} = -g^*_{\sigma\beta} \epsilon^{*\beta\alpha} \frac{\partial u^1}{\partial u^{*\alpha}} \text{ for } \sigma = 1, 2. \tag{3.123}$$

The equations in (3.119) and (3.123) are generalizations of the well–known Cauchy–Riemann differential equations in the theory of complex functions.

If u^{*1} and u^{*2} are isothermal parameters, then $g^*_{\beta\sigma} = \eta \delta^*_{\beta\sigma}$ for $\beta, \sigma = 1, 2$ and $g^* = \eta^2$, hence by (3.119)

$$\frac{\partial u^1}{\partial u^{*1}} = g^*_{1\kappa} \epsilon^{*\kappa\lambda} \frac{\partial u^2}{\partial u^{*\lambda}} = \eta \delta^*_{1\kappa} \epsilon^{*\kappa\lambda} \frac{\partial u^2}{\partial u^{*\lambda}} = \eta \epsilon^{*1\lambda} \frac{\partial u^2}{\partial u^{*\lambda}} = \frac{\eta}{\sqrt{g^*}} \frac{\partial u^2}{\partial u^{*2}} = \frac{\partial u^2}{\partial u^{*2}},$$

and by (3.123)

$$\frac{\partial u^2}{\partial u^{*1}} = -g_{1\beta}^* \epsilon^{*\beta\alpha} \frac{\partial u^1}{\partial u^{*\alpha}} = -\eta \delta_{1\beta}^* \epsilon^{*\beta\alpha} \frac{\partial u^1}{\partial u^{*\alpha}} = -\eta \epsilon^{*1\alpha} \frac{\partial u^1}{\partial u^{*\alpha}} = -\frac{\eta}{\sqrt{g^*}} \frac{\partial u^1}{\partial u^{*2}} = -\frac{\partial u^1}{\partial u^{*2}}.$$

If $u^2(u^{*1}, u^{*2}) = const$ is a family of curves satisfying (3.122), then we obtain the second family $u^1 = const$ of isothermal parameters u^1 and u^2 on S_1 from (3.119). It follows from (3.109) that

$$\eta^2 = g = g^* \left(\frac{\partial(u^{*1}, u^{*2})}{\partial(u^1, u^2)} \right)^2.$$

We show that the Jacobian of the transformation (3.116) does not vanish. From (3.119), we obtain

$$\frac{\partial(u^1, u^2)}{\partial(u^{*1}, u^{*2})} = \frac{\partial u^1}{\partial u^{*1}} \frac{\partial u^2}{\partial u^{*2}} - \frac{\partial u^1}{\partial u^{*2}} \frac{\partial u^2}{\partial u^{*1}} = g_{1\kappa}^* \epsilon^{*\kappa\lambda} \frac{\partial u^2}{\partial u^{*\lambda}} \frac{\partial u^2}{\partial u^{*2}} - g_{2\kappa}^* \epsilon^{*\kappa\lambda} \frac{\partial u^2}{\partial u^{*\kappa}} \frac{\partial u^2}{\partial u^{*1}}$$

$$= g_{11}^* \epsilon^{*12} \left(\frac{\partial u^2}{\partial u^{*2}} \right)^2 + g_{12}^* \epsilon^{*21} \frac{\partial u^2}{\partial u^{*1}} \frac{\partial u^2}{\partial u^{*2}} - g_{21}^* \epsilon^{*12} \frac{\partial u^2}{\partial u^{*2}} \frac{\partial u^2}{\partial u^{*1}}$$

$$- g_{22}^* \epsilon^{*21} \left(\frac{\partial u^2}{\partial u^{*1}} \right)^2$$

$$= \frac{1}{\sqrt{g^*}} \left(g_{22}^* \left(\frac{\partial u^2}{\partial u^{*1}} \right)^2 - 2g_{12}^* \frac{\partial u^2}{\partial u^{*1}} \frac{\partial u^2}{\partial u^{*2}} + g_{11}^* \left(\frac{\partial u^2}{\partial u^{*2}} \right)^2 \right).$$

This quadratic form is positive definite, it can only vanish when both partial derivatives $\partial u^2/\partial u^{*\alpha}$ ($\alpha = 1, 2$) vanish. This cannot happen for an admissible solution of (3.122).

Therefore, we can find a conformal map of a part S_1 of S into the plane. Similarly a local conformal map can be found for any other sufficiently small part of S. All these local maps have to be joined to a global conformal map of S into the plane. This can be achieved by a principle from complex analysis. □

Now we consider *isometric* or *length preserving maps.*

Definition 3.9.13. An admissible map from a surface S to a surface S^* is said to be *isometric* or *length preserving* if the length of any curve γ on S is equal to the length of its image γ^* on S^*.

The following result gives a characterization of isometric maps.

Theorem 3.9.14. *An admissible map from a surface S onto a surface S^* is isometric if and only if*

$$g_{ik}(u^i, u^2) = g_{ik}^*(u^{*1}, u^{*2}) \ (i, k = 1, 2) \tag{3.124}$$

at all corresponding pairs (u^1, u^2) and (u^{*1}, u^{*2}), where g_{ik} and g_{ik}^* $(i, k = 1, 2)$ are the first fundamental coefficients of S and S^*, and S and S^* have the same parameter systems.

Proof. Let u^1 and u^2 be the parameters of S, and u^{*1} and u^{*2} be the parameters of S^*. Since we assume that S and S^* have the same parameter systems, the map from S to S^* is given by $u^{*\alpha} = u^\alpha$ for $\alpha = 1, 2$. If the functions

$$u^\alpha = \phi^\alpha(t) \ (t \in [0, t_1]) \text{ for } \alpha = 1, 2$$

define a curve γ on S, then its image γ^* is given by

$$u^{*\alpha} = \phi^\alpha(t) \ (t \in [0, t_1]) \text{ for } \alpha = 1, 2$$

with the same functions ϕ^α, since S and S^* have the same parameter systems (Remark 3.9.2 (b)). A part $\tilde{\gamma}$ of γ with $u^\alpha = \phi^\alpha(t) \ (t \in [0, t_0])$, where $0 \leq t_0 \leq t_1$ has the arc length

$$s(t_0) = \int_0^{t_0} \sqrt{g_{ik}\left(\phi^1(t), \phi^2(t)\right) \frac{d\phi^i}{dt} \cdot \frac{d\phi^k}{dt}} \, dt \text{ for } 0 \leq t_0 \leq t_1,$$

and the corresponding image $\tilde{\gamma}^*$ has the arc length

$$s^*(t_0) = \int_0^{t_0} \sqrt{g_{ik}^*\left(\phi^1(t), \phi^2(t)\right) \frac{d\phi^i}{dt} \cdot \frac{d\phi^k}{dt}} \, dt \text{ for } 0 \leq t_0 \leq t_1.$$

If $g_{ik}(u^1, u^2) = g_{ik}^*(u^{*1}, u^{*2})$ $(i, k = 1, 2)$ for all pairs (u^1, u^2) and corresponding pairs (u^{*1}, u^{*2}), then it follows that $s(t_0) = s^*(t_0)$.

Conversely if we assume that a curve γ and its image γ^* have the same lengths, and if this also holds for each part $\tilde{\gamma}$ of γ and its image $\tilde{\gamma}^*$, then the integrands of both integrals above must be equal. If the length of any curve on S is to be the same as that of its image on S^*, then the integrands of both integrals above have to be equal for any arbitrary pair of functions ϕ^α and any point t, that is, for all pairs (u^1, u^2) and corresponding pairs (u^{*1}, u^{2*}), we must have

$$g_{ik}(u^i, u^2) = g_{ik}^*(u^{*1}, u^{*2}) \ (i, k = 1, 2).$$

□

Remark 3.9.15. *(a) It follows from Theorems 3.9.5 and 3.9.14 that every isometric map is conformal.*
(b) The map of Example 3.9.3 is isometric.

We are going to show that helicoids and catenoids are isometric.

Example 3.9.16. *Let $a > 0$ be a constant and $D = \mathbb{R} \times (0, 2\pi)$. Then the helicoid given by*

$$\vec{\tilde{x}}(\bar{u}^i) = \{\bar{u}^1 \cos \bar{u}^2, \bar{u}^1 \sin \bar{u}^2, a\bar{u}^2\} \ \left((\bar{u}^1 \, \bar{u}^2) \in D\right)$$

and the catenoid given by

$$\vec{x}(u^i) = \{a \cosh u^1 \cos u^2, a \cosh u^1 \sin u^2, au^1\} \ \left((u^1, u^2) \in D\right)$$

are isometric.

Proof. The first fundamental coefficients $\bar{g}_{ik}$ of the helicoid and g_{ik} of the catenoid are

$$\bar{g}_{11} = 1, \ \bar{g}_{12} = 0, \ \bar{g}_{22} = a^2 + (\bar{u}^1)^2,$$

and

$$g_{11} = g_{22} = a^2 \cosh^2 u^1, \ \text{and} \ g_{12} = 0.$$

We choose suitable parameters u^{*1} and u^{*2} for the helicoid. By Theorem 3.9.14, we must have

$$g_{11}^* = \bar{g}_{11} = \bar{g}_{11} \left(\frac{\partial \bar{u}^1}{\partial u^{*1}} \right)^2 + \bar{g}_{22} \left(\frac{\partial \bar{u}^2}{\partial u^{*1}} \right)^2.$$

Putting

$$\bar{u}^1 = a \sinh u^{*1} \ \text{and} \ \bar{u}^2 = u^{*2}, \tag{3.125}$$

we obtain

$$g_{11}^*(u^{*i}) = a^2 \cosh^2 u^{*1} = \bar{g}_{11}(\bar{u}^i).$$

Obviously the transformation (3.125) is one-to-one. Since the Jacobian also satisfies

$$\frac{\partial(\bar{u}^1, \bar{u}^2)}{\partial(u^{*1}, u^{*2})} = \det \begin{pmatrix} a \cosh u^{*1} & 0 \\ 0 & 1 \end{pmatrix} = a \cosh u^{*1} > 0 \ \text{for all} \ (u^{*1}, u^{+2}) \in D,$$

the transformation (3.125) is admissible.

The parametric representation of the helicoid with respect to the parameters u^{*1} and u^{*2} is given by

$$\vec{x}^*(u^{*i}) = \left\{ a \sinh u^{*1} \cos u^{*2}, a \sinh u^{*1} \sin u^{*2}, a u^{*2} \right\}$$

with the first fundamental coefficients satisfying in the same parameter systems u^{*i} and u^{*2}

$$g_{11}^*(u^{*i}) = a^2 \cosh^2 u^{*1} = g_{11}(u^i), \ g_{12}^*(u^{*i}) = 0 = g_{12}(u^i)$$

and

$$g_{22}^*(u^{*i}) = a^2(1 + \sinh^2 u^{*1}) = a^2 \cosh^2 u^{*1} = g_{22}(u^{*i}).$$

Thus the helicoid and catenoid are isometric by Theorem 3.9.14. $\qquad\square$

Example 3.9.16 is a special case of a more general result which we are going to prove next, namely that every *screw surface* is isometric to some surface of revolution. The following result holds.

Theorem 3.9.17 (Bour). *Every screw surface S can be mapped isometrically onto a surface of revolution RS.*

Proof. We may assume that the surface of revolution RS and the screw surface S are given by the parametric representations

$$\vec{x}(u^i) = \{u^1 \cos u^2, u^1 \sin u^2, h(u^1)\} \tag{3.126}$$

and

$$\vec{x}(\bar{u}^i) = \{\bar{u}^1 \cos \bar{u}^2, \bar{u}^1 \sin \bar{u}^2, \xi(\bar{u}^1) + c\bar{u}^2\}$$

with the first fundamental coefficients

$$g_{11}(u^i) = 1 + (h'(u^1))^2, \ g_{12} = 0, \ g_{22}(u^i) = (u^1)^2,$$

$$\bar{g}_{11}(\bar{u}^i) = 1 + (\xi'(\bar{u}^1))^2, \ \bar{g}_{12}(\bar{u}^i) = c\xi'(\bar{u}^1) \text{ and } \bar{g}_{22}(\bar{u}^i) = (\bar{u}^1)^2 + c^2,$$

and the first fundamental forms

$$ds = \left(1 + (h'(u^1))^2\right)(du^1)^2 + (u^1)(du^2)^2 \tag{3.127}$$

and

$$d\bar{s} = \left(1 + \left(\xi'(\bar{u}^1)\right)^2\right)\left(d\bar{u}^1\right)^2 + 2c\xi'(\bar{u}^1)d\bar{u}^1 d\bar{u}^2 + \left(\left(\bar{u}^1\right)^2 + c^2\right)\left(d\bar{u}^2\right)^2$$

$$= \left(1 + \frac{(\bar{u}^1)^2 (\xi'(\bar{u}^1))^2}{(\bar{u}^1)^2 + c^2}\right)\left(d\bar{u}^1\right)^2 + \left(\left(\bar{u}^1\right)^2 + c^2\right)\left(\frac{c\xi'(\bar{u}^1)}{(\bar{u}^1)^2 + c^2} \cdot d\bar{u}^1 + d\bar{u}^2\right)^2. \tag{3.128}$$

We put

$$du^{*2} = \frac{1}{\eta}\left(\frac{c\xi'(\bar{u}^1)}{(\bar{u}^1)^2 + c^2} \cdot d\bar{u}^1 + d\bar{u}^2\right),$$

where $\eta \neq 0$ is a constant, that is, we use the transformation

$$\bar{u}^1 = u^{*1} \text{ and } \bar{u}^2 = -c \cdot \int \frac{\xi'(u^{*1})}{(u^{*1})^2 + c^2} du^{*1} + \eta u^{*2}.$$

Then u^{*1} and u^{*2} are orthogonal parameters for S^* (as in Example 3.6.1) such that the u^{*1}–lines are helices and the u^{*2}–lines are their orthogonal trajectories. Now it follows from (3.128) that the first fundamental form of the screw surface S^* with respect to the parameters u^{*1} and u^{*2} is given by

$$ds^* = \left(1 + \frac{(u^{*1})^2 (\xi'(u^{*1}))^2}{(u^{*1})^2 + c^2}\right)\left(du^{*1}\right)^2 + \eta^2\left(\left(u^{*1}\right)^2 + c^2\right)\left(du^{*2}\right)^2. \tag{3.129}$$

The surfaces RS and S are isometric by Theorem 3.9.14 if and only if their fundamental forms (3.127) and (3.129) are of the same form. This can be achieved by putting

$$u^2 = u^{*2}, \tag{3.130}$$

$$\left(u^1\right)^2 = \eta^2\left(\left(u^{*1}\right)^2 + c^2\right) \text{ for some constant } \eta \neq 0 \tag{3.131}$$

and

$$\left(1 + (h'(u^1))^2\right) \cdot \left(\frac{du^1}{du^{*1}}\right)^2 = 1 + \frac{(u^{*1})^2 (\xi'(u^{*1}))^2}{(u^{*1})^2 + c^2}. \tag{3.132}$$

$\square$

Remark 3.9.18. *The relations in (3.130), (3.131) and (3.132) enable us to find a screw surface isometric to a given surface of revolution, and conversely, a surface of revolution isometric to a given screw surface.*

If a surface of revolution is given, then we use (3.131) to eliminate the parameter u^1 in (3.132) and then solve (3.132) for $\xi'(u^{*1})$ to find the function ξ of the screw surface.

Conversely, if a screw surface is given, then we use (3.131) to eliminate the parameter u^{*1} in (3.132) and then solve (3.132) for $h'(u^1)$ to find the function h of the surface of revolution.

We apply this method to find screw surfaces which are isometric to a catenoid.

Visualization 3.9.19. *Let $a > 0$ and a catenoid S be given by the parametric representation*

$$\vec{x}(\tilde{u}^i) = \left\{ a \cosh \tilde{u}^1 \cos \tilde{u}^2, a \cosh \tilde{u}^1 \sin \tilde{u}^2, a\tilde{u}^1 \right\} \ for \ (\tilde{u}^1, \tilde{u}^2) \in \tilde{D} = (0, \infty) \times (0, 2\pi).$$

We apply the method described in Remark 3.9.18. Using the transformation

$$\tilde{u}^1 = \cosh^{-1} \left(\frac{u^1}{a} \right) = \log \left(u^1 + \sqrt{(u^1)^2 - a^2} \right) \ for \ u^1 > a \ and \ \tilde{u}^2 = u^2,$$

we obtain the following parametric representation for S in the new parameters

$$\vec{x}(u^i) = \left\{ u^1 \cos u^2, u^1 \sin u^2, a \cdot \cosh^{-1} \left(\frac{u^1}{a} \right) \right\} \ for \ (u^1, u^2) \in D = (a, \infty) \times (0, 2\pi),$$

which is the parametric representation in (3.126) in the proof of Theorem 3.9.17 with $h(u^1) = a \cdot \cosh^{-1}(u^1/a)$ and

$$h'(u^1) = \frac{a}{\sqrt{(u^1)^2 - a^2}} \ and \ 1 + \left(h'(u^1) \right)^2 = \frac{(u^1)^2}{(u^1)^2 - a^2}.$$

First (3.131) yields $u^1 du^1 = \eta^2 u^{*1} du^{*1}$, that is,

$$\frac{du^1}{du^{*1}} = \eta^2 \cdot \frac{u^{*1}}{u^1}.$$

Substituting this in (3.132), eliminating u^1 and solving for $\xi(u^{*1})$, we obtain

$$\left(1 + \left(h'(u^1) \right)^2 \right) \left(\frac{du^1}{du^{*1}} \right) = \eta^4 \cdot \frac{(u^1)^2}{(u^1)^2 - a^2} \cdot \frac{(u^{*1})^2}{(u^1)^2} = \frac{\eta^4 (u^{*1})^2}{\eta^2 \left((u^{*1})^2 + c^2 \right) - a^2}$$

$$= \frac{\eta^2 (u^{*1})^2}{(u^{*1})^2 + c^2 - \frac{a^2}{\eta^2}}$$

$$= 1 + \frac{(u^{*1})^2 \left(\xi' (u^{*1}) \right)^2}{(u^{*1})^2 + c^2} \ for \ u^{*1} > \sqrt{\frac{a^2}{\eta^2} - c^2} \ and \ c < \frac{a}{|\eta|},$$

and

$$\left(\xi'\left(u^{1*}\right)\right)^2 = \frac{\left(u^{*1}\right)^2 + c^2}{\left(u^{*1}\right)^2} \left(\frac{\eta^2 \left(u^{*1}\right)^2}{\left(u^{*1}\right)^2 + c^2 - \frac{a^2}{\eta^2}} - 1\right)$$

$$= \frac{1}{\left(u^{*1}\right)^2} \cdot \frac{\left(u^{*1}\right)^2 + c^2}{\left(u^{*1}\right)^2 + c^2 - \frac{a^2}{\eta^2}} \left(\left(u^{*1}\right)^2 \left(\eta^2 - 1\right) - \left(c^2 - \frac{a^2}{\eta^2}\right)\right).$$

We choose $\eta = 1$ and put $k^2 = a^2 - c^2$ for $a > c$, and obtain

$$\left(\xi'\left(u^{1*}\right)\right)^2) = \frac{\left(\left(u^{*1}\right)^2 + c^2\right) k^2}{\left(u^{*1}\right)^2 \left(\left(u^{*1}\right)^2 - k^2\right)} \text{ for } u^{*1} > k > 0,$$

that is,

$$\xi'\left(u^*\right) = k \cdot \sqrt{\frac{\left(u^{*1}\right)^2 + c^2}{\left(u^{*1}\right)^2 \left(\left(u^{*1}\right)^2 - k^2\right)}}.$$

This yields

$$\xi(u^{*1}) = k \cdot \log\left(\sqrt{\left(u^{*1}\right)^2 + c^2} + \sqrt{\left(u^{*1}\right)^2 - k^2}\right) - c \cdot \tan^{-1}\left(\frac{k}{c} \cdot \sqrt{\frac{\left(u^{*1}\right)^2 + c^2}{\left(u^{*1}\right)^2 - k^2}}\right) + \tilde{d},$$

where $\tilde{d}$ is a constant. We observe that we may choose $\tilde{d} = 0$, since a change in $\tilde{d}$ only results in a movement of the screw surface in the direction of the x^3-axis. For every k with $0 < k \leq a$, that is, for every c with $0 \leq c < a$, we obtain a screw surface S_k with

$$\xi(u^{*1}) = k \cdot \log\left(\sqrt{\left(u^{*1}\right)^2 + c^2} + \sqrt{\left(u^{*1}\right)^2 - k^2}\right) - c \cdot \tan^{-1}\left(\frac{k}{c} \cdot \sqrt{\frac{\left(u^{*1}\right)^2 + c^2}{\left(u^{*1}\right)^2 - k^2}}\right)$$

for $u^{*1} > k$, which is isometric to the catenoid.
If $k = a$, that is, $c = 0$, then we obtain the original catenoid with

$$\xi(u^{*1}) = a \cosh^{-1}\left(\frac{u^{*1}}{a}\right) = \log\left(u^{*1} + \sqrt{(u^{*1})^2 - a^2}\right).$$

If $k = 0$, that is, $c = a$, then $\xi = 0$ and we obtain a helicoid (Figure 3.32).

It turns out that the Gaussian and geodesic curvature are invariant under isometric maps.

Theorem 3.9.20. *The Gaussian curvature K of a surface and the geodesic curvature κ_g of a curve on a surface are invariant under isometric maps.*

Figure 3.32 Isometric maps of Visualization 3.9.19

Proof. The Gaussian curvature K of a surface depends only of the first fundamental coefficients by the theorema egregium, Theorem 3.8.3; also the geodesic curvature κ_g of a curve on a surface only depends on the first fundamental coefficients by Theorem 3.1.8. □

Remark 3.9.21. *Isometric surfaces must necessarily have the same Gaussian curvature in corresponding points. Since the Gaussian curvature of a sphere of radius R is equal to $1/R$ and that of a plane is equal to 0, there is no isometric map between any part of a sphere and a plane; in particular, the stereographic projection is not isometric.*
In general, the equality of the Gaussian curvature is not sufficient for the isometry of surfaces except in the case of surfaces with constant Gaussian curvature.

Theorem 3.9.22. *Under an isometry, the image γ^* of any curve γ of minimal length between the points P_1 and P_2 is also a curve of minimal length between the image points P_1* and P_2^*.*

Proof. If γ^* were not the shortest connection between P_1^* and P_2^*, then there would be a shorter curve $\tilde{\gamma}^*$ between P_1^* and P_2^*. The inverse image $\tilde{\gamma}$ of $\tilde{\gamma}^*$ would then be shorter than γ by isometry, a contradiction to the assumed minimality of γ. □

The class of surfaces that are isometric to a part of the plane is relatively small, namely that of developable surfaces, as we will see in Theorem 3.9.25. The proof uses the fact that developable surfaces can be characterized as surfaces of identically vanishing Gaussian curvature; this is the result of Theorem 3.9.24.
We need the next Lemma in the proof of Theorem 3.9.24.

Lemma 3.9.23. *A curve γ with a parametric representation $\vec{x}(t)$ $(t \in I)$ for which the vectors $\vec{x}''(t)$ and $\vec{x}'(t)$ are linearly dependent for all t is a straight line segment.*

Proof. Let $\vec{x}(t) = \{x^1(t), x^2(t), x^3(t)\}$. Since $\vec{x}'(t) \neq \vec{0}$, there exists a function c such that $(x^k)''(t) = c(t)(x^k)'(t)$ $(k = 1, 2, 3)$ for all $t \in I$. Writing $y^k = (x^k)'$ for $k = 1, 2, 3$, we obtain

$$\frac{(y^k)'(t)}{y^k(t)} = c(t),$$

and by integration

$$y^k(t) = C_k \exp\left(\int c(t)\, dt\right) \text{ for } k = 1, 2, 3,$$

where the C_k are constants of integration. One more integration yields

$$x^k(t^*) = C_k t^* + d_k \text{ with } t^*(t) = \int \exp\left(\int c(t)\, dt\right) dt \text{ for } k = 1, 2, 3,$$

where the d_k are constants of integration. These are the component functions of a straight line segment with t^* as their parameter. □

Theorem 3.9.24. *A surface S of class C^r for $r \geq 2$ is part of a developable surface if and only if its Gaussian curvature K vanishes identically.*

Proof. (i) First we assume that S is a developable surface.

Then its surface normal vectors $\vec{N}$ are constant along each u^1–line by Definition 2.8.14 and consequently its Gauss spherical image is a curve. This implies $K = 0$ by (2.159).

(ii) Now we assume $K = 0$.

Then $K = L/g$ implies $L = 0$. If $L_{ik} = 0$ for all i, k, then S is a plane by Example 2.10.2.

So let

$$L_{11}L_{22} = (L_{12})^2, \tag{3.133}$$

where not all of the coefficients L_{ik} vanish. The differential equations (2.121) for the asymptotic lines on S reduce to

$$L_{11}(du^1)^2 + 2L_{12}du^1du^2 + L_{22}(du^2)^2 = \left(\sqrt{|L_{11}|}du^1 + \sqrt{|L_{22}|}du^2\right)^2 = 0, \tag{3.134}$$

and consequently only one family of asymptotic lines exists; this has to be the case by Remark 2.6.2 (a), when all points are parabolic points. We assume that the parameters u^1 and u^2 of S are chosen such that the asymptotic lines coincide with the u^1–lines. Then it follows from (3.134) that $L_{11}du^1 = 0$, and then $L_{12} = 0$ by (3.133). Now we obtain from the Weingarten equations (2.141) and (2.140)

$$\vec{N}_1 = -L_1^k\vec{x}_k = -g^{kj}L_{j1}\vec{x}_k = -\left(g^{k1}L_{11} + g^{k2}L_{21}\right)\vec{x}_k = \vec{0},$$

hence the surface normal vectors $\vec{N}$ do not depend on u^1. Consequently they are constant along each asymptotic line, which is given by $u^2 = const$ in our parameter system. If these asymptotic lines are straight lines, then the surface is part of a developable surface that is generated by these straight lines. This is indeed the case, since if $L_{11} = L_{12} = 0$ and $L_{22} \neq 0$ the Mainardi–Codazzi equations (3.98) and (3.96) reduce for $i = k = 1$ and $j = 2$ to

$$C_{112} = L_{11;2} - L_{12;2} - \left\{ \begin{matrix} r \\ 11 \end{matrix} \right\} L_{r2} - \left\{ \begin{matrix} r \\ 12 \end{matrix} \right\} L_{r1} = \left\{ \begin{matrix} 2 \\ 11 \end{matrix} \right\} L_{22} = 0$$

and so the Gauss derivation formulae (3.99) yield

$$\vec{x}_{11} = \left\{ \begin{matrix} k \\ 11 \end{matrix} \right\} \vec{x}_k + L_{11}\vec{N} = \left\{ \begin{matrix} 1 \\ 11 \end{matrix} \right\} \vec{x}_1$$

Hence $\vec{x}_{11}''$ and $\vec{x}_1'$ are linearly dependent and the asymptotic lines are straight line segments by Lemma 3.9.23.

□

Theorem 3.9.25. *Any sufficiently small part S of a surface of class C^r for $r \geq 3$ can be mapped isometrically into a plane if and only if S is part of a developable surface.*

Proof. (i) First we assume that S can be mapped isometrically into a plane. Then the Gaussian curvature K of S is equal to that of the plane by Theorem 3.9.20. Since the Gaussian curvature of a plane is identically equal to 0 it follows that $K \equiv 0$ and so S is a part of a developable surface by Theorem 3.9.24.

(ii) Now we show that a sufficiently small part S of a developable surface can be mapped isometrically into a plane.

We introduce geodesic parallel coordinates u^1 and u^2 on S as follows: We choose a geodesic line γ_g as the u^2–line corresponding to $u^1 = 0$, the geodesic lines orthogonal to γ_g as the u^1–lines corresponding to $u^2 = const$, and their orthogonal trajectories as the u^2–lines corresponding to $u^1 = const$. The first fundamental form for a suitable choice of u^1 is given by (3.72) as

$$ds^2 = \left(du^1\right)^2 + g_{22}\left(du^2\right)^2.$$

If we take the arc length along γ_g as the parameter u^2, then we have $ds = du^2$ on γ_g, hence

$$g_{22}(0, u^2) = 1. \tag{3.135}$$

Since γ_g is a geodesic line it follows from (3.21) that

$$\left. \frac{\partial g_{22}}{\partial u^1} \right|_{u^1=0} = 0. \tag{3.136}$$

Since $K = 0$ by Theorem 3.9.24 and S has geodesic parameters, it follows from (3.104) that

$$\frac{\partial^2 \sqrt{g_{22}}}{(\partial u^1)^2} = 0.$$

Integrating twice we obtain

$$\sqrt{g_{22}} = c_1(u^2)u^1 + c_2(u^2).$$

It follows from (3.135) that $c_1 \equiv 1$, and from (3.136) that $c_1(u^2) = 0$. Thus we have $g_{11} = g_{22} = 1$ and $g_{12} = 0$. If u^{*1} and u^{*2} are the Cartesian parameters of the plane, then $u^{*i} = u^i$ for $i = 1, 2$ defines a map which is isometric by Theorem 3.9.14.

□

Remark 3.9.26. *If S is a developable surface then the orthogonal trajectories of geodesic lines are geodesic lines provided at least one of the trajectories is a geodesic line. This property only holds for developable surfaces, for otherwise parts of some other surfaces could be mapped isometrically into the plane in contradiction to Theorems 3.9.26 and 3.9.20.*

Finally, we consider *area preserving* maps.

Definition 3.9.27. An admissible map from a surface S onto a surface S^* is said to be *area preserving* if any part D of S is mapped onto a part D^* of S^* which has the same surface area as D.

We obtain a characterization for area preserving maps similar to the characterizations of isometric and conformal maps in Theorems 3.9.5 and 3.9.14, respectively.

Theorem 3.9.28. *An admissible map from a surface S onto a surface S^* is area preserving if and only if*

$$g = \det(g_{ik}) = g^* = \det(g_{ik}^*) \tag{3.137}$$

at every point on S and its image point on S^*, where g_{ik} and g_{ik}^* $(i, k = 1, 2)$ are the first fundamental coefficients of S and S^*, and S and S^* have the same coordinate systems.

Proof. We consider the integral for the surface area of any part D of S

$$A(D) = \iint\limits_{D} \sqrt{g}\, du^1\, du^2,$$

and the corresponding integral for the surface area of the image D^* on S^*

$$A(D^*) = \iint\limits_{D^*} \sqrt{g^*}\, du^{*1}\, du^{*2}.$$

We conclude as in the proof of Theorem 3.9.5 that the map is area preserving if and only if (3.137) holds. □

The following result is a consequence of Theorems 3.9.5 and 3.9.14.

Theorem 3.9.29. *(a) Every isometric map is area preserving.*
(b) Every area preserving and conformal map is isometric.

Proof. We assume that the surface S and its image S^* have the same parameter systems.

(a) If a map is isometric then we have by (3.124) in Theorem 3.9.14

$$g_{ik}(u^1, u^2) = g_{ik}^*(u^{*1}, u^{*2}) \ (i, k = 1, 2),$$

and so the condition in (3.137) holds. This implies by Theorem 3.9.28 that the map is area preserving.
Thus we have shown Part (a).

(b) Let a map be conformal and area preserving.
Since the map is conformal, we have by (3.109) in Theorem 3.9.5

$$g_{ik}^*(u^{*1}, u^{*2}) = \eta(u^1, u^2) g_{ik}(u^1, u^2) \ (i, k = 1, 2), \text{ for } \eta > 0,$$

hence

$$g^* = \eta^2 g. \tag{3.138}$$

Since the map is also area preserving, it follows from (3.137) in Theorem 3.9.28 that

$$g^* = g. \tag{3.139}$$

Now it follows from (3.138) and (3.139) that $\eta^2 = 1$, hence $\eta = 1$, since $\eta > 0$. Therefore, we have $g_{ik}^* = g_{ik}$ $(i, k = 1, 2)$, and so the map is isometric by (3.124) in Theorem 3.9.14.
Thus we have shown (b).

□

Combining Remark 3.9.15 (a) and Parts (a) and (b) of Theorem 3.9.29, we obtain

Corollary 3.9.30. *A map is isometric if and only if it is conformal and area preserving.*

Example **3.9.31 (Lambert Projection).** *Let the sphere be given by the parametric representation in (3.113) in Example 3.9.8. We put*

$$x_1^* = ru^1 \qquad (-\pi < u^1 < \pi)$$

$$x_2^* = r \sin u^2 \qquad \left(-\frac{\pi}{2} < u^2 < \frac{\pi}{2} \right)$$

and use x^{*1} and $x^*(2)$ as Cartesian coordinates in the plane. This defines the Lambert projection. *Then we have*

$$g_{11}(u^1, u^2) = r^2 \cos^2 u^2, \ g_{12}(u^1, u^2) = 0, \quad g_{22}(u^1, u^2) = r^2,$$
$$g_{11}^*(u^1, u^2) = r^2, \qquad g_{12}^*(u^1, u^2) = 0, \quad g_{22}^*(u^1, u^2) = r^2 \cos u^2.$$

This implies

$$g_{11}g_{22} - g_{12}^2 = g = r^4 \cos^2 u^2 = g^* = g_{11}^* g_{22}^* - g_{12}^{*2},$$

so the Lambert projection is area preserving by Theorem 3.9.28.

Remark 3.9.32. *The Lambert projection can be geometrically generated as follows. Let S and Cyl denote the sphere of radius r, and the circular cylinder tangent to the equator of S and its axis orthogonal to the plane Pl of the equator. First we construct the points $P^* \in Cyl$ as the intersections of the straight lines through $P \in S$ and the point on the axis of the sphere that has the same distance from the plane Pl as the point P (Figure 3.33). Then the points P^* are projected to Pl as in Example 3.9.3. Then S^* is a rectangle with edges of the lengths $2\pi r$ and $2r$. The images of the meridians and parallels of S are straight line segments in S^*; the images of the meridians have equal distances in S^* while the distances of images the parallels in S^* decrease as the parallels approach the poles of S.*

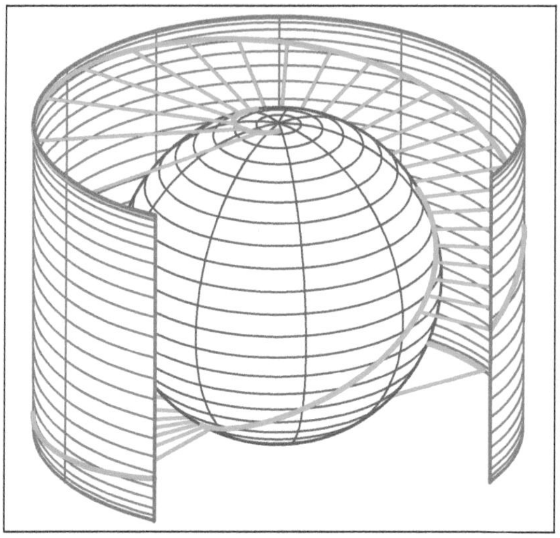

Figure 3.33 Projection of a sphere to a cylinder

We close this section with one more example of an area preserving map.

*Example **3.9.33** (**The Collignon Projection**). Let a sphere S be given by a parametric representation (3.113) in Example 3.9.8. We put*

$$x_1^* = \frac{2\sqrt{2}}{\sqrt{\pi}}\, ru^1 \sin\left(\frac{\pi}{4} - \frac{u^2}{2}\right) \text{ and } x_2^* = r\sqrt{\pi}\left(1 - \sqrt{2}\sin\left(\frac{\pi}{4} - \frac{u^2}{2}\right)\right), \quad (3.140)$$

and introduce x_1^* and x_2^* as Cartesian coordinates in the plane S^*. Then the Collignon projection of S into the plane S^* is given by (3.140).

(a) The Collignon projection is area preserving.

(b) The images in the plane under the Collignon projection of

    (i) the north pole $(0, 0, r)$ of S is the point $(0, r\sqrt{\pi})$;

    (ii) the south pole $(0, 0, -r)$ of S is the straight line segment of length $4r \cdot \sqrt{2\pi}$, parallel to the x^{*1}–axis through $x_2^* = -r\sqrt{\pi}(\sqrt{2} - 1) < 0$;

    (iii) the parallels of S, that is, $u^2 = const$, are straight line segments parallel to the x_1^*–axis;

    (iv) the meridians of S, that is, $u^1 = const$, are straight lines in the $x^{*1}x^{*2}$–plane through the point $(0, r\sqrt{\pi})$.

Proof. (a) We put $u^{*i} = u^i$ $(i = 1, 2)$ and obtain the following parametric representation for the plane S^*

$$\vec{x}^*(u^{*i}) = \left\{\frac{2\sqrt{2}}{\sqrt{\pi}}\, ru^{*1} \sin\left(\frac{\pi}{4} - \frac{u^{*2}}{2}\right), r\sqrt{\pi}\left(1 - \sqrt{2}\sin\left(\frac{\pi}{4} - \frac{u^{*2}}{2}\right)\right)\right\}.$$

We get from (3.113) $g_{11}(u^i) = r^2 \cdot \cos^2 u^2$, $g_{12}(u^i) = 0$, $g_{22}(u^i) = r^2$, and writing $\phi = \phi(u^{*2}) = \pi/4 - u^{*2}/2$,

$$\vec{x}_1^*(u^{*i}) = \left\{ \frac{2\sqrt{2}}{\sqrt{\pi}} r \sin \phi, 0 \right\}, \quad \vec{x}_2^*(u^{*i}) = \left\{ -\frac{\sqrt{2}}{\sqrt{\pi}} r u^{*1} \cos \phi, r \sqrt{\pi} \frac{\sqrt{2}}{2} \cos \phi \right\},$$

$$g_{11}^*(u^{*i}) = \frac{8r^2}{\pi} \sin^2 \phi, \quad g_{12}^*(u^{*i}) = -\frac{4r^2 u^{*1}}{\pi} \sin \phi \cos \phi,$$

$$g_{22}^*(u^{*i}) = r^2 \cos^2 \phi \left(\frac{2(u^{*1})^2}{\pi} + \frac{\pi}{2} \right),$$

$$g^*(u^{*i}) = g_{11}^*(u^{*i}) g_{22}^*(u^{*i}) - g_{12}^{*2}(u^{*i})$$

$$= \frac{16r^4(u^{*1})^2}{\pi^2} \sin^2 \phi \cos^2 \phi + 4r^4 \sin^2 \phi \cos^2 \phi - \frac{16r^4(u^{*1})^2}{\pi^2} \sin^2 \phi \cos^2 \phi$$

$$= r^4 \sin^2 (2\phi) = r^4 \sin^2 \left(u^{*2} - \frac{\pi}{2} \right) = r^4 \cos^2 u^{*2}.$$

Since $u^i = u^{*i}$ for $i = 1, 2$, we obtain $g^*(u^{*i}) = g(u^i)$, and so the Collignon projection is area preserving by Theorem 3.9.28.

(b) We have:

    (i) $u^2 = \pi/2$ for the north pole of S, hence its image is $(0, r\sqrt{\pi})$ by the transformation formulae (3.140);

    (ii) $u^2 = -\pi/2$ for the south pole of S, hence its image is $x_1^* = 2\sqrt{2}\, r u^1/\sqrt{\pi}$, $x_2^* = -r\sqrt{\pi}(\sqrt{2} - 1)$ for $0 \le u^1 \le 2\pi$ by the transformation formulae (3.140); this is a straight line segment of length $4r \cdot \sqrt{2\pi}$, parallel to the x^{*1}–axis through $x_2^* = -r\sqrt{\pi}(\sqrt{2} - 1) < 0$;

    (iii) $u^2 = u_0^2 = const \in (-\pi/2, \pi/2)$ for the parallels; if we put $\alpha = \pi/4 - u_0^2/2$, then their images are $x_1^* = 2\sqrt{2}\, r u^1 \sin \alpha/\sqrt{\pi}$, $x_2^* = r\sqrt{\pi}(1 - \sqrt{2} \sin \alpha)$ for $u^1 \in [0, 2\pi]$ by the transformation formulae (3.140); those are straight line segments parallel to the x_1^*–axis;

    (iv) $u^1 = u_0^1 = const \in [0, 2\pi]$ for the meridians; if we put $\alpha(u^2) = \pi/4 - u^2/2$, then their images are $x_1^* = 2\sqrt{2} \cdot r u_0^1 \sin \alpha(u^2)/\sqrt{\pi}$, $x_2^* = r\sqrt{\pi}(1 - \sqrt{2} \sin \alpha(u^2))$ by the transformation formulae (3.140); we eliminate $\sin \alpha(u^2)$, and obtain $x_1^* = 2r u_0^1/\sqrt{\pi} - 2u_0^1 \cdot x_2^*/\pi$; those are straight lines in the $x^{*1}x^{*2}$–plane through the point $(0, r\sqrt{\pi})$.

    □

3.10   THE GAUSS-BONNET THEOREM

The Gauss–Bonnet theorem, Theorem 3.10.1, is one of the most important theorems in the theory of surfaces. It is an example for results of the global differential geometry. Furthermore this theorem may be considered as an analogue in differential geometry of the Gauss–Green integral theorem in analysis. The common feature of both theorems is that they link integrals over parts of surfaces to integrals over their boundaries.

Theorem 3.10.1 (Gauss–Bonnet).

Let S be a part of a surface with a parametric representation $\vec{x}(u^i)$ $((u^1, u^2) \in D)$ and $\vec{x} \in C^r(D)$ for $r \geq 3$. We assume that the boundary ∂S of S is a simple closed curve γ with a parametric representation $\vec{x}(u^i(s))$ of class $r^* \geq 2$, where s denotes the arc length along γ. If we denote the geodesic curvature of γ by κ_g, the Gaussian curvature of S by K and the surface element of S by dA, then

$$\int_{\partial S} \kappa_g \, ds + \iint_S K \, dA = 2\pi, \tag{3.141}$$

where the orientation of ∂S has to be chosen such that S is to the left of ∂S.

Proof. First we introduce suitable orthogonal parameters for S as follows:

Let γ_0 be an arbitrary, sufficiently small *geodesic circle line* on S with its centre at a point P on S, that is, the points of γ_0 have equal distance from P measured along the geodesic lines through P. Furthermore, let u^1 and u^2 be *geodesic polar coordinates* with their origin in P in the interior of γ_0. Finally, let the boundary curve γ of S be given by $u^1 \equiv 1$ and $u^2 = s$, where s is the arc length along γ.

The existence of such coordinates can be seen as follows:

By Theorem 3.9.12, there is a conformal map F of S into a plane E. Let $S^* = F(S)$, $\gamma^* = F(\gamma)$ and $\gamma_0^* = F(\gamma_0)$. It is well known from complex analysis that the part S^{**} of S^* which is bounded by γ^* and γ_0^* can be mapped by a conformal map G onto an annulus A bounded by two concentric circles lines. Let $\mathcal{C}$ be the family of concentric circles lines in A. Then the family $\mathcal{C}^* = G^{-1}(\mathcal{C})$ of the pre–images of the circle lines in $\mathcal{C}$ is a family of closed curves that cover S^{**} in a one–to–one way. Let $\mathcal{C}^{*\perp}$ be the family of the *orthogonal trajectories* of the curves of $\mathcal{C}^*$. Now the curves in the pre–images $F^{-1}(\mathcal{C}^*)$ and $F^{-1}(\mathcal{C}^{*\perp})$ between γ and γ_0 and the geodesic circles centred at P inside of γ_0 and their orthogonal trajectories are the desired parameter lines.

By (3.21), the geodesic curvature κ_g of γ is given by

$$\kappa_g = \frac{1}{2\sqrt{g_{11}}} \frac{1}{g_{22}} \frac{\partial g_{22}}{\partial u^1} = \frac{1}{\sqrt{g_{11}g_{22}}} \frac{\partial(\sqrt{g_{22}})}{\partial u^1}. \tag{3.142}$$

Since $u^2 = s$ along γ, we have $g_{22} = 1$ on γ, and so by (3.142)

$$\int_\gamma \kappa_g \, ds = \int_\gamma \frac{1}{\sqrt{g_{11}}} \frac{\partial(\sqrt{g_{22}})}{\partial u^1} \, du^2. \tag{3.143}$$

Let S_0 be the domain bounded by γ_0. Since by (3.103)

$$K = -\frac{1}{2\sqrt{g_{11}g_{22}}} \left(\frac{\partial}{\partial u^1} \left(\frac{1}{\sqrt{g_{11}g_{22}}} \frac{\partial g_{22}}{\partial u^1} \right) + \frac{\partial}{\partial u^2} \left(\frac{1}{\sqrt{g_{11}g_{22}}} \frac{\partial g_{11}}{\partial u^2} \right) \right)$$

$$= -\frac{1}{\sqrt{g_{11}g_{22}}} \left(\frac{\partial}{\partial u^1} \left(\frac{1}{\sqrt{g_{11}}} \frac{\partial(\sqrt{g_{22}})}{\partial u^1} \right) + \frac{\partial}{\partial u^2} \left(\frac{1}{\sqrt{g_{22}}} \frac{\partial(\sqrt{g_{11}})}{\partial u^2} \right) \right),$$

this implies

$$\iint\limits_{S \setminus S_0} K \, dA = \iint\limits_{S \setminus S_0} K \sqrt{g} \, du^1 du^2 = -\iint\limits_{S \setminus S_0} \left(\frac{\partial}{\partial u^1} \left(\frac{1}{\sqrt{g_{11}}} \frac{\partial(\sqrt{g_{22}})}{\partial u^1} \right) \right) du^1 du^2 -$$

$$-\iint\limits_{S \setminus S_0} \left(\frac{\partial}{\partial u^2} \left(\frac{1}{\sqrt{g_{22}}} \frac{\partial(\sqrt{g_{11}})}{\partial u^2} \right) \right) du^1 du^2 = I_1 + I_2.$$

The integration with respect to u^2 is along closed curves, so $I_2 = 0$. Furthermore,

$$I_1 = \iint\limits_{S \setminus S_0} K \, dA = -\int\limits_{\gamma} \frac{1}{\sqrt{g_{11}}} \frac{\partial(\sqrt{g_{22}})}{\partial u^1} \, du^2 + \int\limits_{\gamma_0} \frac{1}{\sqrt{g_{11}}} \frac{\partial(\sqrt{g_{22}})}{\partial u^1} \, du^2,$$

and by (3.143)

$$\int\limits_{\gamma} \kappa_g \, ds + \iint\limits_{S} K \, dA = \lim_{\gamma_0 \to P} \int\limits_{\gamma_0} \frac{1}{\sqrt{g_{11}}} \frac{\partial(\sqrt{g_{22}})}{\partial u^1} \, du^2. \tag{3.144}$$

Since both integrals on the left in (3.144) exist, the limit on the right must also exist. We have $g_{11} = 1$ in the interior of γ_0 because of the geodesic polar coordinates, and $\lim_{u^1 \to 0} \partial(\sqrt{g_{22}})/\partial u^1 = 1$. So (3.144) implies (3.141). □

Remark 3.10.2. *(a) The integral $\iint_S K \, dA$ in Theorem 3.10.1 is referred to as the total curvature of S .*
(b) The Gauss–Bonnet theorem may also be applied to parts of surfaces with boundaries that consist of finitely many curves having different tangents in the points where they meet so that the boundary has cusps. In this case, such a boundary curve γ is made smooth by replacing γ in the neighbourhood of each cusp E by a geodesic arc γ_g of radius r such that the directions of its tangents at the points of intersection with the original curve γ coincide with the directions of γ at these points. Then we let r tend to zero. We have

$$\lim_{r \to 0} \int\limits_{\gamma_g} \kappa_g \, ds = \lim_{r \to 0} \int\limits_{\gamma_r} \frac{1}{r} r \, d\varphi = \alpha,$$

where γ_r is an ordinary circular arc of radius r and α denotes the angle of change in the direction of the tangent to γ at E. If γ has cusps $E_1, \ldots, E_n$ with inner angles $\beta_\nu = \pi - \alpha_\nu$ $(\nu = 1, \ldots, n)$, then we have to replace (3.141) by

$$\int\limits_{\gamma} \kappa_g \, ds + \sum_{\nu=1}^{n} \alpha_\nu + \iint\limits_{S} K \, dA = 2\pi \tag{3.145}$$

or

$$(n-2)\pi + \int\limits_{\gamma} \kappa_g \, ds + \iint\limits_{S} K \, dA = \sum_{\nu=1}^{n} \beta_\nu. \tag{3.146}$$

A part of a surface with a connected boundary without double points, which consists of n geodesic line segments and has n cusps, is called a geodesic n–gon. *For such parts of surfaces, we have by (3.146)*

$$\sum_{\nu=1}^{n} \beta_{\nu} = (n-2)\pi + \iint_{S} K \, dA. \tag{3.147}$$

If the Gaussian curvature K has a constant value unequal to zero at each point of a geodesic n–gon, then we obtain for its surface area A(S)

$$A(S) = \frac{1}{K}\left(\sum_{\nu=1}^{n} \beta_{\nu} - (n-2)\pi\right). \tag{3.148}$$

Visualization 3.10.3. *(a) If K = 0, then (3.147) yields the well–known formula for the sum of angles in a planar n–gon*

$$\sum_{\nu=1}^{n} \beta_{\nu} = (n-2)\pi.$$

(b) If n = 3, then (3.147) implies: If the Gaussian curvature of a geodesic triangle is positive throughout, then

$$\sum_{\nu=1}^{3} \beta_{\nu} > \pi,$$

as for geodesic triangles on a sphere (left in Figure 3.34).
If the Gaussian curvature of a geodesic triangle is negative throughout, then

$$\sum_{\nu=1}^{3} \beta_{\nu} < \pi,$$

as for geodesic triangles on a pseudo-sphere (right in Figure 3.34).
The difference between π and the sum of angles in a geodesic triangle is called the excess. *By (3.147), the excess is equal to the total curvature of a triangle.*

We obtain from (3.147) for $n = 0$:

Theorem 3.10.4. *On a surface of class r ≥ 3 with nowhere positive Gaussian curvature, there are no closed geodesic lines such that one of them is a complete boundary of a part of the surface.*

We close this section with an example.

Example 3.10.5. *(a) The simplest example for Theorem 3.10.4 is a plane where the geodesic lines are the straight lines.*
(b) As another example we consider the circular cone. The u^2–lines are closed geodesic lines, but no such circle line is the complete boundary of a part of the surface.

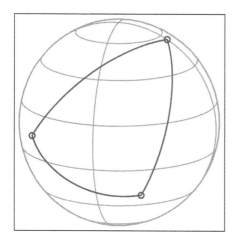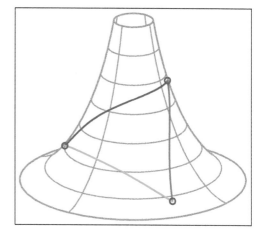

Figure 3.34 Geodesic triangles on a sphere (left) and a pseudo-sphere (right)

3.11 MINIMAL SURFACES

In this section, we study some *minimal surfaces*.

The name minimal surface is connected to the following problem, known as *Plateau's problem*. In 1866, the physicist Plateau posed the problem if, for every closed curve γ, there exists a surface with minimal surface area that has γ as its boundary curve. The answer to this problem is affirmative as we will see in Theorem 3.11.5.

We start with the definition of the notion of minimal surfaces. The term minimal surface will become clear in Theorem 3.11.5.

Definition 3.11.1. A *minimal surface* is a surface with identically vanishing mean curvature

$$H = \frac{1}{2} L_{ik} g^{ik} \equiv 0.$$

Remark 3.11.2. *The Gaussian curvature of a minimal surface cannot be positive.*

Proof. Let κ_1 and κ_2 denote the principal curvatures of the minimal surface. Then it follows from (2.74) in Definition 2.5.17 that $2H = \kappa_1 + \kappa_2 = 0$, and so $\kappa_1 = -\kappa_2$, hence $K = \kappa_1 \kappa_2 \leq 0$. □

Example 3.11.3. *(a) We already know from Visualization 2.5.19 (b) that the catenoid (Figure 2.39) is a minimal surface.*
(b) The helicoid of Visualization 2.5.3 (c) (right in Figure 2.30) is a minimal surface.

Proof. We only have to show Part (b). A helicoid has a parametric representation by Visualization 2.5.3 (c)

$$\vec{x}(u^i) = \{u^2 \cos u^1, u^2 \sin u^1, cu^1\} \ \left((u^1, u^2) \in D = (0, 2\pi) \times I_2 \right),$$

where $c \in \mathbb{R}$ is a constant, and $I_2 \subset \mathbb{R}$ is an interval. It follows that

$$\vec{x}_1 = \{-u^2 \sin u^1, u^2 \cos u^1, c\}, \ \vec{x}_2 = \{\cos u^1, \sin u^1, 0\}, \ g_{12} = 0,$$
$$\vec{N} = \frac{1}{\sqrt{(u^2)^2 + c^2}} \cdot \{-c \sin u^1, c \cos u^1, -u^2\},$$
$$\vec{x}_{11} = -u^2 \{\cos u^1, \sin u^1, 0\}, \ \vec{x}_{22} = \vec{0},$$
$$L_{11} = \vec{N} \bullet \vec{x}_{11} = \frac{cu^2}{\sqrt{(u^2)^2 + c^2}} \cdot (\sin u^1 \cos u^1 - \sin u^1 \cos u^1) = 0,$$
$$L_{22} = \vec{N} \bullet \vec{x}_{22} = 0,$$

and so by (2.72) and (2.74)

$$H = \frac{1}{g} \cdot (g_{22} L_{11} - 2 g_{12} L_{12} + g_{11} L_{22}) = 0.$$

$\square$

First we establish a characterization of minimal surfaces.

Theorem 3.11.4. *A surface S of class $r \geq 3$ is a minimal surface or a sphere if and only if its spherical Gauss map is a conformal map.*

Proof.

(i) First we show that if S is a minimal surface or a sphere then its spherical Gauss map is conformal.

We assume that S is a minimal surface, that is, $H = 0$. Then we have by (2.153) in Definition 2.10.8

$$c_{ik} = -K g_{ik} \text{ for } i, k = 1, 2, \tag{3.149}$$

for the third fundamental coefficients c_{ik} of S, which are the first fundamental coefficients of the image S^* of the spherical Gauss map of S. By Theorem 3.9.5 and Remark 3.11.2, the conditions in (3.149) imply that the spherical Gauss map is conformal.

Trivially the spherical Gauss map of a sphere is conformal.

(ii) Now we show that if the spherical Gauss map of S is conformal, then S is a minimal surface.

We assume that the spherical Gauss map of S is conformal. Then it follows by Theorem 3.9.5 that

$$c_{ik} = \eta(u^i, u^2) g_{ik} \text{ for } i, k = 1, 2,$$

and so by (2.153) in Definition 2.10.8

$$c_{ik} = 2H L_k - K g_{ik} = \eta(u^i, u^2) g_{ik},$$

that is,

$$(K + \eta(u^i, u^2)) g_{ik} = 2H L_{ik} \text{ for } i, k = 1, 2,$$

hence

$$(K + \eta)g_{ik}du^i du^k = 2HL_{ik}du^i du^k. \tag{3.150}$$

We may assume that the second fundamental form does not vanish identically. Otherwise the surface would be a plane by Example 2.10.2, in which case its spherical Gauss image would reduce to a single point and the study of the conformity of the spherical Gauss map would not make sense.

If $K = -\eta < 0$, then $H = 0$ by (3.150) and S is a minimal surface. If we assume $H \neq 0$, then a solution of (3.150) is given by

$$\frac{K + \eta}{2H} = \frac{L_{ik}du^i du^k}{g_{ik}du^i du^2} = \kappa_n,$$

where κ_n denotes the normal curvature and the second equality holds by (2.47) in Remark 2.4.3 (c). Then the normal curvature at any point is the same in every direction and consequently every point of S is an umbilical point, and so S is a sphere by Theorem 2.10.4.

□

The following result explains the term *minimal surface*.

Theorem 3.11.5. *Let S be a surface bounded by a closed curve γ. If the surface area $A(S)$ of S is less than or equal to the surface area of any sufficiently close surface bounded by γ, then $H \equiv 0$ for the mean curvature of S.*

Proof. Let S be given by a parametric representation $\vec{x}(u^j)$ $((u^1, u^2) \in D)$. We consider a family of parallel surfaces given by the parametric representations

$$\vec{x}^*(u^j) = \vec{x}(u^j) + \varepsilon v(u^j)\vec{N}(u^j) \ ((u^1, u^2) \in D).$$

The Weingarten equations (2.141) yield

$$\vec{x}_i^* = \vec{x}_i + \varepsilon \frac{\partial v}{\partial u^i}\vec{N} + \varepsilon v\vec{N}_i = \vec{x}_i - \varepsilon v L_i^k \vec{x}_k + \varepsilon \frac{\partial v}{\partial u^i}\vec{N} \ \text{for } i = 1, 2$$

and

$$g_{ij}^* = \vec{x}_i^* \bullet \vec{x}_j^* = \left(\vec{x}_i - \varepsilon v L_i^k \vec{x}_k + \varepsilon \frac{\partial v}{\partial u^i}\vec{N}\right) \bullet \left(\vec{x}_j - \varepsilon v L_j^m \vec{x}_m + \varepsilon \frac{\partial v}{\partial u^j}\vec{N}\right)$$

$$= \left(\vec{x}_i - \varepsilon v L_i^k \vec{x}_k\right) \bullet \left(\vec{x}_j - \varepsilon v L_j^m \vec{x}_m\right) + \varepsilon^2 \frac{\partial v}{\partial u^i} \cdot \frac{\partial v}{\partial u^j}$$

$$= g_{ij} - \varepsilon v \left(L_j^m g_{im} + L_i^k g_{kj}\right) + \varepsilon^2 \frac{\partial v}{\partial u^i} \cdot \frac{\partial v}{\partial u^j} \ \text{for } i, j = 1, 2.$$

Since by the definition of L_j^m in (2.140)

$$L_j^m g_{im} = g^{ml} L_{lj} g_{im} = \delta_i^l L_{lj} = L_{ij}$$

and similarly $L_i^k g_{kj} = L_{ji} = L_{ij}$, we obtain

$$g_{ij}^* = g_{ij} - 2\varepsilon v L_{ij} + \varepsilon^2(\cdots) \ \text{for } i, j = 1, 2,$$

and

$$g^* = g_{11}^* g_{22}^* - (g_{12}^*)^2$$
$$= \left(g_{11} - 2\varepsilon v L_{11} + \varepsilon^2(\cdots)\right)\left(g_{22} - 2\varepsilon v L_{22} + \varepsilon^2(\cdots)\right) - \left(g_{12} - 2\varepsilon v L_{12} + \varepsilon^2(\cdots)\right)^2$$
$$= g_{11}g_{22} - 2\varepsilon v(g_{11}L_{22} + g_{22}L_{11}) - (g_{12}12 - 4\varepsilon v g_{123}L_{12}) - \varepsilon^2(\cdots)$$
$$= g_{11}g_{22} - g_{12}^2 - 2\varepsilon v\left(g_{11}L_{22} + g_{22}L_{11} - 2g_{12}L_{12}\right) - \varepsilon^2(\cdots),$$

hence by (2.72) in Remark 2.5.16 (a) and (2.74) in Definition 2.5.17

$$g^* = g\left(1 - 4\varepsilon v H\right) + \varepsilon^2(\cdots).$$

This implies

$$\left.\frac{\partial\left(\sqrt{g^*}\right)}{\partial\varepsilon}\right|_{\varepsilon=0} = \left.\frac{\partial\left(\sqrt{g\left(1 - 4\varepsilon v H\right) + \varepsilon^2(\cdots)}\right)}{\partial\varepsilon}\right|_{\varepsilon=0}$$

$$= \left.\frac{-4Hvg + 2\varepsilon(\cdots)}{2\sqrt{g\left(1 - 4\varepsilon v H\right) + \varepsilon^2(\cdots)}}\right|_{\varepsilon=0}$$

$$= -2Hv\sqrt{g}.$$

If the surface area $A(S)$ of S is to be minimal for $\varepsilon = 0$, then we must have

$$\left.\frac{\partial}{\partial\varepsilon}\left(\iint_D \sqrt{g^*}\, du^1 du^2\right)\right|_{\varepsilon=0} = -\iint_D 2Hv\sqrt{g}\, du^1 du^2 = 0$$

for all admissible functions $v(u^j)$. This implies $H = 0$ by the fundamental lemma of calculus of variations, Lemma 3.5.1. □

It is known from elementary geometry that the planar surface, which is bounded by a planar curve of length L and has maximum surface area A, is a disk; the following formula holds

$$A = \pi r^2 = \frac{L^2}{4\pi}, \tag{3.151}$$

where r is the radius of the disk. This result can be used to prove a similar relation between the surface area of a minimal surface and its bounding curve.

Theorem 3.11.6. *Let a part S of a minimal surface of class $r \geq 3$ be bounded by a closed simple curve γ of length L. If the surface area $A(S)$ of S corresponds to the minimum surface area of all surfaces of class r that are bounded by γ then the following estimate holds*

$$A(S) \leq \frac{L^2}{4\pi}.$$

Proof. Let S^* be the cone generated by joining the origin with the points of γ by straight lines. Then S^* is part of a developable surface by Theorem 2.8.16 and so can be mapped isometrically into a plane by Theorem 3.9.25. Furthermore, isometric maps are area preserving by Theorem 3.9.29 (a). Now (3.151) implies

$$A(S^*) \leq \frac{L^2}{4\pi}.$$

This implies the statement, since $A(S)$ is the minimum of the surface area of the part of the surface bounded by γ. ◻

We saw in Example 3.11.3 that catenoids and helicoids are minimal surfaces. It will turn out that they are the only minimal surfaces in the classes of surfaces of revolution and conoids, respectively.

Theorem 3.11.7. *(a) Catenoids are the only minimal surfaces of revolution.*
(b) Helicoids are the only minimal surfaces among the conoids.

Proof. (a) Let S be a surface of revolution given by the parametric representation

$$\vec{x}(u^i) = \{u^1 \cos u^2, u^1 \sin u^2, h(u^1)\} \left((u^1, u^2) \in D\right).$$

Since first and second fundamental coefficients are by (2.24) in Example 2.3.5 and (2.52)–(2.54) in Example 2.4.6

$$g_{11} = 1 + \left(h'(u^1)\right)^2, \; g_{12} = 0, \; g_{22} = \left(u^1\right)^2, \; g = g_{11}g_{22}$$

and

$$L_{11} = \frac{h''(u^1)}{\sqrt{1 + (h'(u^1))^2}}, \; L_{12} = 0 \text{ and } L_{22} = \frac{u^1 h'(u^1)}{\sqrt{1 + (h'(u^1))^2}},$$

the condition for S to be a minimal surface is

$$H = \frac{1}{2g}(L_{11}g_{22} + L_{22}g_{11}) = 0,$$

that is,

$$\frac{L_{11}}{g_{11}} + \frac{L_{22}}{g_{22}} = \frac{1}{\sqrt{1 + (h'(u^1))^2}} \left(\frac{h''(u^1)}{1 + (h'(u^1))^2} + \frac{h'(u^1)}{u^1}\right) = 0.$$

If $h(u^1) = const$, then we obtain a plane parallel to the $x^1 x^2$–plane. If $h'(u^1) \neq 0$, then $H = 0$ is equivalent to

$$\frac{h''(u^1)h'(u^1)}{(h'(u^1))^2 \left(1 + (h'(u^1))^2\right)} = -\frac{1}{u^1}.$$

We put $z = (h')^2$, hence $2h'h'' = z'$ and obtain

$$\frac{z'}{z(1 + z)} = \left(\frac{1}{z} - \frac{1}{1 + z}\right) z' = -\frac{2}{u^1}.$$

Integration yields

$$\log \frac{z}{z+1} = \log \frac{c^2}{(u^1)^2},$$

where $c \neq 0$ is a constant of integration, whence

$$\frac{(h')^2}{1+(h')^2} = \frac{c^2}{(u^1)^2}$$

or

$$h' = \frac{c}{\sqrt{(u^1)^2 - c^2}} \quad \text{for } |u^1| > |c|.$$

Integrating once again we obtain for $c > 0$ and $u^1 > c$

$$h(u^1) = c \cdot \cosh^{-1}\frac{u^1}{c} - d,$$

where d is a constant of integration. We may choose $d = 0$, since a change in the value of d means a translation of S along the x^3–axis. Introducing a new parameters u^{*i} $(i = 1, 2)$ by

$$u^1 = c \cdot \cosh\frac{u^{*1}}{c} \quad \text{and} \quad u^2 = u^{*2},$$

we obtain

$$\vec{x}(u^{*i}) = \left\{ c \cdot \cosh\frac{u^{*1}}{c} \cos u^{*2}, c \cdot \cosh\frac{u^{*1}}{c} \sin u^{*2}, u^{*1} \right\},$$

that is, part of a catenoid.

(b) Let S^* be a conoid given by a parametric representation

$$\vec{x}^*(u^{*i}) = \{u^{*1} \cos u^{*2}, u^{*1} \sin u^{*2}, h(u^{*2})\} \quad \left((u^1, u^2) \in D \subset \mathbb{R}^2\right).$$

We obtain

$$\vec{x}_1^* = \{\cos u^{*2}, \sin u^{*2}, 0\}, \quad \vec{x}_2^* = \{-u^{*1} \sin u^{*2}, u^{*1} \cos u^{*2}, h'(u^{*2})\}$$
$$g_{11}^* = 1, \quad g_{12}^* = 0, \quad g_{22}^* = (u^{*1})^2 + (h'(u^{*2}))^2,$$

$$\vec{x}_{11}^* = \vec{0}, \quad \vec{x}_{12}^* = \{-\sin u^{*2}, \cos u^{*2}, 0\},$$
$$\vec{x}_{22}^* = \{-u^{*1} \cos u^{*2}, -u^{*1} \sin u^{*2}, h''(u^{*2})\},$$

$$\vec{N}^* = \frac{1}{\sqrt{g^*}} \cdot \{h'(u^{*2}) \sin u^{*2}, -h'(u^{*2}) \cos u^{*2}, u^{*1}\},$$

$$L_{11}^* = \vec{N}^* \bullet \vec{x}_{11}^* = 0, \quad L_{12}^* = -\frac{h'(u^{*2})}{\sqrt{g^*}} \quad \text{and} \quad L_{22}^* = \frac{u^{*1} h(u^{*2})}{\sqrt{g^*}}.$$

Now the condition for S^* to be a minimal surface is

$$H = \frac{1}{2g^*} \cdot \left(L_{11}^* g_{22}^{*2} + L_{22}^* g_{11}^*\right) = \frac{u^{*1} h''(u^{*2})}{2(g^*)^{3/2}} = 0,$$

that is, $h''(u^{*2}) = 0$. Hence h must be a linear function of u^{*2} which means that S^* is a helicoid. □

One very important result of this section will be the *Weierstrass formulae* (3.157). They constitute a relation between minimal surfaces and the theory of complex functions.

The following fundamental results, which are also interesting in themselves, are needed for the proof on the Weierstrass formulae.

Theorem 3.11.8. *Let S be a simply connected minimal surface of class $r \geq 3$ with a parametric representation $\vec{x}(u^i)$ $((u^1, u^2) \in D)$ with respect to isothermal parameters. Then the component functions of $\vec{x}$ are harmonic functions, that is, in vector notation*

$$\Delta \vec{x} = \vec{x}_{11} + \vec{x}_{22} = \vec{0},$$

where Δ denotes the Laplace operator.

Proof. The existence of isothermal parameters for S follows from Definition 3.9.9 and Theorem 3.9.12. Then the first fundamental form of S is given by (3.115)

$$ds^2 = \eta(u^1, u^2) \left((du^1)^2 + (du^2)^2 \right)^2, \tag{3.152}$$

where $\eta > 0$, and so $g = g_{11}g_{22} = \eta^2$, hence $g^{11}g^{22} = 1/\eta$. The condition for S to be a minimal surface is now

$$H = \frac{1}{2}g^{ik}L_{ik} = \frac{1}{2\eta}\left(L_{11} + L_{22}\right) = \frac{1}{2\eta}\vec{N} \bullet (\vec{x}_{11} + \vec{x}_{22}) = \frac{1}{2\eta}\vec{N} \bullet \Delta\vec{x} = 0,$$

that is, $\Delta\vec{x}$ is orthogonal to the surface normal vector $\vec{N}$ of S.
It also follows from (3.152) that

$$g_{11} = \vec{x}_1 \bullet \vec{x}_1 = g_{22} = \vec{x}_2 \bullet \vec{x}_2 \text{ and } g_{12} = \vec{x}_1 \bullet \vec{x}_2 = 0 \tag{3.153}$$

and we obtain

$$\frac{1}{2}\frac{\partial g_{11}}{\partial u^1} = \vec{x}_1 \bullet \vec{x}_{11} = \frac{1}{2}\frac{\partial g_{22}}{\partial u^1} = \vec{x}_2 \bullet \vec{x}_{12} \text{ and } \frac{\partial g_{12}}{\partial u^2} = \vec{x}_1 \bullet \vec{x}_{22} + \vec{x}_2 \bullet \vec{x}_{12} = 0.$$

It follows that

$$\vec{x}_1 \bullet \vec{x}_{11} = \vec{x}_2 \bullet \vec{x}_{12} = -\vec{x}_1 \bullet \vec{x}_{22},$$

hence $\vec{x}_1 \bullet \Delta\vec{x} = 0$. Consequently $\Delta\vec{x}$ is orthogonal to $\vec{x}_1$.
If we interchange the partial differentiations in (3.153), then it follows similarly that $\Delta\vec{x}$ is orthogonal to $\vec{x}_2$.
This we have shown that $\Delta\vec{x}$ is orthogonal to $\vec{N}$, $\vec{x}_1$ and $\vec{x}_2$, which implies $\Delta\vec{x} = \vec{0}$. $\square$

We saw in Theorem 3.11.8 that each of the components x^k of the parametric representation

$$\vec{x}(u^i) = \left\{ x^1(u^1, u^2), x^2(u^1, u^2), x^3(u^1, u^2) \right\} \left((u^1, u^2) \in D \right)$$

of a minimal surface of class $r \geq 3$ with respect to isothermal parameters u^1 and u^2 is a solution of the so–called *potential equation*

$$\Delta x^k = \frac{\partial^2 x^k}{(\partial u^1)^2} + \frac{\partial^2 x^k}{(\partial u^2)^2} = 0,$$

that is, the component functions x^k are harmonic functions. By a known result from potential theory, they are the real parts of complex functions z^k of the complex variable $u = u^1 + iu^2$ with $i = \sqrt{-1}$; we write

$$\vec{x} = \mathrm{Re}(\vec{z}), \text{ where } \vec{z}(u) = \vec{x}(u^i) + i\vec{y}(u^i). \tag{3.154}$$

The following result holds.

Theorem 3.11.9. *Let S be a simply connected surface of class $r \geq 3$ with a parametric representation $\vec{x}(u^i)$ ($(u^1, u^2) \in D$) with respect to isothermal parameters. Then S is a minimal surface if and only if there exists a complex analytic function $\vec{z} = \vec{z}(u)$ of $u = u^1 + iu^2$ such that*

$$(\vec{z}')^2 = \left(\frac{d\vec{z}}{du}\right)^2 = 0 \text{ and } \vec{x}(u^i) = \mathrm{Re}\left(\vec{z}(u)\right).$$

Proof. (i) First we assume that S is a minimal surface with a parametric representation $\vec{x}(u^i)$ with respect to isothermal parameters. Then it follows by Theorem 3.11.8 that $\Delta\vec{x} = \vec{0}$. Writing $\vec{z}$ as in (3.154) the Cauchy–Riemann differential equations $\vec{x}_1 = \vec{y}_2$ and $\vec{x}_2 = -\vec{y}_1$ imply

$$\frac{d\vec{z}}{du} = \vec{x}_1 - i\vec{x}_2,$$

and so

$$\left(\frac{d\vec{z}}{du}\right)^2 = \vec{x}_1 \bullet \vec{x}_1 - 2i\vec{x}_1 \bullet \vec{x}_2 - \vec{x}_1 \bullet \vec{x}_2 = g_{11} - 2ig_{12} + g_{22}.$$

Since u^1 and u^2 are isothermal parameters, we have $g_{11} = g_{22}$ and $g_{12} = 0$, and consequently

$$\left(\frac{d\vec{z}}{du}\right)^2 = 0. \tag{3.155}$$

(ii) Conversely we assume that $\vec{x}$ is harmonic and the real part of a complex function $\vec{z}$ that satisfies (3.155). This implies $g_{11} = g_{22} = 0$ and $g_{12} = 0$, that is, u^1 and u^2 are isothermal parameters for S, and $\Delta\vec{x} = \vec{0}$ implies as in the proof of Theorem 3.11.8

$$H = \frac{1}{2\eta}\vec{N} \bullet \Delta\vec{x} = 0.$$

$\square$

Now we can derive the *Weierstrass formulae* which enable us to determine a minimal surface with a parametric representation $\vec{x}(u^i)$ for every given regular function $F \not\equiv 0$ of the complex variable $u = (u^1 + iu^2)$.

We use the notations of Theorem 3.11.9, and assume that f and g are arbitrary regular functions of $u = u^1 + iu^2$ on some domain. If we choose $\vec{z}(u) = \{z^1(u), z^2(u), z^3(u)\}$ such that

$$\frac{dz^1}{du} = f^2 - g^2, \quad \frac{dz^2}{du} = i(f^2 + g^2) \text{ and } \frac{dz^3}{du} = 2fg, \tag{3.156}$$

then

$$\frac{d\vec{z}}{du} \bullet \frac{d\vec{z}}{du} = \left(f^2 - g^2\right)^2 - \left(f^2 + g^2\right)^2 + 4f^2g^2 = 0,$$

that is, (3.155) is satisfied. Integration of the equations in (3.156) yields the following general representation for the minimal surfaces of class $r \geq 3$

$$\begin{cases} x^1 = \text{Re}(z^1) = \text{Re}\left(\int (f^2 - g^2)\,du\right) \\ x^2 = \text{Re}(z^2) = \text{Re}\left(\int i(f^2 + g^2)\,du\right) \\ x^3 = \text{Re}(z^3) = \text{Re}\left(\int 2fg\,du\right). \end{cases} \tag{3.157}$$

Now we assume $f \neq 0$ and $(g/f)' \neq 0$, that is, $fg' - f'g \neq 0$, since if $p \equiv 0$ or $(g/p)' \equiv 0$, then we obtain a part of a plane. We put $t = g/f$, hence

$$du = \frac{f^2}{g'f - f'g} \cdot dt,$$

where $'$ denotes differentiation with respect to u. We also put

$$F(t) = \frac{f^4}{g'f - f'g}.$$

Then $f^2 - g^2 = f^2(1 - g^2/f^2) = (1 - t^2)f^2$ and we obtain from the identity in the first line of (3.157)

$$x^1 = \text{Re}\left(\int (f^2 - g^2)\,du\right) = \text{Re}\left(\int (1 - t^2) \cdot \frac{f^4}{g'f - f'g}\,dt\right)$$
$$= \text{Re}\left(\int (1 - t^2) F(t)\,dt\right).$$

Analogous arguments yield the second and third formulae of the following identities known as the *Weierstrass equations*.

$$\begin{cases} x^1 = \text{Re}\left(\int (1 - t^2)F(t)\,dt\right) \\ x^2 = \text{Re}\left(\int i(1 + t^2)F(t)\,dt\right) \\ x^3 = \text{Re}\left(\int 2tF(t)\,dt\right) \end{cases}. \tag{3.158}$$

Remark 3.11.10. *The Weierstrass formulae (3.158) enable us to find a minimal surface for every regular function F of the complex variable $t = t^1 + it^2$.*
Putting $F = h'''$ then integration by parts yields

$$\left\{ \begin{array}{l} x^1 = \mathrm{Re}\left((1 - t^2)h'' + 2th' - 2h\right) \\ x^2 = \mathrm{Re}\left(i(1 + t^2)h'' - 2ith' + 2ih\right) \\ x^3 = \mathrm{Re}\left(2th'' - 2h'\right)dt). \end{array} \right\} \qquad (3.159)$$

Example 3.11.11 (The helicoid). *If $F(t) = i/(2t^2)$ $(t \neq 0)$ then the Weierstrass formulae (3.158) yield the minimal surface S with the parametric representation*

$$\vec{x}(u^i) = \{x^1(u^1, u^2), x^2(u^1, u^2), x^3(u^1, u^2)\},$$

where $\tan x^3 = x^1/x^2$.

Proof. We write $t = t_1 + it_2$ and obtain from the first formula in (3.158)

$$x^1 = \mathrm{Re}\left(\int (1 - t^2)\frac{i}{2t^2}\,dt\right) = -\mathrm{Re}\left(\frac{i}{2} \cdot \frac{1 + t^2}{t}\right).$$

Since $1/t = \bar{t}/(|t|^2) = (t_1 - it_2)/(t_1^2 + t_2^2)$, we get

$$x^1 = \frac{t_2}{2}\left(1 - \frac{1}{t_1^2 + t_2^2}\right).$$

Similarly the second formula in (3.158) yields

$$x^2 = -\frac{t_1}{2}\left(1 - \frac{1}{t_1^2 + t_2^2}\right).$$

Hence we have

$$\frac{x_1}{x_2} = -\frac{t_2}{t_1}. \qquad (3.160)$$

Furthermore, the third formula in (3.158) yields

$$x^3 = \mathrm{Re}\left(\int \frac{i}{t}\,dt\right) = \mathrm{Re}\left(i\log t\right) = \mathrm{Re}\left(i\left(\log|t| + i\tan^{-1}\left(\frac{t_2}{t_1}\right)\right)\right)$$
$$= -\tan^{-1}\left(\frac{t_2}{t_1}\right),$$

or by (3.160)

$$\tan\left(-x^3\right) = -\tan x^3 = \frac{t_2}{t_1} = -\frac{x^1}{x^2}. \qquad (3.161)$$

We introduce the parameters u^1 and u^2 by

$$x^1 = ru^1 \sin u^2 \text{ and } x^2 = ru^1 \cos u^2.$$

Then (3.161) yields $x^3 = u^2$, that is, the minimal surface is a helicoid. □

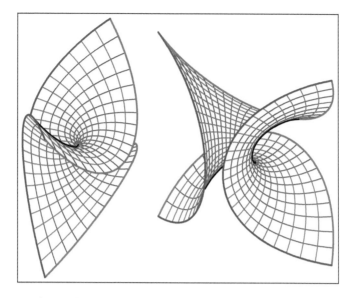

Figure 3.35 Enneper's surface with respect to the parametric representation (3.162)

Visualization 3.11.12 (The Enneper surface). *(a) The choice $F(t) \equiv 1$ in the Weierstrass formulae (3.158) yields* Enneper's minimal surface *ES given by a parametric representation (Figure 3.35)*

$$\vec{x}(u^i) = \left\{ u^1 - \frac{(u^1)^3}{3} + u^1(u^2)^2, u^2 - \frac{(u^2)^3}{3} + u^2(u^1)^2, (u^1)^2 - (u^2)^2 \right\}$$

$$\left((u^1, u^2) \in \mathbb{R}^2 \right). \quad (3.162)$$

(b) We introduce polar coordinates ρ and ϕ in $\mathbb{R}^2$ by

$$u^1(\rho, \phi) = \rho \cos \phi \text{ and } u^2(\rho, \phi) = \rho \sin \phi \text{ for } (\rho, \phi) \in (0, \infty) \times (0, 2\pi).$$

Then Enneper's surface ES has a parametric representation (Figure 3.36)

$$\vec{x}(\rho, \phi) = \left\{ \rho \cos \phi - \frac{\rho^3}{3} \cos(3\phi), \rho \sin \phi + \frac{\rho^3}{3} \sin(3\phi), \rho^2 \cos(2\phi) \right\}$$

$$((\rho, \phi) \in (0, \infty) \times (0, 2\pi)). \quad (3.163)$$

(c) The components $x^k(\rho, \phi)$ of the parametric representation (3.163) satisfy the relation

$$\left(x^1(\rho, \phi) \right)^2 + \left(x^2(\rho, \phi) \right)^2 + \frac{4}{3} \left(x^3(\rho, \phi) \right)^2 = \left(\rho + \frac{\rho^3}{3} \right)^2. \quad (3.164)$$

(d) The lines of self–intersection of Enneper's surface with the parametric representation (3.163) are given by

$$f_1(\rho, \phi) = \cos \phi - \frac{\rho^2}{3} \cos(3\phi) = 0$$

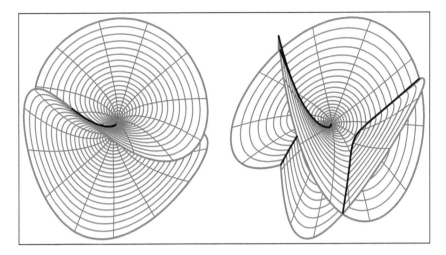

Figure 3.36 Enneper's surface with respect to the parametric representation (3.163)

and

$$f_2(\rho, \phi) = \sin\phi + \frac{\rho^2}{3}\sin(3\phi) = 0;$$

consequently they are in the planes $x = 0$ and $y = 0$, respectively (Figure 3.37).

Proof. (a) The formulae in (3.158) yield for $t = t_1 + it_2$

$$x^1 = \mathrm{Re}\left(\int(1 - t^2)\,dt\right) = \mathrm{Re}\left(t - \frac{t^3}{3}\right)$$

$$= t_1 - \frac{1}{3}\cdot\mathrm{Re}\left((t_1)^3 + 3it_1^2t_2 - 3t_1t_2^2 - it_2^3\right) = t_1 - \frac{t_1^3}{3} + t_1t_2^2,$$

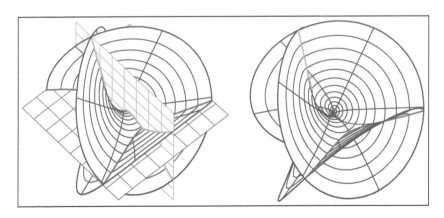

Figure 3.37 Enneper's surface and lines of self-intersection

$$x^2 = \mathrm{Re}\left(\int i(1+t^2)\,dt\right) = \mathrm{Re}\left(i\left(t+\frac{t^3}{3}\right)\right)$$

$$= -t_2 + \frac{1}{3}\mathrm{Re}\cdot\left(i\left((t_1)^3 + 3it_1^2 t_2 - 3t_1 t_2^2 - it_2^3\right)\right) = -t_2 - t_1^2 t_2 + \frac{t_2^3}{3},$$

and

$$x^3 = \mathrm{Re}\left(\int 2t\,dt\right) = \mathrm{Re}\left(t^2\right) = t_1^2 - t_2^2.$$

Introducing new parameters u^1 and u^2 by $t_1 = u^1$ and $t_2 = -u^2$, we obtain the parametric representation in (3.162).

(b) We write $x^k(u^i)$ and $x^k(\rho, \phi)$ ($k = 1, 2, 3$) for the components of the parametric representations of ES in (3.162) and (3.163). Then we obtain

$$x^3(\rho, \phi) = x^3(u^i(\rho, \phi)) = (u^1(\rho, \phi))^2 - (u^2(\rho, \phi))^2$$
$$= \rho^2\left(\cos^2\phi - \sin^2\phi\right) = \rho^2\cos(2\phi).$$

Also using

$$\cos(3\phi) = \mathrm{Re}(\exp(3i\phi)) = \mathrm{Re}\,(\cos\phi + i\sin\phi)^3 = \cos^3\phi - 3\cos\phi\sin^2\phi$$

and

$$\sin(3\phi) = \mathrm{Im}(\exp(3i\phi)) = \mathrm{Im}\,(\cos\phi + i\sin\phi)^3 = -\sin^3\phi + 3\cos^2\phi\cos\phi,$$

we conclude

$$x^1(\rho, \phi) = x^1(u^i(\rho, \phi)) = \rho\cos\phi - \frac{\rho^3}{3}\cos^3\phi + \rho^3\cos\phi\sin^2\phi$$

$$= \rho\cos\phi - \frac{\rho^3}{3}\left(\cos^3\phi - 3\cos\phi\sin^2\phi\right) = \rho\cos\phi - \frac{\rho^3}{3}\cos 3\phi$$

and

$$x^2(\rho, \phi) = x^2(u^i(\rho, \phi)) = \rho\sin\phi - \frac{\rho^3}{3}\sin^3\phi + \rho^3\sin\phi\cos^2\phi$$

$$= \rho\sin\phi - \frac{\rho^3}{3}\left(\sin^3\phi - 3\sin\phi\cos^2\phi\right) = \rho\sin\phi + \frac{\rho^3}{3}\sin(3\phi).$$

(c) We have

$$\left(x^1(\rho, \phi)\right)^2 + \left(x^2(\rho, \phi)\right)^2 + \frac{4}{3}\left(x^3(\rho, \phi)\right)^2 =$$

$$\rho^2\cos^2\phi - \frac{2\rho^4}{3}\cos\phi\cos(3\phi) + \frac{\rho^6}{9}\cos^2(3\phi)$$

$$+ \rho^2 \sin^2 \phi + \frac{2\rho^4}{3} \sin \phi \sin(3\phi) + \frac{\rho^6}{9} \sin^2(3\phi)$$

$$+ \frac{4}{3}\rho^4 \cos^2(2\phi) =$$

$$\rho^2 + \frac{\rho^6}{9} - \frac{2\rho^4}{3}(\cos\phi\cos(3\phi) - \sin\phi\sin(3\phi)) + \frac{4}{3}\rho^4\cos^2(2\phi) =$$

$$\rho^2 + \frac{\rho^6}{9} - \frac{2\rho^4}{3}\cos(4\phi) + \frac{4}{3}\rho^4\cos^2(2\phi) =$$

$$\rho^2 + \frac{\rho^6}{9} - \frac{2\rho^4}{3}\left(\cos^2(2\phi) - \sin^2(2\phi)\right) + \frac{4}{3}\rho^4\cos^2(2\phi) =$$

$$\rho^2 + \frac{\rho^6}{9} - \frac{2\rho^4}{3}\left(2\cos^2(2\phi) - 1\right) + \frac{4}{3}\rho^4\cos^2(2\phi) =$$

$$\rho^2 + \frac{\rho^6}{9} + \frac{2\rho^4}{3} = \left(\rho + \frac{\rho^3}{3}\right)^2.$$

(d)   The points of self–intersection of Enneper's surface given by a the parametric representation (3.163) must satisfy

$$\vec{x}(\rho_1, \phi_1) = \vec{x}(\rho_2, \phi_2),$$

that is,

$$x^k(\rho_1, \phi_1) = x^k(\rho_2, \phi_2) \text{ for } k = 1, 2, 3,$$

and it follows from (3.164) that

$$\rho_1 + \frac{\rho_1^3}{3} = \rho_2 + \frac{\rho_2^3}{3}.$$

Since the function $f : \mathbb{R} \to \mathbb{R}$ with $f(t) = t + t^3/3$ obviously is one–to–one, this implies $\rho_1 = \rho_2 = \rho$. Thus it follows from $x^3(\rho, \phi_1) = x^3(\rho, \phi_2)$, that $\cos(2\phi_1) = \cos(2\phi_2)$, hence $\phi_2 = \pi - \phi_1$ or $\phi_2 = 2\pi - \phi_1$.
If $\phi_2 = \pi - \phi_1$, then $x^1(\rho, \phi_1) = x^1(\rho, \pi - \phi_1)$ implies

$$\cos\phi_1 - \frac{\rho^2}{3}\cos(3\phi_1) = \cos(\pi - \phi_1) - \frac{\rho^2}{3}\cos(3(\pi - \phi_1))$$

$$= -\left(\cos\phi_1 - \frac{\rho^2}{3}\cos(3\phi_1)\right)$$

that is, $x^1(\rho, \phi_1) = -x^1(\rho, \phi_1) = f_1(\rho, \phi_1) = 0$.
If $\phi_2 = 2\pi - \phi_1$, then it can similarly be shown that $x^2(\rho, \phi_1) = -x^2(\rho, \phi_1) = f_2(\rho, \phi_1) = 0$.
It is easy to see that if $\phi_2 = \pi - \phi_1$ or $\phi_2 = 2\pi - \phi_1$, then $\vec{x}(\rho, \phi_1) = \vec{x}(\rho, \phi_2)$.

□

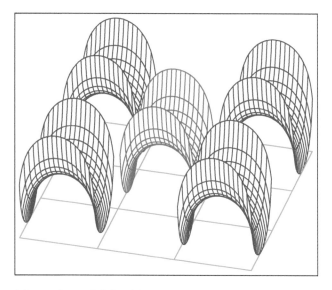

Figure 3.38 Several branches of Scherk's minimal surface

Visualization 3.11.13 (The Scherk surface). *The choice of $F(t) = 2/(1 - t^4)$ ($t \neq \pm 1, \pm i$) in the Weierstrass formulae (3.158) yields* Scherk's minimal surface *with a parametric representation (Figure 3.38)*

$$\vec{x}(u^i) = \left\{ u^1, u^2, \log \left(\frac{\cos u^2}{\cos u^1} \right) \right\} \tag{3.165}$$

for

$$(u^1, u^2) \in D_{j,k} = \left(\frac{(2j-1)\pi}{2}, \frac{(2j+1)\pi}{2} \right) \times \left(\frac{(2k-1)\pi}{2}, \frac{(2k+1)\pi}{2} \right) \quad (k, j \in \mathbb{Z}).$$

Proof. Let $(t \neq \pm 1, \pm i)$.
The first formula in (3.158) yields

$$x^1 = \mathrm{Re} \left(\int 2 \cdot \frac{1 - t^2}{1 - t^4} \, dt \right) = \mathrm{Re} \left(\int 2 \cdot \frac{1}{1 + t^2} \, dt \right) = \mathrm{Re} \left(\int \left(\frac{1}{1 - it} + \frac{1}{1 + it} \right) dt \right)$$

$$= \mathrm{Re} \left(i^{-1} \cdot \mathrm{Log} \left(\frac{1 + it}{1 - it} \right) \right) = \arg \left(\frac{1 + it}{1 - it} \right) + 2k\pi \ (k \in \mathbb{Z}),$$

where Log and $\arg(z)$ denote the complex logarithm and the angle of the complex number z measured anti–clockwise from the positive real axis. We will only consider the principal branch of Log, that is, $k = 0$ in the formula for arg. Writing $t = u + iv$ and $\bar{t} = u - iv$ for the conjugate complex number of t, we obtain

$$\frac{1 + it}{1 - it} = \frac{(1 + it)(1 + i\bar{t})}{|1 - it|^2} = \frac{1 - |t|^2 + 2i\mathrm{Re}(t)}{|1 - it|^2},$$

hence

$$\arg \left(\frac{1 + it}{1 - it} \right) = \tan^{-1} \left(\frac{2u}{1 - (u^2 + v^2)} \right). \tag{3.166}$$

Similarly the second formula in (3.158) yields

$$x^2 = \text{Re}\left(\int 2i \cdot \frac{1+t^2}{1-t^4}\,dt\right) = \text{Re}\left(\int 2i \cdot \frac{dt}{1-t^2}\right) = \text{Re}\left(\int i \cdot \left(\frac{1}{1-t}+\frac{1}{1+t}\right)dt\right)$$

$$= \text{Re}\left(i\left(\cdot\text{Log}\left(\frac{1+t}{1-t}\right)\right)\right) = -\arg\left(\frac{1+t}{1-t}\right).$$

We obtain

$$\frac{1+t}{1-t} = \frac{(1+t)(1-\bar{t})}{|1-t^2|} = \frac{1-|t|^2+2i\text{Im}(t)}{|1-t^2|},$$

hence

$$x^2 = \tan^{-1}\left(\frac{-2v}{1-(u^2+v^2)}\right). \tag{3.167}$$

Finally the third formula in (3.158) yields

$$x^3 = \text{Re}\left(\int \frac{4t}{1-t^4}\,dt\right) = \text{Re}\left(\int 2t\left(\frac{1}{1-t^2}+\frac{1}{1+t^2}\right)dt\right) = \text{Re}\left(\text{Log}\left(\frac{1+t^2}{1-t^2}\right)\right)$$

$$= \log\left|\frac{1+t^2}{1-t^2}\right|. \tag{3.168}$$

We have $t^2 = u^2 - v^2 + 2iuv$ and

$$\left|\frac{1+t^2}{1-t^2}\right|^2 = \frac{(1+t^2)(1+\bar{t}^2)}{(1-t^2)(1-\bar{t}^2)} = \frac{1+|t|^4+\text{Re}\left(t^2+\bar{t}^2\right)}{1+|t|^4-\text{Re}\left(t^2+\bar{t}^2\right)}$$

$$= \frac{1+(u^2+v^2)^2+2(u^2-v^2)}{1+(u^2+v^2)^2-2(u^2-v^2)}. \tag{3.169}$$

Since $\cos^2\varphi = (1+\tan^2\varphi)^{-1}$, we obtain from (3.166) and (3.167)

$$\cos^1 x^1 = \frac{1}{1+\left(\frac{2u}{u^2+v^2-1}\right)^2} = \frac{\left(u^2+v^2-1\right)^2}{\left(u^2+v^2-1\right)^2+4u^2},$$

$$\cos^2 x^2 = \frac{1}{1+\left(\frac{2v}{u^2+v^2-1}\right)^2} = \frac{\left(u^2+v^2-1\right)^2}{\left(u^2+v^2-1\right)^2+4v^2},$$

$$\frac{\cos x^2}{\cos x^1} = \frac{\left(u^2+v^2-1\right)^2+4u^2}{\left(u^2+v^2-1\right)^2+4v^2}$$

$$= \frac{\left(u^2+v^2\right)^2-2\left(u^2+v^2\right)+1+4u^2}{\left(u^2+v^2\right)^2-2\left(u^2+v^2\right)+1+4v^2}$$

$$= \frac{1+\left(u^2+v^2\right)^2+2\left(u^2-v^2\right)}{1+\left(u^2+v^2\right)^2+2\left(v^2-u^2\right)}$$

and by (3.168) and (3.169)

$$x^3 = \frac{1}{2}\cdot\log\left|\frac{1+t^2}{1-t^2}\right| = \log\left(\frac{\cos x^2}{\cos x^1}\right).$$

Finally, if we write $u^1 = u$ and $u^2 = v$, then we obtain the parametric representation (3.165) for $(u^1, u^2) \in D_{0,0}$. □

Example 3.11.14 (The catenoid). *The choice of $F(t) = -1/(2t^2)$ in the Weierstrass formulae (3.158) yields the catenoid with a parametric representation*

$$\vec{x}(u^i) = \left\{ \cosh u^1 \cos u^2, \cosh u^1 \sin u^2, u^1 \right\} \quad \left((u^1, u2) \in \mathbb{R} \times (0, 2\pi) \right). \quad (3.170)$$

Proof. The Weierstrass formulae (3.158) yield

$$x^1 = \mathrm{Re}\left(-\frac{1}{2} \int \left(1 - t^2 \right) \cdot \frac{1}{t^2}\, dt \right) = \frac{1}{2}\mathrm{Re}\left(t + \frac{1}{t} \right)$$

$$x^2 = \mathrm{Re}\left(-\frac{i}{2} \int \left(1 + t^2 \right) \cdot \frac{1}{t^2}\, dt \right) = \frac{1}{2}\mathrm{Re}\left(i \cdot \left(\frac{1}{t} - t \right) \right)$$

and

$$x^3 = \mathrm{Re}\left(\int \frac{-2t}{2t^2}\, dt \right) = -\mathrm{Re}(\mathrm{Log}(t)) = \log\left| \frac{1}{t} \right|.$$

We write $t = \exp\left(u^1 + iu^2 \right)$ and obtain

$$x^1 = \frac{1}{2}\mathrm{Re}\left(\exp\left(u^1 + iu^2 \right) + \exp\left(-u^1 - iu^2 \right) \right) = \frac{1}{2}\left(e^{u^1} + e^{-u^1} \right) \cos u^2$$
$$= \cosh u^1 \cos u^2,$$

$$x^2 = \frac{1}{2}\mathrm{Re}\left(i\exp\left(-u^1 - iu^2 \right) - \exp\left(u^1 + iu^2 \right) \right) = \frac{1}{2}\left(e^{u^1} + e^{-u^1} \right) \sin u^2$$
$$= \cosh u^1 \sin u^2$$

and

$$x^3 = -\log\left(|t| \right) = -\log\left(e^{u^1} \right) = -u^1.$$

Finally, since $\cosh\left(-u^1 \right) = \cosh u^1$, we may replace u^1 by $-u^1$, and obtain the parametric representation (3.170). □

Example 3.11.15 (Conjugate minimal surfaces). *Let S_f and S_g be surfaces given by parametric representations $\vec{x}$ and $\vec{y}$ with the component functions f^k and g^k ($k = 1, 2, 3$). If the functions f^k and g^k satisfy the Cauchy–Riemann differential equations, that is, if*

$$f_1^k = g_2^k \text{ and } f_2^k = -g_1^k \text{ for } k = 1, 2, 3, \quad (3.171)$$

then the component functions are said to be conjugate harmonic.
If $\vec{x}$ and $\vec{y}$ are the isothermal parametric representations of two minimal surfaces such that their component functions are pairwise conjugate harmonic, then the surfaces are said to be conjugate harmonic minimal surfaces.
(a) The helicoid and catenoid are conjugate minimal surfaces.

(b) If two conjugate minimal surfaces with the parametric representations $\vec{x}$ and $\vec{y}$ are given, then the surface with the parametric representation

$$\vec{z}^{(\lambda)} = (\cos \lambda) \cdot \vec{x} + (\sin \lambda) \cdot \vec{y} \tag{3.172}$$

is a minimal surface for each $\lambda \in \mathbb{R}$.

(c) The first fundamental coefficients $g_{ik}^{(\lambda)}$ of each surface with the parametric representation in (3.172) satisfy

$$g_{11}^{(\lambda)} = \vec{x}_1 \bullet \vec{x}_1 = \vec{y}_1 \bullet \vec{y}_1, \ g_{12}^{(\lambda)} = 0 \ and \ g_{22}^{(\lambda)} = \vec{x}_2 \bullet \vec{x}_2 = \vec{y}_2 \bullet \vec{y}_2. \tag{3.173}$$

Proof. We note that (3.171) implies that the component functions are harmonic functions.

(a) The catenoid is a minimal surface by Example 3.11.14; its parametric representation

$$\vec{x}(u^i) = \{\cosh u^1 \cos u^2, \cosh u^1 \sin u^2, u^1\} \ \left((u^1, u^2) \in \mathbb{R} \times (0, 2\pi) \right)$$

is isothermal, since

$$\vec{x}_1 = \{\sinh u^1 \cos u^2, \sinh u^1 \sin u^2, 1\},$$
$$\vec{x}_2 = \{-\cosh u^1 \sin u^2, \cosh u^1 \cos u^2, 0\}$$

and

$$g_{11} = 1 + \sinh^2 u^1 = \cosh^2 u^1 = g_{22} \ and \ g_{12} = 0.$$

The helicoid is a minimal surface by Example 3.11.11; its parametric representation (for $r = 1$ in Example 3.11.11)

$$\vec{y}(u^{*i}) = \{u^{*1} \sin u^{*2}, u^{*1} \cos u^{*2}, u^{*2}\} \ \left((u^{*1}, u^{*2}) \in \mathbb{R} \times (0, 2\pi) \right),$$

however, is not isothermal, since

$$g_{11}^* = 1 \neq 1 + \left(u^{*1} \right)^2 = g_{22}^*.$$

We introduce new parameters u^1 and u^2 by $u^{*1} = \sinh u^1$ and $u^{*2} = -u^2$, and write

$$\vec{y}(u^i) = \vec{y}(u^{*i}(u^j)) = \{-\sinh u^1 \sin u^2, \sinh u^1 \cos u^2, -u^2\}.$$

Then we obtain

$$\vec{y}_1 = \{-\cosh u^1 \sin u^2, \cosh u^1 \sin u^2, 0\} = \vec{x}_2,$$
$$\vec{y}_2 = \{-\sinh u^1 \cos u^2, -\sinh u^1 \cos u^2, -1\} = -\vec{x}_1,$$

hence the component functions of $\vec{x}$ and $\vec{y}$ are harmonic and $\vec{x}(u^i)$ and $\vec{y}(u^i)$ are parametric representations with respect to isothermal parameters.

The catenoid and helicoid are conjugate minimal surfaces.

(b) We assume that $\vec{x}$ and $\vec{y}$ are the parametric representations of two conjugate minimal surfaces, and write g_{ik} and h_{ik} $(i, k = 1, 2)$ for the first fundamental coefficients of $\vec{x}$ and $\vec{y}$, respectively. Since the component functions are conjugate harmonic and the parameters are isothermal, it follows that $\vec{x}_1 = \vec{y}_2$ and $\vec{x}_2 = -\vec{y}_1$, and

$$g_{11} = h_{22}, \ g_{22} = h_{11}, \ g_{12} = -h_{12} = 0 \text{ and } g_{11} = g_{22} = h_{11} = h_{22}.$$

We obtain

$$\vec{z}_1^{(\lambda)} = (\cos \lambda)\vec{x}_1 + (\sin \lambda)\vec{y}_1 \text{ and } \vec{z}_2^{(\lambda)} = (\cos \lambda)\vec{x}_2 + (\sin \lambda)\vec{y}_2,$$

hence

$$\begin{aligned}
g_{11}^{(\lambda)} &= \vec{z}_1^{(\lambda)} \bullet \vec{z}_1^{(\lambda)} = \left(\cos^2 \lambda\right) g_{11} + 2 \left(\cos \lambda \sin \lambda\right) (\vec{x}_1 \bullet \vec{y}_1) + \left(\sin^2 \lambda\right) h_{11} \\
&= \left(\cos^2 \lambda + \sin^2 \lambda\right) g_{11} - 2 \left(\cos \lambda \sin \lambda\right) (\vec{x}_1 \bullet \vec{x}_2) \\
&= g_{11} = h_{11}, \\
g_{12}^{(\lambda)} &= \vec{z}_1^{(\lambda)} \bullet \vec{z}_2^{(\lambda)} \\
&= \left(\cos^2 \lambda\right) g_{12} + \left(\cos \lambda \sin \lambda\right) (\vec{x}_1 \bullet \vec{y}_2 + \vec{x}_2 \bullet \vec{y}_1) + \left(\sin^2 \lambda\right) h_{12} \\
&= (\cos \lambda \sin \lambda) (\vec{x}_1 \bullet \vec{x}_1 - \vec{x}_2 \bullet \vec{x}_2) = (\cos \lambda \sin \lambda) (g_{11} - g_{22}) = 0
\end{aligned}$$

and

$$\begin{aligned}
g_{22}^{(\lambda)} &= \vec{z}_2^{(\lambda)} \bullet \vec{z}_2^{(\lambda)} = \left(\cos^2 \lambda\right) g_{22} + 2 \left(\cos \lambda \sin \lambda\right) (\vec{x}_2 \bullet \vec{y}_2) + \left(\sin^2 \lambda\right) h_{22} \\
&= \left(\cos^2 \lambda + \sin^2 \lambda\right) g_{11} - 2 \left(\cos \lambda \sin \lambda\right) (\vec{x}_1 \bullet \vec{x}_2) \\
&= g_{22} = h_{22}.
\end{aligned}$$

(c) Since $\vec{x}_1 = \vec{y}_2$ and $\vec{x}_2 = -\vec{y}_1$, we have

$$\begin{aligned}
\vec{z}_1^{(\lambda)} \times \vec{z}_2^{(\lambda)} &= ((\cos \lambda)\vec{x}_1 + (\sin \lambda)\vec{y}_1) \times ((\cos \lambda)\vec{x}_2 + (\sin \lambda)\vec{y}_2) \\
&= ((\cos \lambda)\vec{x}_1 - (\sin \lambda)\vec{x}_2) \times ((\cos \lambda)\vec{x}_2 + (\sin \lambda)\vec{x}_1) \\
&= \left((\cos \lambda)^2 + (\sin \lambda)^2\right) (\vec{x}_1 \times \vec{x}_2) = \sqrt{g}\vec{N},
\end{aligned}$$

where $\vec{N}$ is the surface normal vector of the surface given by $\vec{x}(u^i)$. So we have $\vec{N}^{(\lambda)} = \vec{N}$ for the surface normal vector of the surface with $\vec{z}^{(\lambda)}$. Furthermore, we have

$$\vec{z}_{11}^{(\lambda)} = (\cos(\lambda))\vec{x}_{11} + (\sin(\lambda))\vec{y}_{11} = (\cos(\lambda))\vec{x}_{11} - (\sin(\lambda))\vec{x}_{12}$$

and

$$\vec{z}_{22}^{(\lambda)} = (\cos(\lambda))\vec{x}_{22} + (\sin(\lambda))\vec{y}_{22} = (\cos(\lambda))\vec{x}_{22} + (\sin(\lambda))\vec{x}_{12}.$$

Thus the second fundamental coefficients $L_{ik}^{(\lambda)}$ of the surface given by $\vec{z}^{(\lambda)}$ are given by

$$L_{11}^{(\lambda)} = \vec{N}^{(\lambda)} \bullet \vec{z}_{11}^{(\lambda)} = \vec{N} \bullet \left((\cos\lambda)\, \vec{x}_{11} - (\sin\lambda)\, \vec{x}_{12} \right),$$
$$L_{12}^{(\lambda)} = \vec{N}^{(\lambda)} \bullet \vec{z}_{12}^{(\lambda)} = \vec{N} \bullet \left((\cos\lambda)\, \vec{x}_{12} - (\sin\lambda)\, \vec{x}_{22} \right)$$

and

$$L_{22}^{(\lambda)} = \vec{N}^{(\lambda)} \bullet \vec{z}_{22}^{(\lambda)} = \vec{N} \bullet \left((\cos\lambda)\, \vec{x}_{22} + (\sin\lambda)\, \vec{x}_{12} \right).$$

Let H and H^λ denote the mean curvature of the surfaces with $\vec{x}(u^i)$ and $\vec{z}^{(\lambda)}(u^i)$. Then we have by Part (c)

$$2\sqrt{g^{(\lambda)}}\, H^{(\lambda)} = L_{11}^{(\lambda)} g_{22}^{(\lambda)} - 2 L_{12}^{(\lambda)} g_{12}^{(\lambda)} + L_{22}^{(\lambda)} g_{11}^{(\lambda)} = \left(L_{11}^{(\lambda)} + L_{22}^{(\lambda)} \right) g_{11}$$
$$= \left((\cos\lambda)\, L_{11} - (\sin\lambda)\, L_{12} + (\cos\lambda)\, L_{22} + (\sin\lambda)\, L_{12} \right) g_{11}$$
$$= (\cos\lambda)\, (L_{11} g_{22} + L_{22} g_{11}) = 2(\cos\lambda)\sqrt{g}\, H = 0.$$

$\square$

CHAPTER 4

Tensor Algebra and Riemannian Geometry

In this chapter, we redevelop the intrinsic geometry of surfaces independently of their embedding in three-dimensional space $\mathbb{E}^3$, based only on the definition of measurements of lengths in a point set. This concept was originally introduced by *Riemann* in 1854.

For our purpose, we need the concept of a *manifold in n–dimensional space*. We shall obtain surfaces and n–dimensional Euclidean spaces as special cases of the manifolds.

There is one more reason for introducing manifolds. So far we only considered parts of surfaces and not surfaces as a whole. For instance, the whole sphere does not fall under the definition of surfaces given in Chapter 2, since there is no one–to–one correspondence between the whole sphere and a single part of the parameter plane.

Furthermore, we deal with an introduction into *tensor algebra*. This will provide an extremely useful formalism in the solution of problems in Differential Geometry.

In particular, we study

- *the definition of n–dimensional manifolds of class C^k* in Definition 4.1.1 of Section 4.1

- *the transformation formulae between the bases of a finite dimensional vector space, and between the components of vectors with respect to bases* in Proposition 4.2.1

- *spaces V^* of linear functionals on finite-dimensional vector spaces V*, the *isometry of V and V^** in Proposition 4.3.1, the *transformation formulae of the elements of V^* and their components* in Proposition 4.3.3, and the *concepts of contravariant and covariant vectors* in Definition 4.3.4

- *introduction of tensors of the second order* in Definition 4.4.1 and the *transformation formulae for their components with respect to different bases* in 4.4.2

- *symmetric bilinear forms and inner products* in Section 4.5

DOI: 10.1201/9781003370567-4

- *introduction of tensors of arbitrary order* in Definition 4.6.1 the *transformation formulae for their components* in Theorem 4.6.2, the *characterization of tensors by the transformation properties of their components, and the identification of tensors with their components* in Remark 4.6.4; *sums, outer products, contractions* (Definition 4.6.5) and *inner products of tensors*

- *symmetric and anti–symmetric tensors of arbitrary order* in Definition 4.7.1 and the *independence of their properties from the choice of the coordinate system* in Theorem 4.7.2

- *Riemann spaces* in Section 4.8 and the *metric tensor* in Definition 4.8.3

- *the Christoffel symbols in a Riemann space* in Definition 4.9 and *their transformation formulae* in Theorem 4.9.3.

4.1 DIFFERENTIABLE MANIFOLDS

In this section, we give the definition of an n–*dimensional manifold of class* C^k. First we need to recall a few definitions and well–known facts from *topology*.

Definition 4.1.1. (a) Let X be a nonempty set. A family $\mathcal{T}$ of subsets of X is called a *topology for* X, and the members of $\mathcal{T}$ are called *open sets* if $\mathcal{T}$ satisfies the following conditions

(T.1) $\qquad\qquad X, \emptyset \in \mathcal{T}$

(T.2) $\qquad\qquad \bigcup \mathcal{O} \in \mathcal{T}$ for every subcollection $\mathcal{O}$ of $\mathcal{T}$

(T.3) $\qquad\qquad O_1 \cap O_2 \in \mathcal{T}$ for all $O_1, O_2 \in \mathcal{T}$.

The pair $(X, \mathcal{T})$ is called a *topological space*; sometimes we write X for short.
(b) Let S be a set in a topological space $(X, \mathcal{T})$. We say that a subset of S is S–*open* if it is the intersection of S with a set in $\mathcal{T}$. The collection of S–open sets is called the *relative topology of* X *for* S, denoted by $\mathcal{T}_S$. Thus $\mathcal{T}_S = \{O \cap S : O \in \mathcal{T}\}$.
(c) A topological space $(X, \mathcal{T})$ is said to be a *Hausdorff space* if for any two distinct elements $x, x' \in X$ there are open sets O and O' such that $x \in O$, $x' \in O'$ and $O \cap O' = \emptyset$.
(d) Let $(X, \mathcal{T})$ be a topological space. A subset F of X is called *closed* if its complement $X \setminus F$ is open. A point $s \in X$ is called *interior to a set* S if there is an open set O with $s \in O \subset S$, in this case S is called a *neighbourhood of* s.
(e) A *local base at a point* x for a topology $\mathcal{T}$ is a collection Σ of neighbourhoods of x such that given any neighbourhood N of x there is $G \in \Sigma$ with $G \subset N$. A topological space $(X, \mathcal{T})$ is said to be *first countable at* $x \in X$ if it has a countable local base at x; it is said to be *first countable* if it is first countable at each of its points.

Example 4.1.2. *A metric space* (X, d) *has a local base at any of its points* x, *for instance, the collection of all sets* $N_{1/n}(x) = \{x' \in X : d(x', x) < 1/n\}$ $(n = 1, 2, \ldots)$. *Hence every metric space is first countable.*

Definition 4.1.3. Let X and Y be topological spaces. A map $f : X \to Y$ is said to be *continuous at* $x \in X$ if for every neighbourhood G of $f(x)$ in Y the pre–image $f^{-1}(G)$ of G is a neighbourhood of x in X; f is said to be *continuous on a subset S of X* if it is continuous at every point s in S; it is called *continuous* if it is continuous on X. The map f is called a *homeomorphism* if it is one–to–one, onto and continuous and if the inverse map $f^{-1} : Y \to X$ is also continuous.

Now we give the definition of a *manifold*.

Definition 4.1.4. An *n–dimensional manifold M of class C^k* is a first countable Hausdorff space with the following properties:
There exist a family $\mathcal{G} = \{G_\alpha : \alpha \in A\}$ of open sets with $M = \cup_{\alpha \in A} G_\alpha$, a so–called *open covering of M*, and a family $\mathcal{F}$ of homeomorphisms $f_\alpha : G_\alpha \to O_\alpha$ onto open subsets O_α of $\mathbb{R}^n$, and for each pair (α, β) with $G_\alpha \cap G_\beta \neq \emptyset$ there is a map $g_{\alpha,\beta} = f_\beta \circ f_\alpha^{-1} : f_\alpha(G_\alpha \cap G_\beta) \to f_\beta(G_\alpha \cap G_\beta)$ of class C^k with nonvanishing Jacobian.

Remark 4.1.5. *For each $\alpha \in A$, the map f_α in Definition 4.1.4 is a continuous, one–to–one map from the open set G_α of M onto an open subset O_α of $\mathbb{R}^n$ which has a continuous inverse map $f_\alpha^{-1}(O_\alpha) \to G_\alpha$. If $f_\alpha(P) = (\underset{\alpha}{x^1}, \underset{\alpha}{x^2}, \dots, \underset{\alpha}{x^n})$ and $f_\beta(P) = (\underset{\beta}{x^1}, \underset{\beta}{x^2}, \dots, \underset{\beta}{x^n})$ for a point $P \in G_\alpha \cap G_\beta$ then $g_{\alpha,\beta}$ is given by*

$$g_{\alpha,\beta}(f_\alpha(P)) = g_{\alpha,\beta}(\underset{\alpha}{x^1}, \underset{\alpha}{x^2}, \dots, \underset{\alpha}{x^n}) = (\underset{\beta}{x^1}, \underset{\beta}{x^2}, \dots, \underset{\beta}{x^n}).$$

If we write $\underset{\beta}{x^j} = \underset{\beta}{x^j}(\underset{\alpha}{x^i})$ $(j = 1, 2, \dots)$ for the coordinate functions of $g_{\alpha,\beta}$, then the functions $\underset{\beta}{x^j}$ have to have continuous partial derivatives of order k and

$$\frac{\partial \left(\underset{\beta}{x^j} \right)}{\partial \left(\underset{\alpha}{x^j} \right)} \neq 0 \text{ for the Jacobian.}$$

If the functions $\underset{\beta}{x^j}$ are analytic, *that is, if they have power series expansions, then the manifold is said to be* analytic *or of* class C^∞.

Example 4.1.6. *(a) The Euclidean point set $\mathbb{R}^n$ is an n–dimensional manifold of class C^∞. We may choose $\mathcal{G} = \{\mathbb{R}^n\}$ and $\mathcal{F} = \{id\}$ where $id : \mathbb{R}^n \to \mathbb{R}^n$ is the identity map defined by $id(x) = x$ for all $x \in \mathbb{R}^n$.*
(b) Every surface S of class C^r in $\mathbb{R}^3$ with parametric representation $\vec{x}(u^i)$ $((u^1, u^2) \in D)$, where $D \subset \mathbb{R}^2$ is a domain, is a two-dimensional manifold of class C^r. We have $S = \{X \in \mathbb{R}^3 : X = F(u^1, u^2), ((u^1, u^2 \in D)\}$, where the map $F = (f^1, f^2, f^3) : D \to \mathbb{R}^3$ is given by $f^k(u^1, u^2) = x^k(u^1, u^2)$ $(k = 1, 2, 3)$ and satisfies the conditions of Definition 2.1.1. We choose $\mathcal{G} = \{F(D)\}$ and $\mathcal{F} = \{f\}$ with $f = F^{-1}$.
(c) Let $F : \mathbb{R}^3 \to \mathbb{R}$ be a differentiable map and $c \in F(\mathbb{R}^3)$ with $\text{grad } F(X) \neq \vec{0}$. Then $F^{-1}(\{c\})$ is a manifold. If $(\partial F/\partial x^3)(X_0) \neq 0$ for some $X_0 = (x_0^1, x_0^2, x_0^3) \in F^{-1}(\{c\})$, then, by the implicit function theorem, there are a neighbourhood $N = N_{X_0}$ of (x_0^1, x_0^2),

a neighbourhood $V = V_{X_0}$ of x_0^3 and a differentiable map $g = g_{X_0} : N \to V$ with $F(x^1, x^2, g(x^1, x^2)) = c$, and more precisely

$$x \in (N \times V) \cap F^{-1}(\{c\}) \text{ if and only if } x^3 = g(x^1, x^2).$$

Consequently the function f_{X_0} on $(N \times V) \cap F^{-1}(\{c\})$ is given by the projection $P_{X_0}(x^1, x^2, x^3) = (x^1, x^2)$ with the inverse map $P_{X_0}^{-1}(x^1, x^2, g(x^1, x^2))$. Since $F^{-1}(\{c\}) \subset \mathbb{R}^3$, $F^{-1}(\{c\})$ is a metric, hence a Hausdorff space, and is first countable.

Definition 4.1.7. Let M be an n–dimensional manifold of class C^k, $\mathcal{G} = \{G_\alpha : \alpha \in A\}$ and $\tilde{\mathcal{G}} = \{\tilde{G}_\lambda : \lambda \in \tilde{A}\}$ be two open coverings of M, and $f_\alpha : G_\alpha \to O_\alpha$ and $\tilde{f}_\lambda : \tilde{G}_\lambda \to \tilde{O}_\lambda$ be maps that satisfy the conditions of Definition 4.1.4. A transformation from the system $\mathcal{G}$ to the system $\tilde{\mathcal{G}}$ is called an *admissible coordinate transformation* if each of the maps $g_{\alpha,\lambda} : f_\alpha(G_\alpha \cap \tilde{G}_\lambda) \to \tilde{f}_\lambda(G_\alpha \cap \tilde{G}_\lambda)$ is of class C^k or C^∞ and has nonvanishing Jacobian.

Remark 4.1.8. *(a) With the notations of Remark 4.1.5, a transformation from the system $\mathcal{G}$ to the system $\tilde{\mathcal{G}}$ is an admissible coordinate transformation if for each $P \in G_\alpha \cap \tilde{G}_\lambda$, the maps $\underset{\lambda}{\tilde{x}^j} = \underset{\lambda}{\tilde{x}^j}(\underset{\alpha}{x^i})$ are of class C^k (or C^∞) with nonvanishing Jacobian.*
(b) A differentiable manifold has only a topological structure, but no geometric one. The geometric properties will have to be independent of the choice of a coordinate system.

Definition 4.1.9. (a) The continuous image $P(I)$ of an interval $I = [a, b]$ is called a *curve*. A curve is said to be *m times continuously differentiable* ($m \leq k$ if the manifold is of class C^k), if the coordinate functions x^i with respect to some coordinate system are m times continuously differentiable. This definition makes sense, since, by the chain rule, $\tilde{x}^j(t) = \tilde{x}^j(x^i(t))$ is differentiable in any other admissible coordinate system. A *smooth curve* is a curve that is at least once continuously differentiable; a curve that consists of a finite number of smooth parts is called *piecewise smooth*.
(b) Let P be a given point. Two curves through P with coordinates $x^i(t)$ and $y^i(\tau)$ are said to *have the same tangent vector $\vec{t}$ at P* if

$$\frac{dx^i}{dt} = \frac{dy^i}{d\tau} \text{ at } P.$$

Remark 4.1.10. *(a) By the chain rule, the definition of tangent vectors is independent of the choice of the coordinate system.*
(b) Let P be a given point. We define a relation $\mathcal{R}$ on the set $SC(P)$ of all smooth curves through P by $\gamma \mathcal{R} \gamma^*$ if and only if γ and γ^* have the same tangents at P. It is easy to see that $\mathcal{R}$ is an equivalence relation. Now we define the tangent vectors $\vec{t}$ as the equivalence classes $[\vec{t}]$ of $\mathcal{R}$. Let $T(P)$ denote the set of all tangent vectors through P, that is, of all equivalence classes. We define a multiplication with scalars and an addition on $T(P)$. Let $\vec{t} \in T(P)$ and $c \in \mathbb{R}$ be given. If γ is a representative of

the class $\vec{t}$ given by the coordinates $x^i(t)$, then we have for γ^* given by the coordinates $z^i(t) = x^i(ct)$

$$\frac{dz^i}{dt} = c\frac{dx^i}{dt} \text{ at } P,$$

and we define $c \cdot \vec{t}$ as the equivalence class of γ^*. Furthermore, let $\gamma_1(\vec{t})$ and $\gamma_2(\vec{t})$ be two representatives of the class $\vec{t}$ given by the coordinates $x_1^i(t)$ and $x_2^i(t)$, and $\gamma_1(\vec{u})$ and $\gamma_2(\vec{u})$ two representatives of the class $\vec{u}$ given by the coordinates $y_1^i(t)$ and $y_2^i(t)$. Then

$$\frac{dx_1^i}{dt} = \frac{dx_2^i}{dt} \text{ and } \frac{dy_1^i}{dt} = \frac{dy_2^i}{dt} \text{ at } P$$

and so

$$\frac{d}{dt}(x_1^i + x_2^i) = \frac{d}{dt}(y_1^i + y_2^i) \text{ at } P.$$

Consequently we define $\vec{t} + \vec{u}$ to be the equivalence class of the curve given by the coordinates $x_1^i(t) + y_1^i(t)$.

In both cases the definition is independent from the choice of a coordinate system, as can be seen from the chain rule.

It is easy to see that $T(P)$ is a vector space with the operations just defined. Since the curves given by $x^i(t) = t\delta_k^i$ where $\delta_k^k = 1$ and $\delta_k^i = 0$ for $i \neq k$ have the tangent vectors

$$\frac{dx^i}{dt} = \delta_k^i$$

and any tangent vector can be represented as

$$\frac{dy^i}{dt} = a^k\delta_k^i \text{ where } a^k = \frac{dy^k}{dt},$$

the vector space has dimension n.

4.2 TRANSFORMATION OF BASES

In Section 4.1, we saw that we can associate an n–dimensional vector space with every point of a differentiable manifold. Therefore, we take a closer look at vector spaces and first study *transformations of bases*.

Let $H = \{\vec{b}_1, \vec{b}_2, \ldots, \vec{b}_n\}$ be a basis of an n–dimensional vector space V. Then every vector $\vec{x} \in V$ has a unique representation

$$\vec{x} = x^k\vec{b}_k, \tag{4.1}$$

and the real numbers x^k $(k = 1, 2, \ldots, n)$ are called the *components of $\vec{x}$ with respect to the basis H*. If $\tilde{H} = \{\vec{\tilde{b}}_1, \vec{\tilde{b}}_2, \ldots, \vec{\tilde{b}}_n\}$ is another basis, then, by (4.1), each vector $\vec{\tilde{b}}_k$ has a unique representation in terms of the vectors $\vec{b}_i$ and each vector $\vec{b}_j$ has a unique representation in terms of the vectors $\vec{\tilde{b}}_k$

$$\vec{\tilde{b}}_k = a_k^i\vec{b}_i \ (k = 1, 2, \ldots, n) \text{ and } \vec{b}_j = \alpha_j^k\vec{\tilde{b}}_k \ (j = 1, 2, \ldots, n). \tag{4.2}$$

The equations in (4.2) and describe a *basis transformation*.

Proposition 4.2.1. *Let $H = \{\vec{b}_1, \vec{b}_2, \ldots, \vec{b}_n\}$ and $\tilde{H} = \{\vec{\tilde{b}}_1, \vec{\tilde{b}}_2, \ldots, \vec{\tilde{b}}_n\}$ be bases of an n–dimensional vector space V and their transformation be given by (4.2). Then*

$$a^i_k \alpha^l_i = \alpha^i_k a^l_i = \delta^l_k \quad (k, l = 1, 2, \ldots, n), \tag{4.3}$$

and the components x^k and $\tilde{x}^k$ $(k = 1, 2, \ldots, n)$ of a given vector $\vec{x} \in V$ with respect to H and $\tilde{H}$ transform as

$$\tilde{x}^k = \alpha^k_j x^j \ (k = 1, 2, \ldots, n) \text{ and } x^j = a^j_k \tilde{x}^k \ (j = 1, 2, \ldots, n). \tag{4.4}$$

Proof. We have by (4.2)

$$\vec{b}_k = a^i_k \vec{\tilde{b}}_i = a^i_k \alpha^l_i \vec{b}_l \text{ and } \vec{b}_k = \alpha^i_k \vec{\tilde{b}}_i = \alpha^i_k a^l_i \vec{b}_l \ (k = 1, 2, \ldots, n),$$

and this implies (4.3), since the sets H and $\tilde{H}$ are linearly independent. Furthermore,

$$\vec{x} = x^j \vec{b}_j = x^j \alpha^k_j \vec{\tilde{b}}_k = \tilde{x}^k \vec{\tilde{b}}_k = x^k a^j_k \vec{b}_j$$

implies (4.4). $\qquad \square$

Example 4.2.2. *In two-dimensional V^2, let a transformation of bases be defined by $\vec{\tilde{b}}_1 = \vec{b}_1 + \vec{b}_2$ and $\vec{\tilde{b}}_2 = \vec{b}_2$. Then we have*

$$\left(a^i_k \right) = \begin{pmatrix} 1 & 1 \\ 0 & 1 \end{pmatrix} \text{ and } \left(\alpha^i_k \right) = \begin{pmatrix} 1 & -1 \\ 0 & 1 \end{pmatrix}.$$

The components of a vector transform as follows:

$$\tilde{x}^k = \alpha^k_j x^j \ (k = 1, 2), \text{ hence } \tilde{x}^1 = x^1 \text{ and } \tilde{x}^2 = -x_1 + x_2.$$

Example 4.2.3. *Let $\mathbb{E}^n$ be the n–dimensional real Euclidean space with the dot product • defined as usual by*

$$\vec{x} \bullet \vec{y} = \sum_{k=1}^n x^k y^k \text{ for all } \vec{x} = \{x^1, x^2, \ldots, x^n\}, \vec{y} = \{y^1, y^2, \ldots, y^n\} \in \mathbb{E}^n.$$

A basis $H = \{\vec{b}_1, \vec{b}_2, \ldots, \vec{b}_n\}$ is said to be orthonormal *if*

$$\vec{b}_i \bullet \vec{b}_k = \delta_{ik} = \begin{cases} 1 & (i = k) \\ 0 & (i \neq k) \end{cases} \quad (i, k = 1, 2, \ldots, n).$$

Then the formulae in (4.2) transform the orthonormal basis $H = \{\vec{b}_1, \vec{b}_2, \ldots, \vec{b}_n\}$ into the orthonormal basis $\tilde{H} = \{\vec{\tilde{b}}_1, \vec{\tilde{b}}_2, \ldots, \vec{\tilde{b}}_n\}$ if and only if

$$a^j_i = \alpha^i_j \text{ for all } i, j = 1, 2, \ldots, n. \tag{4.5}$$

Proof.

(i) First we assume that the conditions in (4.5) are satisfied.
Then we obtain from $\vec{b}_i \bullet \vec{b}_k = \delta_{ik}$ for $i, k = 1, 2, \ldots, n$ by (4.3) for all $l, m = 1, 2, \ldots, n$

$$\vec{\tilde{b}}_l \bullet \vec{\tilde{b}}_m = a_l^i \vec{b}_i \bullet a_m^k \vec{b}_k = a_l^i a_m^k \vec{b}_i \bullet \vec{b}_k = a_l^i a_m^k \delta_{ik} = \alpha_i^l a_m^k \delta_{ik} = \alpha_k^l a_m^k = \delta_m^l.$$

(ii) Now we assume that the bases H and $\tilde{H}$ are orthonormal and connected by the transformation formulae (4.2). Then we obtain for all $i, j = 1, 2, \ldots, n$

$$\alpha_j^i = \alpha_j^k \delta_{ki} = \alpha_j^k \vec{\tilde{b}}_k \bullet \vec{b}_i = \vec{b}_j \bullet \vec{b}_i = \vec{b}_j \bullet a_i^l \vec{b}_l = a_i^l \delta_{jl} = a_i^j,$$

which is (4.5).

□

4.3   LINEAR FUNCTIONALS AND DUAL SPACES

In this section, we deal with *linear functionals*, *dual spaces* and *transformations of bases of dual spaces*.

Let V be a real vector space. Then a map $f : V \to \mathbb{R}$ is called a *linear functional* if

$$f(a\vec{x} + b\vec{y}) = af(\vec{x}) + bf(\vec{y}) \text{ for all } \vec{x}, \vec{y} \in V \text{ and all } a, b \in \mathbb{R}. \tag{4.6}$$

The set of all linear functionals on V is denoted by V^* and called the *dual space of V*. We define $f + g$ and αf for all $f, g \in V$ and all $\alpha \in \mathbb{R}$ by $(f + g)(\vec{x}) = f(\vec{x}) + g(\vec{x})$ and $(\alpha f)(\vec{x}) = \alpha f(\vec{x})$ $(\vec{x} \in V)$. Then it is easy to see that V^* is a real vector space.

We recall that two vector spaces V and $\tilde{V}$ are said to be *isomorphic* if there exists a linear bijective map between them. The next result is well known from linear algebra.

Proposition 4.3.1. *The vector spaces V and V^* are isomorphic.*

Proof. Let $H = \{\vec{b}_1, \vec{b}_2, \ldots, \vec{b}_n\}$ be a basis of V and $\vec{x} = x^k \vec{b}_k \in V$ be given. We define the maps $f_k : V \to \mathbb{R}$ $(k = 1, 2, \ldots, n)$ by

$$f_k(\vec{y}) = y^k \text{ for all } \vec{y} = y^k \vec{b}_k \in V, \tag{4.7}$$

and put $f = x^k f_k$. Now we define the map T on V by $T(x) = f$. Since obviously $f_k \in V^*$ for each $k = 1, 2, \ldots, n$, it follows that $f \in V^*$, and so $T : V \to V^*$.

(i) First we show that T is linear.
Let $\alpha, \tilde{\alpha} \in \mathbb{R}$ and $\vec{x} = x^k \vec{b}_k, \vec{\tilde{x}} = \tilde{x}^k \vec{b}_k \in V$. Then $\vec{y} = \alpha \vec{x} + \tilde{\alpha} \vec{\tilde{x}} = y^k \vec{b}_k$ with $y^k = \alpha x^k + \tilde{\alpha} \tilde{x}^k$ $(k = 1, 2, \ldots, n)$ and

$$T(\alpha \vec{x} + \tilde{\alpha} \vec{\tilde{x}}) = T(\vec{y}) = y^k f_k = \alpha x^k f_k + \tilde{\alpha} \tilde{x}^k f_k = \alpha T(\vec{x}) + \tilde{\alpha} T(\vec{\tilde{x}}).$$

(ii) Now we show that T is one-to-one.
Let $T(\vec{x}) = T(\tilde{\vec{x}})$, that is,

$$x^j = x^k f_k(\vec{b}_j) = (T(\vec{x}))(\vec{b}_j) = (T(\tilde{\vec{x}})(\vec{b}_j) = \tilde{x}^k f_k(\vec{b}_j) = \tilde{x}^j \ (j = 1, 2, \ldots, n)$$

by the definition of T and the linear functionals f_k $(k = 1, 2, \ldots, n)$, and so
$\vec{x} = x^j \vec{b}_j = \tilde{x}^j \vec{b}_j = \tilde{\vec{x}}$.

(iii) Finally we show that T is onto.
Let $f \in V^*$ be given. We put $x^k = f(\vec{b}_k)$ $(k = 1, 2, \ldots, n)$. Then $\vec{x} = x^k \vec{b}_k \in V$ and, for all $\vec{y} = y^j \vec{b}_j$,

$$(T(\vec{x}))(\vec{y}) = x^k f_k(\vec{y}) = \sum_{k=1}^{n} f(\vec{b}_k) f_k(\vec{y}) = f(\vec{b}_k) y^k = f(\vec{y}),$$

that is, $T(x) = f$.

$\square$

Remark 4.3.2. *Since an n–dimensional vector space V and its dual space V^* are isomorphic, V^* is an n–dimensional vector space. Let $H = \{\vec{b}_1, \vec{b}_2, \ldots, \vec{b}_n\}$ be a basis of V. We define the vectors $\vec{b}^i \in V^*$ $(i = 1, 2, \ldots, n)$ by $\vec{b}^i(\vec{b}_k) = \delta_k^i$ $(i, k = 1, 2, \ldots, n)$. Since $\vec{b}^i \in V^*$ $(i = 1, 2, \ldots, n)$, we have $\vec{b}^i(\vec{x}) = \vec{b}^i(x^k \vec{b}_k) = x^k \vec{b}^i(\vec{b}_k) = x^i$ $(i = 1, 2, \ldots, n)$ for all $\vec{x} = x^k \vec{b}_k \in V$, and $\vec{b}^i$ indeed is defined on all of V. Furthermore, $\lambda_i \vec{b}^i = \vec{0}$ implies $\lambda_i \vec{b}^i(\vec{x}) = \vec{0}$ for all $\vec{x} \in V$, in particular, for $\vec{x} = \vec{b}_k$ $(k = 1, 2, \ldots, n)$, $\lambda_i \vec{b}^i(\vec{b}_k) = \lambda_i \delta_k^i = \lambda_k = 0$ $(k = 1, 2, \ldots, n)$. Thus the set $\mathcal{H} = \{\vec{b}^1, \vec{b}^2, \ldots, \vec{b}^n\}$ is linearly independent. Since V^* is an n–dimensional vector space, $\mathcal{H}$ is a basis of V^*, the so–called* dual basis. *Therefore, every vector $l \in V^*$ has a unique representation $l = l_i \vec{b}^i$, where $l_i = l(\vec{b}_i)$ for $i = 1, 2, \ldots, n$.*

Proposition 4.3.3. *Let $H = \{\vec{b}_1, \vec{b}_2, \ldots, \vec{b}_n\}$ and $\tilde{H} = \{\tilde{\vec{b}}_1, \tilde{\vec{b}}_2, \ldots, \tilde{\vec{b}}_n\}$ be bases of V with the transformation (4.2), and $\mathcal{H} = \{\vec{b}^1, \vec{b}^2, \ldots, \vec{b}^n\}$ and $\tilde{\mathcal{H}} = \{\tilde{\vec{b}}^1, \tilde{\vec{b}}^2, \ldots, \tilde{\vec{b}}^n\}$ be the corresponding dual bases, that is, $\vec{b}^k(\vec{b}_j) = \delta_j^k$ and $\tilde{\vec{b}}^k(\tilde{\vec{b}}_j) = \delta_j^k$ $(k, j = 1, 2, \ldots, n)$. If $l = l_i \vec{b}^i = \tilde{l}_i \tilde{\vec{b}}^i$, then*

$$l_i = \alpha_i^k \tilde{l}_k \ (i = 1, 2, \ldots, n) \ and \ \tilde{l}_k = a_k^i l_i \ (k = 1, 2, \ldots, n). \tag{4.8}$$

Furthermore,

$$\vec{b}^k = a_i^k \tilde{\vec{b}}^i \ (k = 1, 2, \ldots, n) \ and \ \tilde{\vec{b}}^i = \alpha_k^i \vec{b}^k \ (i = 1, 2, \ldots, n). \tag{4.9}$$

Therefore, the components of the vectors of the dual space V^* transform like the vectors of the bases of V, and the vectors of the dual bases transform like the components of the vectors of V.

Proof. Since $l_i = l(\vec{b}_i)$ for $i = 1, 2, \ldots, n$ and $\tilde{l}_k = l(\tilde{\vec{b}}_k)$ for $k = 1, 2, \ldots, n$, the linearity of l implies

$$l_i = l(\vec{b}_i) = l(\alpha_i^k \tilde{\vec{b}}_k) = \alpha_i^k l(\tilde{\vec{b}}_k) = \alpha_i^k \tilde{l}_k \ (i = 1, 2, \ldots, n)$$

and

$$\tilde{l}_k = l(\tilde{\vec{b}}_k) = l(a_k^i \vec{b}_i) = a_k^i l(\vec{b}_i) = a_k^i l_i \ (k = 1, 2, \ldots, n),$$

which proves (4.8).

Furthermore, we obtain

$$\vec{b}^i = \vec{b}^i(\vec{b}_k)\vec{b}^k = \vec{b}^i(\alpha_k^l \tilde{\vec{b}}_l)\vec{b}^k = \alpha_k^l \vec{b}^i(\tilde{\vec{b}}_l)\vec{b}^k = \alpha_k^l \delta_l^i \vec{b}^k = \alpha_k^i \vec{b}^k \ (i = 1, 2, \ldots, n)$$

and

$$\vec{b}^k = \vec{b}^k(\vec{b}_i)\tilde{\vec{b}}^i = \vec{b}^k(a_i^l \tilde{\vec{b}}_l)\tilde{\vec{b}}^i = a_i^l \vec{b}^k(\tilde{\vec{b}}_l)\tilde{\vec{b}}^i = a_i^l \delta_l^k \tilde{\vec{b}}^i = a_i^k \tilde{\vec{b}}^i \ (k = 1, 2, \ldots, n),$$

which proves (4.9). □

Definition 4.3.4. The vectors of V and V^* are called *contravariant* and *covariant* vectors, respectively. The components of contravariant vectors have a superscript, and those of covariant vectors have a subscript.

Remark 4.3.5. *For basis transformations given by*

$$\tilde{\vec{b}}_k = a_k^i \vec{b}_i \ (k = 1, 2, \ldots, n) \ and \ \vec{b}_j = \alpha_j^k \tilde{\vec{b}}_k \ (j = 1, 2, \ldots, n),$$

the components x^k of a contravariant vector and the components x_k of a covariant vector transform as follows

$$\tilde{x}^k = \alpha_j^k x^j \ (k = 1, 2, \ldots, n) \qquad x^j = a_k^j \tilde{x}^k \ (j = 1, 2, \ldots, n)$$
$$\tilde{x}_k = a_k^j x_j \ (k = 1, 2, \ldots, n) \quad and \ x_j = \alpha_j^k \tilde{x}_k \ (j = 1, 2, \ldots, n).$$

4.4 TENSORS OF SECOND ORDER

Now we generalize the concepts of contravariant and covariant vectors.

Let X and Y be two vector spaces. A function $B : X \times Y \to \mathbb{R}$ is called a *bilinear form* if for every fixed vector $\vec{y}$ in Y, the map $B(\cdot, \vec{y}) : X \to \mathbb{R}$ and for every fixed vector $\vec{x}$ in X, the map $B(\vec{x}, \cdot) : Y \to \mathbb{R}$ are linear functionals. Here, we are interested in those cases where X and Y are equal to an n–dimensional vector space V or its dual space. There are three different cases

(a) $X = Y = V$, (b) $X = Y = V^*$, (c) $X = V$ and $Y = V^*$.

These three cases lead to the so–called *tensors of second order*.

Definition 4.4.1. (a) A bilinear form B on $X = Y = V$ is called a *covariant tensor of second order*.
(b) A bilinear form B on $X = Y = V^*$ is called a *contravariant tensor of second order*.
(c) A bilinear form B on $X = V$ and $Y = V^*$ is called a *mixed tensor of second order*.

Proposition 4.4.2. *Let $H = \{\vec{b}_1, \vec{b}_2, \ldots, \vec{b}_n\}$ and $\tilde{H} = \{\vec{\tilde{b}}_1, \vec{\tilde{b}}_2, \ldots, \vec{\tilde{b}}_n\}$ be bases of V, and $\mathcal{H} = \{\vec{b}^1, \vec{b}^2, \ldots, \vec{b}^n\}$ and $\tilde{\mathcal{H}} = \{\vec{\tilde{b}}^1, \vec{\tilde{b}}^2, \ldots, \vec{\tilde{b}}^n\}$ be bases of V^* with the transformations (4.2) and (4.9).*

(a) *Every covariant tensor B of second order can be written as*

$$B(\vec{x}, \vec{y}) = b_{ik} x^i y^k \quad (\vec{x} = x^i \vec{b}_i, \vec{y} = y^i \vec{b}_i \in V), \tag{4.10}$$

where $b_{ik} = B(\vec{b}_i, \vec{b}_k)$ for all $i, k = 1, 2, \ldots, n$, and conversely (4.10) defines a covariant tensor of second order. The components b_{ik} of B satisfy the transformation formulae

$$\tilde{b}_{jl} = a^i_j a^k_l b_{ik} \text{ and } b_{jl} = \alpha^i_j \alpha^k_l \tilde{b}_{ik} \quad (j, l = 1, 2, \ldots, n). \tag{4.11}$$

(b) *Every contravariant tensor B of second order can be written as*

$$B(\vec{x}, \vec{y}) = b^{ik} x_i y_k \quad (\vec{x} = x_i \vec{b}^i, \vec{y} = y_i \vec{b}^i \in V), \tag{4.12}$$

where $b^{ik} = B(\vec{b}^i, \vec{b}^k)$ for all $i, k = 1, 2, \ldots, n$, and conversely (4.12) defines a covariant tensor of second order. The components b^{ik} of B satisfy the transformation formulae

$$\tilde{b}^{jl} = \alpha^j_i \alpha^l_k b^{ik} \text{ and } b^{jl} = a^j_i a^l_k \tilde{b}^{ik} \quad (j, l = 1, 2, \ldots, n). \tag{4.13}$$

(c) *Every mixed tensor B of second order can be written as*

$$B(\vec{x}, \vec{y}) = b^k_i x^i y_k \quad (\vec{x} = x^i \vec{b}_i, \vec{y} = y_i \vec{b}^i \in V), \tag{4.14}$$

where $b^k_i = B(\vec{b}_i, \vec{b}^k)$ for all $(i, k = 1, 2, \ldots, n)$, and conversely (4.14) defines a mixed tensor of second order. The components b^k_i of B satisfy the transformation formulae

$$\tilde{b}^l_j = a^i_j \alpha^l_k b^k_i \text{ and } b^l_j = \alpha^i_j a^l_k \tilde{b}^k_i \quad (j, l = 1, 2, \ldots, n). \tag{4.15}$$

Proof.

(a) Let B be a covariant tensor of second order and $\vec{x} = x^i \vec{b}_i, \vec{y} = y^i \vec{b}_i \in V$. Then
$B(\vec{x}, \vec{y}) = B(x^i \vec{b}_i, y^k \vec{b}_k) = x^i y^k B(\vec{b}_i, \vec{b}_k) = x^i y^k b_{ik}$.
The converse part is trivial.
Furthermore, we have by (4.2)

$$\tilde{b}_{jl} = B(\vec{\tilde{b}}_j, \vec{\tilde{b}}_l) = B(a^i_j \vec{b}_i, a^k_l \vec{b}_k) = a^i_j a^k_l B(\vec{b}_i, \vec{b}_k) = a^i_j a^k_l b_{ik} \quad (j, l = 1, 2, \ldots, n)$$

and

$$b_{jl} = B(\vec{b}_j, \vec{b}_l) = B(\alpha^i_j \vec{\tilde{b}}_i, \alpha^k_l \vec{\tilde{b}}_k) = \alpha^i_j \alpha^k_l B(\vec{\tilde{b}}_i, \vec{\tilde{b}}_k) = \alpha^i_j \alpha^k_l \tilde{b}_{ik} \quad (j, l = 1, 2, \ldots, n).$$

(b) Let B be a contravariant tensor of second order and $\vec{x} = x_i \vec{b}^i, \vec{y} = y_i \vec{b}^i \in V^*$.
Then $B(\vec{x}, \vec{y}) = B(x_i \vec{b}^i, y_k \vec{b}^k) = x_i y_k B(\vec{b}^i, \vec{b}^k) = x_i y_k b^{ik}$.
The converse part is trivial.
Furthermore, we have by (4.9),

$$\tilde{b}^{jl} = B(\vec{\tilde{b}}^j, \vec{\tilde{b}}^l) = B(\alpha_i^j \vec{b}^i, \alpha_k^l \vec{b}^k) = \alpha_i^j \alpha_k^l B(\vec{b}^i, \vec{b}^k) = \alpha_i^j \alpha_k^l b^{ik} \quad (j, l = 1, 2, \ldots, n)$$

and

$$b^{jl} = B(\vec{b}^j, \vec{b}^l) = B(a_i^j \vec{\tilde{b}}^i, a_k^l \vec{\tilde{b}}^k) = a_i^j a_k^l B(\vec{\tilde{b}}^i, \vec{\tilde{b}}^k) = a_i^j a_k^l \tilde{b}^{ik} \quad (j, l = 1, 2, \ldots, n).$$

(c) Let B be a mixed tensor of second order and $\vec{x} = x^i \vec{b}_i \in V$, $\vec{y} = y_i \vec{b}^i \in V^*$.
Then $B(\vec{x}, \vec{y}) = B(x^i \vec{b}_i, y_k \vec{b}^k) = x^i y_k B(\vec{b}_i, \vec{b}^k) = x^i y_k b_i^k$.
The converse part is trivial.
Furthermore, we have by (4.2) and (4.9),

$$\tilde{b}_j^l = B(\vec{\tilde{b}}_j, \vec{\tilde{b}}^l) = B(a_j^i \vec{b}_i, \alpha_k^l \vec{b}^k) = a_j^i \alpha_k^l B(\vec{b}_i, \vec{b}^k) = a_j^i \alpha_k^l b_i^k \quad (j, l = 1, 2, \ldots, n)$$

and

$$b_j^l = B(\vec{b}_j, \vec{b}^l) = B(\alpha_j^i \vec{\tilde{b}}_i, a_k^l \vec{\tilde{b}}^k) = \alpha_j^i a_k^l B(\vec{\tilde{b}}_i, \vec{\tilde{b}}^k) = \alpha_j^i a_k^l \tilde{b}_i^k \quad (j, l = 1, 2, \ldots, n).$$

$\square$

Example 4.4.3. *(a) Products of the form $B(\vec{x}, \vec{y}) = l(\vec{x}) m(\vec{y})$ of linear functionals
l and m are covariant tensors of second order. We have for the components $b_{ik} =
B(\vec{b}_i, \vec{b}_k) = l(\vec{b}_i) m(\vec{b}_k) = l_i m_k$ for $i, k = 1, 2, \ldots, n$, and all products of the components
of the two covariant vectors appear in the formula.*
*(b) Let $\vec{x} = \{x^1, x^2, \ldots, x^n\}, \vec{y} = \{y^1, y^2, \ldots, y^n\} \in \mathbb{V}^n$ be two contravariant vectors.
We put $b^{ik} = x^i y^k$ $(i, k = 1, 2, \ldots, n)$. Then a contravariant tensor B of second order
is defined by $B(l, m) = b^{ik} l_i m_k$ $(l, m \in (\mathbb{V}^n)^*)$. The linearity is trivial. Furthermore,
the bilinear form is independent of the basis, since*

$$\tilde{b}^{jr} \tilde{l}_j \tilde{m}_r = b^{ik} \alpha_i^j \alpha_k^r a_j^s a_r^t l_s m_t = b^{ik} \delta_i^s \delta_k^t l_s m_t = b^{ik} l_i m_k.$$

(c) The outer product $\vec{x} \wedge \vec{y}$ of two contravariant vectors is defined by

$$s^{ik} = (\vec{x} \wedge \vec{y})^{ik} = x^i y^k - x^k y^i \quad (i, k = 1, 2, \ldots, n). \tag{4.16}$$

Its tensor properties can be shown by the study of the transformation of the components

$$\tilde{s}^{jr} = \tilde{x}^j \tilde{y}^r - \tilde{x}^r \tilde{y}^j = \alpha_i^j \alpha_k^r (x^i y^k - x^k y^i) = \alpha_i^j \alpha_k^r s^{ik} \quad (j, r = 1, 2, \ldots, n).$$

4.5 SYMMETRIC BILINEAR FORMS AND INNER PRODUCTS

Since *symmetric covariant tensors of second order*, that is, tensors that satisfy $B(\vec{x}, \vec{y}) = B(\vec{y}, \vec{x})$ for all vectors $\vec{x}$ and $\vec{y}$, are of special interest, we shall study them more closely. The corresponding quadratic form is positive definite, that is, $B(\vec{x}, \vec{x}) > 0$ for all $\vec{x} \neq \vec{0}$. We write $\vec{x} \bullet \vec{y} = B(\vec{x}, \vec{y})$, and call $\vec{x} \bullet \vec{y}$ *inner product* and $\vec{x} \bullet \vec{x}$ the corresponding *square of the length*. If we choose a basis $\{\vec{b}_1, \ldots, \vec{b}_n\}$, then

$$g_{ik} = \vec{b}_i \bullet \vec{b}_k \quad (i, k = 1, 2, \ldots, n).$$

It is a well–known fact from linear algebra that $\det(g_{ik}) \neq 0$ and that the inverse matrix (g^{ik}) exists, that is, $g^{ik} = g^{ki}$ $(i, k = 1, 2, \ldots, n)$ and

$$g_{ij} g^{jk} = \delta_i^k \quad (i, k = 1, 2, \ldots, n). \tag{4.17}$$

Thus an inner product which is *adjoint* to the first one is defined by

$$l \cdot m = g^{ik} l_i m_k.$$

The map $x^i \mapsto g_{ik} x^k$ $(i = 1, 2, \ldots, n)$ from a space into its dual space is invertible by (4.17), and the inverse map is given by $m_i \mapsto g^{ik} m_k$ $(i = 1, 2, \ldots, n)$.

If $\vec{z}_{(1)}, \vec{z}_{(2)}, \ldots, \vec{z}_{(n)}$ are *orthonormal vectors*, that is, $\vec{z}_{(i)} \bullet \vec{z}_{(k)} = \delta_{(ik)}$ $(i, k = 1, 2, \ldots, n)$ then

$$g^{ik} = \delta^{rs} z^i_{(r)} z^k_{(s)} \quad (i, k = 1, 2, \ldots, n). \tag{4.18}$$

The set $Z = \{\vec{z}_{(1)}, \vec{z}_{(2)}, \ldots, \vec{z}_{(n)}\}$ is a basis of the space, and so every vector $\vec{x}$ with the representation $\vec{x} = x^i \vec{b}_i$ can also be represented as

$$\vec{x} = \xi^i \vec{x}_{(i)} = \xi^i z^k_{(i)} \vec{b}_k.$$

We put

$$b^i_j = \delta^{rs} z^i_{(r)} z^k_{(s)} g_{kj} \quad (i, j = 1, 2, \ldots, n).$$

Then (4.18) will be proved, once we have shown $b^i_j = \delta^i_j$ $(i, j = 1, 2, \ldots, n)$. But

$$b^i_j = \delta^{rs} z^i_{(r)} z^k_{(s)} g_{kj} \xi^l z^i_{(l)} = \delta^{rs} \delta_{sl} z^i_{(r)} \xi^l = z^i_{(l)} \xi^l = x^i$$

implies $b^i_j = \delta^i_j$.

4.6 TENSORS OF ARBITRARY ORDER

In this section, we generalize the concepts of the previous sections.

Definition 4.6.1. A multilinear form T of r contravariant and s covariant vectors is called *r–contravariant, s–covariant tensor* or *(r, s)–tensor*. Once basis vectors $\vec{b}_k$ and $\vec{b}^i$ of the vector space and the dual space have been chosen, the components of the tensor are obtained from

$$T^{i_1 \ldots i_s}_{k_1 \ldots k_r} = T\left(\vec{b}^{i_1}, \ldots, \vec{b}^{i_s}; \vec{b}_{k_1}, \ldots, \vec{b}_{k_r}\right).$$

We put $T = 0$ if the value of T is equal to 0 for arbitrary arguments. Obviously this is the case if and only if all components of the tensor are equal to 0 with respect to some basis.

Theorem 4.6.2. *Any superscript and any subscript transforms with a transformation of bases as the components of a contravariant and of a covariant vector, respectively. Conversely this rule of transformation is sufficient for the components of an invariant multilinear form.*

Proof.

(i) We assume that the transformations of the bases are given by

$$\vec{\tilde{b}}^{i_1} = \alpha^{i_1}_{j_1} \vec{b}^{j_1}, \ldots, \vec{\tilde{b}}^{i_s} = \alpha^{i_s}_{j_s} \vec{b}^{j_s} \text{ and } \vec{\tilde{b}}_{k_1} = a^{l_1}_{k_1} \vec{b}_{l_1}, \ldots, \vec{\tilde{b}}_{k_r} = a^{l_r}_{k_r} \vec{b}_{l_r}.$$

Then we have

$$\tilde{T}^{i_1 \ldots i_s}_{k_1 \ldots k_r} = T\left(\vec{\tilde{b}}^{i_1}, \ldots, \vec{\tilde{b}}^{i_s}; \vec{\tilde{b}}_{k_1}, \ldots, \vec{\tilde{b}}_{k_r} \right)$$

$$= \alpha^{i_1}_{j_1} \cdots \alpha^{i_s}_{j_s} a^{l_1}_{k_1} \cdots a^{l_r}_{k_r} T\left(\vec{b}^{j_1}, \ldots, \vec{b}^{j_s}; \vec{b}_{l_1}, \ldots, \vec{b}_{l_r} \right)$$

$$= \alpha^{i_1}_{j_1} \cdots \alpha^{i_s}_{j_s} a^{l_1}_{k_1} \cdots a^{l_r}_{k_r} T^{j_1 \ldots j_s}_{l_1 \ldots l_r}.$$

This proves the stated rule of transformation.

(ii) Let $x^{(1)}_{i_1}, \ldots, x^{(s)}_{i_s}$ and $x^{k_1}_{(1)}, \ldots, x^{k_r}_{(r)}$ be the components of s covariant vectors $\vec{x}^{(1)}, \ldots, \vec{x}^{(s)}$ and r contravariant vectors $\vec{x}_{(1)}, \ldots, \vec{x}_{(r)}$, respectively. We put

$$T(\vec{x}^{(1)}, \ldots, \vec{x}^{(s)}; \vec{x}_{(1)}, \ldots, \vec{x}_{(r)}) = T^{i_1 \ldots i_s}_{k_1 \ldots k_r} x^{(1)}_{i_1} \cdots x^{(s)}_{i_s} x^{k_1}_{(1)} \cdots x^{k_r}_{(r)}.$$

Then this obviously defines a multilinear form and

$$\tilde{T}^{m_1 \ldots m_s}_{n_1 \ldots n_r} \tilde{x}^{(1)}_{m_1} \cdots \tilde{x}^{(s)}_{m_s} \tilde{x}^{n_1}_{(1)} \cdots \tilde{x}^{n_r}_{(r)}$$

$$= \alpha^{m_1}_{j_1} \cdots \alpha^{m_s}_{j_s} a^{l_1}_{n_1} \cdots a^{l_r}_{n_r} T^{j_1 \ldots j_s}_{l_1 \ldots l_r} a^{i_1}_{m_1} \cdots a^{i_s}_{m_s} x^{(1)}_{i_1} \cdots x^{(s)}_{i_s} \alpha^{n_1}_{t_1} \ldots \alpha^{n_r}_{t_r} x^{t_1}_{(1)} \cdots x^{t_r}_{(r)}$$

$$= T^{j_1 \ldots j_s}_{l_1 \ldots l_r} \delta^{i_1}_{j_1} \cdots \delta^{i_s}_{j_s} \delta^{l_1}_{t_1} \cdots \delta^{l_r}_{t_r} x^{(1)}_{i_1} \cdots x^{(s)}_{i_s} x^{t_1}_{(1)} \cdots x^{t_r}_{(r)}$$

$$= T^{i_1 \ldots i_s}_{t_1 \ldots t_r} x^{(1)}_{i_1} \cdots x^{(s)}_{i_s} x^{t_1}_{(1)} \cdots x^{t_r}_{(r)}.$$

$\square$

Example 4.6.3. *Let $T^{j_1 \ldots j_s}_{i_1 \ldots i_r}$, $\tilde{T}^{j_1 \ldots j_s}_{i_1 \ldots i_r}$ and $\tilde{\tilde{T}}^{j_1 \ldots j_s}_{i_1 \ldots i_r}$ that satisfy the transformation formulae*

$$\tilde{T}^{j_1 \ldots j_s}_{i_1 \ldots i_r} = \alpha^{j_1}_{l_1} \cdots \alpha^{j_s}_{l_s} a^{m_1}_{i_1} \cdots a^{m_r}_{i_r} T^{l_1 \ldots l_s}_{m_1 \ldots m_r} \tag{4.19}$$

and

$$\tilde{\tilde{T}}^{j_1 \ldots j_s}_{i_1 \ldots i_r} = \beta^{j_1}_{l_1} \cdots \beta^{j_s}_{l_s} b^{m_1}_{i_1} \cdots b^{m_r}_{i_r} T^{l_1 \ldots l_s}_{m_1 \ldots m_r}. \tag{4.20}$$

Then we obtain the following transformation formulae between $\tilde{T}^{j_1...j_s}_{i_1...i_r}$ and $\widetilde{\tilde{T}}^{j_1...j_s}_{i_1...i_r}$ as follows:

$$
\begin{aligned}
T^{l_1...l_s}_{m_1...m_r} &= \delta^{n_1}_{m_1} \cdots \delta^{n_r}_{m_r} \delta^{l_1}_{t_1} \cdots \delta^{l_s}_{t_s} T^{t_1...t_s}_{n_1...n_r} \\
&= \alpha^{p_1}_{m_1} a^{n_1}_{p_1} \cdots \alpha^{p_r}_{m_r} a^{n_r}_{p_r} \alpha^{q_1}_{t_1} a^{l_1}_{q_1} \cdots \alpha^{q_s}_{t_s} a^{l_s}_{q_s} T^{t_1...t_s}_{n_1...n_r} \\
&= \alpha^{p_1}_{m_1} \cdots \alpha^{p_r}_{m_r} a^{l_1}_{q_1} \cdots a^{l_s}_{q_s} \alpha^{q_1}_{t_1} \cdots \alpha^{q_s}_{t_s} a^{n_1}_{p_1} \cdots a^{n_r}_{p_r} T^{t_1...t_s}_{n_1...n_r} \\
&= \alpha^{p_1}_{m_1} \cdots \alpha^{p_r}_{m_r} a^{l_1}_{q_1} \cdots a^{l_s}_{q_s} \tilde{T}^{q_1...q_s}_{p_1...p_r},
\end{aligned}
$$

hence

$$
\widetilde{\tilde{T}}^{j_1...j_s}_{i_1...i_r} = \beta^{j_1}_{l_1} \cdots \beta^{j_s}_{l_s} b^{m_1}_{i_1} \cdots b^{m_r}_{i_r} \alpha^{p_1}_{m_1} \cdots \alpha^{p_r}_{m_r} a^{l_1}_{q_1} \cdots a^{l_s}_{q_s} \tilde{T}^{q_1...q_s}_{p_1...p_r}.
$$

Remark 4.6.4. *(a) Covariant vectors are $(1,0)$–tensors, contravariant vectors are $(0,1)$–tensors, and scalars are $(0,0)$–tensors.*
(b) Tensors are characterized by the transformation properties of their components. Therefore, we frequently identify tensors with their components.
(c) The sum of two tensors T and U of the same structure with the components

$$
T^{i_1...i_s}_{k_1...k_r} \quad and \quad U^{i_1...i_s}_{k_1...k_r}
$$

is defined by

$$
V^{i_1...i_s}_{k_1...k_r} = (T+U)^{i_1...i_s}_{k_1...k_r} = T^{i_1...i_s}_{k_1...k_r} + U^{i_1...i_s}_{k_1...k_r};
$$

it is an (r,s)–tensor.
(d) The outer product $V = T * U$ of an (r,s)–tensor T and an (r',s')–tensor U with the components

$$
T^{i_1...i_s}_{k_1...k_r} \quad and \quad U^{l_1...l_{s'}}_{m_1...m_{r'}}
$$

is defined by

$$
V^{i_1...i_s l_1...l_{s'}}_{k_1...k_r m_1...m_{r'}} = T^{i_1...i_s}_{k_1...k_r} \cdot U^{l_1...l_{s'}}_{m_1...m_{r'}};
$$

it is an $(r+r', s+s')$–tensor. Multiplication and addition are commutative

$$
T * U = U * T \quad and \quad T + U = U + T.
$$

Furthermore, the distributive and associative laws hold, for instance,

$$
V * (U+T) = V * U + V * T \ and \ T * (U * V) = (T * U) * V.
$$

This follows from the corresponding laws for real numbers. The tensors of the same structure make up a vector space.

Proof. We only need to show the statement concerning the outer product of two tensors.
Let T and U be (r,s)– and (r',s')–tensors with the components $T^{l_1...l_s}_{m_1...m_r}$ and $U^{p_1...p_{s'}}_{q_1...q_{r'}}$ that satisfy the transformation formulae

$$
\tilde{T}^{j_1...j_s}_{i_1...i_r} = \alpha^{j_1}_{l_1} \cdots \alpha^{j_s}_{l_s} a^{m_1}_{i_1} \cdots a^{m_r}_{i_r} T^{l_1...l_s}_{m_1...m_r}
$$

and

$$\widetilde{U}^{n_1\ldots n_{s'}}_{k_1\ldots k_{r'}} = \alpha^{n_1}_{p_1} \cdots \alpha^{n_{s'}}_{p_{s'}} a^{q_1}_{k_1} \cdots a^{q_{r'}}_{k_{r'}} U^{p_1\ldots p_{s'}}_{q_1\ldots q_{r'}}.$$

Then we obtain for the transformation formulae of the outer product $V = T * U$

$$
\begin{aligned}
\widetilde{V}^{j_1\ldots j_s n_1\ldots n_{r'}}_{i_1\ldots i_r k_1\ldots k_{r'}} &= \widetilde{T}^{j_1\ldots j_s}_{i_1\ldots i_r} \widetilde{U}^{n_1\ldots n_{s'}}_{k_1\ldots k_{r'}} \\
&= \alpha^{j_1}_{l_1} \cdots \alpha^{j_s}_{l_s} a^{m_1}_{i_1} \cdots a^{m_r}_{i_r} T^{l_1\ldots l_s}_{m_1\ldots m_r} \alpha^{n_1}_{p_1} \cdots \alpha^{n_{s'}}_{p_{s'}} a^{q_1}_{k_1} \cdots a^{q_{r'}}_{k_{r'}} U^{p_1\ldots p_{s'}}_{q_1\ldots q_{r'}} \\
&= \alpha^{j_1}_{l_1} \cdots \alpha^{j_s}_{l_s} \alpha^{n_1}_{p_1} \cdots \alpha^{n_{s'}}_{p_{s'}} a^{m_1}_{i_1} \cdots a^{m_r}_{i_r} a^{q_1}_{k_1} \cdots a^{q_{r'}}_{k_{r'}} V^{l_1\ldots l_s p_1\ldots p_{s'}}_{m_1\ldots m_r q_1\ldots q_{r'}}.
\end{aligned}
$$

$\square$

Definition 4.6.5. Let T be a tensor with the components $T^{i_1\ldots i_s}_{k_1\ldots k_r}$. The *contraction* of T is defined by

$$T^{i_1\ldots i_{m-1},i_{m+1}\ldots i_s}_{k_1\ldots k_{n-1},k_{n+1}\ldots k_r} = T^{i_1\ldots i_{m-1}j i_{m+1}\ldots i_s}_{k_1\ldots k_{n-1}j k_{n+1}\ldots k_r}.$$

Remark 4.6.6. *The contraction of an (r,s)–tensor yields an $(r-1, s-1)$–tensor.*

Proof. We obtain for the transformation formulae of the contraction of an (r,s)–tensor T

$$
\begin{aligned}
\widetilde{V}^{j_1\ldots j_{q-1}j_{q+1}\ldots j_s}_{i_1\ldots i_{p-1}i_{p+1}\ldots i_r} &= \widetilde{T}^{j_1\ldots j_{q-1},j_{q+1}\ldots j_s}_{i_1\ldots i_{p-1},i_{p+1}\ldots i_r} = \widetilde{T}^{j_1\ldots j_{q-1}t j_{q+1}\ldots j_s}_{i_1\ldots i_{p-1}t i_{p+1}\ldots i_r} \\
&= \alpha^{j_1}_{l_1} \cdots \alpha^{j_{q-1}}_{l_{q-1}} \alpha^{t}_{l_q} \alpha^{j_{q+1}}_{l_{q+1}} \cdots \alpha^{j_s}_{l_s} a^{n_1}_{i_1} \cdots a^{n_{p-1}}_{i_{p-1}} a^{n_p}_{t} a^{n_{p+1}}_{i_{p+1}} \cdots a^{n_r}_{i_r} T^{l_1\ldots l_s}_{n_1\ldots n_r} \\
&= \alpha^{j_1}_{l_1} \cdots \alpha^{j_{q-1}}_{l_{q-1}} \alpha^{j_{q+1}}_{l_{q+1}} \cdots \alpha^{j_s}_{l_s} a^{n_1}_{i_1} \cdots a^{n_{p-1}}_{i_{p-1}} a^{n_{p+1}}_{i_{p+1}} \cdots a^{n_r}_{i_r} \alpha^{t}_{l_q} a^{n_p}_{t} T^{l_1\ldots l_{q-1}l_q l_{q+1}\ldots l_s}_{n_1\ldots n_{p-1}n_p n_{p+1}\ldots n_r} \\
&= \alpha^{j_1}_{l_1} \cdots \alpha^{j_{q-1}}_{l_{q-1}} \alpha^{j_{q+1}}_{l_{q+1}} \cdots \alpha^{j_s}_{l_s} a^{n_1}_{i_1} \cdots a^{n_{p-1}}_{i_{p-1}} a^{n_{p+1}}_{i_{p+1}} \cdots a^{n_r}_{i_r} \delta^{n_p}_{l_q} T^{l_1\ldots l_{q-1}l_q l_{q+1}\ldots l_s}_{n_1\ldots n_{p-1}n_p n_{p+1}\ldots n_r} \\
&= \alpha^{j_1}_{l_1} \cdots \alpha^{j_{q-1}}_{l_{q-1}} \alpha^{j_{q+1}}_{l_{q+1}} \cdots \alpha^{j_s}_{l_s} a^{n_1}_{i_1} \cdots a^{n_{p-1}}_{i_{p-1}} a^{n_{p+1}}_{i_{p+1}} \cdots a^{n_r}_{i_r} T^{l_1\ldots l_{q-1}n_p l_{q+1}\ldots l_s}_{n_1\ldots n_{p-1}n_p n_{p+1}\ldots n_r} \\
&= \alpha^{j_1}_{l_1} \cdots \alpha^{j_{q-1}}_{l_{q-1}} \alpha^{j_{q+1}}_{l_{q+1}} \cdots \alpha^{j_s}_{l_s} a^{n_1}_{i_1} \cdots a^{n_{p-1}}_{i_{p-1}} a^{n_{p+1}}_{i_{p+1}} \cdots a^{n_r}_{i_r} V^{l_1\ldots l_{q-1}l_{q+1}\ldots l_s}_{n_1\ldots n_{p-1}n_{p+1}\ldots n_r}.
\end{aligned}
$$

$\square$

Example 4.6.7. *(a) The contraction of a $(1,1)$–tensor with the components T^i_k yields a scalar*

$$T_{,} = T^i_i = T^1_1 + T^2_2 + \cdots + T^n_n.$$

(b) Contracting a $(3,2)$–tensor T with the components $T^{i_1 i_2}_{k_1 k_2 k_3}$ with respect to k_2 and i_1 we obtain the $(2,1)$–tensor with the components

$$T^{,i_2}_{k_1,k_3} = T^{j i_2}_{k_1 j k_3}.$$

Definition 4.6.8. The outer product of two tensors with contraction with respect to indices from different factors is called *inner product of the two tensors*.

Remark 4.6.9. *The inner product of an (r,s)–tensor and an (r',s')–tensor is an $(r+r'-1, s+s'-1)$–tensor.*

Proof. The statement is an immediate consequence of Remarks 4.6.4 (d) and 4.6.6.

□

Example 4.6.10. *(a) We consider the first fundamental coefficients g_{ik} and g^{lm}. The outer product is*

$$c_{ik}^{lm} = g_{ik} g^{lm},$$

and for $k = l$, we obtain the contraction

$$c_{i,}^{\;\;m} = c_{ik}^{km} = g_{ik} g^{km} = \delta_i^m.$$

(b) We consider the values g^{ik} and L_{lm}. The values g^{ik} transform like the components of a tensor, and the values L_{lm} also transform like the components of a tensor, provided the transformation of the coordinates preserves the orientation of the surface normal vectors. We obtain the outer product

$$c_{lm}^{ik} = g^{ik} L_{lm},$$

and the contractions

$$d_m^k = c_{,m}^k = g^{ik} L_{im} \text{ and } d_{,}^{} = d_k^k = g^{ik} L_{ik} = 2H.$$

4.7 SYMMETRIC AND ANTI-SYMMETRIC TENSORS

In this section, we study symmetric and anti–symmetric tensors.

Definition 4.7.1. A tensor is called *symmetric* in two of its indices, if it is invariant with respect to their interchange; it is called *anti–symmetric* if it changes sign.

Theorem 4.7.2. *Symmetry and anti–symmetry of tensors are independent of the choice of the coordinate system.*

Proof. We show that

$$T_{i_1 i_2 \ldots i_r}^{j_1 \ldots j_s} = T_{i_2 i_1 \ldots i_r}^{j_1 \ldots j_s} \text{ implies } \widetilde{T}_{i_1 i_2 \ldots i_r}^{j_1 \ldots j_s} = \widetilde{T}_{i_2 i_1 \ldots i_r}^{j_1 \ldots j_s}.$$

We have

$$\widetilde{T}_{i_1 i_2 \ldots i_r}^{j_1 \ldots j_s} = \alpha_{l_1}^{j_1} \alpha_{l_2}^{j_2} \cdots \alpha_{l_s}^{j_s} a_{i_1}^{m_1} a_{i_2}^{m_2} \cdots a_{i_r}^{m_r} T_{m_1 m_2 \ldots m_r}^{l_1 l_2 \ldots l_s}$$
$$= \alpha_{l_1}^{j_1} \alpha_{l_2}^{j_2} \cdots \alpha_{l_s}^{j_s} a_{i_2}^{m_2} a_{i_1}^{m_1} \cdots a_{i_r}^{m_r} T_{m_2 m_1 \ldots m_r}^{l_1 l_2 \ldots l_s} = \widetilde{T}_{i_2 i_1 \ldots i_r}^{j_1 \ldots j_s}.$$

□

Remark 4.7.3. *Any (r, s)–tensor may be written as the sum of a symmetric and an anti–symmetric tensor, for instance,*

$$T_{ik} = \frac{1}{2} \left(T_{ik} + T_{ki} \right) + \frac{1}{2} \left(T_{ik} - T_{ki} \right).$$

Example **4.7.4.** *The first fundamental coefficients* g_{ik} *are the components of a symmetric tensor. The values* ϵ_{ik} *with*

$$\epsilon_{11} = \epsilon_{22} = 0 \quad and \quad \epsilon_{12} = -\epsilon_{21} = \sqrt{g}$$

are the components of an anti–symmetric tensor.

Definition 4.7.5. Two tensors with components A_{ik} and B^{jl} and non–vanishing determinants are called *conjugated* if

$$A_{ik}B^{kl} = \delta_i^l.$$

Example **4.7.6.** *The values* g_{ik} *and* g^{jl} *are the components of conjugated tensors.*

4.8 RIEMANN SPACES

The concept of *Riemann spaces* is needed in the measurement of lengths of curves on a manifold. We shall see that we require a covariant symmetric tensor $g_{ik}(x)$ on the manifold. Then the length of a segment of a curve given by a parametric representation $\vec{x}(t)$ ($t \in [t_1, t_2]$) will be defined by

$$s(t_2) - s(t_1) = \int\limits_{t_1}^{t_2} \sqrt{g_{ik}\frac{dx^i}{dt}\frac{d^k}{dt}}dt.$$

Let a coordinate system S with the coordinates (x^i) be given in some neighbourhood of a point P_0 of an n–dimensional manifold and $P_0 = (\underset{0}{x^1}, \underset{0}{x^2}, \dots, \underset{0}{x^n})$. Then a special basis is defined for the tangent space $T(P_0)$ at P_0. The vectors $\vec{b}_i$ of the basis of $T(P_0)$ are the tangents to the curves $\gamma_{(i)}$ with the parametric representations

$$\vec{x}_{(i)}(t) = \left\{\underset{0}{x^1}, \dots, \underset{0}{x^{i-1}}, \underset{0}{x^i} + t, \underset{0}{x^{i+1}}, \dots, \underset{0}{x^n}\right\} \quad (i = 1, 2, \dots, n). \tag{4.21}$$

If we choose another coordinate system $\tilde{S}$ with coordinates $(\tilde{x}^i)$ then this yields another basis for the tangent space $T(P_0)$ at the point P_0, and we are going to find the corresponding basis transformation according to (4.2).

Let γ be an arbitrary curve with the parametric representations

$$\vec{x}(t) = \{x^1(t), \dots, x^n(t)\} \text{ and } \vec{\tilde{x}}(t) = \{\tilde{x}^1(t), \dots, \tilde{x}^n(t)\}$$

in the coordinate systems S and $\tilde{S}$ with the coordinates (x^i) and $(\tilde{x}^i)$. Since $x^i(t) = x^i(\tilde{x}^k(t))$ ($i = 1, 2, \dots, n$), the chain rule yields

$$\frac{dx^i}{dt} = \frac{\partial x^i}{\partial \tilde{x}^k}\frac{d\tilde{x}^k}{dt} \quad (i = 1, 2, \dots, n). \tag{4.22}$$

Since the tangent vector at P_0 is independent of the bases $\{\vec{b}_1, \dots, \vec{b}_n\}$ and $\{\vec{\tilde{b}}_1, \dots, \vec{\tilde{b}}_n\}$, we must have

$$\vec{b}_i\frac{dx^i}{dt} = \vec{\tilde{b}}_k\frac{d\tilde{x}^k}{dt}.$$

Substituting (4.21) and (4.2), we obtain

$$\vec{b}_i \frac{\partial x^i}{\partial \tilde{x}^k} \frac{d\tilde{x}^k}{dt} = \vec{b}_k \frac{d\tilde{x}^k}{dt} = a_k^i \vec{b}_i \frac{d\tilde{x}^k}{dt}.$$

Since this holds for arbitrary $d\tilde{x}^k/dt$, we must have

$$a_k^i = \frac{\partial x^i}{\partial \tilde{x}^k} \; (i, k = 1, 2, \ldots, n). \tag{4.23}$$

Similarly, we obtain for the inverse

$$\alpha_j^k = \frac{\partial \tilde{x}^k}{\partial x^j} \; (j, k = 1, 2, \ldots, n).$$

Example 4.8.1. *Let S be a surface in $\mathbb{R}^3$ with a parametric representation $\vec{x}(u^i)$. Then the tangent vectors $\vec{b}_i$ above are the partial derivatives $\vec{x}_i$, hence the tangent vectors of the u^i–lines. We define $g_{ik} = \vec{x}_i \bullet \vec{x}_k$ $(i, k = 1, 2)$. Then the following formulae of transformation hold*

$$\tilde{g}_{ik} = g_{lm} \frac{\partial u^l}{\partial \tilde{u}^i} \frac{\partial u^m}{\partial \tilde{u}^k} \; (i, k = 1, 2).$$

Example 4.8.2. *We consider the tensors $\tilde{T}_{i_1 \ldots i_r}^{j_1 \ldots j_s}$ and $\tilde{\tilde{T}}_{i_1 \ldots i_r}^{j_1 \ldots j_s}$ of Example 4.6.3 that satisfy the transformation formulae in (4.19) and (4.20) in the special case, where*

$$\alpha_i^k = \frac{\partial \tilde{x}^k}{\partial x^i}, \; a_i^k = \frac{\partial x^k}{\partial \tilde{x}^i}, \; \beta_i^k = \frac{\partial \tilde{\tilde{x}}^k}{\partial x^i} \text{ and } b_i^k = \frac{\partial x^k}{\partial \tilde{\tilde{x}}^i} \text{ for } i, k = 1, 2, \ldots, n.$$

Then

$$\tilde{T}_{i_1 \ldots i_r}^{j_1 \ldots j_s} = \frac{\partial \tilde{x}^{j_1}}{\partial x^{l_1}} \cdots \frac{\partial \tilde{x}^{j_s}}{\partial x^{l_r}} \frac{\partial x^{m_1}}{\partial \tilde{x}^{i_1}} \cdots \frac{\partial x^{m_r}}{\partial \tilde{x}^{i_r}} T_{m_1 \ldots m_r}^{l_1 \ldots l_s}$$

and

$$\tilde{\tilde{T}}_{i_1 \ldots i_r}^{j_1 \ldots j_s} = \frac{\partial \tilde{\tilde{x}}^{j_1}}{\partial x^{l_1}} \cdots \frac{\partial \tilde{\tilde{x}}^{j_s}}{\partial x^{l_s}} \frac{\partial x^{m_1}}{\partial \tilde{\tilde{x}}^{i_1}} \cdots \frac{\partial x^{m_r}}{\partial \tilde{\tilde{x}}^{i_r}} T_{m_1 \ldots m_r}^{l_1 \ldots l_s}$$

imply

$$\begin{aligned}
\tilde{\tilde{T}}_{i_1 \ldots i_r}^{j_1 \ldots j_s} &= \frac{\partial \tilde{\tilde{x}}^{j_1}}{\partial x^{l_1}} \cdots \frac{\partial \tilde{\tilde{x}}^{j_s}}{\partial x^{l_s}} \frac{\partial x^{m_1}}{\partial \tilde{\tilde{x}}^{i_1}} \cdots \frac{\partial x^{m_r}}{\partial \tilde{\tilde{x}}^{i_r}} \frac{\partial \tilde{x}^{p_1}}{\partial x^{m_1}} \cdots \frac{\partial \tilde{x}^{p_r}}{\partial x^{m_r}} \frac{\partial x^{l_1}}{\partial \tilde{x}^{q_1}} \cdots \frac{\partial x^{l_s}}{\partial \tilde{x}^{q_s}} \tilde{T}_{p_1 \ldots p_r}^{q_1 \ldots q_s} \\
&= \frac{\partial \tilde{\tilde{x}}^{j_1}}{\partial x^{l_1}} \frac{\partial x^{l_1}}{\partial \tilde{x}^{q_1}} \cdots \frac{\partial \tilde{\tilde{x}}^{j_s}}{\partial x^{l_s}} \frac{\partial x^{l_s}}{\partial \tilde{x}^{q_s}} \frac{\partial \tilde{x}^{p_1}}{\partial x^{m_1}} \frac{\partial x^{m_1}}{\partial \tilde{\tilde{x}}^{i_1}} \cdots \frac{\partial \tilde{x}^{p_r}}{\partial x^{m_r}} \frac{\partial x^{m_r}}{\partial \tilde{\tilde{x}}^{i_r}} \tilde{T}_{p_1 \ldots p_r}^{q_1 \ldots q_s} \\
&= \frac{\partial \tilde{\tilde{x}}^{j_1}}{\partial \tilde{x}^{q_1}} \cdots \frac{\partial \tilde{\tilde{x}}^{j_s}}{\partial \tilde{x}^{q_s}} \frac{\partial \tilde{x}^{p_1}}{\partial \tilde{\tilde{x}}^{i_1}} \cdots \frac{\partial \tilde{x}^{p_r}}{\partial \tilde{\tilde{x}}^{i_r}} \tilde{T}_{p_1 \ldots p_r}^{q_1 \ldots q_s}.
\end{aligned}$$

Definition 4.8.3. Let G be a symmetric tensor in $\mathbb{R}^n$ with the components $g_{ik} = \vec{b}_i \bullet \vec{b}_k$ $(i, k = 1, 2, \ldots, n)$ in a coordinate system S with the coordinates (x^i), where the vectors $\vec{b}_i$ $(i = 1, 2, \ldots, n)$ are the tangent vectors defined above. We assume that the functions g_{ik} are continuous. Then G is called *metric tensor*.

Let γ be a curve in $\mathbb{R}^n$ with the parametric representations

$$\vec{x}(t) = \{x^1(t), x^2(t), \ldots, x^n(t)\} \text{ and } \vec{\tilde{x}}(t) = \{\tilde{x}^1(t), \tilde{x}^2(t), \ldots, \tilde{x}^n(t)\}$$

in the coordinate systems S and $\tilde{S}$ with the coordinates (x^i) and $(\tilde{x}^i)$. Then

$$\tilde{g}_{ik} = \frac{d\tilde{x}^i}{dt} \frac{d\tilde{x}^k}{dt} = g_{lm} \frac{\partial x^l}{\partial \tilde{x}^i} \frac{\partial x^m}{\partial \tilde{x}^k} \frac{\partial \tilde{x}^i}{\partial x^j} \frac{dx^j}{dt} \frac{\partial \tilde{x}^k}{\partial x^n} \frac{dx^n}{dt}$$

$$= g_{lm} \delta^l_j \delta^m_n \frac{dx^j}{dt} \frac{dx^n}{dt} = g_{jn} \frac{dx^j}{dt} \frac{dx^n}{dt} = g_{ik} \frac{dx^i}{dt} \frac{dx^k}{dt}.$$

We define the length s by

$$s = s(t_1, t_2) = \int_{t_1}^{t_2} \sqrt{e \cdot g_{ik} \frac{dx^i}{dt} \frac{dx^k}{dt}} \, dt,$$

where the *indicator function* e is defined such that

$$e \cdot g_{ik} \frac{dx^i}{dt} \frac{dx^k}{dt} \geq 0, \text{ that is, } e = \text{sign}\left(g_{ik} \frac{dx^i}{dt} \frac{dx^k}{dt}\right).$$

Thus a curve may be split into arcs with $e > 0$, $e = 0$ and $e < 0$. The length is independent of both the coordinate system and the parameters.

Definition 4.8.4. Let S and $\tilde{S}$ be two coordinate systems in $\mathbb{R}^n$, G be a metric tensor with the components $g_{ik}(x^j)$ and $\tilde{g}_{ik}(\tilde{x}^j)$ $(i, k = 1, 2, \ldots, n)$ and s be a length defined by

$$s = \int_{t_1}^{t_2} \sqrt{e \cdot g_{ik} \frac{dx^i}{dt} \frac{dx^k}{dt}} \, dt.$$

Then $\mathbb{R}^n$ with this length is called *Riemann space*.

Example 4.8.5 (**Euclidean $\mathbb{E}^n$**).
Let S be a Cartesian coordinate system and the components of the metric tensor be given by $g_{ik} = \delta_{ik}$ $(i, k = 1, 2, \ldots, n)$. For a curve γ with a parametric representation $\vec{x}(t) = \{x^1(t), x^2(t), \ldots, x^n(t)\}$, we have

$$s = \int_{t_1}^{t_2} \sqrt{e \cdot \delta_{ik} \frac{dx^i}{dt} \frac{dx^k}{dt}} \, dt = \int_{t_1}^{t_2} \sqrt{\sum_{i=1}^n \left(\frac{dx^i}{dt}\right)^2} \, dt,$$

and the transformation formulae

$$\tilde{g}_{ik} = \tilde{\delta}_{ik} = g_{ml} \frac{\partial x^m}{\partial \tilde{x}^i} \frac{\partial \tilde{x}^l}{\partial \tilde{x}^k} = \sum_{m=1}^n \frac{\partial x^m}{\partial \tilde{x}^i} \frac{\partial x^m}{\partial \tilde{x}^k} \quad (i, k = 1, 2, \ldots, n).$$

Example 4.8.6 (Surfaces in $\mathbb{E}^3$).
Let a surface in $\mathbb{R}^3$ be given by some parametric representation $\vec{x}(u^\alpha) = \vec{x}(u^1, u^2)$. A curve given by $(u^1(t), u^2(t))$ in the parameter plane generates a curve with a parametric representation $\vec{x}(t) = \vec{x}(u^1(t), u^2(t))$ on the surface. By Example 4.8.5, the arc length along this curve is given by

$$s = \int_{t_1}^{t_2} \sqrt{\sum_{k=1}^{3} \left(\frac{dx^k}{dt}\right)^2}\, dt = \int_{t_1}^{t_2} \sqrt{\sum_{k=1}^{3} \frac{\partial x^k}{\partial u^\alpha} \frac{du^\alpha}{dt} \frac{\partial x^k}{\partial u^\beta} \frac{du^\beta}{dt}}\, dt$$

$$= \int_{t_1}^{t_2} \sqrt{\vec{x}_\alpha \bullet \vec{x}_\beta \frac{du^\alpha}{dt} \frac{du^\beta}{dt}}\, dt = \int_{t_1}^{t_2} \sqrt{g_{\alpha\beta} \frac{du^\alpha}{dt} \frac{du^\beta}{dt}}\, dt.$$

We have for parameter transformations

$$\tilde{g}_{\alpha\beta} = \frac{\partial u^\gamma}{\partial \tilde{u}^\alpha} \frac{\partial u^\delta}{\partial \tilde{u}^\beta} g_{\gamma\delta}.$$

Let $\tilde{S}$ be some other coordinate system in $\mathbb{E}^3$ with

$$\tilde{x}^i = \tilde{x}^i(x^j) \quad and \quad \tilde{x}^i(u^\alpha) = \tilde{x}^i(x^j(u^\alpha)) \ (i = 1, 2, 3).$$

The length of a curve with a parametric representation

$$\vec{\tilde{x}}(t) = \left\{ \tilde{x}^1(x^j(u^\alpha(t))), \tilde{x}^2(x^j(u^\alpha(t))), \tilde{x}^3(x^j(u^\alpha(t))) \right\}$$

is given by

$$s = \int_{t_1}^{t_2} \sqrt{\tilde{g}_{ik} \frac{d\tilde{x}^i}{dt} \frac{d\tilde{x}^k}{dt}}\, dt, \ where \ \tilde{g}_{ik} = \sum_{m=1}^{3} \frac{\partial x^m}{\partial \tilde{x}^i} \frac{\partial x^m}{\partial \tilde{x}^k} \ (i, k = 1, 2).$$

This implies

$$\tilde{g}_{ik} \frac{d\tilde{x}^i}{dt} \frac{d\tilde{x}^k}{dt} = \sum_{m=1}^{3} \frac{\partial x^m}{\partial \tilde{x}^i} \frac{\partial x^m}{\partial \tilde{x}^k} \frac{\partial \tilde{x}^i}{\partial x^j} \frac{\partial x^j}{\partial u^\alpha} \frac{du^\alpha}{dt} \frac{\partial \tilde{x}^k}{\partial x^l} \frac{\partial x^l}{\partial u^\beta} \frac{du^\beta}{dt}$$

$$= \sum_{m=1}^{3} \frac{\partial x^m}{\partial \tilde{x}^i} \frac{\partial \tilde{x}^i}{\partial x^j} \frac{\partial x^m}{\partial \tilde{x}^k} \frac{\partial \tilde{x}^k}{\partial x^l} \frac{\partial x^j}{\partial u^\alpha} \frac{\partial x^l}{\partial u^\beta} \frac{du^\alpha}{dt} \frac{du^\beta}{dt}$$

$$= \sum_{m=1}^{3} \delta_j^m \delta_l^m \frac{\partial x^j}{\partial u^\alpha} \frac{\partial x^l}{\partial u^\beta} \frac{du^\alpha}{dt} \frac{du^\beta}{dt}$$

$$= \sum_{m=1}^{3} \frac{\partial x^m}{\partial u^\alpha} \frac{\partial x^m}{\partial u^\beta} \frac{du^\alpha}{dt} \frac{du^\beta}{dt} = g_{\alpha\beta} \frac{du^\alpha}{dt} \frac{du^\beta}{dt},$$

hence

$$\tilde{g}_{ik} \frac{d\tilde{x}^i}{dt} \frac{\tilde{x}^k}{dt} = g_{\alpha\beta} \frac{du^\alpha}{dt} \frac{du^\beta}{dt}.$$

4.9   THE CHRISTOFFEL SYMBOLS

In this section, we study the *Christoffel symbols* which we introduce in a formal way.

Definition 4.9.1. Let a metric tensor G with the components g_{ik} and its conjugated tensor be given in a Riemann space. We define the *first* and *second Christoffel symbols* by

$$[ijk] = \frac{1}{2}(g_{ik,j} - g_{ij,k} + g_{jk,i}) = \frac{1}{2}\left(\frac{\partial g_{ik}}{\partial x^j} - \frac{\partial g_{ij}}{\partial x^k} + \frac{\partial g_{jk}}{\partial x^i}\right)$$

and

$$\Gamma_{ij}^k = \left\{\begin{matrix} k \\ ij \end{matrix}\right\} = g^{kl}[ijl].$$

Remark 4.9.2. *We note that the values Γ_{ij}^k are not the components of a tensor as we will see in the transformation formulae (4.28) below.*

Theorem 4.9.3. *The following formulae hold*

$$[ikl] = g_{lj}\left\{\begin{matrix} j \\ ik \end{matrix}\right\}, \quad [ijk] = [jik], \quad \left\{\begin{matrix} k \\ ij \end{matrix}\right\} = \left\{\begin{matrix} k \\ ji \end{matrix}\right\}, \tag{4.24}$$

$$\frac{\partial g_{ik}}{\partial x^j} = [ijk] + [kji], \tag{4.25}$$

$$\frac{\partial \tilde{g}_{ij}}{\partial \tilde{x}^k} = \left(\frac{\partial^2 x^l}{\partial \tilde{x}^i \partial \tilde{x}^k}\frac{\partial x^m}{\partial \tilde{x}^j} + \frac{\partial x^l}{\partial \tilde{x}^i}\frac{\partial^2 x^m}{\partial \tilde{x}^j \partial \tilde{x}^k}\right)g_{lm} \tag{4.26}$$
$$+ \frac{\partial x^l}{\partial \tilde{x}^i}\frac{\partial x^m}{\partial \tilde{x}^j}\frac{\partial g_{lm}}{\partial x^n}\frac{\partial x^n}{\partial \tilde{x}^k},$$

$$\widetilde{[ijk]} = \left(\frac{\partial x^l}{\partial \tilde{x}^i}\frac{\partial x^n}{\partial \tilde{x}^j}[lnm] + \frac{\partial^2 x^l}{\partial \tilde{x}^i \partial \tilde{x}^j}g_{lm}\right)\frac{\partial x^m}{\partial \tilde{x}^k}, \tag{4.27}$$

$$\widetilde{\left\{\begin{matrix} k \\ ij \end{matrix}\right\}} = \left(\frac{\partial x^p}{\partial \tilde{x}^i}\frac{\partial x^q}{\partial \tilde{x}^j}\left\{\begin{matrix} m \\ pq \end{matrix}\right\} + \frac{\partial^2 x^m}{\partial \tilde{x}^i \partial \tilde{x}^j}\right)\frac{\partial \tilde{x}^k}{\partial x^m} \tag{4.28}$$

and

$$\frac{\partial^2 x^p}{\partial \tilde{x}^i \partial \tilde{x}^j} = \frac{\partial x^p}{\partial \tilde{x}^k}\widetilde{\left\{\begin{matrix} k \\ ij \end{matrix}\right\}} - \frac{\partial x^l}{\partial \tilde{x}^i}\frac{\partial x^m}{\partial \tilde{x}^j}\left\{\begin{matrix} p \\ lm \end{matrix}\right\}. \tag{4.29}$$

Proof. The proof is analogous to that of Lemma 3.1.4 with the exception of the identities in (4.26) and (4.29). The transformation formula

$$\tilde{g}_{ij} = \frac{\partial x^l}{\partial \tilde{x}^i}\frac{\partial x^m}{\partial \tilde{x}^j}g_{lm}$$

and the product and chain rules together imply

$$\frac{\partial \tilde{g}_{ij}}{\partial \tilde{x}^k} = \left(\frac{\partial^2 x^l}{\partial \tilde{x}^i \partial \tilde{x}^k}\frac{\partial x^m}{\partial \tilde{x}^j} + \frac{\partial x^l}{\partial \tilde{x}^i}\frac{\partial^2 x^m}{\partial \tilde{x}^j \partial \tilde{x}^k}\right)g_{lm} + \frac{\partial x^l}{\partial \tilde{x}^i}\frac{\partial x^m}{\partial \tilde{x}^j}\frac{\partial g_{lm}}{\partial x^n}\frac{\partial x^n}{\partial \tilde{x}^k}.$$

It follows from (4.28) that

$$
\frac{\partial x^p}{\partial \tilde{x}^k} \widetilde{\begin{Bmatrix} k \\ ij \end{Bmatrix}} = \left(\frac{\partial x^l}{\partial \tilde{x}^i} \frac{\partial x^q}{\partial \tilde{x}^j} \begin{Bmatrix} m \\ lq \end{Bmatrix} + \frac{\partial^2 x^m}{\partial \tilde{x}^i \partial \tilde{x}^j} \right) \frac{\partial \tilde{x}^k}{\partial x^m} \frac{\partial x^p}{\partial \tilde{x}^k}
$$

$$
= \left(\frac{\partial x^l}{\partial \tilde{x}^i} \frac{\partial x^q}{\partial \tilde{x}^j} \begin{Bmatrix} m \\ lq \end{Bmatrix} + \frac{\partial^2 x^m}{\partial \tilde{x}^i \partial \tilde{x}^j} \right) \delta_m^p
$$

$$
= \frac{\partial x^l}{\partial \tilde{x}^i} \frac{\partial x^q}{\partial \tilde{x}^j} \begin{Bmatrix} p \\ lq \end{Bmatrix} + \frac{\partial^2 x^p}{\partial \tilde{x}^i \partial \tilde{x}^j},
$$

hence

$$
\frac{\partial^2 x^p}{\partial \tilde{x}^i \partial \tilde{x}^j} = \frac{\partial x^p}{\partial \tilde{x}^k} \widetilde{\begin{Bmatrix} k \\ ij \end{Bmatrix}} - \frac{\partial x^l}{\partial \tilde{x}^i} \frac{\partial x^q}{\partial \tilde{x}^j} \begin{Bmatrix} p \\ lq \end{Bmatrix}.
$$

The identity in (4.29) now follows if we rename the index of summation q by m. □

CHAPTER 5

Tensor Analysis

Tensors first appeared in the theory of surfaces. Typical examples were the first and second fundamental coefficients. A tensor of a given order was assigned to every point of a manifold. We also saw that certain derivatives of tensors were needed in various problems of differential geometry. It seems natural to try and define derivatives of a tensor in a way such that the result again is a tensor. This guarantees the independence of a derivative of the choice of a coordinate system.

In particular, we study

- *covariant derivatives of contravariant, covariant vectors and* $(1,1)$*–tensors introduced* in Definitions 5.1.2, 5.1.3 and 5.1.4

- *covariant derivatives of* (r,s)*–tensors* in Definition 5.2.1, *their basic properties* in Theorems 5.2.2 and 5.2.4, and the *Ricci identity for the covariant derivatives of the metric tensor* in Theorem 5.2.6

- *the mixed Riemann tensor of curvature in a Riemann space with metric tensor* g_{ik} in Definition 5.3.1 and the *interchange of the order of its covariant differentiation in Ricci's identity,* Theorem 5.3.2

- *the Bianchi identities for the derivatives of the mixed Riemann tensor of curvature and the covariant Riemann tensor of curvature* in Theorem 5.4.1

- *the Beltrami differentiator of first order in a Riemann space* in Definition 5.5.1, the *divergence of contravariant and covariant vectors* in Definitions 5.5.3 and 5.5.6, *and some properties and a few cases of special interest* in Theorem 5.5.5, Remarks 5.5.2, 5.5.4, 5.5.7, and Example 5.5.8

- *a geometric meaning of the covariant differentiation, the Levi–Cività parallelism* in Definition 5.6.3

- *the fundamental theorem of the theory of surfaces,* Theorem 5.7

- *a geometric interpretation of the Riemann tensor of curvature*

- *spaces with vanishing tensor of curvature, the existence and uniqueness of autoparallel curves in Riemann spaces* in Theorem 5.9.3

DOI: 10.1201/9781003370567-5

- *an extension of Frenet's formulae for curves in Riemann spaces* in Theorem 5.10.1

- *Riemann normal coordinates and the curvature of spaces and the Bertrand–Puiseux theorem,* Theorem 5.11.1.

5.1 COVARIANT DIFFERENTIATION

In this section, we deal with the following problem: differentiation should turn a tensor into a tensor and, in special cases, coincide with partial differentiation.

Example 5.1.1. *(a) Partial differentiation of a $(0,0)$–tensor, that is, of a scalar, yields a covariant vector.*
Let F be a $(0,0)$–tensor with the components $F(x^1, x^2, \ldots, x^n)$, that is, the values of a real–valued function on $\mathbb{R}^n$. Furthermore let $F^*(x^{*1}, x^{*2}, \ldots, x^{*n})$ be its components in another coordinate system. Then

$$F^*(x^{*1}, \ldots, x^{*n}) = F(x^1(x^{*1}, \ldots, x^{*n}), \ldots, x^n(x^{*1}, \ldots, x^{*n}))$$

and

$$F_i^* = \frac{\partial F^*}{\partial x^{*i}} = \frac{\partial F}{\partial x^k}\frac{\partial x^k}{\partial x^{*i}} = \frac{\partial x^k}{\partial x^{*i}}F_k \text{ for all } i.$$

Thus the components of the partial derivative of a $(0,0)$–tensor transform in the same way as the components of a covariant vector.
(b) Partial differentiation of a contravariant tensor T with the components T^i and T^{*i} in the coordinate systems S and S^* yields

$$\frac{\partial T^{*i}}{\partial x^{*j}} = \frac{\partial}{\partial x^{*j}}\left(\frac{\partial x^{*i}}{\partial x^m}T^m\right) = \frac{\partial x^n}{\partial x^{*j}}\frac{\partial x^{*i}}{\partial x^m}\frac{\partial T^m}{\partial x^n} + \frac{\partial^2 x^{*i}}{\partial x^m \partial x^n}\frac{\partial x^n}{\partial x^{*j}}T^m \text{ for all } i \text{ and } j. \quad (5.1)$$

Only the first term of the result on the right hand side in (5.1) satisfies the law of transformation for the components of a tensor. The problem, however, is in the second term.

We try to add certain terms to the partial derivative of a tensor such that the result will transform like a tensor.
First we consider a contravariant vector T with the components T^i and T^{*i} in the coordinate systems S and S^*, and write for the new kind of derivative

$$T^i_{,j} = \frac{\partial T^i}{\partial x^j} + \Lambda^i_{jk}T^k \text{ and } T^{*i}_{,j} = \frac{\partial T^{*i}}{\partial x^{*j}} + \Lambda^{*i}_{jk}T^{*k} \text{ for all } i \text{ and } j \quad (5.2)$$

with the values Λ^i_{jk} and Λ^{*i}_{jk} yet to be determined.
Since the terms $T^i_{,j}$ are to transform like the components of a tensor, that is, satisfy

$$T^{*i}_{,j} = \frac{\partial x^{*i}}{\partial x^m}\frac{\partial x^n}{\partial x^{*j}}T^m_{,n} \text{ for all } i \text{ and } j, \quad (5.3)$$

we have to choose the terms Λ^i_{jk} and Λ^{*i}_{jk} such that identities in (5.3) hold. Substituting (5.2) into (5.3), we obtain

$$\frac{\partial T^{*i}}{\partial x^{*j}} + \Lambda^{*i}_{jk} T^{*k} = T^{*i}_{,j} = \frac{\partial x^{*i}}{\partial x^m} \frac{\partial x^n}{\partial x^{*j}} T^m_{,n} = \frac{\partial x^{*i}}{\partial x^m} \frac{\partial x^n}{\partial x^{*j}} \left(\frac{\partial T^m}{\partial x^n} + \Lambda^m_{nk} T^k \right) \text{ for all } i \text{ and } j.$$

Now this, (5.2) and (5.1) together yield

$$T^{*i}_{,j} = \frac{\partial T^{*i}}{\partial x^{*j}} + \Lambda^{*i}_{jk} T^{*k} = \frac{\partial x^{*i}}{\partial x^m} \frac{\partial x^n}{\partial x^{*j}} \frac{\partial T^m}{\partial x^n} + \frac{\partial^2 x^{*i}}{\partial x^m \partial x^n} \frac{\partial x^n}{\partial x^{*j}} T^m + \Lambda^{*i}_{jl} \frac{\partial x^{*l}}{\partial x^k} T^k$$

$$= \frac{\partial x^{*i}}{\partial x^m} \frac{\partial x^n}{\partial x^{*j}} \left(\frac{\partial T^m}{\partial x^n} + \Lambda^m_{nk} T^k \right)$$

for all i and j. This implies

$$\frac{\partial^2 x^{*i}}{\partial x^m \partial x^n} \frac{\partial x^n}{\partial x^{*j}} T^m + \frac{\partial x^{*l}}{\partial x^k} T^k \Lambda^{*i}_{jl} = \frac{\partial x^{*i}}{\partial x^m} \frac{\partial x^n}{\partial x^{*j}} \Lambda^m_{nk} T^k \text{ for all } i \text{ and } j.$$

If we rename the index of summation m in the first sum on the left hand side to k, then we obtain

$$\left(\frac{\partial^2 x^{*i}}{\partial x^k \partial x^n} \frac{\partial x^n}{\partial x^{*j}} + \frac{\partial x^{*l}}{\partial x^k} \Lambda^{*i}_{jl} - \frac{\partial x^{*i}}{\partial x^m} \frac{\partial x^n}{\partial x^{*j}} \Lambda^m_{nk} \right) T^k = 0 \text{ for all } i \text{ and } j.$$

The last identity has to hold true for all values of T^k. Thus

$$\frac{\partial^2 x^{*i}}{\partial x^k \partial x^n} \frac{\partial x^n}{\partial x^{*j}} + \frac{\partial x^{*l}}{\partial x^k} \Lambda^{*i}_{jl} - \frac{\partial x^{*i}}{\partial x^m} \frac{\partial x^n}{\partial x^{*j}} \Lambda^m_{nk} = 0 \text{ for all } i, j \text{ and } k.$$

This implies

$$\frac{\partial x^{*j}}{\partial x^p} \left(\frac{\partial^2 x^{*i}}{\partial x^k \partial x^n} \frac{\partial x^n}{\partial x^{*j}} + \frac{\partial x^{*l}}{\partial x^k} \Lambda^{*i}_{jl} - \frac{\partial x^{*i}}{\partial x^m} \frac{\partial x^n}{\partial x^{*j}} \Lambda^m_{nk} \right)$$

$$= \frac{\partial^2 x^{*i}}{\partial x^k \partial x^p} + \frac{\partial x^{*l}}{\partial x^k} \frac{\partial x^{*j}}{\partial x^p} \Lambda^{*i}_{jl} - \frac{\partial x^{*i}}{\partial x^m} \Lambda^m_{pk} = 0$$

for all i, k and p. We have by identity (4.29)

$$-\frac{\partial x^{*l}}{\partial x^k} \frac{\partial x^{*j}}{\partial x^p} \Lambda^{*i}_{jl} + \frac{\partial x^{*i}}{\partial x^m} \Lambda^m_{pk} = \frac{\partial x^{*i}}{\partial x^m} \left\{ \begin{matrix} m \\ pk \end{matrix} \right\} - \frac{\partial x^{*l}}{\partial x^k} \frac{\partial x^{*j}}{\partial x^p} \left\{ \begin{matrix} i \\ jl \end{matrix} \right\}^* \text{ for all } i, k \text{ and } p,$$

and we may choose

$$\Lambda^k_{ij} = \left\{ \begin{matrix} k \\ ij \end{matrix} \right\} = \Gamma^k_{ij} \text{ for all } i, j \text{ and } k.$$

This gives rise to the following definition.

Definition 5.1.2. The *covariant derivative of a contravariant vector* with the components T^i is defined by

$$T^i_{,j} = \frac{\partial T^i}{\partial x^j} + \left\{ \begin{matrix} i \\ jk \end{matrix} \right\} T^k \text{ for all } i \text{ and } j.$$

We introduce the *covariant differentiation of covariant vectors* by requiring that the product rule holds.

Let T be a covariant vector with the components T_i and U be a contravariant vector with the components U^i. Then we require for the covariant differentiation by Definition 5.1.2

$$(T_i U^i)_{;j} = T_{i;j} U^i + T_i U^i_{;j} = T_{i;j} U^i + T_i \left(\frac{\partial U^i}{\partial x^j} + \left\{ \begin{matrix} i \\ jk \end{matrix} \right\} U^k \right) \text{ for all } j.$$

Since $T_i U^i$ is a scalar, the result of its covariant differentiation is equal to its partial derivative by Example 5.1.1 (a). Thus we obtain

$$\frac{\partial T_i}{\partial x^j} U^i + T_i \frac{\partial U^i}{\partial x^j} = (T_i U^i)_{,j} = T_{i;j} U^i + T_i \frac{\partial U^i}{\partial x^j} + T_i \left\{ \begin{matrix} i \\ jk \end{matrix} \right\} U^k \text{ for all } j,$$

and consequently

$$\frac{\partial T_i}{\partial x^j} U^i = T_{i;j} U^i + T_i \left\{ \begin{matrix} i \\ jk \end{matrix} \right\} U^k \text{ or } \frac{\partial T_k}{\partial x^j} U^k = T_{k;j} U^k + T_i \left\{ \begin{matrix} i \\ jk \end{matrix} \right\} U^k \text{ for all } j.$$

Since the last identity has to hold true for arbitrary contravariant tensors U, we must have

$$T_{k;j} = \frac{\partial T_k}{\partial x^j} - T_i \left\{ \begin{matrix} i \\ jk \end{matrix} \right\} \text{ for all } k \text{ and } j.$$

This gives rise to the following definition.

Definition 5.1.3. The *covariant derivative of a covariant vector T* with the components T_i is defined by

$$T_{i;j} = \frac{\partial T_i}{\partial x^j} - \left\{ \begin{matrix} k \\ ij \end{matrix} \right\} T_k \text{ for all } i \text{ and } j.$$

Before we generalize the definition of the covariant differentiation to arbitrary tensors, we consider the covariant differentiation of a $(1,1)$–tensor T with the components T_i^k. The following transformation formulae hold for all i nd k

$$T_i^{*k} = \frac{\partial x^m}{\partial x^{*i}} \frac{\partial x^{*k}}{\partial x^n} T_m^n.$$

Partial differentiation yields

$$\frac{\partial T_i^{*k}}{\partial x^{*j}} = \frac{\partial x^m}{\partial x^{*i}} \frac{\partial x^{*k}}{\partial x^n} \frac{\partial T_m^n}{\partial x^l} \frac{\partial x^l}{\partial x^{*j}} + T_m^n \left(\frac{\partial^2 x^m}{\partial x^{*j} \partial x^{*i}} \frac{\partial x^{*k}}{\partial x^n} + \frac{\partial^2 x^{*k}}{\partial x^r \partial x^n} \frac{\partial x^r}{\partial x^{*j}} \frac{\partial x^m}{\partial x^{*i}} \right)$$

for all i, j and k. Applying (4.29), we eliminate the second order partial derivatives

$$\frac{\partial^2 x^m}{\partial x^{*j} \partial x^{*i}} \frac{\partial x^{*k}}{\partial x^n} + \frac{\partial x^m}{\partial x^{*i}} \frac{\partial^2 x^{*k}}{\partial x^r \partial x^n} \frac{\partial x^r}{\partial x^{*j}}$$
$$= \frac{\partial x^m}{\partial x^{*l}} \left\{ \begin{matrix} l \\ ji \end{matrix} \right\}^* \frac{\partial x^{*k}}{\partial x^n} - \frac{\partial x^l}{\partial x^{*i}} \frac{\partial x^q}{\partial x^{*j}} \left\{ \begin{matrix} m \\ lq \end{matrix} \right\} \frac{\partial x^{*k}}{\partial x^n}$$

$$+ \frac{\partial x^m}{\partial x^{*i}} \frac{\partial x^{*k}}{\partial x^l} \left\{ {l \atop rn} \right\} \frac{\partial x^r}{\partial x^{*j}} - \frac{\partial x^{*l}}{\partial x^r} \frac{\partial x^{*q}}{\partial x^n} \left\{ {k \atop lq} \right\}^* \frac{\partial x^m}{\partial x^{*i}} \frac{\partial x^r}{\partial x^{*j}}$$

for all i, j, k, m and n. Thus

$$\frac{\partial T_i^{*k}}{\partial x^{*j}} - T_m^n \frac{\partial x^m}{\partial x^{*l}} \frac{\partial x^{*k}}{\partial x^n} \left\{ {l \atop ji} \right\}^* + T_m^n \frac{\partial x^m}{\partial x^{*i}} \frac{\partial x^{*q}}{\partial x^n} \frac{\partial x^{*l}}{\partial x^r} \frac{\partial x^r}{\partial x^{*j}} \left\{ {k \atop lq} \right\}^*$$

$$= \frac{\partial T_i^{*k}}{\partial x^{*j}} - T_m^n \frac{\partial x^m}{\partial x^{*l}} \frac{\partial x^{*k}}{\partial x^n} \left\{ {l \atop ji} \right\}^* + T_m^n \frac{\partial x^m}{\partial x^{*i}} \frac{\partial x^{*q}}{\partial x^n} \delta_j^{*l} \left\{ {k \atop lq} \right\}^*$$

$$= \frac{\partial T_i^{*k}}{\partial x^{*j}} - T_l^{*k} \left\{ {l \atop ij} \right\}^* + T_i^{*q} \left\{ {k \atop jq} \right\}^*$$

$$= \frac{\partial T_m^n}{\partial x^l} \frac{\partial x^l}{\partial x^{*j}} \frac{\partial x^m}{\partial x^{*i}} \frac{\partial x^{*k}}{\partial x^n} - T_m^n \frac{\partial x^l}{\partial x^{*i}} \frac{\partial x^q}{\partial x^{*j}} \frac{\partial x^{*k}}{\partial x^n} \left\{ {m \atop lq} \right\} + T_m^n \frac{\partial x^r}{\partial x^{*j}} \frac{\partial x^m}{\partial x^{*i}} \frac{\partial x^{*k}}{\partial x^l} \left\{ {l \atop rn} \right\}$$

and interchanging n and l in the last term

$$= \left(\frac{\partial T_m^n}{\partial x^l} \frac{\partial x^l}{\partial x^{*j}} \frac{\partial x^m}{\partial x^{*i}} - T_m^n \frac{\partial x^l}{\partial x^{*i}} \frac{\partial x^q}{\partial x^{*j}} \left\{ {m \atop lq} \right\} + T_m^l \frac{\partial x^r}{\partial x^{*j}} \frac{\partial x^m}{\partial x^{*i}} \left\{ {n \atop rl} \right\} \right) \frac{\partial x^{*k}}{\partial x^n}$$

and interchanging m and l in the second term

$$= \left(\frac{\partial T_m^n}{\partial x^l} \frac{\partial x^l}{\partial x^{*j}} - T_l^n \frac{\partial x^q}{\partial x^{*j}} \left\{ {l \atop mq} \right\} + T_m^l \frac{\partial x^r}{\partial x^{*j}} \left\{ {n \atop rl} \right\} \right) \frac{\partial x^m}{\partial x^{*i}} \frac{\partial x^{*k}}{\partial x^n}$$

and changing l to r in the first term and q to r in the second term

$$= \left(\frac{\partial T_m^n}{\partial x^r} - T_l^n \left\{ {l \atop mr} \right\} + T_m^l \left\{ {n \atop rl} \right\} \right) \frac{\partial x^r}{\partial x^{*j}} \frac{\partial x^m}{\partial x^{*i}} \frac{\partial x^{*k}}{\partial x^n}$$

for all i, j and k. Consequently, we have

$$\frac{\partial T_i^{*k}}{\partial x^{*j}} - T_l^{*k} \left\{ {l \atop ij} \right\}^* + T_i^{*q} \left\{ {k \atop jq} \right\}^*$$

$$= \frac{\partial x^r}{\partial x^{*j}} \frac{\partial x^m}{\partial x^{*i}} \frac{\partial x^{*k}}{\partial x^n} \left(\frac{\partial T_m^n}{\partial x^r} - T_l^n \left\{ {l \atop mr} \right\} + T_m^l \left\{ {n \atop rl} \right\} \right)$$

for all i, j and k. Thus the terms

$$T_{m;r}^n = \frac{\partial T_m^n}{\partial x^r} - T_l^n \left\{ {l \atop mr} \right\} + T_m^l \left\{ {n \atop rl} \right\} \quad \text{for all } m, n \text{ and } r$$

satisfy the transformation formulae of a tensor.

Definition 5.1.4. The *covariant derivative of a $(1,1)$–tensor* with the components T_m^n is defined by

$$T_{m;r}^n = \frac{\partial T_m^n}{\partial x^r} - T_l^n \left\{ {l \atop mr} \right\} + T_m^l \left\{ {n \atop rl} \right\} \quad \text{for all } m, n \text{ and } r.$$

Remark 5.1.5. *We summarize. The covariant derivatives of a contravariant vector, a covariant vector and a* $(1,1)$*–tensor with the components* T^i, T_i *and* T^i_j *are defined by*

$$T^i_{;k} = \frac{\partial T^i}{\partial x^k} + \left\{ \begin{matrix} i \\ kl \end{matrix} \right\} T^l \qquad \text{for all } i \text{ and } k,$$

$$T_{i;k} = \frac{\partial T_i}{\partial x^k} - \left\{ \begin{matrix} l \\ ik \end{matrix} \right\} T_l \qquad \text{for all } i \text{ and } k,$$

and

$$T^i_{j;k} = \frac{\partial T^i_j}{\partial x^k} - T^i_l \left\{ \begin{matrix} l \\ jk \end{matrix} \right\} + T^l_j \left\{ \begin{matrix} i \\ kl \end{matrix} \right\} \quad \text{for all } i, j \text{ and } k.$$

5.2 THE COVARIANT DERIVATIVE OF AN (R, S)–TENSOR

In this section, we extend the concept of covariant derivatives to the general case of (r, s)–tensors. Furthermore we shall prove the *Ricci identity*.

Definition 5.2.1. The *covariant derivative of an* (r, s)*–tensor* T with the components

$$T^{j_1 \ldots j_s}_{i_1 \ldots i_r}$$

is defined by

$$\left\{ \begin{aligned} T^{j_1 \ldots j_s}_{i_1 \ldots i_r; k} &= \frac{\partial T^{j_1 \ldots j_s}_{i_1 \ldots i_r}}{\partial x^k} - \sum_{l=1}^{r} T^{j_1 \ldots j_s}_{i_1 \ldots i_{l-1} m i_{l+1} \ldots i_r} \left\{ \begin{matrix} m \\ i_l k \end{matrix} \right\} \\ &\quad + \sum_{n=1}^{s} T^{j_1 \ldots j_{n-1} p j_{n+1} \ldots j_s}_{i_1 \ldots i_r} \left\{ \begin{matrix} j_n \\ pk \end{matrix} \right\}. \end{aligned} \right. \tag{5.4}$$

Theorem 5.2.2. *The covariant derivative of an* (r, s)*–tensor is an* $(r+1, s)$*–tensor. Covariant differentiation increases the number of covariant indices by one.*

Proof. The proof is similar to that in the case of a $(1,1)$–tensor.
Let T be an (r, s)–tensor with the components $T^{j_1 \ldots j_s}_{i_1 \ldots i_r}$ and $T^{*j_1 \ldots j_s}_{i_1 \ldots i_r}$ in the coordinate systems S and S^*, respectively. We put

$$A_1 = \frac{\partial T^{*j_1 \ldots j_s}_{i_1 \ldots i_r}}{\partial x^{*k}},$$

$$A_2 = \sum_{l=1}^{r} T^{*j_1 \ldots j_s}_{i_1 \ldots i_{l-1} m i_{l+1} \ldots i_r} \left\{ \begin{matrix} m \\ i_l k \end{matrix} \right\}^*$$

and

$$A_3 = \sum_{n=1}^{s} T^{*j_1 \ldots j_{n-1} p j_{n+1} \ldots j_s}_{i_1 \ldots i_r} \left\{ \begin{matrix} j_n \\ pk \end{matrix} \right\}^*.$$

so that

$$T^{*j_1\ldots j_s}_{i_1\ldots i_r;k} = A_1 - A_2 + A_3 \tag{5.5}$$

by Definition 5.2.1. Using the transformation formula for the components of an (r,s)-tensor, we obtain

$$A_1 = \frac{\partial}{\partial x^{*k}}\left(\frac{\partial x^{m_1}}{\partial x^{*i_1}}\cdots\frac{\partial x^{m_r}}{\partial x^{*i_r}}\frac{\partial x^{*j_1}}{\partial x^{l_1}}\cdots\frac{\partial x^{*j_s}}{\partial x^{l_s}}T^{l_1\ldots l_s}_{m_1\ldots m_r}\right)$$

$$= \frac{\partial x^{m_1}}{\partial x^{*i_1}}\cdots\frac{\partial x^{m_r}}{\partial x^{*i_r}}\frac{\partial x^\alpha}{\partial x^{*k}}\frac{\partial x^{*j_1}}{\partial x^{l_1}}\cdots\frac{\partial x^{*j_s}}{\partial x^{l_s}}\frac{\partial T^{l_1\ldots l_s}_{m_1\ldots m_r}}{\partial x^\alpha}$$

$$+ T^{l_1\ldots l_s}_{m_1\ldots m_r}\left(\sum_{\alpha=1}^{r}\frac{\partial^2 x^{m_\alpha}}{\partial x^{*i_\alpha}\partial x^{*k}}\frac{\partial x^{m_1}}{\partial x^{*i_1}}\cdots\frac{\partial x^{m_{\alpha-1}}}{\partial x^{*i_{\alpha-1}}}\frac{\partial x^{m_{\alpha+1}}}{\partial x^{*i_{\alpha+1}}}\cdots\frac{\partial x^{m_r}}{\partial x^{*i_r}}\frac{\partial x^{*j_1}}{\partial x^{l_1}}\cdots\frac{\partial x^{*j_s}}{\partial x^{l_s}}\right.$$

$$\left.+ \sum_{\beta=1}^{s}\frac{\partial^2 x^{*j_\beta}}{\partial x^{l_\beta}\partial x^\gamma}\frac{\partial x^{m_1}}{\partial x^{*i_1}}\cdots\frac{\partial x^{m_r}}{\partial x^{*i_r}}\frac{\partial x^{*j_1}}{\partial x^{l_1}}\cdots\frac{\partial x^{*j_{\beta-1}}}{\partial x^{l_{\beta-1}}}\frac{\partial x^{*j_{\beta+1}}}{\partial x^{l_{\beta+1}}}\cdots\frac{\partial x^{*j_s}}{\partial x^{l_s}}\frac{\partial x^\gamma}{\partial x^{*k}}\right).$$

We have by (4.29)

$$A_{11} = T^{l_1\ldots l_s}_{m_1\ldots m_r}\sum_{\alpha=1}^{r}\frac{\partial^2 x^{m_\alpha}}{\partial x^{*i_\alpha}\partial x^{*k}}\frac{\partial x^{m_1}}{\partial x^{*i_1}}\cdots\frac{\partial x^{m_{\alpha-1}}}{\partial x^{*i_{\alpha-1}}}\frac{\partial x^{m_{\alpha+1}}}{\partial x^{*i_{\alpha+1}}}\cdots\frac{\partial x^{m_r}}{\partial x^{*i_r}}\frac{\partial x^{*j_1}}{\partial x^{l_1}}\cdots\frac{\partial x^{*j_s}}{\partial x^{l_s}}$$

$$= T^{l_1\ldots l_s}_{m_1\ldots m_r}\sum_{\alpha=1}^{r}\frac{\partial x^{m_\alpha}}{\partial x^{*\beta}}\left\{\begin{matrix}\beta\\i_\alpha k\end{matrix}\right\}^*\frac{\partial x^{m_1}}{\partial x^{*i_1}}\cdots\frac{\partial x^{m_{\alpha-1}}}{\partial x^{*i_{\alpha-1}}}\frac{\partial x^{m_{\alpha+1}}}{\partial x^{*i_{\alpha+1}}}\cdots\frac{\partial x^{m_r}}{\partial x^{*i_r}}\frac{\partial x^{*j_1}}{\partial x^{l_1}}\cdots\frac{\partial x^{*j_s}}{\partial x^{l_s}}$$

$$- T^{l_1\ldots l_s}_{m_1\ldots m_r}\sum_{\alpha=1}^{r}\frac{\partial x^\gamma}{\partial x^{*i_\alpha}}\frac{\partial x^\delta}{\partial x^{*k}}\left\{\begin{matrix}m_\alpha\\\gamma\delta\end{matrix}\right\}\frac{\partial x^{m_1}}{\partial x^{*i_1}}\cdots\frac{\partial x^{m_{\alpha-1}}}{\partial x^{*i_{\alpha-1}}}\frac{\partial x^{m_{\alpha+1}}}{\partial x^{*i_{\alpha+1}}}\cdots\frac{\partial x^{m_r}}{\partial x^{*i_r}}\frac{\partial x^{*j_1}}{\partial x^{l_1}}\cdots\frac{\partial x^{*j_s}}{\partial x^{l_s}}$$

$$= \sum_{\alpha=1}^{r}T^{l_1\ldots l_s}_{m_1\ldots m_{\alpha-1}m_\alpha m_{\alpha+1}\ldots m_r}\frac{\partial x^{m_1}}{\partial x^{*i_1}}\cdots\frac{\partial x^{m_{\alpha-1}}}{\partial x^{*i_{\alpha-1}}}\frac{\partial x^{m_\alpha}}{\partial x^{*\beta}}\frac{\partial x^{m_{\alpha+1}}}{\partial x^{*i_{\alpha+1}}}\cdots\frac{\partial x^{m_r}}{\partial x^{*i_r}}\cdot$$

$$\cdot\frac{\partial x^{*j_1}}{\partial x^{l_1}}\cdots\frac{\partial x^{*j_s}}{\partial x^{l_s}}\left\{\begin{matrix}\beta\\i_\alpha k\end{matrix}\right\}^*$$

$$- \left(\sum_{l=1}^{r}T^{l_1\ldots l_s}_{m_1\ldots m_{l-1}mm_{l+1}\ldots m_r}\left\{\begin{matrix}m\\m_l\delta\end{matrix}\right\}\right)\frac{\partial x^{m_1}}{\partial x^{*i_1}}\cdots\frac{\partial x^{m_r}}{\partial x^{*i_r}}\frac{\partial x^{*j_1}}{\partial x^{l_1}}\cdots\frac{\partial x^{*j_s}}{\partial x^{l_s}}\frac{\partial x^\delta}{\partial x^{*k}},$$

where we change the indices of summation m_α and γ to m and m_l, respectively, in the second sum. An application of the transformation formula in the first sum yields

$$A_{11} = A_2 - \frac{\partial x^{m_1}}{\partial x^{*i_1}}\cdots\frac{\partial x^{m_r}}{\partial x^{*i_r}}\frac{\partial x^\alpha}{\partial x^{*k}}\frac{\partial x^{*j_1}}{\partial x^{l_1}}\cdots\frac{\partial x^{*j_s}}{\partial x^{l_s}}\sum_{l=1}^{r}T^{l_1\ldots l_s}_{m_1\ldots m_{l-1}mm_{l+1}\ldots m_r}\left\{\begin{matrix}m\\m_l\alpha\end{matrix}\right\}.$$

$$\tag{5.6}$$

Similarly it follows that

$$A_{22} = T^{l_1\ldots l_s}_{m_1\ldots m_r}\sum_{\beta=1}^{s}\frac{\partial^2 x^{*j_\beta}}{\partial x^{l_\beta}\partial x^\gamma}\frac{\partial x^{m_1}}{\partial x^{*i_1}}\cdots\frac{\partial x^{m_r}}{\partial x^{*i_r}}\frac{\partial x^{*j_1}}{\partial x^{l_1}}\cdots\frac{\partial x^{*j_{\beta-1}}}{\partial x^{l_{\beta-1}}}\frac{\partial x^{*j_{\beta+1}}}{\partial x^{l_{\beta+1}}}\cdots\frac{\partial x^{*j_s}}{\partial x^{l_s}}\frac{\partial x^\gamma}{\partial x^{*k}}$$

$$= T^{l_1,\ldots,l_s}_{m_1\ldots m_r}\sum_{\beta=1}^{s}\frac{\partial x^{m_1}}{\partial x^{*i_1}}\cdots\frac{\partial x^{m_r}}{\partial x^{*i_r}}\frac{\partial x^{*j_1}}{\partial x^{l_1}}\cdots\frac{\partial x^{*j_{\beta-1}}}{\partial x^{l_{\beta-1}}}\frac{\partial x^{*j_\beta}}{\partial x^\delta}\frac{\partial x^{*j_{\beta+1}}}{\partial x^{l_{\beta+1}}}\cdots\frac{\partial x^{*j_s}}{\partial x^{l_s}}\frac{\partial x^\gamma}{\partial x^{*k}}\left\{\begin{matrix}\delta\\l_\beta\gamma\end{matrix}\right\}$$

$$- T^{l_1 \ldots l_s}_{m_1 \ldots m_r} \sum_{\beta=1}^{s} \frac{\partial x^{m_1}}{\partial x^{*i_1}} \cdots \frac{\partial x^{m_r}}{\partial x^{*i_r}} \frac{\partial x^{*j_1}}{\partial x^{l_1}} \cdots \frac{\partial x^{*j_{\beta-1}}}{\partial x^{l_{\beta-1}}} \frac{\partial x^{*t}}{\partial x^{l_\beta}} \frac{\partial x^{*j_{\beta+1}}}{\partial x^{l_{\beta+1}}} \cdots \frac{\partial x^{*j_s}}{\partial x^{l_s}} \cdot$$

$$\cdot \begin{Bmatrix} j_\beta \\ tu \end{Bmatrix}^* \frac{\partial x^{*u}}{\partial x^\gamma} \frac{\partial x^\gamma}{\partial x^{*k}}$$

$$= \left(\sum_{\beta=1}^{s} T^{l_1 \ldots l_{\beta-1} p l_{\beta+1} \ldots l_s}_{m_1 \ldots m_r} \begin{Bmatrix} l_\beta \\ p\gamma \end{Bmatrix} \right) \frac{\partial x^{m_1}}{\partial x^{*i_1}} \cdots \frac{\partial x^{m_r}}{\partial x^{*i_r}} \frac{\partial x^{*j_1}}{\partial x^{l_1}} \cdots \frac{\partial x^{*j_s}}{\partial x^{l_s}} \frac{\partial x^\gamma}{\partial x^{*k}}$$

$$- \sum_{\beta=1}^{s} \frac{\partial x^{m_1}}{\partial x^{*i_1}} \cdots \frac{\partial x^{m_r}}{\partial x^{*i_r}} \frac{\partial x^{*j_1}}{\partial x^{l_1}} \cdots \frac{\partial x^{*j_{\beta-1}}}{\partial x^{l_{\beta-1}}} \frac{\partial x^{*t}}{\partial x^{l_\beta}} \frac{\partial x^{*j_{\beta+1}}}{\partial x^{l_{\beta+1}}} \cdots \frac{\partial x^{*j_s}}{\partial x^{l_s}} T^{l_1 \ldots l_s}_{m_1 \ldots m_r} \begin{Bmatrix} j_\beta \\ tk \end{Bmatrix}^*.$$

Therefore,

$$A_{22} = \frac{\partial x^{m_1}}{\partial x^{*i_1}} \cdots \frac{\partial x^{m_r}}{\partial x^{*i_r}} \frac{\partial x^{*j_1}}{\partial x^{l_1}} \cdots \frac{\partial x^{*j_s}}{\partial x^{l_s}} \frac{\partial x^\alpha}{\partial x^{*k}} \left(\sum_{\beta=1}^{s} T^{l_1 \ldots l_{\beta-1} p l_{\beta+1} \ldots l_s}_{m_1 \ldots m_r} \begin{Bmatrix} l_\beta \\ p\alpha \end{Bmatrix} \right) - A_3. \quad (5.7)$$

Finally, we obtain from (5.5), (5.6) and (5.7)

$$T^{*j_1 \ldots j_s}_{i_1 \ldots i_r; k} = \frac{\partial x^{m_1}}{\partial x^{*i_1}} \cdots \frac{\partial x^{m_r}}{\partial x^{*i_r}} \frac{\partial x^\alpha}{\partial x^{*k}} \frac{\partial x^{*j_1}}{\partial x^{l_1}} \cdots \frac{\partial x^{*j_s}}{\partial x^{l_s}}$$

$$\left(\frac{\partial T^{l_1 \ldots l_s}_{m_1 \ldots m_r}}{\partial x^\alpha} - \sum_{l=1}^{r} T^{l_1 \ldots l_s}_{m_1 \ldots m_{l-1} m m_{l+1} \ldots m_r} \begin{Bmatrix} m \\ m_l \alpha \end{Bmatrix} + \right.$$

$$\left. + \sum_{\beta=1}^{s} T^{l_1 \ldots l_{\beta-1} p l_{\beta+1} \ldots l_s}_{m_1 \ldots m_r} \begin{Bmatrix} l_\beta \\ p\alpha \end{Bmatrix} \right)$$

$$= \frac{\partial x^{m_1}}{\partial x^{*i_1}} \cdots \frac{\partial x^{m_r}}{\partial x^{*i_r}} \frac{\partial x^\alpha}{\partial x^{*k}} \frac{\partial x^{*j_1}}{\partial x^{l_1}} \cdots \frac{\partial x^{*j_s}}{\partial x^{l_s}} T^{l_1 \ldots l_s}_{m_1 \ldots m_r; k},$$

and the components of the covariant derivative of an (r, s)–tensor satisfy the transformation formulae of an $(r + 1, s)$–tensor. □

Example 5.2.3. *The covariant derivatives of $(2, 0)$–, $(1, 1)$– and $(0, 2)$–tensors T are given by*

$$T_{ij;k} = \frac{\partial T_{ij}}{\partial x^k} - T_{mj} \begin{Bmatrix} m \\ ik \end{Bmatrix} - T_{im} \begin{Bmatrix} m \\ jk \end{Bmatrix}, \quad (5.8)$$

$$T^j_{i;k} = \frac{\partial T^j_i}{\partial x^k} - T^j_m \begin{Bmatrix} m \\ ik \end{Bmatrix} + T^m_i \begin{Bmatrix} j \\ mk \end{Bmatrix} \quad (5.9)$$

and

$$T^{ij}_{;k} = \frac{\partial T^{ij}}{\partial x^k} + T^{mj} \begin{Bmatrix} m \\ ik \end{Bmatrix} + T^{im} \begin{Bmatrix} j \\ mk \end{Bmatrix}. \quad (5.10)$$

Theorem 5.2.4. *(a) Let T and U be tensors with the components $T_{i_1...i_r}^{j_1...j_s}$ and $U_{i_1...i_r}^{j_1...j_s}$. Then the covariant derivative of the sum and difference of T and U taken termwise, that is,*

$$\left(T_{i_1...i_r}^{j_1...j_s} + U_{i_1...i_r}^{j_1...j_s} \right)_{;k} = T_{i_1...i_r;k}^{j_1...j_s} + U_{i_1...i_r;k}^{j_1...j_s}.$$

(b) Let T and U be tensors with the components $T_{i_1...i_r}^{j_1...j_s}$ and $U_{m_1...m_q}^{n_1...n_p}$. Then the product rule holds for the covariant differentiation of the outer product of T and U, that is,

$$\left(T_{i_1...i_r}^{j_1...j_s} * U_{m_1...m_q}^{n_1...n_p} \right)_{;k} = T_{i_1...i_r;k}^{j_1...j_s} * U_{m_1...m_q}^{n_1...n_p} + T_{i_1...i_r}^{j_1...j_s} * U_{m_1...m_q}^{n_1...n_p;k}.$$

Proof. (a) Let

$$V_{i_1...i_r}^{j_1...j_s} = T_{i_1...i_r}^{j_1...j_s} + U_{i_1...i_r}^{j_1...j_s}.$$

Then it follows by Definition 5.2.1 that

$$
\begin{aligned}
V_{i_1...i_r;k}^{j_1...j_s} &= \frac{\partial V_{i_1...i_r}^{j_1...j_s}}{\partial x^k} \quad - \sum_{l=1}^{r} V_{i_1...i_{l-1}mi_{l+1}...i_r}^{j_1...j_s} \left\{ \begin{matrix} m \\ i_l k \end{matrix} \right\} + \\
&\quad + \sum_{n=1}^{s} V_{i_1...i_r}^{j_1...j_{n-1}pj_{n+1}...j_s} \left\{ \begin{matrix} j_n \\ pk \end{matrix} \right\} \\
&= \frac{\partial T_{i_1...i_r}^{j_1...j_s}}{\partial x^k} \quad - \sum_{l=1}^{r} T_{i_1...i_{l-1}mi_{l+1}...i_r}^{j_1...j_s} \left\{ \begin{matrix} m \\ i_l k \end{matrix} \right\} + \\
&\quad + \sum_{n=1}^{s} T_{i_1...i_r}^{j_1...j_{n-1}pj_{n+1}...j_s} \left\{ \begin{matrix} j_n \\ pk \end{matrix} \right\} \\
&\quad + \frac{\partial U_{i_1...i_r}^{j_1...j_s}}{\partial x^k} \quad - \sum_{l=1}^{r} U_{i_1...i_{l-1}mi_{l+1}...,i_r}^{j_1...j_s} \left\{ \begin{matrix} m \\ i_l k \end{matrix} \right\} + \\
&\quad + \sum_{n=1}^{s} U_{i_1...i_r}^{j_1...j_{n-1}pj_{n+1}...j_s} \left\{ \begin{matrix} j_n \\ pk \end{matrix} \right\} \\
&= T_{i_1...i_r;k}^{j_1...j_s} \quad + U_{i_1...i_r;k}^{j_1...j_s}.
\end{aligned}
$$

(b) Let

$$V_{i_1...i_r m_1...m_q}^{j_1...j_s n_1...n_p} = T_{i_1...i_r}^{j_1...j_s} \cdot U_{m_1...m_q}^{n_1...n_q}.$$

Then we have

$$
\begin{aligned}
V_{i_1...i_r m_1...m_q;k}^{j_1...j_s n_1...n_p} &= \frac{\partial V_{i_1...i_r m_1...m_q}^{j_1,...,j_s n_1...n_p}}{\partial x^k} \\
&\quad - \sum_{l=1}^{r} V_{i_1...i_{l-1}mi_{l+1}...i_r m_1...m_q}^{j_1...j_s n_1...n_p} \left\{ \begin{matrix} m \\ i_l k \end{matrix} \right\} \\
&\quad - \sum_{l=1}^{q} V_{i_1...i_r m_1...m_{l-1}mm_{l+1}...m_q}^{j_1...j_s n_1...n_p} \left\{ \begin{matrix} m \\ m_l k \end{matrix} \right\} \\
&\quad + \sum_{l=1}^{s} V_{i_1...i_r m_1...m_q}^{j_1...j_{l-1}tj_{l+1}...j_s n_1...n_p} \left\{ \begin{matrix} j_l \\ tk \end{matrix} \right\} \\
&\quad + \sum_{l=1}^{p} V_{i_1...i_r m_1...m_q}^{j_1...j_s n_1...n_{l-1}tn_{l+1}...n_p} \left\{ \begin{matrix} n_l \\ tk \end{matrix} \right\}
\end{aligned}
$$

$$= \left(\frac{\partial T^{j_1 \ldots j_s}_{i_1 \ldots i_r}}{\partial x^k} - \sum_{l=1}^{r} T^{j_1, \ldots, j_s}_{i_1 \ldots i_{l-1} m i_{l+1} \ldots i_r} \left\{ \begin{matrix} m \\ i_l k \end{matrix} \right\} \right.$$

$$\left. + \sum_{l=1}^{s} T^{j_1 \ldots j_{l-1} t j_{l+1} \ldots j_s}_{i_1 \ldots i_r} \left\{ \begin{matrix} j_l \\ t k \end{matrix} \right\} \right) \cdot U^{n_1 \ldots n_p}_{m_1 \ldots m_q}$$

$$+ \left(\frac{\partial U^{n_1 \ldots n_p}_{m_1 \ldots m_q}}{\partial x^k} - \sum_{l=1}^{q} U^{n_1 \ldots n_p}_{m_1 \ldots m_{l-1} m m_{l+1} \ldots m_q} \left\{ \begin{matrix} m \\ m_l k \end{matrix} \right\} \right.$$

$$\left. + \sum_{l=1}^{p} U^{n_1 \ldots n_{l-1} t n_{l+1} \ldots n_p}_{m_1 \ldots m_q} \left\{ \begin{matrix} n_l \\ t k \end{matrix} \right\} \right) \cdot T^{j_1 \ldots j_s}_{i_1 \ldots i_r}$$

$$= T^{j_1 \ldots j_s}_{i_1 \ldots i_r; k} \cdot U^{n_1 \ldots n_p}_{m_1 \ldots m_q} + T^{j_1 \ldots j_s}_{i_1 \ldots i_r} \cdot U^{n_1 \ldots n_p}_{m_1 \ldots m_q; k}.$$

□

Example **5.2.5.** *We have by Theorem 5.2.4 (b)*

$$(T_{ij} U^{mn})_{;k} = T_{ij;k} U^{mn} + T_{ij} U^{mn}_{;k}.$$

We put $m = j$ and obtain the contraction

$$\left(T_{ij} U^{jn} \right)_{;k} = T_{ij;k} U^{jn} + T_{ij} U^{jn}_{;k}.$$

Theorem 5.2.6 (Ricci).
The covariant derivatives of the tensors with the components g_{ik}, g^{ik} and δ^k_i are the zero tensor.

Proof. We have by (5.8), (4.24) and (4.26)

$$g_{ij;k} = \frac{\partial g_{ij}}{\partial x^k} - g_{mj} \left\{ \begin{matrix} m \\ ik \end{matrix} \right\} - g_{im} \left\{ \begin{matrix} m \\ jk \end{matrix} \right\} = \frac{\partial g_{ij}}{\partial x^k} - [ikj] - [jki]$$

$$= [ikj] + [jki] - [ikj] - [jki] = 0.$$

Furthermore we obtain by (5.10)

$$g^{ij}_{;k} = \frac{\partial g^{ij}}{\partial x^k} + g^{mj} \left\{ \begin{matrix} i \\ mk \end{matrix} \right\} + g^{im} \left\{ \begin{matrix} j \\ mk \end{matrix} \right\}. \tag{5.11}$$

Finally $g_{ij} g^{jm} = \delta^m_i$ implies

$$\frac{\partial g_{ij}}{\partial x^k} g^{jm} + g_{ij} \frac{\partial g^{jm}}{\partial x^k} = 0$$

and

$$\frac{\partial g_{ij}}{\partial x^k} g^{jm} g^{li} + g^{li} g_{ij} \frac{\partial g^{jm}}{\partial x^k} = \frac{\partial g_{ij}}{\partial x^k} g^{jm} g^{li} + \frac{\partial g^{lm}}{\partial x^k} = 0.$$

This yields by (4.25) and (4.24)

$$\frac{\partial g^{lm}}{\partial x^k} = -\frac{\partial g_{ij}}{\partial x^k} g^{jm} g^{li} = -[ikj] g^{jm} g^{li} - [jki] g^{jm} g^{li}$$

$$= - \left\{ \begin{matrix} m \\ ik \end{matrix} \right\} g^{li} - \left\{ \begin{matrix} l \\ jk \end{matrix} \right\} g^{jm}.$$

If we replace l and m by i and j and rename the indices of summation i and j by m then we conclude

$$\frac{\partial g^{ij}}{\partial x^k} = -\left\{\begin{matrix} j \\ mk \end{matrix}\right\} g^{im} - \left\{\begin{matrix} i \\ mk \end{matrix}\right\} g^{mj}.$$

Now $g^{ij}_{;k} = 0$ is an immediate consequence of (5.11).
By Example 5.2.5 and what we have just shown, we conclude

$$\delta^j_{i;k} = \left(g_{im}g^{mj}\right)_{;k} = g_{im;k}g^{mj} + g_{im}g^{mj}_{;k} = 0.$$

$\square$

Example 5.2.7. *The second order covariant derivative of a scalar F is given by*

$$F_{;ik} = (F_{;i})_{;k} = \left(\frac{\partial F}{\partial x^i}\right)_{;k} = \frac{\partial^2 F}{\partial x^i \partial x^k} - \left\{\begin{matrix} j \\ ik \end{matrix}\right\} \frac{\partial F}{\partial x^j}.$$

Thus Gauss's formulae of differentiation may be written as

$$x^l_{;ik} = \frac{\partial^2 x^l}{\partial x^i \partial x^k} - \left\{\begin{matrix} j \\ ik \end{matrix}\right\} \frac{\partial x^l}{\partial x^j} = \left\{\begin{matrix} j \\ ik \end{matrix}\right\} \frac{\partial x^l}{\partial x^j} + L_{ik}N^l - \left\{\begin{matrix} j \\ ik \end{matrix}\right\} \frac{\partial x^l}{\partial x^j}$$

$$= L_{ik}N^l.$$

Furthermore, the Mainardi–Codazzi identities

$$\left\{\begin{matrix} r \\ ik \end{matrix}\right\} L_{rj} - \left\{\begin{matrix} r \\ ij \end{matrix}\right\} L_{rk} + \frac{\partial L_{ik}}{\partial u^j} - \frac{\partial L_{ij}}{\partial u^k} = 0$$

may be written as

$$L_{ik;j} = L_{ij;k}.$$

5.3 THE INTERCHANGE OF ORDER FOR COVARIANT DIFFERENTIATION AND RICCI'S IDENTITY

In general, the second order covariant derivative depends on the order of differentiation. A trivial exception is the case where the covariant derivative reduces to the normal partial derivative, because all Christoffel symbols vanish identically, as is the case with respect to Cartesian coordinates.

Definition 5.3.1. In a Riemann space with the metric tensor g_{ik}, the *mixed Riemann tensor of curvature* is defined by

$$R^j_{ilk} = \frac{\partial}{\partial x^k}\left(\left\{\begin{matrix} j \\ il \end{matrix}\right\}\right) - \frac{\partial}{\partial x^l}\left(\left\{\begin{matrix} j \\ ik \end{matrix}\right\}\right) + \left\{\begin{matrix} m \\ il \end{matrix}\right\}\left\{\begin{matrix} j \\ mk \end{matrix}\right\} - \left\{\begin{matrix} m \\ ik \end{matrix}\right\}\left\{\begin{matrix} j \\ ml \end{matrix}\right\}.$$

Now we prove the important *Ricci identity*.

Theorem 5.3.2 (Ricci identity).

Let T be an (r,s)–tensor with components $T^{i_1\ldots i_s}_{j_1\ldots j_r}$. Then

$$
\begin{aligned}
T^{i_1\ldots i_s}_{j_1\ldots j_r;lk} - T^{i_1\ldots i_s}_{j_1\ldots j_r;kl} &= \sum_{\rho=1}^{r} T^{i_1\ldots i_s}_{j_1\ldots j_{\rho-1}mj_{\rho+1}\ldots j_r} R^m_{j_\rho kl} \\
&\quad - \sum_{\sigma=1}^{s} T^{i_1\ldots i_{\sigma-1}mi_{\sigma+1}\cdots i_s}_{j_1\ldots j_r} R^{i_\sigma}_{mkl}.
\end{aligned}
\tag{5.12}
$$

The identity in (5.12) is called Ricci's identity.

Proof.

(i) First we prove the theorem for a covariant vector T with the components T_i. We have

$$
T_{i;l} = \frac{\partial T_i}{\partial x^l} - T_j \left\{ {j \atop il} \right\},
$$

and

$$
\begin{aligned}
T_{i;lk} &= \frac{\partial}{\partial x^k}(T_{i;l}) - T_{m;l}\left\{ {m \atop ik} \right\} - T_{i;m}\left\{ {m \atop lk} \right\} \\
&= \frac{\partial^2 T_i}{\partial x^k \partial x^l} - \frac{\partial}{\partial x^k}\left(T_j \left\{ {j \atop il} \right\}\right) - \left(\frac{\partial T_m}{\partial x^l} - T_n \left\{ {n \atop ml} \right\}\right)\left\{ {m \atop ik} \right\} \\
&\quad - \left(\frac{\partial T_i}{\partial x^m} - T_n \left\{ {n \atop im} \right\}\right)\left\{ {m \atop lk} \right\}
\end{aligned}
$$

and similarly

$$
\begin{aligned}
T_{i;kl} &= \frac{\partial^2 T_i}{\partial x^l \partial x^k} - \frac{\partial}{\partial x^l}\left(T_j \left\{ {j \atop ik} \right\}\right) - \left(\frac{\partial T_m}{\partial x^k} - T_n \left\{ {n \atop mk} \right\}\right)\left\{ {m \atop il} \right\} \\
&\quad - \left(\frac{\partial T_i}{\partial x^m} - T_n \left\{ {n \atop im} \right\}\right)\left\{ {m \atop kl} \right\}.
\end{aligned}
$$

Thus

$$
\begin{aligned}
T_{i;lk} - T_{i;kl} &= T_j\left(\frac{\partial}{\partial x^l}\left(\left\{ {j \atop ik} \right\}\right) - \frac{\partial}{\partial x^k}\left(\left\{ {j \atop il} \right\}\right) + \left\{ {j \atop ml} \right\}\left\{ {m \atop ik} \right\}\right. \\
&\qquad \left. - \left\{ {j \atop mk} \right\}\left\{ {m \atop il} \right\}\right) \\
&\quad - \frac{\partial T_m}{\partial x^l}\left\{ {m \atop ik} \right\} - \frac{\partial T_i}{\partial x^m}\left\{ {m \atop lk} \right\} + \frac{\partial T_m}{\partial x^k}\left\{ {m \atop il} \right\} + \frac{\partial T_i}{\partial x^m}\left\{ {m \atop kl} \right\} \\
&\quad + \frac{\partial T_j}{\partial x^l}\left\{ {j \atop ik} \right\} - \frac{\partial T_j}{\partial x^k}\left\{ {j \atop il} \right\} \\
&= T_j\left(\frac{\partial}{\partial x^l}\left(\left\{ {j \atop ik} \right\}\right) - \frac{\partial}{\partial x^k}\left(\left\{ {j \atop il} \right\}\right) + \left\{ {j \atop ml} \right\}\left\{ {m \atop ik} \right\}\right. \\
&\qquad \left. - \left\{ {j \atop mk} \right\}\left\{ {m \atop il} \right\}\right)
\end{aligned}
$$

$$- \frac{\partial T_j}{\partial x^l} \left\{ {j \atop ik} \right\} + \frac{\partial T_j}{\partial x^k} \left\{ {j \atop il} \right\} + \frac{\partial T_j}{\partial x^l} \left\{ {j \atop ik} \right\} - \frac{\partial T_j}{\partial x^k} \left\{ {j \atop il} \right\}$$

$$= T_j \left(\frac{\partial}{\partial x^l} \left(\left\{ {j \atop ik} \right\} \right) - \frac{\partial}{\partial x^k} \left(\left\{ {j \atop il} \right\} \right) + \left\{ {j \atop ml} \right\} \left\{ {m \atop ik} \right\} \right.$$
$$\left. - \left\{ {j \atop mk} \right\} \left\{ {m \atop il} \right\} \right)$$

$$= T_j R^j_{ikl}. \tag{5.13}$$

(ii) Now we show the assertion for a $(2,0)$–tensor T with the components T_{ri}. We have

$$T_{ri;l} = \frac{\partial T_{ri}}{\partial x^l} - T_{mi} \left\{ {m \atop rl} \right\} - T_{rm} \left\{ {m \atop il} \right\},$$

$$T_{ri;lk} = \frac{\partial}{\partial x^k} (T_{ri;l}) - T_{mi;l} \left\{ {m \atop rk} \right\} - T_{rm;l} \left\{ {m \atop il} \right\} - T_{ri;m} \left\{ {m \atop lk} \right\},$$

$$T_{ri;kl} = \frac{\partial}{\partial x^l} (T_{ri;k}) - T_{mi;k} \left\{ {m \atop rl} \right\} - T_{rm;k} \left\{ {m \atop ik} \right\} - T_{ri;m} \left\{ {m \atop kl} \right\},$$

hence

$$T_{ri;lk} - T_{ri;kl} = - \frac{\partial T_{mi}}{\partial x^k} \left\{ {m \atop rl} \right\} - T_{mi} \frac{\partial}{\partial x^k} \left(\left\{ {m \atop lr} \right\} \right) - \frac{\partial T_{rm}}{\partial x^k} \left\{ {m \atop il} \right\}$$
$$- T_{rm} \frac{\partial}{\partial x^k} \left(\left\{ {m \atop il} \right\} \right)$$

$$+ \frac{\partial T_{mi}}{\partial x^l} \left\{ {m \atop rk} \right\} + T_{mi} \frac{\partial}{\partial x^l} \left(\left\{ {m \atop rk} \right\} \right) + \frac{\partial T_{rm}}{\partial x^l} \left\{ {m \atop ik} \right\}$$
$$+ T_{rm} \frac{\partial}{\partial x^l} \left(\left\{ {m \atop ik} \right\} \right)$$

$$- \frac{\partial T_{mi}}{\partial x^l} \left\{ {m \atop rk} \right\} + T_{si} \left\{ {s \atop ml} \right\} \left\{ {m \atop rk} \right\} + T_{ms} \left\{ {s \atop il} \right\} \left\{ {m \atop rk} \right\}$$

$$+ \frac{\partial T_{mi}}{\partial x^k} \left\{ {m \atop rl} \right\} - T_{si} \left\{ {s \atop mk} \right\} \left\{ {m \atop rl} \right\} - T_{ms} \left\{ {s \atop ik} \right\} \left\{ {m \atop rl} \right\}$$

$$- \frac{\partial T_{rm}}{\partial x^l} \left\{ {m \atop ik} \right\} + T_{sm} \left\{ {s \atop rl} \right\} \left\{ {m \atop ik} \right\} + T_{rs} \left\{ {s \atop ml} \right\} \left\{ {m \atop ik} \right\}$$

$$+ \frac{\partial T_{rm}}{\partial x^k} \left\{ {m \atop il} \right\} - T_{sm} \left\{ {s \atop rk} \right\} \left\{ {m \atop il} \right\} - T_{rs} \left\{ {s \atop mk} \right\} \left\{ {m \atop il} \right\}$$

$$= T_{mi} \left(\frac{\partial}{\partial x^l} \left(\left\{ {m \atop rk} \right\} \right) - \frac{\partial}{\partial x^k} \left(\left\{ {m \atop rl} \right\} \right) \right)$$

$$+ T_{si} \left(\begin{Bmatrix} s \\ ml \end{Bmatrix} \begin{Bmatrix} m \\ rk \end{Bmatrix} - \begin{Bmatrix} s \\ mk \end{Bmatrix} \begin{Bmatrix} m \\ rl \end{Bmatrix} \right)$$

$$+ T_{rm} \left(\frac{\partial}{\partial x^l} \left(\begin{Bmatrix} m \\ ik \end{Bmatrix} \right) - \frac{\partial}{\partial x^k} \left(\begin{Bmatrix} m \\ il \end{Bmatrix} \right) \right)$$

$$+ T_{rs} \left(\begin{Bmatrix} s \\ ml \end{Bmatrix} \begin{Bmatrix} m \\ ik \end{Bmatrix} - \begin{Bmatrix} s \\ mk \end{Bmatrix} \begin{Bmatrix} m \\ il \end{Bmatrix} \right) + A,$$

where

$$A = T_{ms} \left(\begin{Bmatrix} s \\ il \end{Bmatrix} \begin{Bmatrix} m \\ rk \end{Bmatrix} - \begin{Bmatrix} s \\ ik \end{Bmatrix} \begin{Bmatrix} m \\ rl \end{Bmatrix} \right) + T_{sm} \left(\begin{Bmatrix} s \\ rl \end{Bmatrix} \begin{Bmatrix} m \\ ik \end{Bmatrix} - \begin{Bmatrix} s \\ rk \end{Bmatrix} \begin{Bmatrix} m \\ il \end{Bmatrix} \right)$$

$$= T_{ms} \left(\begin{Bmatrix} s \\ il \end{Bmatrix} \begin{Bmatrix} m \\ rk \end{Bmatrix} - \begin{Bmatrix} s \\ ik \end{Bmatrix} \begin{Bmatrix} m \\ rl \end{Bmatrix} + \begin{Bmatrix} m \\ rl \end{Bmatrix} \begin{Bmatrix} s \\ ik \end{Bmatrix} - \begin{Bmatrix} m \\ rk \end{Bmatrix} \begin{Bmatrix} s \\ il \end{Bmatrix} \right) = 0.$$

If we interchange the indices s and m in the terms of T_{si} and T_{rs}, then we obtain

$$T_{ri;lk} - T_{ri;kl} = T_{mi} \left(\frac{\partial}{\partial x^l} \left(\begin{Bmatrix} m \\ rk \end{Bmatrix} \right) - \frac{\partial}{\partial x^k} \left(\begin{Bmatrix} m \\ rl \end{Bmatrix} \right) + \begin{Bmatrix} m \\ sl \end{Bmatrix} \begin{Bmatrix} s \\ rk \end{Bmatrix} \right.$$
$$\left. - \begin{Bmatrix} m \\ sk \end{Bmatrix} \begin{Bmatrix} s \\ rl \end{Bmatrix} \right)$$

$$+ T_{rm} \left(\frac{\partial}{\partial x^l} \left(\begin{Bmatrix} m \\ ik \end{Bmatrix} \right) - \frac{\partial}{\partial x^k} \left(\begin{Bmatrix} m \\ il \end{Bmatrix} \right) + \begin{Bmatrix} m \\ sl \end{Bmatrix} \begin{Bmatrix} s \\ ik \end{Bmatrix} \right.$$
$$\left. - \begin{Bmatrix} m \\ sk \end{Bmatrix} \begin{Bmatrix} s \\ il \end{Bmatrix} \right)$$

$$= T_{mi} R^m_{rkl} - T_{rm} R^m_{ikl}.$$

(iii) Now we show the general formulae for $(r,0)$–tensors

$$T_{m_1 \ldots m_r; lk} - T_{m_1 \ldots m_r; kl} = \sum_{\rho=1}^{r} T_{m_1 \ldots m_{\rho-1} s m_{\rho+1} \ldots m_r} R^s_{m_\rho lk}. \qquad (5.14)$$

We have by (5.4)

$$T_{i_1 \ldots i_r; l} = \frac{\partial T_{i_1 \ldots i_r}}{\partial x^l} - \sum_{\rho=1}^{r} T_{i_1 \ldots i_{\rho-1} m i_{\rho+1} \ldots i_r} \begin{Bmatrix} m \\ i_\rho l \end{Bmatrix},$$

$$T_{i_1 \ldots i_r; lk} = \frac{\partial}{\partial x^k} \left(T_{i_1 \ldots i_r; l} \right) - \sum_{\rho=1}^{r} T_{i_1 \ldots i_{\rho-1} m i_{\rho+1} \ldots i_r; k} \begin{Bmatrix} m \\ i_\rho l \end{Bmatrix} - T_{i_1 \ldots i_r; m} \begin{Bmatrix} m \\ lk \end{Bmatrix}$$

$$= \frac{\partial^2}{\partial x^k \partial x^l} \left(T_{i_1 \ldots i_r} \right) - \frac{\partial}{\partial x^k} \left(\sum_{\rho=1}^{r} T_{i_1 \ldots i_{\rho-1} m i_{\rho+1} \ldots i_r} \begin{Bmatrix} m \\ i_\rho l \end{Bmatrix} \right)$$

$$- \sum_{\rho=1}^{r} \frac{\partial}{\partial x^l} \left(T_{i_1 \ldots i_{\rho-1} m i_{\rho+1} \ldots i_r} \right) \begin{Bmatrix} m \\ i_\rho k \end{Bmatrix}$$

$$+ \sum_{\rho=2}^{r} \sum_{\tau=1}^{\rho-1} T_{i_1 \ldots i_{\tau-1} u i_{\tau+1} \ldots i_{\rho-1} m i_{\rho+1} \ldots i_r} \begin{Bmatrix} m \\ i_\rho k \end{Bmatrix} \begin{Bmatrix} u \\ i_\tau l \end{Bmatrix}$$

$$+ \sum_{\rho=1}^{r-1} \sum_{\tau=\rho+1}^{r} T_{i_1 \ldots i_{\rho-1} m i_{\rho+1} \ldots i_{\tau-1} u i_{\tau+1} \ldots i_r} \begin{Bmatrix} m \\ i_\rho k \end{Bmatrix} \begin{Bmatrix} u \\ i_\tau l \end{Bmatrix}$$

$$+ \sum_{\rho=1}^{r} T_{i_1 \ldots i_{\rho-1} u i_{\rho+1} \ldots i_r} \begin{Bmatrix} m \\ i_\rho k \end{Bmatrix} \begin{Bmatrix} u \\ ml \end{Bmatrix} - T_{i_1 \ldots i_r ; m} \begin{Bmatrix} m \\ lk \end{Bmatrix}$$

$$- \frac{\partial}{\partial x^m} (T_{i_1 \ldots i_r ; m}) \begin{Bmatrix} m \\ lk \end{Bmatrix} - \sum_{\rho=1}^{r} T_{i_1 \ldots i_{\rho-1} u i_{\rho+1} \ldots i_r} \begin{Bmatrix} u \\ i_\rho m \end{Bmatrix} \begin{Bmatrix} m \\ lk \end{Bmatrix}$$

$$= \frac{\partial^2}{\partial x^k \partial x^l} (T_{i_1 \ldots i_r}) - \sum_{\rho=1}^{r} \frac{\partial}{\partial x^k} (T_{i_1 \ldots i_{\rho-1} m i_{\rho+1} \ldots i_r}) \begin{Bmatrix} m \\ i_\rho l \end{Bmatrix}$$

$$- \sum_{\rho=1}^{r} T_{i_1 \ldots i_{\rho-1} m i_{\rho+1} \ldots i_r} \frac{\partial}{\partial x^k} \left(\begin{Bmatrix} m \\ i_\rho l \end{Bmatrix} \right)$$

$$- \sum_{\rho=1}^{r} \frac{\partial}{\partial x^l} (T_{i_1 \ldots i_{\rho-1} m i_{\rho+1} \ldots i_r}) \begin{Bmatrix} m \\ i_\rho k \end{Bmatrix}$$

$$+ \sum_{\rho=2}^{r} \sum_{\tau=1}^{\rho-1} T_{i_1 \ldots i_{\tau-1} u i_{\tau+1} \ldots i_{\rho-1} m i_{\rho+1} \ldots i_r} \begin{Bmatrix} m \\ i_\rho k \end{Bmatrix} \begin{Bmatrix} u \\ i_\tau l \end{Bmatrix}$$

$$+ \sum_{\rho=1}^{r} T_{i_1 \ldots i_{\rho-1} u i_{\rho+1} \ldots i_r} \begin{Bmatrix} m \\ i_\rho k \end{Bmatrix} \begin{Bmatrix} u \\ ml \end{Bmatrix}$$

$$+ \sum_{\rho=1}^{r-1} \sum_{\tau=\rho+1}^{r} T_{i_1 \ldots i_{\rho-1} m i_{\rho+1} \ldots i_{\tau-1} u i_{\tau+1} \ldots i_r} \begin{Bmatrix} m \\ i_\rho k \end{Bmatrix} \begin{Bmatrix} u \\ i_\tau l \end{Bmatrix} - T_{i_1 \ldots i_r ; m} \begin{Bmatrix} m \\ lk \end{Bmatrix}$$

$$- \frac{\partial}{\partial x^m} (T_{i_1 \ldots i_r}) \begin{Bmatrix} m \\ lk \end{Bmatrix} - \sum_{\rho=1}^{r} T_{i_1 \ldots i_{\rho-1} u i_{\rho+1} \ldots i_r} \begin{Bmatrix} u \\ i_\rho m \end{Bmatrix} \begin{Bmatrix} m \\ lk \end{Bmatrix}.$$

We put

$$A_{kl} = \frac{\partial}{\partial x^m} (T_{i_1 \ldots i_r}) \begin{Bmatrix} m \\ lk \end{Bmatrix} - \sum_{\rho=1}^{r} T_{i_1 \ldots i_{\rho-1} u i_{\rho+1} \ldots i_r} \begin{Bmatrix} u \\ i_\rho m \end{Bmatrix} \begin{Bmatrix} m \\ lk \end{Bmatrix}$$

and

$$B_{kl} = \sum_{\rho=1}^{r} \frac{\partial}{\partial x^k} (T_{i_1 \ldots i_{\rho-1} m i_{\rho+1} \ldots i_r}) \begin{Bmatrix} m \\ i_\rho l \end{Bmatrix} + \sum_{\rho=1}^{r} \frac{\partial}{\partial x^l} (T_{i_1 \ldots i_{\rho-1} m i_{\rho+1} \ldots i_r}) \begin{Bmatrix} m \\ i_\rho k \end{Bmatrix},$$

and obtain

$$A_{kl} - A_{lk} = \frac{\partial}{\partial x^m} (T_{i_1 \ldots i_r}) \begin{Bmatrix} m \\ lk \end{Bmatrix} - \sum_{\rho=1}^{r} T_{i_1 \ldots i_{\rho-1} u i_{\rho+1} \ldots i_r} \begin{Bmatrix} u \\ i_\rho m \end{Bmatrix} \begin{Bmatrix} m \\ lk \end{Bmatrix}$$

$$- \frac{\partial}{\partial x^m} (T_{i_1 \ldots i_r}) \begin{Bmatrix} m \\ kl \end{Bmatrix} - \sum_{\rho=1}^{r} T_{i_1 \ldots i_{\rho-1} u i_{\rho+1} \ldots i_r} \begin{Bmatrix} u \\ i_\rho m \end{Bmatrix} \begin{Bmatrix} m \\ kl \end{Bmatrix} = 0$$

and

$$B_{kl} - B_{lk} = \sum_{\rho=1}^{r} \frac{\partial}{\partial x^k} \left(T_{i_1...i_{\rho-1}mi_{\rho+1}...i_r}\right) \begin{Bmatrix} m \\ i_\rho l \end{Bmatrix}$$

$$+ \sum_{\rho=1}^{r} \frac{\partial}{\partial x^l} \left(T_{i_1...i_{\rho-1}mi_{\rho+1}...i_r}\right) \begin{Bmatrix} m \\ i_\rho k \end{Bmatrix}$$

$$- \sum_{\rho=1}^{r} \frac{\partial}{\partial x^l} \left(T_{i_1...i_{\rho-1}mi_{\rho+1}...i_r}\right) \begin{Bmatrix} m \\ i_\rho k \end{Bmatrix}$$

$$- \sum_{\rho=1}^{r} \frac{\partial}{\partial x^k} \left(T_{i_1...i_{\rho-1}mi_{\rho+1}...i_r}\right) \begin{Bmatrix} m \\ i_\rho l \end{Bmatrix} = 0.$$

It follows that

$$T_{i_1...i_r;kl} - T_{i_1...i_r;kl} = -B_{kl} - \sum_{\rho=1}^{r} T_{i_1...i_{\rho-1}mi_{\rho+1}...i_r} \frac{\partial}{\partial x^k} \left(\begin{Bmatrix} m \\ i_\rho l \end{Bmatrix} \right)$$

$$+ \sum_{\rho=2}^{r} \sum_{\tau=1}^{\rho-1} T_{i_1...i_{\tau-1}ui_{\tau+1}...i_{\rho-1}mi_{\rho+1}...i_r} \begin{Bmatrix} m \\ i_\rho k \end{Bmatrix} \begin{Bmatrix} u \\ i_\tau l \end{Bmatrix}$$

$$+ \sum_{\rho=1}^{r} T_{i_1...i_{\rho-1}ui_{\rho+1}\cdots i_r} \begin{Bmatrix} m \\ i_\rho k \end{Bmatrix} \begin{Bmatrix} u \\ ml \end{Bmatrix}$$

$$+ \sum_{\rho=1}^{r-1} \sum_{\tau=\rho+1}^{r} T_{i_1...i_{\rho-1}mi_{\rho+1}...i_{\tau-1}ui_{\tau+1}...i_r} \begin{Bmatrix} m \\ i_\rho k \end{Bmatrix} \begin{Bmatrix} u \\ i_\tau l \end{Bmatrix} - A_{kl}$$

$$+ B_{lk} + \sum_{\rho=1}^{r} T_{i_1...i_{\rho-1}mi_{\rho+1}...i_r} \frac{\partial}{\partial x^l} \left(\begin{Bmatrix} m \\ i_\rho k \end{Bmatrix} \right)$$

$$- \sum_{\rho=2}^{r} \sum_{\tau=1}^{\rho-1} T_{i_1...i_{\tau-1}ui_{\tau+1}...i_{\rho-1}mi_{\rho+1}...i_r} \begin{Bmatrix} m \\ i_\rho l \end{Bmatrix} \begin{Bmatrix} u \\ i_\tau k \end{Bmatrix}$$

$$- \sum_{\rho=1}^{r} T_{i_1...i_{\rho-1}ui_{\rho+1}\cdots i_r} \begin{Bmatrix} m \\ i_\rho l \end{Bmatrix} \begin{Bmatrix} u \\ mk \end{Bmatrix}$$

$$- \sum_{\rho=1}^{r-1} \sum_{\tau=\rho+1}^{r} T_{i_1...i_{\rho-1}mi_{\rho+1}...i_{\tau-1}ui_{\tau+1}...i_r} \begin{Bmatrix} m \\ i_\rho l \end{Bmatrix} \begin{Bmatrix} u \\ i_\tau k \end{Bmatrix} + A_{lk}$$

$$= \sum_{\rho=1}^{r} T_{i_1...i_{\rho-1}mi_{\rho+1}...i_r} \left(\frac{\partial}{\partial x^l} \begin{Bmatrix} m \\ i_\rho k \end{Bmatrix} - \frac{\partial}{\partial x^k} \begin{Bmatrix} m \\ i_\rho l \end{Bmatrix} \right)$$

$$+ \sum_{\rho=1}^{r} T_{i_1...i_{\rho-1}ui_{\rho+1}\cdots i_r} \left(\begin{Bmatrix} m \\ i_\rho k \end{Bmatrix} \begin{Bmatrix} u \\ ml \end{Bmatrix} - \begin{Bmatrix} m \\ i_\rho l \end{Bmatrix} \begin{Bmatrix} u \\ mk \end{Bmatrix} \right)$$

$$+ \sum_{\rho=2}^{r} \sum_{\tau=1}^{\rho-1} T_{i_1...i_{\tau-1}ui_{\tau+1}...i_{\rho-1}mi_{\rho+1}...i_r} \cdot$$

$$\left(\begin{Bmatrix} m \\ i_\rho k \end{Bmatrix} \begin{Bmatrix} u \\ i_\tau l \end{Bmatrix} - \begin{Bmatrix} m \\ i_\rho l \end{Bmatrix} \begin{Bmatrix} u \\ i_\tau k \end{Bmatrix} \right)$$

$$+ \sum_{\rho=1}^{r-1} \sum_{\tau=\rho+1}^{r} T_{i_1 \dots i_{\rho-1} m i_{\rho+1} \dots i_{\tau-1} u i_{\tau+1} \dots i_r} \cdot$$

$$\left(\left\{ \begin{matrix} m \\ i_\rho k \end{matrix} \right\} \left\{ \begin{matrix} u \\ i_\tau l \end{matrix} \right\} - \left\{ \begin{matrix} m \\ i_\rho l \end{matrix} \right\} \left\{ \begin{matrix} u \\ i_\tau k \end{matrix} \right\} \right).$$

We put

$$C = \sum_{\rho=1}^{r} T_{i_1 \dots i_{\rho-1} m i_{\rho+1} \dots i_r} \left(\frac{\partial}{\partial x^l} \left\{ \begin{matrix} m \\ i_\rho k \end{matrix} \right\} - \frac{\partial}{\partial x^k} \left\{ \begin{matrix} m \\ i_\rho l \end{matrix} \right\} \right)$$

$$+ \sum_{\rho=1}^{r} T_{i_1 \dots i_{\rho-1} u i_{\rho+1} \dots i_r} \left(\left\{ \begin{matrix} m \\ i_\rho k \end{matrix} \right\} \left\{ \begin{matrix} u \\ ml \end{matrix} \right\} - \left\{ \begin{matrix} m \\ i_\rho l \end{matrix} \right\} \left\{ \begin{matrix} u \\ mk \end{matrix} \right\} \right)$$

and obtain interchanging the indices of summation u and m in the second sum

$$C = \sum_{\rho=1}^{r} T_{i_1 \dots i_{\rho-1} m i_{\rho+1} \dots i_r} \left(\frac{\partial}{\partial x^l} \left\{ \begin{matrix} m \\ i_\rho k \end{matrix} \right\} - \frac{\partial}{\partial x^k} \left\{ \begin{matrix} m \\ i_\rho l \end{matrix} \right\} + \right.$$

$$\left. \left\{ \begin{matrix} u \\ i_\rho k \end{matrix} \right\} \left\{ \begin{matrix} m \\ ul \end{matrix} \right\} - \left\{ \begin{matrix} u \\ i_\rho l \end{matrix} \right\} \left\{ \begin{matrix} m \\ uk \end{matrix} \right\} \right)$$

$$= \sum_{\rho=1}^{r} T_{i_1 \dots i_{\rho-1} m i_{\rho+1} \dots i_r} R^{m}_{i_\rho kl}$$

by Definition 5.3.1. We also write D as follows and change the order of summation in the second sum

$$D = \sum_{\rho=2}^{r} \sum_{\tau=1}^{\rho-1} T_{i_1 \dots i_{\rho-1} u i_{\rho+1} \dots i_{\tau-1} m i_{\tau+1} \dots i_r} \cdot$$

$$\left(\left\{ \begin{matrix} m \\ i_\rho k \end{matrix} \right\} \left\{ \begin{matrix} u \\ i_\tau l \end{matrix} \right\} - \left\{ \begin{matrix} m \\ i_\rho l \end{matrix} \right\} \left\{ \begin{matrix} u \\ i_\tau k \end{matrix} \right\} \right)$$

$$+ \sum_{\rho=1}^{r-1} \sum_{\tau=\rho+1}^{r} T_{i_1 \dots i_{\tau-1} m i_{\tau+1} \dots i_{\rho-1} u i_{\rho+1} \dots i_r} \cdot$$

$$\left(\left\{ \begin{matrix} m \\ i_\rho k \end{matrix} \right\} \left\{ \begin{matrix} u \\ i_\tau l \end{matrix} \right\} - \left\{ \begin{matrix} m \\ i_\rho l \end{matrix} \right\} \left\{ \begin{matrix} u \\ i_\tau k \end{matrix} \right\} \right)$$

$$= \sum_{\rho=2}^{r} \sum_{\tau=1}^{\rho-1} T_{i_1 \dots i_{\rho-1} u i_{\rho+1} \dots i_{\tau-1} m i_{\tau+1} \dots i_r} \cdot$$

$$\left(\left\{ \begin{matrix} m \\ i_\rho k \end{matrix} \right\} \left\{ \begin{matrix} u \\ i_\tau l \end{matrix} \right\} - \left\{ \begin{matrix} m \\ i_\rho l \end{matrix} \right\} \left\{ \begin{matrix} u \\ i_\tau k \end{matrix} \right\} \right)$$

$$+ \sum_{\tau=2}^{r} \sum_{\rho=1}^{\tau-1} T_{i_1 \dots i_{\tau-1} m i_{\tau+1} \dots i_{\rho-1} u i_{\rho+1} \dots i_r} \cdot$$

$$\left(\left\{ \begin{matrix} m \\ i_\rho k \end{matrix} \right\} \left\{ \begin{matrix} u \\ i_\tau l \end{matrix} \right\} - \left\{ \begin{matrix} m \\ i_\rho l \end{matrix} \right\} \left\{ \begin{matrix} u \\ i_\tau k \end{matrix} \right\} \right)$$

and change the indices of summation from τ to ρ and u to m in the second sum to obtain

$$D = \sum_{\rho=2}^{r} \sum_{\tau=1}^{\rho-1} T_{i_1 \ldots i_{\rho-1} u i_{\rho+1} \ldots i_{\tau-1} m i_{\tau+1} \ldots i_r} \cdot$$

$$\left(\left\{ \begin{matrix} m \\ i_\rho k \end{matrix} \right\} \left\{ \begin{matrix} u \\ i_\tau l \end{matrix} \right\} - \left\{ \begin{matrix} m \\ i_\rho l \end{matrix} \right\} \left\{ \begin{matrix} u \\ i_\tau k \end{matrix} \right\} \right)$$

$$- \sum_{\rho=2}^{r} \sum_{\tau=1}^{\rho-1} T_{i_1 \ldots i_{\rho-1} u i_{\rho+1} \ldots i_{\tau-1} m i_{\tau+1} \ldots i_r} \cdot$$

$$\left(\left\{ \begin{matrix} u \\ i_\tau k \end{matrix} \right\} \left\{ \begin{matrix} m \\ i_\rho l \end{matrix} \right\} - \left\{ \begin{matrix} u \\ i_\tau l \end{matrix} \right\} \left\{ \begin{matrix} m \\ i_\rho k \end{matrix} \right\} \right) = 0.$$

Finally we have

$$T_{i_1 \ldots i_r; kl} - T_{i_1 \ldots i_r; kl} = C + D = \sum_{\rho=1}^{r} T_{i_1 \ldots i_{\rho-1} m i_{\rho+1} \ldots i_r} R^{m}_{i_\rho kl}.$$

(iv) The general formulae in (5.12) could be established in a similar way. Here we prove the formulae for the covariant derivative of a contravariant vector and a $(0,2)$–tensor by applying suitable contractions.

We have $g^{ij}_{;k} = 0$ by the Ricci Theorem, Theorem 5.2.6. Putting $T_l = g_{lm} T^m$, we conclude $T^i = \delta^i_m T^m = g^{il} g_{lm} T^m = g^{il} T_l$ and

$$T^i_{;jk} = \left(g^{il} T_l \right)_{;jk} = \left(\left(g^{il} T_l \right)_{;j} \right)_{;k} = \left(g^{il}_{;j} T_l + g^{il} T_{l;j} \right)_{;k}$$

$$= g^{il}_{;k} T_{l;j} + g^{il} T_{l;jk} = g^{il} T_{l;jk}.$$

This implies together with (5.13)

$$T^i_{;jk} - T^i_{;kj} = g^{il} (T_{l;jk} - T_{l;kj}) = g^{il} R^{m}_{lkj} T_m.$$

We put $R_{ilkj} = g_{im} R^{m}_{lkj}$. Then it follows that $R^{m}_{lkj} = g^{mr} R_{rlkj}$ and

$$T^i_{;jk} - T^i_{;kj} = g^{il} g^{mr} R_{rlkj} T_m = g^{il} g^{mr} T_m R_{rlkj} = g^{il} T^r R_{rlkj}$$

$$= -g^{il} T^r R_{lrkj} = -T^r g^{il} R_{lrkj} = -T^r R^{i}_{rkj}.$$

Similarly we obtain for a $(0,2)$–tensor

$$T^{il}_{;jk} - T^{il}_{;kj} = -T^{rl} R^{i}_{rkj} - T^{ir} R^{l}_{rkj}$$

and now Ricci's identity (5.12) for arbitrary tensors follows.

$\square$

Theorem 5.3.3. *The covariant derivative of a covariant vector in $\mathbb{V}^3$ is a symmetric tensor if and only if the vector itself is the gradient of a scalar of class $r \geq 2$.*

Proof. Let T be a covariant vector with the components T_i. Then we obtain for the covariant derivative $T_{i;k}$ by Definition 5.1.3 and the fact that $\left\{{j \atop ik}\right\} = \left\{{j \atop ki}\right\}$ for all i, j and k

$$T_{i;k} - T_{k;i} = \frac{\partial T_i}{\partial x^k} - T_j \left\{{j \atop ik}\right\} - \left(\frac{\partial T_k}{\partial x^i} - T_j \left\{{j \atop ki}\right\}\right) = \frac{\partial T_i}{\partial x^k} - \frac{\partial T_k}{\partial x^i} \qquad (5.15)$$

for all i and k. The last two terms in (5.15) are the components of the curl in $\mathbb{V}^3$ of a vector with the components T_i $(i = 1, 2, 3)$. The curl of a vector in $\mathbb{V}^3$ vanishes if and only if the vector is the gradient of a scalar Φ, that is, $T_i = \Phi_i$ for $i = 1, 2, 3$. Thus the identities in (5.15) vanish if and only if

$$\Phi_{;ik} - \Phi_{;ki} = \frac{\partial}{\partial x^k}\left(\frac{\partial \Phi}{\partial x^i}\right) - \frac{\partial}{\partial x^i}\left(\frac{\partial \Phi}{\partial x^k}\right) = 0 \text{ for all } i \text{ and } k.$$

□

5.4 BIANCHI'S IDENTITIES FOR THE COVARIANT DERIVATIVE OF THE TENSORS OF CURVATURE

In this section, we establish the *Bianchi identities* for the derivatives of the mixed Riemann tensor of curvature and the covariant Riemann tensor of curvature.

Theorem 5.4.1 (Bianchi identities).
The Riemann tensors of curvature satisfy the following identities for all i, j, k, l and m

$$R^i_{jkl;m} + R^i_{jmk;l} + R^i_{jlm;k} = 0 \qquad (5.16)$$

and

$$R_{ijkl;m} + R_{ijmk;l} + R_{ijlm;k} = 0. \qquad (5.17)$$

The identities in (5.16) and (5.17) are called the Bianchi identities.

Proof. (i) First we show:
  In any n–dimensional Riemann space R there exists a coordinate system such that the Christoffel symbols vanish in a given point.
  Let S be a coordinate system in R, $P_0 = (\underset{0}{x^1}, \underset{0}{x^2}, \dots, \underset{0}{x^n}) \in R$ be a point, and $\left\{{i \atop jk}\right\}_0$ be the Christoffel symbols at P_0 with respect to S. We define a new coordinate system S^* by

$$x^i = x^{*i} + \underset{0}{x^i} - \frac{1}{2}\left\{{i \atop jk}\right\}_0 x^{*j} x^{*k} \qquad (i = 1, 2, \dots, n). \qquad (5.18)$$

Then P_0 is the origin of S^*, and differentiation yields

$$\frac{\partial x^i}{\partial x^{*l}} = \delta^i_l - \frac{1}{2}\left\{{i \atop jk}\right\}_0 \delta^j_l x^{*k} - \frac{1}{2}\left\{{i \atop jk}\right\}_0 x^{*j} \delta^k_l$$

$$= \delta_l^i - \frac{1}{2}\left\{{i \atop lk}\right\}_0 x^{*k} - \frac{1}{2}\left\{{i \atop jl}\right\}_0 x^{*j} = \delta_l^i - \left\{{i \atop lk}\right\}_0 x^{*k}$$

for all i and l. Thus we have

$$\left.\frac{\partial x^i}{\partial x^{*l}}\right|_{P_0} = \delta_l^i \text{ for all } i \text{ and } l,$$

and the transformation in (5.18) is admissible in a neighbourhood of P_0. Furthermore, we obtain

$$\frac{\partial^2 x^i}{\partial x^{*l}\partial x^{*m}} = -\left\{{i \atop lk}\right\}_0 \delta_m^k = -\left\{{i \atop lm}\right\}_0 \tag{5.19}$$

for all i, l and m. The transformation formulae in (4.28) and (5.19) yield at the point P_0

$$\left\{{i \atop jk}\right\}_0^* = \left\{{l \atop jk}\right\}_0^* \frac{\partial x^i}{\partial x^{*l}} = \left.\left(\frac{\partial x^p}{\partial x^{*j}}\frac{\partial x^q}{\partial x^{*k}}\left\{{m \atop pq}\right\} + \frac{\partial^2 x^m}{\partial x^{*j}\partial x^{*k}}\right)\frac{\partial x^{*l}}{\partial x^m}\frac{\partial x^i}{\partial x^{*l}}\right|_{P_0}$$

$$= \left(\delta_j^p \delta_k^q\left\{{m \atop pq}\right\}_0 - \left\{{m \atop jk}\right\}_0\right)\delta_m^i = \left\{{i \atop jk}\right\}_0 - \left\{{i \atop jk}\right\}_0 = 0$$

for all i, j and k.

(ii) Now we show the identities in (5.16).

If we introduce a coordinate system S such that the Christoffel symbols vanish in some fixed point P_0 (as in Part (i) of the proof), then we have at P_0 by the definition of the terms R^i_{jkl} in Definition 5.3.1

$$R^i_{jkl}(P_0) = \left.\left(\frac{\partial}{\partial x^l}\left(\left\{{i \atop jk}\right\}\right) - \frac{\partial}{\partial x^k}\left(\left\{{i \atop jl}\right\}\right)\right)\right|_{P_0}$$

and

$$R^i_{jkl;m}(P_0) = \left.\left(\frac{\partial}{\partial x^m}\left(\frac{\partial}{\partial x^l}\left(\left\{{i \atop jk}\right\}\right) - \frac{\partial}{\partial x^k}\left(\left\{{i \atop jl}\right\}\right)\right)\right)\right|_{P_0}$$

$$= \left.\left(\frac{\partial^2}{\partial x^l\partial x^m}\left(\left\{{i \atop jk}\right\}\right) - \frac{\partial^2}{\partial x^k\partial x^m}\left(\left\{{i \atop jl}\right\}\right)\right)\right|_{P_0}.$$

Replacing k, l and m by m, k and l, respectively, we obtain

$$R^i_{jmk;l}(P_0) = \left.\left(\frac{\partial^2}{\partial x^k\partial x^l}\left(\left\{{i \atop jm}\right\}\right) - \frac{\partial^2}{\partial x^m\partial x^l}\left(\left\{{i \atop jk}\right\}\right)\right)\right|_{P_0},$$

and replacing m, k and l by l, m and k, respectively, we obtain

$$R^i_{jlm;\,k}(P_0) = \left(\frac{\partial^2}{\partial x^m \partial x^k}\left(\left\{\begin{matrix} i \\ jl \end{matrix}\right\}\right) - \frac{\partial^2}{\partial x^l \partial x^k}\left(\left\{\begin{matrix} i \\ jm \end{matrix}\right\}\right)\right)\Bigg|_{P_0}.$$

Therefore, we obtain for all i, k, j, l and m

$$R^i_{jkl;\,m}(P_0) + R^i_{jmk;\,l}(P_0) + R^i_{jlm;\,k}(P_0)$$

$$= \left(\frac{\partial^2}{\partial x^l \partial x^m}\left(\left\{\begin{matrix} i \\ jk \end{matrix}\right\}\right) - \frac{\partial^2}{\partial x^k \partial x^m}\left(\left\{\begin{matrix} i \\ jl \end{matrix}\right\}\right)\right)\Bigg|_{P_0}$$

$$+ \left(\frac{\partial^2}{\partial x^k \partial x^l}\left(\left\{\begin{matrix} i \\ jm \end{matrix}\right\}\right) - \frac{\partial^2}{\partial x^m \partial x^l}\left(\left\{\begin{matrix} i \\ jk \end{matrix}\right\}\right)\right)\Bigg|_{P_0}$$

$$+ \left(\frac{\partial^2}{\partial x^m \partial x^k}\left(\left\{\begin{matrix} i \\ jl \end{matrix}\right\}\right) - \frac{\partial^2}{\partial x^l \partial x^k}\left(\left\{\begin{matrix} i \\ jm \end{matrix}\right\}\right)\right)\Bigg|_{P_0} = 0.$$

Thus we have proved the identities in (5.16) at P_0 for our special choice of the coordinate system. Since the left hand side in (5.16) is the sum of the components of a tensor and therefore a component of a tensor itself and vanishes at P_0 with respect to the coordinate system S, it must also vanish with respect to any coordinate system. Finally, since P_0 was an arbitrary point, the identities in (5.16) must hold in any point.

(iii) Finally, we show the identities in (5.17).
We have by the Ricci identity, Theorem 5.2.6,

$$R^i_{jkl;\,m} = \left(g^{ir}R_{rjkl}\right)_{;\,m} = g^{ir}R_{rjkl;\,m},$$

and the identities in (5.17) follow from (5.16).

$$\square$$

5.5   BELTRAMI'S DIFFERENTIATORS

In this section, we define the *Beltrami differentiator of first order* in a Riemann space, and study a few special cases of interest.

Definition 5.5.1. Let Φ and Ψ be two given real–valued functions of class $r \geq 2$ in a Riemann space. We define the *Beltrami differentiator of first order* $\triangledown$ by

$$\triangledown(\Phi, \Psi) = g^{ik}\Phi_{;\,i}\Psi_{;\,k},$$

and we write $\triangledown\Phi = \triangledown(\Phi, \Phi)$, for short.

Remark 5.5.2. *(a) Since Φ and Ψ are scalars, we have*

$$\Phi_{;\,i} = \frac{\partial\Phi}{\partial x^i}\text{ for all }i\text{ and }\Psi_{;\,k} = \frac{\partial\Psi}{\partial x^k}\text{ for all }k,$$

hence

$$\nabla(\Phi, \Psi) = g^{ik} \frac{\partial \Phi}{\partial x^i} \frac{\partial \Psi}{\partial x^k} \ and \ \nabla \Phi = g^{ik} \frac{\partial \Phi}{\partial x^i} \frac{\partial \Phi}{\partial x^k}.$$

Thus $\nabla \Phi$ is the square of the length of the gradient of Φ.
We have for Cartesian coordinates $g^{11} = g^{22} = g = 1$ and $g^{12} = 0$, hence

$$\nabla \Phi = \sum_{i=1}^{n} \left(\frac{\partial \Phi}{\partial x^i} \right)^2.$$

(b) Curves given by $\Phi(x^i) = const$ are called level lines *of the function Φ. If the values ξ^i are the components of the tangent vectors to the level lines, then*

$$\frac{\partial \Phi}{\partial x^i} \xi^i = 0.$$

This means that the gradient is orthogonal to the level lines; it is in the direction of maximum change of the function.
(c) If the collections of curves given by the equations

$$\Phi(x^i) = const \ and \ \Psi(x^i) = const$$

are members of a net, then the angle γ between the gradients of the functions is equal to the angle between the curves, since the gradient is always orthogonal to the corresponding curve, hence

$$\cos \gamma = \frac{\nabla(\Phi, \Psi)}{\sqrt{(\nabla \Phi \bullet \nabla \Psi)}}.$$

The collections of curves given by $\Phi(x^i) = const$ and $\Psi(x^i) = const$ make up an orthogonal net if and only if

$$\nabla(\Phi, \Psi) \equiv 0.$$

Next we define the *divergence of a contravariant vector.*

Definition 5.5.3. The *divergence of a contravariant vector T* with the components T^k is defined by

$$\operatorname{div} T^k = T^k_{;k} \ for \ all \ k.$$

Remark 5.5.4. *Being the contraction of a $(1,1)$–tensor the divergence of a contravariant vector is a scalar. We have by Definition 5.1.2*

$$T^k_{;k} = \frac{\partial T^k}{\partial x^k} + \left\{ \begin{matrix} k \\ kl \end{matrix} \right\} T^l \ for \ all \ k.$$

Theorem 5.5.5. *We have*

$$\operatorname{div} T^k = \frac{1}{\sqrt{g}} \frac{\partial}{\partial x^k} \left(\sqrt{g} T^k \right), \tag{5.20}$$

where $g = \det(g_{ij})$.

Proof.

(i) First we show

$$\left\{ \begin{matrix} i \\ ik \end{matrix} \right\} = \frac{\partial(\log \sqrt{g})}{\partial x^k}. \tag{*}$$

We conclude from $g = \sum_{(k_1,\ldots,k_n)} \text{sign}(k_1,\ldots,k_n) g_{1k_1} \cdots g_{nk_n}$ that

$$\frac{\partial g}{\partial x^k} = \frac{\partial}{\partial x^k} \left(\sum_{(k_1,\ldots,k_n)} \text{sign}(k_1,\ldots,k_n) g_{1k_1} \cdots g_{nk_n} \right)$$

$$= \sum_{j=1}^{n} \sum_{(k_1,\ldots,k_n)} \text{sign}(k_1,\ldots,k_n) g_{1k_1} \cdots g_{j-1,k_{j-1}} \frac{\partial}{\partial x^k} \left(g_{jk_j} \right) g_{j+1,k_{j+1}} \cdots g_{nk_n}$$

$$= \sum_{j=1}^{n} \sum_{l=1}^{n} \frac{\partial g_{jl}}{\partial x^k} G_{jl},$$

where the terms G_{jl} denote the adjoints of the determinant g. We conclude from $g^{jl} = G_{jl}/g$ for all j and l that by (4.25) and (4.24)

$$\frac{\partial g}{\partial x^k} = g \cdot g^{jl} \frac{\partial g_{jl}}{\partial x^k} = g \cdot g^{jl} ([jkl] + [lkj])$$

$$= g \left(g^{jl}[jkl] + g^{jl}[jkl] \right) = g \left(g^{il}[ikl] + g^{il}[ikl] \right) = 2g \left\{ \begin{matrix} i \\ ik \end{matrix} \right\}.$$

Finally, we obtain

$$\frac{\partial(\log \sqrt{g})}{\partial x^k} = \frac{1}{2g} \frac{\partial g}{\partial x^k} = \left\{ \begin{matrix} i \\ ik \end{matrix} \right\},$$

which is (*).

(ii) Now we prove the identity in (5.20).
We have

$$\text{div } T^k = T^k_{;k} = \frac{\partial T^k}{\partial x^k} + \left\{ \begin{matrix} k \\ kj \end{matrix} \right\} T^j = \frac{\partial T^k}{\partial x^k} + \frac{1}{2g} \frac{\partial g}{\partial x^j} T^j$$

and

$$\frac{1}{\sqrt{g}} \frac{\partial}{\partial x^k} \left(\sqrt{g} T^k \right) = \frac{\partial T^k}{\partial x^k} + \frac{1}{\sqrt{g}} T^k \frac{\partial(\sqrt{g})}{\partial x^k} = \frac{\partial T^k}{\partial x^k} + \frac{1}{2g} T^k \frac{\partial g}{\partial x^k}.$$

This shows (5.20).

$\square$

Now we define the *divergence of a covariant vector* and the *Laplacian of a real-valued function*.

Definition 5.5.6. (a) The *divergence of a covariant vector* T with the components T_k is defined by

$$\text{div } T_k = g^{ki} T_{k;i}.$$

(b) Let F be a real–valued function. Then the partial derivatives

$$F_i = \frac{\partial F}{\partial x^i}$$

are the components of a covariant vector. Then the *Laplacian operator* $\triangle$ is defined by

$$\triangle F = \text{div } F_i.$$

Remark 5.5.7. *(a) We have for conjugate vectors, that is, for vectors that satisfy* $T_k = g_{kj}T^j$,

$$\text{div } T^k = T^k_{;k} = \left(g^{kj}T_j\right)_{;k} = g^{kj}T_{j;k} = g^{jk}T_{j;k} = \text{div } T_k.$$

(b) We obtain from Theorem 5.5.5

$$\triangle F = \text{div } F_i = \text{div } F^i = \frac{1}{\sqrt{g}}\frac{\partial}{\partial x^i}\left(\sqrt{g}F^i\right) = \frac{1}{\sqrt{g}}\frac{\partial}{\partial x^i}\left(\sqrt{g}g^{ik}F_k\right).$$

Example 5.5.8. *Let S denote the Cartesian coordinate system of $\mathbb{R}^3$ and S^* be the system of cylindrical coordinates defined by*

$$x^1 = x^{*1}\cos x^{*2}, \ x^2 = x^{*1}\sin x^{*2} \ and \ x^3 = x^{*3}.$$

Since $g_{ik} = \delta_{ik}$, we have with respect to the system S

$$\triangle F = \sum_{k=1}^{3}\frac{\partial^2 F}{(\partial x^k)^2}.$$

We obtain for $\triangle^*$ with respect to the system S^* from the transformation formulae

$$g^*_{ik} = \frac{\partial x^m}{\partial x^{*i}}\frac{\partial x^n}{\partial x^{*k}}g_{mn} = \sum_{m=1}^{3}\frac{\partial x^m}{\partial x^{*i}}\frac{\partial x^m}{\partial x^{*k}}.$$

Since

$$\left(\frac{\partial x^i}{\partial x^{*j}}\right) = \begin{pmatrix} \cos x^{*2} & \sin x^{*2} & 0 \\ -x^{*1}\sin x^{*2} & x^{*1}\cos x^{*2} & 0 \\ 0 & 0 & 1 \end{pmatrix},$$

$$g^*_{11} = g^*_{33} = 1, \ g^*_{22} = (x^{*1})^2 \ and \ g^*_{ik} = 0 \ otherwise$$

and

$$\left(g^{*ik}\right) = \begin{pmatrix} 1 & 0 & 0 \\ 0 & \left(\dfrac{1}{x^{*1}}\right)^2 & 0 \\ 0 & 0 & 1 \end{pmatrix},$$

it follows that

$$\triangle^* F = \frac{1}{\sqrt{g^*}}\frac{\partial}{\partial x^{*i}}\left(\sqrt{g^*}g^{*ik}\frac{\partial F}{\partial x^{*k}}\right) = \frac{1}{\sqrt{g^*}}\frac{\partial}{\partial x^{*i}}\left(\sqrt{g^*}g^{*ii}\frac{\partial F}{\partial x^{*i}}\right)$$

$$= \frac{1}{x^{*1}}\left(\frac{\partial}{\partial x^{*1}}\left(x^{*1}\frac{\partial F}{\partial x^{*1}}\right) + \frac{\partial}{\partial x^{*2}}\left(x^{*1}\frac{1}{(x^{*1})^2}\frac{\partial F}{\partial x^{*2}}\right) + \frac{\partial}{\partial x^{*3}}\left(x^{*1}\frac{\partial F}{\partial x^{*3}}\right)\right)$$

$$= \frac{1}{x^{*1}}\frac{\partial F}{\partial x^{*1}} + \frac{\partial^2 F}{(\partial x^{*1})^2} + \frac{1}{(x^{*1})^2}\frac{\partial^2 F}{(\partial x^{*2})^2} + \frac{\partial^2 F}{(\partial x^{*3})^2}.$$

5.6 A GEOMETRIC MEANING OF THE COVARIANT DIFFERENTIATION, THE LEVI–CIVITÀ PARALLELISM

In this section, we give a geometric interpretation of the covariant differentiation. This will lead to the concept of *Levi–Cività parallelism*.

We recall the concept of a *surface vector* (Definition 3.7.1).

Definition 5.6.1. Let S be a surface in $\mathbb{R}^3$ with a parametric representation $\vec{x}(u^i)$ $((u^1, u^2) \in D)$, P be a point on S, and $T(P)$ be the tangent plane of S at P. Then a unit vector $\vec{z}$ in the tangent plane is called *surface vector of S at P*. We shall always write $\vec{z} = \{\xi^1, \xi^2\}$, where the values ξ^i $(i = 1, 2)$ are the components of the vector $\vec{z}$ along the vectors $\vec{x}_i$ $(i = 1, 2)$.

If we move the surface vectors of a plane in a parallel way along a straight line in the plane, that is, along a geodesic line in the plane, then the angles between the vectors and the tangents of the geodesic line are constant.

Similarly, we require for the parallel movement of surface vectors of arbitrary surfaces along geodesic lines that the angles between the surface vectors and the geodesic line are constant. This leads to the following result for the components ξ^i $(i = 1, 2)$ of the surface vectors under parallel movement along a geodesic line.

Theorem 5.6.2. *The components ξ^α $(\alpha = 1, 2)$ of a surface vector $\vec{z} = \{\xi^1, \xi2\}$ under parallel movement along a geodesic line given by $(u^1(s), u^2(s))$ satisfy the differential equations*

$$\frac{\delta \xi^\alpha}{\delta s} = \frac{d\xi^\alpha}{ds} + \xi^\beta \left\{ \begin{matrix} \alpha \\ \beta \gamma \end{matrix} \right\} \dot{u}^\gamma = 0 \ (\alpha = 1, 2). \tag{5.21}$$

Proof. Let the surface vectors at the points of the geodesic line be given by $\vec{z}(s) = \xi^\alpha(s)\vec{x}_\alpha(u^\beta(s))$, $\vec{x}(s)$ be a parametric representation of the geodesic line, s be the arc length along the geodesic line and

$$\dot{\vec{x}}(s) = \dot{u}^\beta(s)\vec{x}_\beta(u^\gamma(s)) \neq \vec{z}(s).$$

We have $g_{\alpha\beta}\xi^\alpha\xi^\beta = 1$ for all s. The surface vectors are parallel, if their angles with the vectors $\dot{\vec{x}}(s)$ are constant. This means that

$$\frac{d}{ds}\left(g_{\alpha\beta}\xi^\alpha\dot{u}^\beta\right) = 0$$

or

$$\frac{\partial g_{\alpha\beta}}{\partial u^\gamma}\xi^\alpha\dot{u}^\beta\dot{u}^\gamma + g_{\alpha\beta}\dot{\xi}^\alpha\dot{u}^\beta + g_{\alpha\beta}\xi^\alpha\ddot{u}^\beta = 0.$$

We rename β by τ in the last term above and apply (4.25) to obtain

$$([\alpha\gamma\beta] + [\beta\gamma\alpha])\,\xi^\alpha\dot{u}^\beta\dot{u}^\gamma + g_{\alpha\beta}\dot{\xi}^\alpha\dot{u}^\beta + g_{\alpha\tau}\xi^\alpha\ddot{u}^\tau =$$
$$\left(([\alpha\gamma\beta] + [\beta\gamma\alpha])\,\xi^\alpha\dot{u}^\gamma + g_{\alpha\beta}\dot{\xi}^\alpha\right)\dot{u}^\beta + g_{\alpha\tau}\xi^\alpha\ddot{u}^\tau = 0.$$

Using the identities in (4.24), we obtain

$$[\beta\gamma\alpha] = g_{\alpha\tau} \left\{ \begin{matrix} \tau \\ \beta\gamma \end{matrix} \right\} \text{ and } \xi^\alpha[\alpha\gamma\beta] = \xi^\delta[\delta\gamma\beta] = \xi^\delta g_{\alpha\beta} \left\{ \begin{matrix} \alpha \\ \delta\gamma \end{matrix} \right\},$$

and conclude

$$
\begin{aligned}
0 &= \left(([\alpha\gamma\beta] + [\beta\gamma\alpha]) \xi^\alpha \dot{u}^\gamma + g_{\alpha\beta}\dot{\xi}^\alpha \right) \dot{u}^\beta + g_{\alpha\tau}\xi^\alpha \ddot{u}^\tau \\
&= \left(\xi^\delta g_{\alpha\beta} \left\{ \begin{matrix} \alpha \\ \delta\gamma \end{matrix} \right\} \dot{u}^\gamma + \xi^\alpha g_{\alpha\tau} \left\{ \begin{matrix} \tau \\ \beta\gamma \end{matrix} \right\} \dot{u}^\gamma + g_{\alpha\beta}\dot{\xi}^\alpha \right) \dot{u}^\beta + g_{\alpha\tau}\xi^\alpha \ddot{u}^\tau \\
&= g_{\alpha\beta} \left(\dot{\xi}^\alpha + \xi^\delta \left\{ \begin{matrix} \alpha \\ \delta\gamma \end{matrix} \right\} \dot{u}^\gamma \right) \dot{u}^\beta + g_{\alpha\tau}\xi^\alpha \left(\ddot{u}^\tau + \left\{ \begin{matrix} \tau \\ \beta\gamma \end{matrix} \right\} \dot{u}^\beta \dot{u}^\gamma \right) \\
&= g_{\alpha\beta} \left(\dot{\xi}^\alpha + \xi^\delta \left\{ \begin{matrix} \alpha \\ \delta\gamma \end{matrix} \right\} \dot{u}^\gamma \right) \dot{u}^\beta,
\end{aligned}
$$

since $(u^1(s), u^2(s))$ is a geodesic line. Therefore, we have

$$g_{\alpha\beta} \left(\dot{\xi}^\alpha + \xi^\delta \left\{ \begin{matrix} \alpha \\ \delta\gamma \end{matrix} \right\} \dot{u}^\gamma \right) \dot{u}^\beta = 0. \tag{5.22}$$

Now we show that the vector with the components

$$\frac{\delta\xi^\alpha}{\delta s} = \dot{\xi}^\alpha + \xi^\delta \left\{ \begin{matrix} \alpha \\ \delta\gamma \end{matrix} \right\} \dot{u}^\gamma \ (\alpha = 1, 2)$$

is the zero vector.

It follows from $g_{\alpha\beta}\xi^\alpha\xi^\beta = 1$ that

$$\left(g_{\alpha\beta}\xi^\alpha\xi^\beta \right)_{;\sigma} = g_{\alpha\beta;\sigma}\xi^\alpha\xi^\beta + g_{\alpha\beta}\xi^\alpha_{;\sigma}\xi^\beta + g_{\alpha\beta}\xi^\alpha\xi^\beta_{;\sigma} = 2g_{\alpha\beta}\xi^\alpha_{;\sigma}\xi^\beta = 0,$$

since $g_{\alpha\beta;\sigma} = 0$ for all α and β by Theorem 5.2.6.

Furthermore, we have

$$\frac{\delta\xi^\alpha}{\delta s} = \frac{\partial\xi^\alpha}{\partial u^\sigma}\dot{u}^\sigma + \xi^\delta \left\{ \begin{matrix} \alpha \\ \delta\sigma \end{matrix} \right\} \dot{u}^\sigma = \left(\frac{\partial\xi^\alpha}{\partial u^\sigma} + \xi^\delta \left\{ \begin{matrix} \alpha \\ \delta\sigma \end{matrix} \right\} \right) \dot{u}^\sigma = \xi^\alpha_{;\sigma}\dot{u}^\sigma \ (\alpha = 1, 2),$$

hence

$$g_{\alpha\beta}\xi^\alpha_{;\sigma}\xi^\beta\dot{u}^\sigma = g_{\alpha\beta}\frac{\delta\xi^\alpha}{\delta s}\xi^\beta = 0. \tag{5.23}$$

This means that the vector with the components $\frac{\delta\xi^\alpha}{\delta s}$ is orthogonal both to the geodesic line by (5.22) and to the surface vector $\vec{x}_\alpha\xi^\alpha$ by (5.23). Since this is an arbitrary vector, we must have

$$\frac{\delta\xi^\alpha}{\delta s} = \xi^\alpha_{;\sigma}\dot{u}^\sigma = \frac{d\xi^\alpha}{ds} + \xi^\delta \left\{ \begin{matrix} \alpha \\ \delta\gamma \end{matrix} \right\} \dot{u}^\gamma = 0 \ (\alpha = 1, 2). \tag{5.24}$$

□

Now we extend the concept of parallel movement along geodesic lines to parallel movement along arbitrary curves.

Definition 5.6.3. The parallel movement of a surface vector with the components ξ^α along an arbitrary curve given by $(u^1(s), u^2(s))$ is defined by

$$\frac{\delta \xi^\alpha}{\delta s} = \xi^\alpha_{;\sigma} \dot{u}^\sigma = \frac{d\xi^\alpha}{ds} + \xi^\delta \left\{ {\alpha \atop \delta\gamma} \right\} \dot{u}^\gamma = 0 \quad (\alpha = 1, 2); \tag{5.25}$$

this is called *parallel movement by Levi–Cività*.

Remark 5.6.4. *(a) Since $\frac{\delta \xi^\alpha}{\delta s} = \xi^\alpha_{;\sigma} \dot{u}^\sigma$, we may write (5.25) as $\xi^\alpha_{;\sigma} \dot{u}^\sigma = 0$ for $\alpha = 1, 2$. Thus the contravariant surface vectors $\vec{z}(s)$ with the components $\xi^\alpha(s)$ are parallel in the sense of Levi–Cività with respect to a curve on a surface given by $(u^1(s), u^2(s))$ if and only if the components of the contravariant vector vanish, which is the contraction of the covariant derivative of $\vec{z}(s)$ and the contravariant vector with the components $\dot{u}^\alpha(s)$.*
(b) If t is an arbitrary parameter of the curve then (5.25) obviously becomes

$$\frac{d\xi^\alpha}{dt} + \xi^\delta \left\{ {\alpha \atop \delta\gamma} \right\} \frac{du^\gamma}{dt} = 0 \quad (\alpha = 1, 2). \tag{5.26}$$

From $\xi^\alpha = g^{\alpha\nu} \xi_\nu$ and $g^* = \det(g^{\mu\nu}) \neq 0$, we obtain the differential equations for a covariant surface vector

$$\frac{d\xi_\nu}{dt} - \xi_\delta \left\{ {\delta \atop \nu\gamma} \right\} \frac{du^\gamma}{dt} = 0 \quad (\nu = 1, 2).$$

(c) The Christoffel symbols vanish for the Cartesian coordinates of the plane. Then (5.26) reduces to

$$\frac{d\xi^\alpha}{dt} = 0 \quad (\alpha = 1, 2).$$

Thus the components of the surface vectors are constant. This is the situation in the case described at the beginning of this section.
(d) By (5.25), the parallel movement is an intrinsic property of the surface.
(e) The parallel movement is independent of the choice of the parameters of the surface, since the left hand side of (5.25) is a tensor.
(f) The tangents of a geodesic line γ are parallel to each other with respect to γ, since, for $\xi^\alpha = \dot{u}^\alpha$ $(\alpha = 1, 2)$, we have

$$\frac{\delta \xi^\alpha}{\delta s} = \ddot{u}^\alpha + \left\{ {\alpha \atop \delta\gamma} \right\} \dot{u}^\delta \dot{u}^\gamma = 0 \quad (\alpha = 1, 2).$$

This property is referred to as the autoparallelism of geodesic lines.

Example 5.6.5. *(a) We consider the circular cylinder with a parametric representation*

$$\vec{x}(u^i) = \{r \cos u^2, r \sin u^2, u^1\}.$$

All Christoffel symbols vanish and the parallel movement by Levi–Cività is given by

$$\frac{d\xi^\alpha}{dt} = 0, \text{ that is, } \xi^\alpha(t) = \xi^\alpha(0) \quad (\alpha = 1, 2)$$

for all $t \geq 0$. Thus the movement is independent of the choice of the curve along which it takes place.

(b) We consider the circular cone with a parametric representation

$$\vec{x}(u^i) = \left\{ \frac{1}{\sqrt{a}} u^1 \cos u^2, \frac{1}{\sqrt{a}} u^1 \sin u^2, u^1 \right\},$$

where $a > 0$ is a constant. Putting $\alpha = \sqrt{a/(a+1)}$, we obtain

$$g_{11} = \frac{a+1}{a} = \frac{1}{\alpha^2}, \; g_{12} = 0 \text{ and } g_{22} = \frac{(u^1)^2}{a},$$

and by (3.47) at the beginning of Section 3.3, the first differential equation in (5.26) reduces to

$$\frac{d\xi^1}{dt} = \frac{1}{g_{11}} \frac{dg_{22}}{du^1} \xi^2 \frac{du^2}{dt} = \frac{\alpha^2 u^1}{a} \xi^2 \frac{u^2}{dt}.$$

The choice of

$$\xi^2 = \frac{1}{\sqrt{g_{22}}} \sqrt{1 - g_{11}(\xi^1)^2} = \frac{\sqrt{a}}{\alpha u^1} \sqrt{\alpha^2 - (\xi^1)^2}$$

yields

$$\frac{d\xi^1}{dt} = \frac{1}{\sqrt{a+1}} \sqrt{\alpha^2 - (\xi^1)^2} \frac{du^2}{dt}.$$

This implies

$$\int \frac{d\xi^1}{\sqrt{\alpha^2 - (\xi^1)^2}} = \arcsin \frac{\xi^1}{\alpha} = \frac{u^2(t)}{\sqrt{a+1}} + c,$$

where $c \in \mathbb{R}$ is a constant.

If the curve for the movement is given by $(u^1(t), u^2(t))$ and we choose the initial conditions

$$u^2(0) = 0, \; \xi^1(0) = \frac{\sin \Theta_0}{\sqrt{g_{11}(u^1_0)}} \text{ and } \xi^2(0) = \frac{\cos \Theta_0}{\sqrt{g_{22}(u^1_0)}},$$

where Θ_0 is the angle between the u^2–line at $t = 0$ and the surface vector, then

$$\xi^1(t) = \alpha \sin \left(\frac{u^2(t)}{\sqrt{a+1}} + \Theta_0 \right)$$

and

$$\xi^2(t) = \frac{\sqrt{a}}{u^1(t)} \sqrt{1 - \sin^2 \left(\frac{u^2(t)}{\sqrt{a+1}} + \Theta_0 \right)} = \frac{\sqrt{a}}{u^1(t)} \left| \cos \left(\frac{u^2(t)}{\sqrt{a+1}} + \Theta_0 \right) \right|.$$

5.7   THE FUNDAMENTAL THEOREM FOR SURFACES

In this section, we prove the *fundamental theorem of the theory of surfaces*, which is the analogue for surfaces of the fundamental theorem of curves, Theorem 1.11.1. It roughly states that a surface is uniquely defined by its first and second fundamental coefficients. Thus the first and second fundamental coefficients g_{ik} and L_{ik} play a similar role for surfaces as the curvature and torsion κ and τ in the case of curves.

We need the following result from the theory of partial differential equations in the proof of the fundamental theorem for surfaces.

Theorem 5.7.1. *Let a system of partial differential equations*

$$\frac{\partial y^i}{\partial x^\alpha} = f_\alpha^i(x^\beta; y^k) \ (i = 1, 2, \ldots, n; \alpha = 1, 2, \ldots, m) \tag{5.27}$$

be given. Furthermore, let $D \subset \mathbb{R}^{m+n}$ be a domain and $f^i \in C^2(D)$ $(i = 1, 2, \ldots, n)$. The system is said to be totally integrable *if given any point $(x_0^\beta, y_0^k) \in D$ there exists at least one solution $y^i(x^\alpha) \in C^2$ of (5.27) which satisfies the initial conditions*

$$y^i(x_0^\alpha) = y_0^i \quad (i = 1, 2, \ldots, n). \tag{5.28}$$

(a) *Given initial conditions there is at most one solution of the system (5.27) (uniqueness).*

(b) *The system (5.27) is totally integrable if and only if, for*

$$J_{\alpha\beta}^i = f_{\alpha\beta}^i - f_{\beta\alpha}^i + f_{\alpha k}^i f_\beta^k - f_{\beta k}^i f_\alpha^k \quad (i = 1, 2, \ldots, n; \alpha, \beta = 1, 2, \ldots, m), \tag{5.29}$$

we have

$$J_{\alpha\beta}^i = 0 \ (i = 1, 2, \ldots, n; \alpha, \beta = 1, 2, \ldots, m). \tag{5.30}$$

Proof.

(i) First we show the necessity of condition in (5.30) for the total integrability of the system (5.27).

Let $y^i(x^\alpha)$ $(i = 1, 2, \ldots, n)$ be a solution of the system (5.27) which is two times continuously differentiable and satisfies the initial conditions in (5.28). Then we may interchange the order of differentiation, and so

$$y_{\alpha\beta}^i - y_{\beta\alpha}^i = 0 \ (i = 1, 2, \ldots, n; \alpha, \beta = 1, 2, \ldots, m).$$

Now we have

$$\begin{aligned} y_{\alpha\beta}^i - y_{\beta\alpha}^i &= \frac{\partial}{\partial x^\beta}\left(f_\alpha^i(x^\beta; y^k)\right) - \frac{\partial}{\partial x^\alpha}\left(f_\beta^i(x^\beta; y^k)\right) \\ &= f_{\alpha\beta}^i - f_{\beta\alpha}^i + f_{\alpha k}^i \frac{\partial y^k}{\partial x^\beta} - f_{\beta k}^i \frac{\partial y^k}{\partial x^\alpha} \\ &= f_{\alpha\beta}^i - f_{\beta\alpha}^i + f_{\alpha k}^i f_\beta^k - f_{\beta k}^i f_\alpha^k = J_{\alpha\beta}^i = 0 \end{aligned}$$

for $i = 1, 2, \ldots, n$ and $\alpha, \beta = 1, 2, \ldots, m$.

This shows the necessity of the conditions in (5.30).

(ii) Now we show the sufficiency of the conditions in (5.30) for the total integrability of the system (5.27) and the uniqueness of the solution.
We consider the system

$$\frac{dy^i}{dt} = f^i_\alpha(x^\beta(t); y^k)\frac{dx^\alpha}{dt} \quad (i = 1, 2, \ldots, n) \tag{5.31}$$

of ordinary differential equations. Since

$$\frac{dy^i}{dt} = \frac{\partial y^i}{\partial x^\alpha}\frac{dx^\alpha}{dt} \quad (i = 1, 2, \ldots, n),$$

any solution $y^i(x^\beta)$ $(i = 1, 2, \ldots, n)$ of (5.27) taken along the curve given by $x^\beta(t)$ $(\beta = 1, 2, \ldots, m)$, that is, any solution with $y^i(t) = y^i(x^\beta(t))$ has to be a solution of the system (5.31). By the uniqueness theorem for systems of ordinary differential equations, there is one and only one solution $y^i(t)$ with $y^i(0) = y^i_0$ $(i = 1, 2, \ldots, n)$ for $x^\beta(0) = x^\beta_0$ $(\beta = 1, 2, \ldots, m)$. This contains the uniqueness statement in Part (a) of the theorem, since any two solutions that satisfy the same initial conditions have to have the same values at arbitrary points that are joined with x^β_0 by a curve. Thus they have to be equal.

To prove the existence of a solution of the system (5.27), we have to use (5.30). It suffices to prove that the integration of (5.31) along different curves given by $x^\beta(t)$ and $x^{*\beta}(t)$ with

$$x^\beta(0) = x^{*\beta}(0) = x^\beta_0 \text{ and } x^\beta(1) = x^{*\beta}(1) = x^\beta_1 \ (\beta = 1, 2, \ldots, m)$$

always leads to the same value $y^i(x^\beta)$, whenever $y^i(0) = y^i_0$ $(i = 1, 2, \ldots, n)$ is chosen as an initial condition. The functions $y^i(x^\beta)$ $(i = 1, 2, \ldots, n)$ defined in some neighbourhood of x^β_0 $(\beta = 1, 2, \ldots, m)$ then indeed satisfy the differential equations in (5.27). We consider all curves given by

$$x^\beta(t; \varepsilon) = (1 - \varepsilon)x^\beta(t) + \varepsilon x^{*\beta}(t) \ (0 \le \varepsilon \le 1; \beta = 1, 2, \ldots, m)$$

of class C^2 (as are x^β and $x^{*\beta}$), and which satisfy

$$x^\beta_0 = x^\beta(0, \varepsilon) \text{ and } x^\beta_1 = x^\beta(1; \varepsilon) \ (\beta = 1, 2, \ldots, m).$$

Let $y^i(t; \varepsilon)$ $(i = 1, 2, \ldots, n)$ denote the corresponding solutions of (5.31) that satisfy

$$y^i(0; \varepsilon) = y^i_0 \ (i = 1, 2, \ldots, n).$$

Since the right hand side in (5.31) is differentiable with respect to ε, each $y^i(t; \varepsilon)$ is also differentiable with respect to ε. Thus we may write

$$z^i(t; \varepsilon) = \frac{dy^i}{d\varepsilon} - f^i_\alpha(x^\beta(t; \varepsilon); y^k(t; \varepsilon))\frac{dx^\alpha}{d\varepsilon} \quad (i = 1, 2, \ldots, n). \tag{5.32}$$

Now (5.31) implies

$$
\begin{aligned}
\frac{dz^i}{dt} &= \frac{d}{dt}\left(\frac{dy^i}{d\varepsilon} - f^i_\alpha(x^\beta(t;\varepsilon); y^k(t;\varepsilon))\frac{dx^\alpha}{d\varepsilon}\right) \\
&= \frac{d}{d\varepsilon}\left(\frac{dy^i}{dt}\right) - \left(f^i_{\alpha\beta}\frac{dx^\beta}{dt} + f^i_{\alpha k}\frac{dy^k}{dt}\right)\frac{dx^\alpha}{d\varepsilon} - f^i_\alpha\frac{d}{dt}\left(\frac{dx^\alpha}{d\varepsilon}\right) \\
&= \frac{d}{d\varepsilon}\left(f^i_\alpha(x^\beta(t;\varepsilon); y^k(t;\varepsilon))\frac{dx^\alpha}{dt}\right) - f^i_{\alpha\beta}\frac{dx^\beta}{dt}\frac{dx^\alpha}{d\varepsilon} - f^i_{\alpha k}\frac{dy^k}{dt}\frac{dx^\alpha}{d\varepsilon} \\
&\quad - f^i_\alpha\frac{d}{dt}\left(\frac{dx^\alpha}{d\varepsilon}\right) \\
&= f^i_{\alpha\beta}\frac{dx^\beta}{d\varepsilon}\frac{dx^\alpha}{dt} + f^i_{\alpha k}\frac{dy^k}{d\varepsilon}\frac{dx^\alpha}{dt} + f^i_\alpha\frac{d}{dt}\left(\frac{dx^\alpha}{d\varepsilon}\right) - f^i_{\alpha\beta}\frac{dx^\beta}{dt}\frac{dx^\alpha}{d\varepsilon} \\
&\quad - f^i_{\alpha k}\frac{dy^k}{dt}\frac{dx^\alpha}{d\varepsilon} f^i_\alpha\frac{d}{dt}\left(\frac{dx^\alpha}{d\varepsilon}\right) \quad (i = 1, 2, \ldots, n),
\end{aligned}
$$

hence

$$
\frac{dz^i}{dt} = f^i_{\alpha k}\frac{dy^k}{d\varepsilon}\frac{dx^\alpha}{dt} + f^i_{\alpha\beta}\frac{dx^\beta}{d\varepsilon}\frac{dx^\alpha}{dt} - f^i_{\alpha k}\frac{dy^k}{dt}\frac{dx^\alpha}{d\varepsilon} - f^i_{\alpha\beta}\frac{dx^\beta}{dt}\frac{dx^\alpha}{d\varepsilon} \tag{5.33}
$$

for $i = 1, 2, \ldots, n$. We obtain by (5.32), (5.29) and (5.33)

$$
\begin{aligned}
f^i_{\alpha k}z^k\frac{dx^\alpha}{dt} &+ J^i_{\alpha\beta}\frac{dx^\alpha}{dt}\frac{dx^\beta}{d\varepsilon} \\
&= f^i_{\alpha k}\frac{dy^k}{d\varepsilon}\frac{dx^\alpha}{dt} - f^i_{\alpha k}f^k_\beta\frac{dx^\beta}{d\varepsilon}\frac{dx^\alpha}{dt} + f^i_{\alpha\beta}\frac{dx^\alpha}{dt}\frac{dx^\beta}{d\varepsilon} - f^i_{\beta\alpha}\frac{dx^\alpha}{dt}\frac{dx^\beta}{d\varepsilon} \\
&\quad + f^i_{\alpha k}f^k_\beta\frac{dx^\alpha}{dt}\frac{dx^\beta}{d\varepsilon} - f^i_{\beta k}f^k_\alpha\frac{dx^\alpha}{dt}\frac{dx^\beta}{d\varepsilon}
\end{aligned}
$$

by interchanging the indices of summation α and β in the last term in the second line

$$
= f^i_{\alpha k}\frac{dy^k}{d\varepsilon}\frac{dx^\alpha}{dt} + f^i_{\alpha\beta}\frac{dx^\beta}{d\varepsilon}\frac{dx^\alpha}{dt} - f^i_{\alpha\beta}\frac{dx^\beta}{dt}\frac{dx^\alpha}{d\varepsilon} - f^i_{\beta k}f^k_\alpha\frac{dx^\alpha}{dt}\frac{dx^\beta}{d\varepsilon}
$$

by observing that $f^k_\alpha dx^\alpha/dt = dy^k/dt$ and then interchanging the indices of summation β and α in the last term

$$
\begin{aligned}
&= f^i_{\alpha k}\frac{dy^k}{d\varepsilon}\frac{dx^\alpha}{dt} + f^i_{\alpha\beta}\frac{dx^\beta}{d\varepsilon}\frac{dx^\alpha}{dt} - f^i_{\alpha\beta}\frac{dx^\beta}{dt}\frac{dx^\alpha}{d\varepsilon} - f^i_{\alpha k}\frac{dy^k}{dt}\frac{dx^\alpha}{d\varepsilon} \\
&= \frac{dz^i}{dt}
\end{aligned}
$$

for $i = 1, 2, \ldots, n$, that is,

$$
\frac{dz^i}{dt} = f^i_{\alpha k}z^k\frac{dx^\alpha}{dt} + J^i_{\alpha\beta}\frac{dx^\alpha}{dt}\frac{dx^\beta}{dt} \quad (i = 1, 2, \ldots, n).
$$

The assumption $J^i_{\alpha\beta} = 0$ for $i = 1, 2, \ldots, n$ now yields

$$\frac{dz^i}{dt} = f^i_{\alpha k} z^k \frac{dx^\alpha}{dt} \quad (i = 1, 2, \ldots, n). \tag{5.34}$$

Since $z^i(0; \varepsilon) = 0$ $(i = 1, 2, \ldots, n)$ by the initial conditions, $z^i(t; \varepsilon) = 0$ $(i = 1, 2, \ldots, n)$ are the only solutions of (5.34) by the uniqueness theorem. This and (5.32) together imply

$$\frac{dy^i}{d\varepsilon} = 0 \text{ at } t = 1, \text{ since } \frac{dx^\beta}{d\varepsilon} = 0 \text{ at } t = 1.$$

Therefore, $y^i(\underset{1}{x^\beta})$ is independent of the choice of the curve.

□

Now we are able to prove the fundamental theorem of the theory of surfaces.

Theorem 5.7.2 (Fundamental Theorem of the Theory of Surfaces).
Let $g^*_{ik}(u^1, u^2)$ and $L^*_{ik}(u^1, u^2)$ be given symmetric tensors of class C^2 and C^1, respectively, that satisfy the conditions

$$R^*_{hikj} = L^*_{ik}L^*_{hj} - L^*_{ij}L^*_{hk}, \tag{5.35}$$

$$L^*_{ik;\,j} = L^*_{ij;\,k} \tag{5.36}$$

and

$$g^*_{ik}\xi^i\xi^k > 0 \text{ for all } \xi^i \neq 0.. \tag{5.37}$$

Then there is – up to movements – one and only one surface with a parametric representation $\vec{x}(u^1, u^2)$ of class C^3 the fundamental coefficients of which satisfy

$$g_{ik} = g^*_{ik} \text{ and } L_{ik} = L^*_{ik}$$

in suitable coordinates.

Proof.

(i)   First we show that the system

$$\vec{x}^*_{i;\,k} = L^*_{ik}\vec{N}^*, \tag{5.38}$$

$$\vec{N}^*_i = -L^{*r}_i \vec{x}^*_r \tag{5.39}$$

of the Gauss and Weingarten equations for the derivatives ((3.99) with Definition 5.1.3, and (2.141)) has one and only one system of solutions $\vec{x}^*_i$ and $\vec{N}^*$ that satisfies the given initial conditions

$$\vec{x}^*(\underset{0}{u^1}, \underset{0}{u^2}) = \underset{0}{\vec{x}}, \ \vec{x}^*_i(\underset{0}{u^1}, \underset{0}{u^2}) = \underset{0}{\vec{x}_i} \ (i = 1, 2) \text{ and } \vec{N}^*(\underset{0}{u^1}, \underset{0}{u^2}) = \underset{0}{\vec{N}},$$

where

$$\vec{x}_i \underset{0}{\bullet} \vec{x}_k = g_{ik}^*(u^1, u^2) \ (i, k = 1, 2), \ \vec{N} \underset{0}{\bullet} \vec{x} = 0 \text{ and } \vec{N}^2 = 1.$$

To prove this, we have to show that the conditions of integrability in Theorem 5.7.1 are satisfied. These conditions follow from the fact that the order of the second derivatives of $\vec{N}^*$ and $\vec{x}_i^*$ may be interchanged. We have by the Ricci identity (5.13)

$$\vec{x}_{i;kj}^* - \vec{x}_{i;jk}^* = -R_{ikj}^{*m} \vec{x}_m^*. \tag{5.40}$$

On the other hand (5.38), (5.39), (5.36) and (5.35) together imply

$$\begin{aligned}
\vec{x}_{i;kj}^* - \vec{x}_{i;jk}^* &= (L_{ik;j}^* - L_{ij;k}^*)\vec{N}^* + L_{ik}^* \vec{N}_j^* - L_{ij}^* \vec{N}_k^* \\
&= (L_{ik;j}^* - L_{ij;k}^*)\vec{N}^* - (L_{ik}^* L_j^{*m} - L_{ij}^* L_k^{*m})\vec{x}_m^* \\
&= 0 \cdot \vec{N}^* - \left(L_{ik}^* g^{*ml} L_{lj}^* - L_{ij}^* g^{*ml} L_{lk}^* \right) \vec{x}_m^* \\
&= -g^{*ml}(L_{ik}^* L_{lj}^* - L_{ij}^* L_{lk}^*)\vec{x}_m^* \\
&= -g^{*ml} R_{likj}^* \vec{x}_m^* = -R_{ikj}^{*m} \vec{x}_m^*.
\end{aligned}$$

Consequently the condition in (5.40) is satisfied. Furthermore, we have

$$\vec{N}_{i;k}^* = \frac{\partial \vec{N}_i^*}{\partial u^k} - \left\{ {m \atop ik} \right\} \vec{N}_m^* \text{ and } \vec{N}_{k;i}^* = \frac{\partial \vec{N}_k^*}{\partial u^i} - \left\{ {m \atop ki} \right\} \vec{N}_m^*,$$

and $\left\{ {m \atop ik} \right\} = \left\{ {m \atop ki} \right\}$ implies

$$\vec{N}_{i;k}^* - \vec{N}_{k;i}^* = \vec{N}_{ik}^* - \vec{N}_{ki}^* = \vec{0} \ (i, k = 1, 2). \tag{5.41}$$

On the other hand (5.39), (5.38), Theorem 5.2.6 and (5.36) together imply

$$\begin{aligned}
\vec{N}_{i;k}^* - \vec{N}_{k;i}^* &= -\left(L_{i;k}^{*r} - L_{k;i}^{*r} \right) \vec{x}_r^* - \left(L_i^{*r} \vec{x}_{r;k}^* - L_k^{*r} \vec{x}_{r;i}^* \right) \\
&= -\left(\left(g^{*rj} L_{ij}^* \right)_{;k} - \left(g^{*rj} L_{kj}^* \right)_{;i} \right) \vec{x}_r^* - (L_i^{*r} L_{rk}^* - L_k^{*r} L_{ri}^*) \vec{N}^* \\
&= -g^{*rj} \left(L_{ij;k}^* - L_{kj;i}^* \right) \vec{x}_r^* - (L_i^{*r} L_{rk}^* - L_k^{*r} L_{ri}^*) \vec{N}^* \\
&= 0 \cdot \vec{x}_r^* - (L_i^{*r} L_{rk}^* - L_k^{*r} L_{ri}^*) \vec{N}^* = \left(g^{*lr} L_{li}^* L_{rk}^* - g^{*lr} L_{lk}^* L_{ri}^* \right) \vec{N}^*
\end{aligned}$$

(with the interchange of the indices l and r in the last term)

$$= g^{*lr} \left(L_{li}^* L_{rk}^* - L_{rk}^* L_{li}^* \right) \vec{N}^* = \vec{0}.$$

Consequently the conditions in (5.41) are satisfied.

(ii) Now we show that there exists a function $\vec{x}$ of class C^3 with

$$\frac{\partial \vec{x}}{\partial u^i} = \vec{x}_i^* \text{ for } i = 1, 2, \text{ and } \vec{x}(u^1, u^2) = \vec{x}.$$

We have by the symmetry of L_{ik}^*

$$\vec{x}_{i;\,k}^* - \vec{x}_{k;\,i}^* = (L_{ik}^* - L_{ki}^*)\vec{N}^* = \vec{0} \ (i, k = 1, 2),$$

and so

$$\frac{\partial \vec{x}^{*i}}{\partial u^k} - \frac{\partial \vec{x}^{*k}}{\partial u^i} = \vec{0} \ (i, k = 1, 2).$$

The fact that the function is of class C^3 follows from (5.38) and (5.39). This step shows that there is at most one surface when an initial point and the vectors $\vec{x}_i$ and $\underset{0}{\vec{N}}$ are given at this point. Furthermore, the surface has to be given by the function $\vec{x}$. (Values without $*$ will refer to this surface.)

(iii) Now we show $g_{ik} = g_{ik}^*$ and $\vec{N} = \vec{N}^*$.
It follows from (5.38) and (5.39) that

$$\left\{ \begin{array}{l} (\vec{x}_i \bullet \vec{x}_j)_{;\,k} = \vec{x}_{i;\,k} \bullet \vec{x}_j + \vec{x}_i \bullet \vec{x}_{j;\,k} = L_{ik}^*(\vec{N}^* \bullet \vec{x}_j) + L_{jk}^*(\vec{N}^* \bullet \vec{x}_i) \\ (\vec{N}^* \bullet \vec{x}_j)_{;\,k} = \vec{N}_k^* \bullet \vec{x}_j + \vec{N}^* \bullet \vec{x}_{j;\,k} = -L_k^{*r}(\vec{x}_r \bullet \vec{x}_j) + L_{jk}^*(\vec{N}^*)^2 \\ \text{and} \\ \left((\vec{N}^*)^2\right)_{;\,k} = -2L_k^{*r}(\vec{N}^* \bullet \vec{x}_r). \end{array} \right\} \quad (5.42)$$

We put

$$\phi_{ij} = \vec{x}_i \bullet \vec{x}_j, \ \phi_j = \vec{N}^* \bullet \vec{x}_j \text{ and } \phi = (\vec{N}^*)^2. \quad (5.43)$$

Then (5.42) is a system of partial differential equations for the functions ϕ_{ij}, ϕ_j and ϕ. This system again has at most a unique solution that satisfies the given initial conditions. By construction, the functions ϕ_{ij}, ϕ_j and ϕ in (5.42) are solutions for the initial conditions at $(u^1, u^2) = \underset{0}{u} = (\underset{0}{u^1}, \underset{0}{u^2})$,

$$\phi_{ij}(\underset{0}{u}) = g_{ij}(\underset{0}{u}), \ \phi_j(\underset{0}{u}) = 0 \text{ and } \phi(\underset{0}{u}) = 1.$$

On the other hand, an application of Theorem 5.2.6 shows that $\phi_{ij} = g_{ij}^*$, $\phi_j = 0$ and $\phi = 1$ is a system of solutions for these initial conditions. Using the terms ϕ_{ij}, ϕ_j and ϕ, we obtain in (5.42)

$$\phi_{ij;\,k} = L_{ik}^*\phi_j + L_{jk}^*\phi_i, \ \phi_{j;\,k} = -L_k^{*r}\phi_{rj} + L_{jk}^*\phi \text{ and } \phi_{;\,k} = -2L_k^{*r}\phi_r.$$

Since $\phi_{ij} = g_{ij}^*$, $\phi_j = 0$ and $\phi = 1$, and $\phi_{ij;\,k} = g_{ij;\,k}^* = 0$ by Theorem 5.2.6, we conclude

$$\phi_{ij;\,k} = 0 = L_{ik}^* \cdot 0 + L_{jk}^* \cdot 0$$
$$\phi_{j;\,k} = 0 = -L_k^{*r}g_{rj}^* + L_{jk}^* = -L_{kj}^* + L_{jk}^* = -L_{jk}^* + L_{jk}^*$$

and

$$\phi_{;\,k} = 0 = -2L_k^{*r}\phi_r = -2 \cdot 0.$$

The two solutions consequently coincide.
This proves (iii).

(iv) Now we show $L_{ik} = L_{ik}^*$.

The Gauss equations for the derivatives imply that the surface satisfies $\vec{x}_{i;\,k} = L_{ik}\vec{N}$. Now it follows from $\vec{N}^* = \vec{N}$ and $\vec{x}_{i;\,k}^* = L_{ik}^*\vec{N}^*$ that $L_{ik} = L_{ik}^*$.

(v) Finally, the fact that $\vec{x}$ is of class C^3 is an immediate consequence of the equations for the derivatives.

$\square$

We give two applications of Theorem 5.7.2.

Example 5.7.3. *If the fundamental coefficients of a surface S satisfy*

$$g_{11} = g_{22} = 1, \; g_{12} = 0, \; L_{11} = -1 \; and \; L_{12} = L_{22} = 0,$$

then S is a circular cylinder of radius $r = 1$.

Proof. We obtain

$$L_1^1 = g^{1j}L_{j1} = -1, \; L_1^2 = g^{2j}L_{j1} = 0, \; L_2^2 = g^{2j}L_{j2} = 0$$

and, by (3.47) at the beginning of Section 3.3, $\left\{ {i \atop jk} \right\} = 0$ for $i, j, k = 1, 2$. So the Gauss and Weingarten equations for the derivatives (3.99) and (2.141) reduce to

$$\vec{x}_{11} = -\vec{N}, \tag{5.44}$$

$$\vec{x}_{12} = \vec{x}_{22} = \vec{0}, \tag{5.45}$$

$$\vec{N}_1 = \vec{x}_1 \tag{5.46}$$

and

$$\vec{N}_2 = \vec{0}. \tag{5.47}$$

It follows from (5.44) and (5.46) that $\vec{x}_{111} = -\vec{N}_1 = -\vec{x}_1$ and integration with respect to u^1 yields

$$\vec{x}(u^i) = \vec{a}(u^2)\cos u^1 + \vec{b}(u^2)\sin u^1 + \vec{c}(u^2), \tag{5.48}$$

where $\vec{a}$, $\vec{b}$ and $\vec{c}$ are vector–valued functions of u^2. Since $\vec{x}_{12} = \vec{0}$ in (5.45), we obtain

$$\vec{x}_{12} = -\vec{a}'(u^2)\sin u^1 + \vec{b}'(u^2)\cos u^1 = \vec{0}.$$

This $\vec{a}'(u^2) = \vec{b}'(u^2) = \vec{0}$, so $\vec{a}(u^2) = \vec{a}$ and $\vec{b}(u^2) = \vec{b}$, where $\vec{a}$ and $\vec{b}$ are constant vectors. Finally $\vec{x}_{22} = \vec{0}$ in (5.45) implies $\vec{x}_{22} = \vec{c}''(u^2) = \vec{0}$, hence $\vec{c}(u^2) = \vec{c}_1 u^2 + \vec{d}$ with constant vectors $\vec{c}_1$ and $\vec{d}$. Thus we have

$$\vec{x}(u^i) = \vec{a}\cos u^1 + \vec{b}\sin u^1 + \vec{c}_1 u^2 + \vec{d}. \tag{5.49}$$

Since $\vec{x}_1 = -\vec{a}\sin u^1 + \vec{b}\cos u^1$ and $\vec{x}_2 = \vec{c}_1$, it follows from

$$g_{11} = \vec{a}^2\sin^2 u^1 + \vec{b}^2\cos^2 u^1 - 2\vec{a} \bullet \vec{b}\sin u^1 \cos u^1 = 1,$$

$$g_{12} = -\vec{a} \bullet \vec{c}_1 \sin u^1 + \vec{b} \bullet \vec{c}_1 \cos u^1 = 0$$

and

$$g_{22} = \vec{c}_1^2 = 1$$

that $\vec{c}_1^2 = 1$ and

$$
\begin{array}{ll}
\text{for } u^1 = 0: & \vec{b} \bullet \vec{c}_1 = 0,\ \vec{b}^2 = 1, \\
\text{for } u^1 = \pi/2: & \vec{a}^2 = 1,\ \vec{a} \bullet \vec{c}_1 = 0 \\
\text{and } \text{for } u^1 = \pi/4: & \vec{a} \bullet \vec{b} = 0.
\end{array}
$$

Consequently (5.49) is a parametric representation of a circular cylinder of radius 1. □

Example 5.7.4. *If the fundamental coefficients of a surface S satisfy*

$$g_{11} = 1,\ g_{12} = 0,\ g_{22} = \sin^2 u^1, \quad L_{11} = 1,\ L_{12} = 0 \ and\ L_{22} = \sin^2 u^1 \ (u^1 \in (0, \pi)),$$

then S is a sphere of radius 1.

Proof. We obtain from (3.47) at the beginning of Section 3.3

$$\left\{ \begin{matrix} 1 \\ 11 \end{matrix} \right\} = \left\{ \begin{matrix} 1 \\ 12 \end{matrix} \right\} = \left\{ \begin{matrix} 1 \\ 21 \end{matrix} \right\} = 0,\ \left\{ \begin{matrix} 1 \\ 22 \end{matrix} \right\} = -\sin u^1 \cos u^1,$$

$$\left\{ \begin{matrix} 2 \\ 11 \end{matrix} \right\} = \left\{ \begin{matrix} 2 \\ 22 \end{matrix} \right\} = 0 \ and\ \left\{ \begin{matrix} 2 \\ 12 \end{matrix} \right\} = \left\{ \begin{matrix} 2 \\ 21 \end{matrix} \right\} = \cot u^1.$$

Furthermore, we have

$$g^{11} = \frac{g_{22}}{g} = 1,\ g^{12} = 0,\ g^{22} = \frac{g_{11}}{g} = \frac{1}{\sin^2 u^1},$$

$$L_1^1 = g^{11} L_{11} = 1,\ L_2^1 = g^{11} L_{12} = 0 \ and\ L_2^2 = g^{22} L_{22} = 1.$$

Therefore, the Gauss and Weingarten equations for the derivatives (3.99) and (2.141) reduce to

$$\vec{x}_{11} = \left\{ \begin{matrix} r \\ 11 \end{matrix} \right\} \vec{x}_r + L_{11}\vec{N} = \vec{N}, \tag{5.50}$$

$$\vec{x}_{12} = \left\{ \begin{matrix} r \\ 12 \end{matrix} \right\} \vec{x}_r + L_{12}\vec{N} = \cot u^1 \vec{x}_2, \tag{5.51}$$

$$\vec{x}_{22} = \left\{ \begin{matrix} r \\ 22 \end{matrix} \right\} \vec{x}_r + L_{22}\vec{N} = -\sin u^1 \cos u^1 \vec{x}_1 + \sin^2 u^1 \vec{N}, \tag{5.52}$$

$$\vec{N}_1 = -L_1^i \vec{x}_i = -\vec{x}_1 \tag{5.53}$$

and

$$\vec{N}_2 = -L_2^i \vec{x}_i = -\vec{x}_2. \tag{5.54}$$

First (5.50) and (5.53) together imply

$$\vec{x}_{111} = \vec{N}_1 = -\vec{x}_1.$$

Integration with respect to u^1 yields

$$\vec{x}(u^i) = \vec{a}(u^2) \sin u^1 + \vec{b}(u^2) \cos u^1 + \vec{c}(u^2), \tag{5.55}$$

where $\vec{a}$, $\vec{b}$ and $\vec{c}$ are vector–valued functions of u^2. We obtain from this and (5.51)

$$\vec{x}_{12} = \vec{a}' \cos u^1 - \vec{b}' \sin u^1 = \vec{x}_2 \cot u^1 = \vec{a}' \cos u^1 + \vec{b}' \cos u^1 \cot u^1 + \vec{c}' \cot u^1$$

and so $\vec{b}'(\sin u^1 + \cos u^1 \cot u^1) = -\vec{c}' \cot u^1$ or equivalently $\vec{b}' = -\vec{c}' \cos u^1$. Since $\vec{b}'$ and $\vec{c}'$ are functions that depend on u^2 only, we must have $\vec{b}' = \vec{c}' = \vec{0}$, and so $\vec{b}$ and $\vec{c}$ are constant vectors. Furthermore (5.55), (5.52) and (5.50) together imply

$$\begin{aligned}
\vec{x}_{22} = \vec{a}'' \sin u^1 &= -\sin u^1 \cos u^1 \vec{x}_1 + \sin^2 u^1 \vec{x}_{11} \\
&= -(\sin u^1 \cos u^1)(\vec{a} \cos u^1 - \vec{b} \sin u^1) + \sin^2 u^1(-\vec{a} \sin u^1 - \vec{b} \cos u^1) \\
&= -\vec{a} \sin u^1(\cos^2 u^1 + \sin^2 u^1) + \vec{b} \sin u^1(\sin u^1 \cos u^1 - \sin u^1 \cos u^1) \\
&= -\vec{a} \sin u^1,
\end{aligned}$$

that is, $\vec{a}'' = -\vec{a}$, and so $\vec{a} = \vec{d} \cos u^2 + \vec{e} \sin u^2$ with constant vectors $\vec{d}$ and $\vec{e}$. Therefore, we have

$$\vec{x}(u^i) = (\vec{d} \cos u^2 + \vec{e} \sin u^2) \sin u^1 + \vec{b} \cos u^1 + \vec{c}$$

and furthermore,

$$\vec{x}_2 \bullet \vec{x}_2 = g_{22} = \sin^2 u^2 = \left(-\vec{d} \sin u^2 + \vec{e} \cos u^2\right)^2 \sin^2 u^1$$

or equivalently,

$$1 = \vec{d}^2 \sin^2 u^2 - 2(\vec{d} \bullet \vec{e}) \sin u^2 \cos u^2 + \vec{e}^2 \cos^2 u^2.$$

For $u^2 = 0, \pi/2, \pi/4$, we conclude $\vec{e}^2 = \vec{d}^2 = 1$ and $\vec{d} \bullet \vec{e} = 0$. Now

$$\begin{aligned}
\vec{x}_1 \bullet \vec{x}_2 = g_{12} = 0 &= \\
&= \left((\vec{d} \cos u^2 + \vec{e} \sin u^2) \cos u^1 - \vec{b} \sin u^1\right) \bullet \left(-\vec{d} \sin u^2 + \vec{e} \cos u^2\right) \sin u^1 \\
&= -\vec{d}^2 \sin u^1 \sin u^2 \cos u^1 \cos u^2 + \vec{e}^2 \sin u^1 \sin u^2 \cos u^1 \cos u^2 \\
&\quad + \vec{b} \bullet \vec{d} \sin^2 u^1 \sin u^2 - \vec{b} \bullet \vec{e} \sin^2 u^1 \cos u^2 \\
&= (\vec{b} \bullet \vec{d}) \sin^2 u^1 \sin u^2 - (\vec{e} \bullet \vec{b}) \sin^2 u^1 \cos u^2
\end{aligned}$$

implies $\vec{b} \bullet \vec{d} = \vec{e} \bullet \vec{b} = 0$. Finally

$$\vec{x}_1 \bullet \vec{x}_1 = g_{11} = 1$$

$$= \left((\vec{d} \cos u^2 + \vec{e} \sin u^2) \cos u^1 - \vec{b} \sin u^1 \right)^2$$
$$= \cos^2 u^1 \cos^2 u^2 + \cos^2 u^1 \sin^2 u^2 + \vec{b}^2 \sin^2 u^1$$
$$= \cos^2 u^1 + \vec{b}^2 \sin^2 u^1$$

implies $\vec{b}^2 = 1$. Therefore, the vectors $\vec{d}$, $\vec{e}$ and $\vec{b}$ are orthonormal. Thus we have

$$\|\vec{x} - \vec{c}\|^2 = \sin^2 u^1 \cos^2 u^2 + \sin^2 u^1 \sin^2 u^2 + \cos^2 u^1 = 1,$$

and the surface is a sphere of radius 1, centred at the point C with the position vector $\vec{c}$. □

5.8 A GEOMETRIC MEANING OF THE RIEMANN TENSOR OF CURVATURE

In this section, we give a geometric interpretation of the Riemann tensor of curvature.

Let γ be a curve with a parametric representation $\vec{x}(t)$ in a Riemann space and T_0 be the tangent space to γ at $\vec{x} = \vec{x}(t_0)$. Then we may move the tangent space T_0 along γ by applying the parallel movement by Levi–Cività to every surface vector in T_0. This means, we assign to every surface vector with the components ξ^i in T_0 the solutions $\xi^i(t)$ $(i = 1, 2, \ldots, n)$ of the system of differential equations

$$\frac{\delta \xi^i}{\delta t} = \xi^i_{;j} \frac{dx^j}{dt} = 0 \text{ with the initial conditions } \xi^i(t_0) = \underset{0}{\xi^i} \ (i = 1, 2, \ldots, n) \quad (5.56)$$

(Remark 5.6.4 (b), (5.26)).

Let t be a fixed parameter of the curve γ and T_t be the tangent space to γ at $\vec{x}(t)$. The parallel movement, which in general depends on the curve and the Christoffel symbols, defines a map

$$A(\gamma; t) : T_0 \to T_t \text{ by } \xi^i(t) = A^i_k(\gamma; t) \xi^k \ (i = 1, 2, \ldots, n).$$

The map $A(\gamma; t)$ is linear, since the system (5.56) is linear in the components ξ^i.

Now let γ be a piecewise smooth curve. Then we obtain a linear map by joining the linear maps from the tangent planes at the initial points to the tangent planes at the end points of the smooth parts. If the end point of γ coincides with its initial point then we obtain a linear map $A(\gamma)$ from T_0 onto itself. Using vector notation, we may write

$$\vec{z}^* = A(\gamma)\vec{z} \text{ for all surface vectors } \vec{z} \in T_0.$$

Let γ_1 and γ_2 be piecewise smooth curves with initial and end points in a given point P. Then we define the product curve $\gamma = \gamma_2 \gamma_1$ to be the curve that results from first passing along γ_1 and then along γ_2. Furthermore, let γ^{-1} be the curve that has the opposite orientation of γ. The corresponding linear maps from the tangent plane $T(P)$ at P onto itself clearly satisfy

$$A(\gamma_2 \gamma_1) = A(\gamma_2) \circ A(\gamma_1) \text{ and } A(\gamma^{-1}) = (A(\gamma))^{-1}.$$

Proposition 5.8.1. *Let M be a differentiable manifold which is* path connected, *that is, any two points of M can be joined by a curve on M. Furthermore, let $P \in M$ be a point and $\mathcal{H}(P)$ denote the set of all maps $A(\gamma)$ from $T(P)$ onto itself. Then $\mathcal{H}(P)$ is a group, the so–called* holonomy group *of connectedness at the point P.*

Proof. Let $A, B \in \mathcal{H}(P)$, that is, $A = A(\gamma_1)$ and $B = A(\gamma_2)$ for two curves γ_1 and γ_2 with initial and end points in P. Then $BA = A(\gamma_2\gamma_1) = A(\gamma_2) \circ A(\gamma_1) \in \mathcal{H}(P)$. Obviously, we have

$$A(\gamma_3(\gamma_2\gamma_1)) = A(\gamma_3) \circ A(\gamma_2\gamma_1) = A(\gamma_3) \circ (A(\gamma_2) \circ A(\gamma_1)) = (A(\gamma_3) \circ A(\gamma_2)) \circ A(\gamma_1)$$
$$= A(\gamma_3\gamma_2) \circ A(\gamma_1) = A((\gamma_3\gamma_2)\gamma_1).$$

Thus the law of associativity is satisfied.

The identity map E defined by $E(\vec{z}) = \vec{z}$ for all surface vectors $\vec{z} \in T(P)$ is an element of $\mathcal{H}(P)$, since $E = A(\gamma) \circ A(\gamma^{-1}) = A(\gamma\gamma^{-1})$. Furthermore, we have $E \circ A(\gamma) = A(\gamma)$ for all $A(\gamma) \in \mathcal{H}(P)$. So E is the neutral element of $\mathcal{H}(P)$.

Finally, let $A \in \mathcal{H}(P)$. Then $A = A(\gamma)$ for some curve γ. We put $A^{-1} = A(\gamma^{-1})$, and obtain $A^{-1} \in \mathcal{H}(P)$ and $AA^{-1} = A(\gamma\gamma^{-1}) = E$. □

Now we consider special curves γ_ε that are similar to parallelograms with one vertex at a point P. We assume that P is the origin of the coordinate system and that the Christoffel symbols vanish at P, that is, $\left\{ {i \atop jk} \right\}(0) = 0$ for all i, j and k. Such a choice of a coordinate system is possible by the proof of Theorem 5.4.1. Let η^i and ζ^i be the components of two linearly independent vectors $\vec{y}$ and $\vec{z}$. We define the curve γ_ε by

$$\vec{x}(t) = \begin{cases} t\varepsilon\vec{y} & (t \in [0,1]) \\ \varepsilon\vec{y} + (t-1)\varepsilon\vec{z} & (t \in [1,2]) \\ \varepsilon(\vec{y} + \vec{z}) - (t-2)\varepsilon\vec{y} & (t \in [2,3]) \\ \varepsilon\vec{z}(1 - (t-3)) & (t \in [3,4]), \end{cases}$$

or using the components

$$x^i(t) = \begin{cases} t\varepsilon\eta^i & (t \in [0,1]) \\ \varepsilon\eta^i + (t-1)\varepsilon\zeta^i & (t \in [1,2]) \\ \varepsilon(\eta^i + \zeta^i) - (t-2)\varepsilon\eta^i & (t \in [2,3]) \\ \varepsilon\zeta^i(1 - (t-3)) & (t \in [3,4]). \end{cases}$$

The following result shows that the second order approximation of the change of a surface vector by parallel movement along γ_ε depends on the Riemann tensor of curvature at P.

Theorem 5.8.2. *The Riemann tensor of curvature is a measure for the change of a surface vector under parallel movement along very small parallelograms γ_ε. More precisely, the linear map $A(\gamma_\varepsilon)$ is given by*

$$A_k^i(\gamma_\varepsilon) = \delta_i^k + \varepsilon^2 R_{kjl}^i(0)\eta^j\zeta^l + \varepsilon^3(\cdots). \tag{5.57}$$

This relation is independent of the choice of a coordinate system, since it is invariant.

Proof. Let ξ^i be the components of a surface vector $\vec{x}$ at P that is to be moved along γ_ε in a parallel way. The Taylor expansion yields

$$\xi^i(t+1) = \xi^i(t) + \frac{d\xi^i(t)}{dt} + \frac{1}{2} \cdot \frac{d^2\xi^i(t)}{dt^2} + \cdots. \tag{5.58}$$

We write

$$\left\{ \begin{matrix} i \\ jk \end{matrix} \right\}_{,l} = \frac{\partial}{\partial x^l} \left(\left\{ \begin{matrix} i \\ jk \end{matrix} \right\} \right).$$

By the parallelism, we have

$$\frac{d\xi^i(t)}{dt} = -\left\{ \begin{matrix} i \\ jk \end{matrix} \right\}(t)\xi^k(t)\frac{dx^j(t)}{dt} \tag{5.59}$$

and

$$\frac{d^2\xi^i(t)}{dt^2} = -\left\{ \begin{matrix} i \\ jk \end{matrix} \right\}_{,l}(t)\xi^k(t)\frac{dx^j(t)}{dt}\frac{dx^l(t)}{dt} + \left\{ \begin{matrix} i \\ jk \end{matrix} \right\}(t)\left\{ \begin{matrix} k \\ rs \end{matrix} \right\}(t)\xi^r(t)\frac{dx^j(t)}{dt}\frac{dx^s(t)}{dt}. \tag{5.60}$$

We neglect terms in ε of order higher than 2. The Taylor expansion yields

$$\left\{ \begin{matrix} i \\ jk \end{matrix} \right\}(t) = \left\{ \begin{matrix} i \\ jk \end{matrix} \right\}(0) + \left\{ \begin{matrix} i \\ jk \end{matrix} \right\}_{,l}(0)\Delta x^l + \cdots = \left\{ \begin{matrix} i \\ jk \end{matrix} \right\}_{,l}(0)\Delta x^l + \cdots. \tag{5.61}$$

The change of the components ξ^i under the parallel movement along γ_ε is given by

$$\xi^i(4) - \xi^i(0) = \sum_{m=1}^{4} (\xi^i(m) - \xi^i(m-1)). \tag{5.62}$$

We obtain from (5.59) and (5.60)

$$\frac{d\xi^i(0)}{dt} = -\left\{ \begin{matrix} i \\ jk \end{matrix} \right\}(0)\xi^k(0)\frac{dx^j(0)}{dt} = 0 \tag{5.63}$$

and

$$\frac{d^2\xi^i(0)}{dt} = -\left\{ \begin{matrix} i \\ jk \end{matrix} \right\}_{,l}(0)\xi^k(0)\varepsilon^2\eta^j\eta^l. \tag{5.64}$$

Now we put

$$f^{(1)}(m) = \frac{1}{\varepsilon^2}\frac{dx^j(m)}{dt}x^l(m) \text{ and } f^{(2)}(m) = \frac{1}{\varepsilon^2}\frac{dx^j(m)}{dt}\frac{dx^l(m)}{dt} \quad (m=1,2,3).$$

We have by (5.59) and (5.61) for $m \geq 1$

$$\frac{d\xi^i(m)}{dt} = -\left\{ \begin{matrix} i \\ jk \end{matrix} \right\}(m)\xi^k(m)\frac{dx^j(m)}{dt} = -\left\{ \begin{matrix} i \\ jk \end{matrix} \right\}_{,l}(0)\varepsilon^2 f^{(1)}(m)\xi^k(m) + \varepsilon^3(\cdots),$$

and by (5.60) and (5.61)

$$\frac{d^2\xi^i(m)}{dt^2} = -\begin{Bmatrix} i \\ jk \end{Bmatrix}_{,l}(m)\xi^k(m)\frac{dx^j(m)}{dt}\frac{dx^l(m)}{dt}$$

$$+\begin{Bmatrix} i \\ jk \end{Bmatrix}(m)\begin{Bmatrix} k \\ rs \end{Bmatrix}(m)\xi^r(m)\frac{dx^j(m)}{dt}\frac{dx^s(m)}{dt}$$

$$= -\varepsilon^2\begin{Bmatrix} i \\ jk \end{Bmatrix}_{,l}(0)f^{(2)}(m)\xi^k(m)+$$

$$+\left(\begin{Bmatrix} i \\ jk \end{Bmatrix}_{,l}(0)x^l(m) + \cdots\right)\xi^r(m)\frac{dx^j(m)}{dt}\frac{dx^s(m)}{dt}\begin{Bmatrix} k \\ rs \end{Bmatrix}(m)$$

$$= -\varepsilon^2\begin{Bmatrix} i \\ jk \end{Bmatrix}_{,l}(0)f^{(2)}(m)\xi^k(m) + \varepsilon^3(\cdots).$$

Applying (5.58), we obtain

$$\frac{d\xi^i(m)}{dt} = -\varepsilon^2\begin{Bmatrix} i \\ jk \end{Bmatrix}_{,l}(0)f^{(1)}(m)\left(\xi^k(m-1) + \frac{d\xi^k(m-1)}{dt} + \frac{1}{2}\frac{d^2\xi^k(m-1)}{dt^2} + \cdots\right)$$

$$+ \varepsilon^3(\cdots) = \cdots$$

$$\cdots = -\varepsilon^2\begin{Bmatrix} i \\ jk \end{Bmatrix}_{,l}(0)f^{(1)}(m)\xi^k(0) + \varepsilon^3(\cdots),$$

and similarly

$$\frac{d^2\xi^i(m)}{dt^2} = -\varepsilon^2\begin{Bmatrix} i \\ jk \end{Bmatrix}_{,l}(0)f^{(2)}(m)\xi^k(0) + \varepsilon^3(\cdots).$$

This yields

$$\frac{d\xi^i(1)}{dt} = -\varepsilon^2\begin{Bmatrix} i \\ jk \end{Bmatrix}_{,l}(0)\xi^k(0)\eta^l\zeta^j \qquad \frac{d^2\xi^i(1)}{dt^2} = -\varepsilon^2\begin{Bmatrix} i \\ jk \end{Bmatrix}_{,l}(0)\xi^k(0)\zeta^l\zeta^j$$

$$\frac{d\xi^i(2)}{dt} = \varepsilon^2\begin{Bmatrix} i \\ jk \end{Bmatrix}_{,l}(0)\xi^k(0)\eta^j(\eta^l + \zeta^l) \qquad \frac{d^2\xi^i(2)}{dt^2} = -\varepsilon^2\begin{Bmatrix} i \\ jk \end{Bmatrix}_{,l}(0)\xi^k(0)\eta^l\eta^j$$

$$\frac{d\xi^i(3)}{dt} = \varepsilon^2\begin{Bmatrix} i \\ jk \end{Bmatrix}_{,l}(0)\xi^k(0)\zeta^j\zeta^l \qquad \frac{d^2\xi^i(3)}{dt^2} = -\varepsilon^2\begin{Bmatrix} i \\ jk \end{Bmatrix}_{,l}(0)\xi^k(0)\zeta^l\zeta^j.$$

Now it follows from this, (5.62) and (5.63) that

$$\frac{1}{\varepsilon^2}\left(\xi^i(4) - \xi^i(0)\right) = \frac{1}{\varepsilon^2}\sum_{m=0}^{3}\left(\xi^i(m+1) - \xi^i(m)\right)$$

$$= \frac{1}{\varepsilon^2}\sum_{m=0}^{3}\left(\frac{d\xi^i(m)}{dt} + \frac{1}{2}\cdot\frac{d^2\xi^i(m)}{dt^2}\right)$$

$$= \begin{Bmatrix} i \\ jk \end{Bmatrix}_{,l}(0)\xi^k(0)\left(0 - \eta^l\zeta^j + \eta^j(\eta^l + \zeta^l) + \zeta^i\zeta^l + \right.$$

$$+\frac{1}{2}\left(-\eta^j\eta^l - \zeta^l\zeta^i - \eta^l\eta^j - \zeta^l\zeta^j\right)\right) + \varepsilon(\cdots)$$

$$= \left\{\begin{matrix} i \\ jk \end{matrix}\right\}_{,l}(0)\xi^k(0)\left(-\eta^l\zeta^j + \eta^j(\eta^l + \zeta^l) + \right.$$

$$\left. +\zeta^j\zeta^l - \eta^j\eta^l - \zeta^j\zeta^l\right) + \varepsilon(\cdots)$$

$$= \left\{\begin{matrix} i \\ jk \end{matrix}\right\}_{,l}(0)\xi^k(0)\left(\eta^j\zeta^l - \eta^l\zeta^j\right) + \varepsilon(\cdots)$$

$$= \left(\left\{\begin{matrix} i \\ jk \end{matrix}\right\}_{,l}(0) - \left\{\begin{matrix} i \\ lk \end{matrix}\right\}_{,j}(0)\right)\xi^k(0)\eta^j\zeta^l + \varepsilon(\cdots)$$

$$= R^i_{kjl}(0)\xi^k(0)\eta^j\zeta^l + \varepsilon(\cdots)$$

by the definition of the Riemann tensor of curvature R^i_{kjl}, since $\left\{\begin{matrix} i \\ jk \end{matrix}\right\}(0) = 0$. Thus the map $A(\gamma_\varepsilon)$ is given by (5.57).

The smaller we choose ε in (5.57) the better the difference $\xi^i(4) - \xi^i(0)$ is approximated by the term that contains the tensor of curvature.

The invariance of the relation in (5.57) follows from the fact that it is a relation between the components of tensors. □

5.9   SPACES WITH VANISHING TENSOR OF CURVATURE

In this section, we study when the tensor of curvature in a Riemann space vanishes identically.

In Euclidean $\mathbb{E}^n$, the metric tensor in Cartesian coordinates is given by $g_{ik} = \delta_{ik}$. This implies $\left\{\begin{matrix} i \\ jk \end{matrix}\right\} = 0$ and consequently $R^i_{ljk} = 0$. Therefore, we have for a parallel vector field along an arbitrary curve with parameter t

$$\frac{\delta\xi^i}{\delta t} = \frac{d\xi^i}{dt} = 0 \ (i = 1, 2, \ldots, n).$$

Thus the parallel movement is independent of the choice of the curve.

The next result shows that a Riemann space with a tensor of curvature that vanishes at a point is locally the same as Euclidean $\mathbb{E}^n$ in the sense that parallel movements locally are independent of the choice of a curve.

Theorem 5.9.1. *Let P_0 be a point in a Riemann space such that $R^i_{ljk} = 0$ at P_0 for the components of the tensor of curvature. Then there is a coordinate system S^* with coordinates x^{*i} in some neighbourhood N of P_0 such that $\left\{\begin{matrix} i \\ jk \end{matrix}\right\}^* = 0$ on N. Therefore, parallel vectors with the components*

$$\xi^{*i}(\underset{1}{x^{*j}}) \ and \ \xi^{*i}(\underset{2}{x^{*j}})$$

at any two distinct points

$$P_1 = (\underset{1}{x^{*1}}, \underset{1}{x^{*2}}, \ldots, \underset{1}{x^{*n}}), P_2 = (\underset{2}{x^{*1}}, \underset{2}{x^{*2}}, \ldots, \underset{2}{x^{*n}}) \in N$$

have the same components. The parallel movement is independent of the choice of a curve.

Proof. First we show that if $R^i_{ljk} = 0$ then there is a neighbourhood N_1 of P_0 such that for every vector with the components ξ^i_0 there exists a vector field with the components $\xi^i(x)$ in N_1 such that $\xi^i(\underset{0}{x}) = \xi^i_0$ and which consists of parallel vectors independent of the choice of a curve. It is necessary and sufficient for this that the system

$$\frac{\partial \xi^i}{\partial x^k} = - \begin{Bmatrix} i \\ lk \end{Bmatrix} \xi^l \ (i, k = 1, 2, \dots, n)$$

of differential equations has a solution with $\xi^i(\underset{0}{x}) = \xi^i_0$. But this is true, since the corresponding condition of integrability in Theorem 5.7.1 is $R^i_{ljk} = 0$. This shows that the parallel movement is independent of the curve.

Now we choose those n vector fields among the parallel fields for which

$$\xi^i_{(k)} = \delta^i_k \ (i, k = 1, 2, \dots, n). \tag{5.65}$$

Since the functions $\xi^i_{(k)}$ are continuous, there is a neighbourhood N_2 of P_0 such that

$$\det(\xi^i_{(k)}) \neq 0 \text{ on } N_2. \tag{5.66}$$

Now we consider the system

$$\frac{\partial x^i}{\partial x^{*k}} = \xi^i_{(k)}(x) \tag{5.67}$$

of differential equations. Since the Christoffel symbols are symmetric and

$$\frac{\partial^2 x^i}{\partial x^{*k} \partial x^{*j}} = \frac{\partial}{\partial x^l} \left(\xi^i_{(k)}(x) \right) \frac{\partial x^l}{\partial x^{*j}} = \frac{\partial \xi^i_{(k)}}{\partial x^l} \xi^l_{(j)} = - \begin{Bmatrix} i \\ sl \end{Bmatrix} \xi^s_{(k)} \xi^l_{(j)},$$

the system (5.67) satisfies the condition of integrability. Thus there exist solutions $x^i(x^{*k})$ of (5.67) in some neighbourhood N of P_0. It follows from (5.66) that the inverse functions $x^{*k}(x^i)$ exist. They define an admissible transformation of coordinates. We have for the vector fields with respect to the new coordinates by the transformation formula for the components of contravariant vectors

$$\xi^l_{(k)} = \frac{\partial x^l}{\partial x^{*i}} \xi^{*i}_{(k)}.$$

Using (5.67), we obtain

$$\xi^l_{(i)} \xi^{*i}_{(k)} = \frac{\partial x^l}{\partial x^{*i}} \xi^{*i}_{(k)} = \xi^l_{(k)},$$

and so, by (5.66), $\xi^{*i}_{(k)} = \delta^i_k$ on N. Since the vector fields are parallel, we conclude

$$0 = \frac{\partial \xi^{*i}_{(k)}}{\partial x^j} + \begin{Bmatrix} i \\ jr \end{Bmatrix}^* \xi^{*r}_{(k)} = 0 + \begin{Bmatrix} i \\ jr \end{Bmatrix}^* \delta^r_k = \begin{Bmatrix} i \\ jk \end{Bmatrix}^*.$$

Thus the Christoffel symbols with respect to the new coordinates vanish on all of N. □

We saw in Remark 5.6.4 (f) that the tangents of a geodesic line are parallel to each other. Therefore, we may define *autoparallel curves*.

Definition 5.9.2. A curve in a Riemann space is called an *autoparallel curve* if its tangents are parallel in the sense of Levi–Cività.

The next results establishes the local existence of an autoparallel curve through every point of a Riemman space in each direction.

Theorem 5.9.3. *There is an autoparallel curve through every point P of a Riemann space in each direction. The autoparallel curves are uniquely defined in a neighbourhood of P.*

Proof. For sufficiently small t, the system

$$\frac{d^2 x^i}{dt^2} + \left\{ \begin{matrix} i \\ jk \end{matrix} \right\} \frac{dx^j}{dt} \frac{dx^k}{dt} = 0 \ (i = 1, 2, \ldots, n) \tag{5.68}$$

has a unique solution $x^i(t)$ that satisfies the conditions

$$x^i(t_0) = \underset{0}{x^i} \text{ and } \frac{dx^i}{dt}(t_0) = \underset{0}{\xi^i} \ (i = 1, 2, \ldots, n).$$

It remains to be shown that the same solution satisfies the initial conditions

$$x^i(t_1) = \underset{0}{x^i} \text{ and } \frac{dx^i}{dt}(t_1) = \lambda \cdot \underset{0}{\xi^i} \ (i = 1, 2, \ldots, n).$$

Indeed, for each i, $x^i(t) = x^i(\lambda(t - t_1) + t_0)$ is a solution for these initial conditions. This is the only solution by the uniqueness theorem for differential equations, since the new curve differs from the old one only by the choice of parameters. □

Now let $t = t(\tau)$ be a parameter transformation for an autoparallel curve. Then $x^i(t) = x^i(\tau(t))$ implies

$$\frac{dx^i}{dt} = \frac{dx^i}{d\tau} \frac{d\tau}{dt}, \quad \frac{d^2 x^i}{dt^2} = \frac{d^2 x^i}{d\tau^2} \left(\frac{d\tau}{dt} \right)^2 + \frac{dx^i}{d\tau} \frac{d^2 \tau}{dt^2},$$

and so by (5.68)

$$\frac{d^2 x^i}{d\tau^2} = \frac{1}{(d\tau/dt)^2} \left(\frac{d^2 x^i}{dt^2} - \frac{dx^i}{d\tau} \frac{d^2 \tau}{dt^2} \right)$$

$$= \frac{1}{(d\tau/dt)^2} \left(- \left\{ \begin{matrix} i \\ jk \end{matrix} \right\} \frac{dx^j}{d\tau} \frac{dx^k}{d\tau} \left(\frac{d\tau}{dt} \right)^2 - \frac{dx^i}{d\tau} \frac{d^2 \tau}{dt^2} \right).$$

Thus (5.68) transforms to

$$\frac{d^2 x^i}{d\tau^2} + \left\{ \begin{matrix} i \\ jk \end{matrix} \right\} \frac{dx^j}{d\tau} \frac{dx^k}{d\tau} = - \left(\frac{dx^i}{d\tau} \frac{d^2 \tau}{dt^2} \right) \frac{1}{(d\tau/dt)^2}. \tag{5.69}$$

A parameter τ for which the right hand side in (5.69) vanishes is called a *natural parameter of the autoparallel curves*. Since $dx^i/d\tau \neq 0$ for admissible parameter transformations, natural parameters τ are characterized by the condition $d^2\tau/dt^2 = 0$. This yields the following result.

Theorem 5.9.4. *If t is a natural parameter of an autoparallel curve, then all other natural parameters τ satisfy*

$$\tau = c_0 t + c_1,$$

where $c_0 \neq 0$ and c_1 are constants.

We close this section with the differential equations for autoparallel curves.

Theorem 5.9.5. *A solution of the system*

$$\frac{d^2 x^i}{d\tau^2} + \left\{ \begin{array}{c} i \\ jk \end{array} \right\} \frac{dx^j}{d\tau} \frac{dx^k}{d\tau} = -\mu(\tau) \frac{dx^i}{d\tau} \quad (i = 1, 2, \ldots, n)$$

is an autoprallel curve.

Proof. We introduce a new parameter t by

$$\frac{d^2\tau}{dt^2} = \mu(\tau) \left(\frac{d\tau}{dt} \right)^2.$$

The assertion now follows from (5.69). □

5.10   AN EXTENSION OF FRENET'S FORMULAE

In Section 1.7, we proved Frenet's formulae in Euclidean $\mathbb{E}^3$ which gave the derivatives of the vectors of the trihedron of a curve in terms of the tangent, principal normal and binormal vectors. Now we study an extension of Frenet's formulae for curves in a Riemann space.

Theorem 5.10.1. *Let γ be a curve with a parametric representation $\vec{x}(s) = \{x^1(s), x^2(s), \ldots, x^n(s)\}$ of class C^{n+1} in a Riemann space with the metric tensor g_{ik}, and s be the arc length along γ. We define the vectors $\vec{x}_{(k)} = \{\xi^1_{(k)}, \xi^2_{(k)}, \ldots, \xi^n_{(k)}\}$ for $k = 1, 2, \ldots, n$ by*

$$\xi^i_{(k)} = \begin{cases} \dfrac{dx^i}{ds} & (k = 1) \\[2ex] \xi^i_{(k)} = \dfrac{\delta \xi^i_{(k-1)}}{\delta s} = \dfrac{d\xi^i_{(k-1)}}{ds} + \left\{ \begin{array}{c} i \\ jl \end{array} \right\} \xi^l_{(k-1)} \dfrac{dx^j}{ds} & (2 \leq k \leq n) \end{cases} \quad (i = 1, 2, \ldots, n),$$

and assume that the vectors $\vec{x}_{(k)}$ are linearly independent.
Let $\vec{y}_{(k)} = \{\eta^1_{(k)}, \eta^2_{(k)}, \ldots, \eta^n_{(k)}\}$ $(k = 1, 2, \ldots, n)$ denote the vectors obtained from the vectors $\vec{x}_{(k)}$ $(k = 1, 2, \ldots, n)$ by the Gram–Schmidt orthonormalization process. We put

$$\kappa_{(j)} = g_{ik} \frac{\delta \eta^i_{(j)}}{\delta s} \eta^k_{(j+1)} \quad (j = 1, 2, \ldots, n-1).$$

Then we have

$$
\begin{cases}
\dfrac{\delta \eta_{(1)}^i}{\delta s} = & \kappa_{(1)} \eta_{(2)}^i \\[2mm]
\dfrac{\delta \eta_{(2)}^i}{\delta s} = & -\kappa_{(1)} \eta_{(1)}^i + \kappa_{(2)} \eta_{(3)}^i \\[1mm]
\cdots & \cdots \\[1mm]
\dfrac{\delta \eta_{(j)}^i}{\delta s} = & -\kappa_{(j-1)} \eta_{(j-1)}^i + \kappa_{(j)} \eta_{(j+1)}^i \\[1mm]
\cdots & \cdots \\[1mm]
\dfrac{\delta \eta_{(n)}^i}{\delta s} = & -\kappa_{(n-1)} \eta_{(n-1)}^i
\end{cases}
\tag{5.70}
$$

for $i = 1, 2, \ldots, n$.

Proof. The vectors $\vec{y}_{(k)}$ are linear combinations of the vectors $\vec{x}_{(1)}, \vec{x}_{(2)}, \ldots, \vec{x}_{(k)}$ by construction. Furthermore, $\delta \eta_{(k)}^i / \delta s$ is a linear combination of $\xi_{(1)}^i, \xi_{(2)}^i, \ldots, \xi_{(k+1)}^i$. Since each component $\xi_{(j)}^i$ can be given in terms of $\eta_{(r)}^i$ for $r \le j$, we may write

$$
\frac{\delta \eta_{(k)}^i}{\delta s} = \sum_{j=1}^{k+1} a_{kj} \eta_{(j)}^i
\tag{5.71}
$$

with the coefficients a_{kj} yet to be determined. The orthonormality of the vectors $\vec{y}_{(k)}$ implies $g_{ik} \eta_{(j)}^i \eta_{(l)}^k = \delta_{jl}$. From Ricci's theorem, Theorem 5.2.6, we conclude

$$
0 = \delta_{jl;m} = (g_{ik} \eta_{(j)}^i \eta_{(l)}^k)_{;m} = g_{ik} \eta_{(j);m}^i \eta_{(l)}^k + g_{ik} \eta_{(j)}^i \eta_{(l);m}^k.
$$

Multiplying this by dx^m/ds and summing with respect to m, we obtain from (5.71), since

$$
\eta_{(j);m}^i \frac{dx^m}{ds} = \frac{\delta \eta_{(j)}^i}{\delta s},
$$

$$
0 = g_{ik} \frac{\delta \eta_{(j)}^i}{\delta s} \eta_{(l)}^k + g_{ik} \eta_{(j)}^i \frac{\delta \eta_{(l)}^k}{\delta s} = g_{ik} \sum_{s=1}^{k+1} a_{js} \eta_{(s)}^i \eta_{(l)}^k + g_{ik} \sum_{s=1}^{k+1} a_{ls} \eta_{(s)}^k \eta_{(j)}^i
$$

$$
= \sum_{s=1}^{k+1} a_{js} g_{ik} \eta_{(s)}^i \eta_{(l)}^k + \sum_{s=1}^{k+1} a_{ls} g_{ik} \eta_{(s)}^k \eta_{(j)}^i = \sum_{s=1}^{k+1} (a_{js} \delta_{sl} + a_{ls} \delta_{sj})
$$

$$
= a_{jl} + a_{lj}.
$$

Consequently, the matrix (a_{jl}) is skew symmetric. Since $a_{lj} = 0$ for $j \ge l + 2$, at most the entries $a_{j,j-1}$ and $a_{j,j+1}$ are unequal to zero. We put $\kappa_{(j)} = a_{j,j+1}$. Then we obtain

$$
\kappa_{(j)} = g_{ik} \frac{\delta \eta_{(j)}^i}{\delta s} \eta_{(j+1)}^k \quad (j = 1, 2, \ldots, n - 1).
$$

This shows the identities in (5.70). The values $\kappa_{(j)}$ are invariants connected with the curve, since all terms in their definitions are components of vectors. $\qquad \square$

Remark 5.10.2. *(a) In Euclidean* $\mathbb{E}^3$, *we have* $\left\{ {}^{i}_{jk} \right\} = 0$ *for all* i, j *and* k, *and so*

$$\vec{x}_{(1)} = \frac{d\vec{x}}{ds} = \vec{v}_1 = \vec{y}_{(1)} \ and \ \vec{x}_{(2)} = \frac{d\vec{x}_{(1)}}{ds}.$$

Since $\vec{x}_{(1)} \bullet \vec{x}_{(2)} = 0$, *we may choose*

$$\vec{y}_{(2)} = \frac{\vec{x}_{(2)}}{\|\vec{x}_{(2)}\|} = \vec{v}_2 \ and \ \vec{y}_{(3)} = \vec{y}_{(1)} \times \vec{y}_{(2)} = \vec{v}_1 \times \vec{v}_2 = \vec{v}_3.$$

Furthermore, we have

$$\kappa_{(1)} = g_{ik} \frac{\delta \eta^i_{(1)}}{\delta s} \eta^k_{(2)} = \sum_{k=1}^{3} \dot{v}_1^k v_2^k = \dot{\vec{v}}_1 \bullet \vec{v}_2,$$

hence $\kappa_{(1)} = \kappa$, *the curvature, and*

$$\kappa_{(2)} = g_{ik} \frac{\delta \eta^i_{(2)}}{\delta s} \eta^k_{(3)} = \dot{\vec{v}}_2 \bullet \vec{v}_3,$$

hence $\kappa_{(2)} = \tau$, *the torsion. Thus the identities in (5.70) reduce to Frenet's formulae of Section 1.7.*

(b) If S *is a surface in Euclidean* $\mathbb{E}^3$, *then there is only the first identity in (5.70) and* $\kappa_{(1)} = \kappa_g$, *the geodesic curvature. We know from (3.14) in the proof of Theorem 3.1.8 that*

$$\kappa_g \vec{t} = \left(\ddot{u}^i + \left\{ {}^{i}_{jk} \right\} \dot{u}^j \dot{u}^k \right) \vec{x}_i,$$

and so the assertion follows from the definition of the components $\xi^i_{(2)}$.

5.11 RIEMANN NORMAL COORDINATES AND THE CURVATURE OF SPACES

Cartesian coordinates play an important role in Euclidean space. Very useful coordinates can be introduced in Riemann spaces. They have many properties similar to those of Cartesian coordinates and are referred to as *Riemann normal coordinates*.

Let P_0 be a point in a Riemann space with the metric tensor g_{ik} and let the coordinates x^i $(i = 1, 2, \ldots, n)$ be *geodesic in* $P_0 = (x) = (x^1, x^2, \ldots, x^n)$, that is, $\left\{ {}^{i}_{jk} \right\}(x) = 0$. There is one and only one geodesic line through P_0 in the direction of a unit vector with the components ξ^i $(i = 1, 2, \ldots, n)$. Let $\vec{x}(s) = \{x^1(s), x^2(s), \ldots, x^n(s)\}$ be a parametric representation of the geodesic line γ with $x^i(0) = x^i_0$, $\dot{x}^i(0) = \xi^i$ $(i = 1, 2, \ldots, n)$ and s be the arc length along γ. The Taylor expansion yields

$$x^i(s) = x^i_0 + s\dot{x}^i(0) + \frac{s^2}{2}\ddot{x}^i(0) + \frac{s^3}{3!}\dddot{x}^i(0) + \cdots \quad (i = 1, 2, \ldots, n). \tag{5.72}$$

The differential equations for geodesic lines

$$\ddot{x}^i = -\left\{{i \atop jk}\right\}\dot{x}^j\dot{x}^k \; (i = 1, 2, \ldots, n)$$

yield

$$\dddot{x}^i = -\left\{{i \atop jk}\right\}_{,l}\dot{x}^l\dot{x}^j\dot{x}^k - \left\{{i \atop jk}\right\}\ddot{x}^j\dot{x}^k - \left\{{i \atop jk}\right\}\dot{x}^j\ddot{x}^k$$

$$= -\left\{{i \atop jk}\right\}_{,l}\dot{x}^l\dot{x}^j\dot{x}^k - 2\left\{{i \atop jk}\right\}\ddot{x}^j\dot{x}^k,$$

$$\ddot{x}^i(0) = -\left\{{i \atop jk}\right\}(\underset{0}{x})\xi^j\xi^k = 0$$

and

$$\dddot{x}^i(0) = -\left\{{i \atop jk}\right\}_{,l}(\underset{0}{x})\xi^l\xi^j\xi^k.$$

for $(i = 1, 2, \ldots, n)$. Therefore (5.72) becomes

$$x^i(s) = \underset{0}{x}^i + s\xi^i - \frac{1}{3!}s^3\left\{{i \atop jk}\right\}_{,l}(\underset{0}{x})\xi^l\xi^j\xi^k + s^4(\cdots) \; (i = 1, 2, \ldots, n)$$

for sufficiently small s. We put $x^{*i} = s\xi^i$. Then a coordinate transformation is given in a neighbourhood of P_0 by

$$x^i = \underset{0}{x}^i + x^{*i} - \frac{1}{3!}\left\{{i \atop jk}\right\}_{,l}(\underset{0}{x})x^{*j}x^{*k}x^{*l} + \cdots \; (i = 1, 2, \ldots, n).$$

Furthermore, we have

$$\frac{\partial x^i}{\partial x^{*m}} = \delta_m^i - \frac{1}{3!}\left\{{i \atop jk}\right\}_{,l}(\underset{0}{x})\delta_m^{*j}x^{*k}x^{*l} - \frac{1}{3!}\left\{{i \atop jk}\right\}_{,l}(\underset{0}{x})x^{*j}\delta_m^{*k}x^{*l}$$

$$- \frac{1}{3!}\left\{{i \atop jk}\right\}_{,l}(\underset{0}{x})x^{*j}x^{*k}\delta_m^{*l} + \cdots \; (i, m = 1, 2, \ldots).$$

Since we have $x^{*i}(\underset{0}{x}^i) = 0$ at $x^i = \underset{0}{x}^i$ by the definition of x^{*i}, it follows that the Jacobian of the transformation is equal to one at $(x) = (\underset{0}{x})$, hence unequal to zero in some neighbourhood of P_0, and so the transformation is locally admissible.

The differential equations for the geodesic lines with respect to the new coordinates are

$$\ddot{x}^{*i} + \left\{{i \atop jk}\right\}^* x^{*j}x^{*k} = 0 \; (i = 1, 2, \ldots, n).$$

We put $x^{*i} = s\xi^i$ $(i = 1, 2, \ldots, n)$, and obtain

$$\left\{{i \atop jk}\right\}^* \xi^i\xi^k = 0 \; (i = 1, 2, \ldots, n), \tag{5.73}$$

and multiplication with s yields

$$\left\{ \begin{matrix} i \\ jk \end{matrix} \right\}^* (x^*) x^{*j} x^{*k} = 0 \ (i = 1, 2, \ldots, n).$$

The geodesic lines through P_0 are given by the linear equations

$$x^{*i} = \xi^i s \quad (i = 1, 2, \ldots, n)$$

with respect to the so–called *Riemann normal coordinates* x^{*i} *with centre at* P_0, in analogy of the case of straight lines in Euclidean space.

The following result shows that the Gaussian curvature of a surface may easily be found by measuring lengths on the surface. Furthermore, its proof contains a very illustrating proof for the *theorema egregium*, Theorem 3.8.3.

Theorem 5.11.1 (Bertrand–Puiseux (1848)).

Let x^{*1} and x^{*2} be Riemann normal coordinates with centre at a point P for a two-dimensional Riemann space. We introduce geodesic polar coordinates u^1 and u^2 by

$$x^{*1} = u^1 \cos u^2 \text{ and } x^{*2} = u^1 \sin u^2,$$

where u^1 is the distance from P measured along the geodesic lines through P and u^2 is the angle between a geodesic line and a given direction. If K_0 denotes the Gaussian curvature at P and $L(r)$ is the circumference of a geodesic circle of radius r, that is, the length of the u^2–line for $u^1 = r$, then

$$K_0 = \lim_{r \to 0} \frac{3(2\pi r - L(r))}{\pi r^3}. \tag{5.74}$$

Proof. We obtain from the transformation formulae

$$g_{22} = g^*_{ik} \frac{\partial x^{*i}}{\partial u^2} \frac{\partial x^{*k}}{\partial u^2} = g^*_{11} \left(\frac{\partial x^{*1}}{\partial u^2} \right)^2 + 2 g^*_{12} \frac{\partial x^{*1}}{\partial u^2} \frac{\partial x^{*2}}{\partial u^2} + g^*_{22} \left(\frac{\partial x^{*2}}{\partial u^2} \right)^2$$

$$= (u^1)^2 \left(g^*_{11} \sin^2 u^2 - 2 g^*_{12} \sin u^2 \cos u^2 + g^*_{22} \cos^2 u^2 \right).$$

This implies

$$\lim_{u^1 \to 0} g_{22} = 0.$$

Furthermore, we have

$$g_{22,1} = \frac{2}{u^1} g_{22} + (u^1)^2 \left(\left(g^*_{11,1} \frac{\partial x^{*1}}{\partial u^1} + g^*_{11,2} \frac{\partial x^{*2}}{\partial u^1} \right) \sin^2 u^2 \right.$$

$$\left. - 2 \left(g^*_{12,1} \frac{\partial x^{*1}}{\partial u^1} + g^*_{12,2} \frac{\partial x^{*2}}{\partial u^1} \right) \sin u^2 \cos u^2 + \left(g^*_{22,1} \frac{\partial x^{*1}}{\partial u^1} + g^*_{22,2} \frac{\partial x^{*2}}{\partial u^1} \right) \cos^2 u^2 \right).$$

Since (5.73) holds for all directions at P, it follows that $\left\{ \begin{matrix} i \\ jk \end{matrix} \right\}^* (P_0) = 0$ for all i, j and k, and so $g^*_{ij,k}(P) = 0$ for all i, j and k by (4.24) and (4.25). Now

$$\left(\sqrt{g_{22}} \right)_{,1} = \frac{1}{2} \frac{g_{22,1}}{\sqrt{g_{22}}}$$

implies

$$\lim_{u^1 \to 0} \left(\sqrt{g_{22}}\right)_{,1} = 0.$$

Since it holds in geodesic coordinates that

$$\frac{\partial^2 \left(\sqrt{g_{22}}\right)}{(\partial u^1)^2} = -K\sqrt{g_{22}}$$

for the Gaussian curvature K by (3.104) of Lemma 3.8.5, we also have

$$\frac{\partial^2 \left(\sqrt{g_{22}}\right)}{(\partial u^1)^2} \to 0 \; (u^1 \to 0)$$

and

$$\frac{\partial^3 \left(\sqrt{g_{22}}\right)}{(\partial u^1)^3} = -\frac{\partial K}{\partial u^1}\sqrt{g_{22}} - K(\sqrt{g_{22}})_{,1} \to -K_0 \; (u^1 \to 0),$$

where K_0 is the Gaussian curvature at the point P. The Taylor expansion at P yields

$$\sqrt{g_{22}} = u^1 - \frac{1}{6}K_0(u^1)^3 + (u^1)^4(\cdots)$$

and

$$g_{22} = (u^1)^2 - \frac{1}{3}K_0(u^1)^4 + (u^1)^5(\cdots).$$

Finally, we have for the circumference of the geodesic circle corresponding to $u^1 = r$

$$L(r) = \int_0^{2\pi} \sqrt{g_{22}}\, du^2 = 2\pi r - \frac{1}{3}K_0\pi r^3 + r^4(\cdots),$$

hence (5.74) follows. □

Now we deal with the *curvature of space*.
Let $\vec{x}$ and $\vec{y}$ be two vectors with the components ξ^i and η^i $(i = 1, 2, \ldots, n)$ in the tangent space at P_0. We introduce Riemann normal coordinates x^{*i} $(i = 1, 2, \ldots, n)$ with the centre at P_0 such that $\vec{x}$ and $\vec{y}$ are the directions of the x^{*1}– and x^{*2}–lines, respectively. The geodesic lines with the tangent vectors $\alpha\vec{x} + \beta\vec{y}$ at P_0 span a surface in a neighbourhood of P_0 which is given by the equations

$$x^{*3} = \cdots = x^{*n} = 0.$$

The Gaussian curvature of this surface which depends on x^{*1} and x^{*2} only is given by

$$K = \frac{g_{2j}R^j_{112}}{g} \; \text{(cf. (3.101) in Theorem 3.8.3).}$$

We write $R_{ijkl} = g_{ir}R^r_{jkl}$ and use the fact that $R^i_{hjk} + R^i_{hkj} = 0$ which is an immediate consequence of the definition of the tensor of curvature to conclude

$$K = -\frac{R_{2121}}{g_{11}g_{22} - g_{12}g_{21}} = -\frac{R_{2121}\xi^1\eta^2\xi^1\eta^2}{(g_{11}g_{22} - g_{12}g_{21})\xi^1\eta^2\xi^1\eta^2}. \tag{5.75}$$

We have in arbitrary coordinates

$$K = \frac{R_{ijkl}\eta^i\eta^k\xi^j\xi^l}{(g_{il}g_{jk} - g_{ik}g_{jl})\eta^i\eta^k\xi^j\xi^l},$$ (5.76)

since this invariant term reduces to (5.75) in the special coordinates. The term in (5.76) is the Gaussian curvature of the two-dimensional surface spanned by the geodesic lines through P_0 in the directions that are linear combinations of the given vectors $\vec{x}$ and $\vec{y}$ in the tangent space at P_0; it is called the *curvature of the space at P_0 in the two-dimensional direction spanned by $\vec{x}$ and $\vec{y}$.*

The *Ricci tensor* R_{jk} is a contraction of the tensor of curvature, namely

$$R_{jk} = R^i_{jki},$$

and the *scalar of curvature* R is defined by

$$R = g^{jl}R_{jl}.$$

The Ricci tensor and the scalar of curvature can be interpreted as certain means of the Gaussian curvature. Let $\vec{x}_{(1)}, \vec{x}_{(2)}, \ldots, \vec{x}_{(n)}$ be orthonormal vectors with the components $\xi^i_{(k)}$ $(i = 1, 2, \ldots, n)$ at a point P with the coordinates $(x) = (x^1, x^2, \ldots, x^n)$, and let $K(r,s)$ be the curvature in (5.76) which corresponds to the direction of the plane spanned by the vectors $\vec{x}_{(r)}$ and $\vec{x}_{(s)}$. Then we obtain

$$K(r,s) = R_{ijkl}\xi^i_{(s)}\xi^l_{(s)}\xi^j_{(r)}\xi^k_{(r)}.$$ (5.77)

We have by (4.18)

$$g^{il} = \sum_{s=1}^{n} \xi^i_{(s)}\xi^l_{(s)},$$

and so, by the definition of the Ricci tensor,

$$\sum_{s=1}^{n} K(1,s) = \sum_{s=1}^{n} R_{ijkl}\xi^i_{(s)}\xi^l_{(s)}\xi^j_{(1)}\xi^k_{(1)} = g^{il}R_{ijkl}\xi^j_{(1)}\xi^k_{(1)}$$

$$= g^{il}g_{ir}R^r_{jkl}\xi^j_{(1)}\xi^k_{(1)} = \delta^l_r R^r_{jkl}\xi^j_{(1)}\xi^k_{(1)} = R^l_{jkl}\xi^j_{(1)}\xi^k_{(1)}$$

$$= R_{jk}\xi^j_{(1)}\xi^k_{(1)}.$$

Since $K(1,1) = 0$, the term

$$\frac{1}{n-1}R_{jk}\xi^j_{(1)}\xi^k_{(1)}$$

is the arithmetic mean of the curvatures corresponding to $n-1$ orthogonal planes to $\vec{x}_{(1)}$. The mean is independent of the choice of the vectors $\vec{x}_{(2)}, \ldots, \vec{x}_{(n)}$. Furthermore, we have

$$\sum_{r,s=1}^{n} K(r,s) = \sum_{r=1}^{n} R_{jk}\xi^j_{(r)}\xi^k_{(r)} = g^{jk}R_{jk} = R.$$

Consequently $R/(n(n-1))$ is equal to the arithmetic mean of all curvatures that correspond to all possible directions of planes that can be constructed from n orthonormal vectors. In particular, for $n = 2$, there is only one direction of a plane, and

$$K = \frac{R}{2} = \frac{1}{2} g^{ik} g^{il} R_{ijkl}.$$

Bibliography

[1] A. Alotaibi, E. Malkowsky, and F. Özger. Some notes on matrix mappings and the Hausdorff measure of noncompactness. *Filomat*, 28(5):1057–1072, 2014.

[2] F. Başar, B. Altay, and E. Malkowsky. Matrix transformations on some sequence spaces related to strong Cesàro summability and boundedness. *Publ. Math. Debrecen*, 73(1–2):193–213, 2008.

[3] F. Başar, B. Altay, and E. Malkowsky. Matrix transformations on the matrix domains of strongly C_1–summable and bounded sequences. *Appl. Math. Comput.*, 211:255–264, 2009.

[4] B. de Malafosse and E. Malkowsky. On the Banach algebra $(w_\infty(\Lambda), w_\infty(\Lambda))$ and applications to the solvability of matrix equations in $w_\infty(\Lambda)$. *Publ. Math. Debrecen*, 85(12):197–217, 2014.

[5] B. de Malafosse and E. Malkowsky. On the spectra of the operator $B(\tilde{r}, \tilde{s})$ mapping in $(w_\infty(\lambda))_a$ and $(w_0(\lambda))_a$ where λ is a nondecreasing exponentially bounded sequence. *Mat. Vesnik*, 72(1):30–42, 2020.

[6] I. Djolović and E. Malkowsky. The Hausdorff measure of noncompactness of operators in the matrix domains of triangles in the spaces of C_1 summable and bounded sequences. *Appl. Math. Comput.*, 216::1122–1130, 2010.

[7] I. Djolović and E. Malkowsky. On matrix mappings into some strong Cesàro sequence spaces. *Appl. Math. Comput.*, 218:6155–6163, 2012.

[8] I. Djolović and E. Malkowsky. Banach algebras of matrix transformations between some sequence spaces related to Λ strong convergence and boundedness. *Appl. Math. Comput.*, 219:8779–8789, 2013.

[9] I. Djolović and E. Malkowsky. Generalization of some results on $p\alpha$–duals. *Banach J. Math. Anal.*, 8(2):124–130, 2014.

[10] I. Djolović and E. Malkowsky. A note on some results related to infinite matrices on weighted ℓ_1 spaces. *Funct. Anal. Approx. Comput.*, 8(1):1–5, 2016.

[11] I. Djolović and E. Malkowsky. Compactness of multiplication, composition and weighted composition operators between some classical sequence spaces: A new approach. *Rocky Mountain J. Math.*, 47(8):2545–2564, 2017.

[12] I. Djolović, E. Malkowsky, and K. Petković. Two methods for the characterization of compact operators between BK spaces. *Banach J. Math. Anal.*, 9(3):1–13, 2015.

[13] I. Djolović, E. Malkowsky, and K. Petković. New approach to some results related to mixed norm spaces. *Filomat*, 30(1):83–88, 2016.

[14] I. Djolović, K. Petković, and E. Malkowsky. Matrix mappings and general bounded linear operators on the space bv. *Math. Slovaca*, 68(2):405–414, 2018.

[15] Ć. Dolićanin and A. B. Antonevich. *Dynamical Systems Generated by Linear Maps*. Academic Mind, State University of Novi Pazar; Springer Verlag, Berlin Heidelberg, 2014.

[16] Ć. Dolićanin, B. Iričanin, M. Vujisić, and P. Osmokrović. Monte carlo simulations of proton and ion beam irradiation on titanium dioxide memristors. In *PIERS 2010 XI'AN: Progress in Electromagnetics Research Symposium Proceedings*, volume 1–2, pages 1209–1214, 2010.

[17] Ć. Dolićanin, E. Malkowsky, and V. Veličković. *Diferencijalna Geometrija i Njena Vizuelizacija*. Državni Univerzitet u Novom Pazaru; Akademska Misao, Beograd, 2021. ISBN: 978-86-81506-09-7.

[18] Ć. Dolićanin, K. Stanković, D. Dolićanin, and B. Lončar. Statistical treatment of nuclear counting results. *Nucl. Technol. Radiat. Prot.*, 26(2):164–170, 2011.

[19] Ć. Dolićanin and A. A. Tikhonov. On dynamical equations in s–parameters for rigid body attitude motion. In *International Conference on Mechanics. Seventh Polyakhovs Reading*, 2015.

[20] K. Endl. *Kreative Computergrafik*. Würfel–Verlag, 1986.

[21] K. Endl and R. Endl. *Computergraphik 1 – Eine Software zur Geometrie in Turbo–Pascal*. Würfel–Verlag, 1989.

[22] K. Endl and R. Endl. *Computergraphik 2 – Eine Software zur Geometrie objektorientierter Programmierung mit Turbo–Pascal*. Würfel–Verlag, 1991.

[23] K. Endl and E. Malkowsky. *Aufgaben zur analytischen Geometrie und linearen Algebra mit einer Einführung in die 3D–Computergrafik in Turbo–Pascal*. VDI–Verlag Düsseldorf, 1987.

[24] J. H. Eschenburg and J. Jost. *Differentialgeometrie und Minimalflächen*. Springer Spektrum, 2014.

[25] M. Failing. *Entwicklung numerischer Algorithmen zur computergrafischen Darstellung spezieller Probleme der Differentialgeometrie und Kristallographie*. Shaker Verlag, Aachen, Germany, 1996.

[26] M. Failing and E. Malkowsky. Ein effizienter Nullstellenalgorithmus zur computergrafischen Darstellung spezieller Kurven und Flächen. *Mitt. Math. Sem. Giessen*, 229:11–28, 1996.

[27] W. Gander. *Learning MATLAB: A Problem Solving Approach.* Springer, 2015.

[28] A. Gray. *Modern Differential Geometry of Curves and Surfaces.* CRC Press, 1993.

[29] A. Gray. *Modern Differential Geometry of Curves and Surfaces with Mathematica.* CRC Press, second edition, 1997.

[30] A. Gray, E. Abbena, and S. Salamon. *Modern Differential Geometry of Curves and Surfaces with Mathematica.* Studies in Advanced Mathematics. Chapman and Hall, CRC Press, 2006.

[31] H. W. Guggenheimer. *Differential Geometry.* Dover Publications, 1977.

[32] W. Haack. *Differential–Geometrie.* Wolfenbütteler Verlagsanstalt G.m.b.H., Wolfenbüttel und Hannover, 1949.

[33] S. Helgason. *Differential Geometry and Symmetric Spaces.* in Press, New York, 1962.

[34] K. Hildebrandt and K. Polthier. Generalized shape operators on polyhedral surfaces. *Comput. Aided Geom. Design*, 28(5):321–343, 2011.

[35] T. Hoffmann, M. Kilian, K. Leschke, and F. Martin, editors. *Minimal Surfaces: Integrable Systems and Visualisation.* Springer International Publishing, 2021.

[36] M. M. M. Jaradat, Z. Mustafa, A. H. Ansari, S. Chandok, and Ć. Dolićanin. Some approximate fixed point results and application on graph theory for partial $(h-F)$–generalized convex contraction mappings with special class of functions on complete metric space. *Journal of Nonlinear Sciences and Applications*, 10(4):1695–1708, 2017.

[37] J. Jost. *Differentialgeometrie und Minimalflächen.* Springer–Verlag, Berlin, Heidelberg, 1994.

[38] E. Kacapor, T. Atanacković, and Ć. Dolićanin. Optimal shape and first integrals for inverted compressed column. *Mathematics*, 8(3):334, 2020.

[39] W. Klingenberg. *Eine Vorlesung über Differentialgeometrie.* Springer–Verlag, 1973.

[40] P. Knežević, D. Šumarac, Z. Perović, Ć. Dolićanin, and Z. Burzić. A preisach model for monotonic tension response of structural mild steel with damage. *Periodica Polytechnica – Civil Engineering*, 64(1):296–303, 2020.

[41] S. Kobayashi. *Differential Geometry of Curves and Surfaces.* Springer, 2019.

[42] E. Kreyszig. *Differential Geometry.* Dover Publications, Inc., New York, first edition, 1991.

[43] W. Kühnel. *Differentialgeometrie, Kurven–Flächen–Mannigfaltigkeiten.* Springer Spektrum, 2013.

[44] C. Lange and K. Polthier. Anisotropic smoothing of point sets. *Comput. Aided Geom. Design*, 22(7):680–692, 2005.

[45] D. Laugwitz. *Differentialgeometrie.* B. G. Teubner, Stuttgart, 1977.

[46] M. M. Lipschutz. *Schaum's Outline of Differential Geometry.* McGraw–Hill Education, first edition, 1969.

[47] D. Majerek. Application of Geogebra for teaching mathematics. *Adv. Sci. Technol. Res. J.*, 8(24):51–54, 2014.

[48] M. Majewsky. *Getting Started with MuPAD.* Springer–Verlag Berlin Heidelberg, 2005.

[49] E. Malkowsky. Computergrafik in der Differentialgeometrie, eine offene Software zur differentialgeometrie. In *Technologie–Vermittlungs–Agentur Berlin, Dokumentationsband zum Hochschul–Computer–Forum*, 7–9, 1993.

[50] E. Malkowsky. An open software in OOP for computer graphics and some applications in differential geometry. In *Proceedings of the 20th South African Symposium on Numerical Mathematics*, pages 51–80, 1994.

[51] E. Malkowsky. An open sofware in OOP for computer graphics in differential geometry, the basic concepts. *ZAMM Z. Angew. Math. Mech.*, 76(1):467–468, 1996.

[52] E. Malkowsky. Visualisation and animation in mathematics. In *Proceedings of the 2nd German–Serbian Summer School of Modern Mathematical Physics, Kopaonik, Yugoslavia*, 2002. electronic publication.

[53] E. Malkowsky. Visualisation and animation in mathematical education. In *Proceedings of the DAAD Workshop Bauinformatik, Budva, Montenegro, 12–14, September, 2003*, 2003. on CD.

[54] E. Malkowsky. Visualisation and animation in mathematics and physics. *Facta Univ. Ser. Autom. Control Robot.*, 4(16):43–54, 2004.

[55] E. Malkowsky. Visualisation and animation in mathematics and physics. In *Proceedings of the Institute of Mathematics of NAS of Ukraine*, volume 50, pages 1415–1422, 2004.

[56] E. Malkowsky. FK spaces, their duals and the visualisation of neighbourhoods. *Appl. Sci. (APPS)*, 8(1):112–127, 2006.

[57] E. Malkowsky. The graphical representation of neighbourhoods in certain topologies. In *Proceedings of the Conference Contemporary Geometry and Related Topics, Belgrade (2006)*, pages 305–324, 2006.

[58] E. Malkowsky. Some compact operators on the Hahn space. *Sci. Res. Comm.*, 1(1):doi: 10.52460/src.2021.001, 2021.

[59] E. Malkowsky and F. Başar. A survey on some paranormed sequence spaces. *Filomat*, 31(4):1099–1122, 2017.

[60] E. Malkowsky and I. Djolović. Compact operators into the spaces of strongly C_1 summable and bounded sequences. *Nonlinear Anal.*, 74(11):3736–3750, 2011.

[61] E. Malkowsky, G. V. Milovanović, V. Rakočević, and O. Tuğ. The roots of polynomials and the operator Δ_i^3 on the Hahn sequence space h. *Comp. Appl. Math.*, 40(222):https://doi.org/10.1007/s40314–021–01611–6, 2021.

[62] E. Malkowsky and W. Nickel. *Computergrafik in der Differentialgeometrie, ein Arbeitsbuch für Studenten, inklusive objektorientierter Software.* Vieweg–Verlag, Wiesbaden, Germany, Braunschweig, 1993.

[63] E. Malkowsky and F. Özger. A note on some sequence spaces of weighted means. *Filomat*, 26(3), 2012.

[64] E. Malkowsky, F. Özger, and V. Veličković. Some spaces related to Cesàro sequence spaces and an application to crystallography. *MATCH Commun. Math. Comput Chem.*, 70(3):867–884, 2013.

[65] E. Malkowsky, F. Özger, and V. Veličković. Matrix transformations on mixed paranorm spaces. *Filomat*, 31(10):2957–2966, 2017.

[66] E. Malkowsky, F. Özger, and V. Veličković. Some mixed paranorm spaces. *Filomat*, 31(4):1079–1098, 2017.

[67] E. Malkowsky and V. Rakočević. On matrix domains of triangles. *Appl. Math. Comput.*, 189:1146–1163, 2007.

[68] E. Malkowsky, V. Rakočević, and O. Tuğ. Compact operators on the Hahn space. *Monatsh. Math.*, 196:519–551, 2021.

[69] E. Malkowsky, V. Rakočević, and V. Veličković. Bounded linear and compact operators between the Hahn space and spaces of strongly summable and bounded sequences. *Bull. Sci. Math. Nat. Sci. Math.*, 45:25–41, 2020.

[70] E. Malkowsky and V. Veličković. An application of functional analysis in computer graphics and crystallography. In *Proceedings of the Conference YUINFO 2000, Kopaonik, Yugoslavia*, 2000. on CD.

[71] E. Malkowsky and V. Veličković. Computer graphics in geometry and differential geometry. In *Proceedings of the Conference moNGeometrija 2000*, pages 188–198, 2000.

[72] E. Malkowsky and V. Veličković. Visualisation of differential geometry. In *Proceedings of the YUSNM 2000 Conference*, 2000.

[73] E. Malkowsky and V. Veličković. Graphics in differential geometry, an approach in OOP. In *Proceedings of the International DYNET Workshop on Modern Programming Concepts and Their Applications in Engineering, Niš, Yugoslavia, 22–24 November, 2001*, pages 1–20, 2001.

[74] E. Malkowsky and V. Veličković. On the Gaussian and mean curvature of certain surfaces. *Novi Sad J. Math.*, 31(1):65–74, 2001.

[75] E. Malkowsky and V. Veličković. Potential surfaces and their graphical representations. *Filomat*, 15:47–54, 2001.

[76] E. Malkowsky and V. Veličković. Representation of parallel and focal surfaces. In *Proceedings of the Conference YUINFO 2001, Kapaonik*, 2001. on CD.

[77] E. Malkowsky and V. Veličković. Some curves and surfaces of given curvature and their graphical representations. In *Proceedings of a Workshop on Computational Intelligence and Information Technologies, Niš, Yugoslavia, 20–21 June, 2001*, pages 9–15, 2001.

[78] E. Malkowsky and V. Veličković. Some geometric properties of screw surfaces and exponential cones. In *Proceeding of the 10th Congress of Yugoslav Mathematicians, Belgrade, 2001*, pages 395–399, 2001.

[79] E. Malkowsky and V. Veličković. Visualisation of differential geometry. *Facta Univ. Ser. Autom. Control Robot.*, 3(11):127–134, 2001.

[80] E. Malkowsky and V. Veličković. The representation of the modulus and argument of complex functions. In *Proceedings of the Conference YUINFO 2002, Kopaonik, Yugoslavia*, 2002. on CD.

[81] E. Malkowsky and V. Veličković. A software for the visualisation of differential geometry. *Visual Mathematics*, 4(1), 2002. Electronic publication, http://www.mi.sanu.ac.rs/vismath/malkovsky/index.htm.

[82] E. Malkowsky and V. Veličković. Visualisation and animation in mathematics. In *Proceedings of the Conference YUINFO 2003, Kopaonik, Yugoslavia*, 2003. on CD.

[83] E. Malkowsky and V. Veličković. Visualisation of isometric maps. *Filomat*, 17:107–116, 2003.

[84] E. Malkowsky and V. Veličković. Analytic transformations between surfaces with animations. In *In Proceedings of the Institute of Mathematics of NAS of Ukraine*, volume 50, pages 1496–1501, 2004.

[85] E. Malkowsky and V. Veličković. Visualisation and animation in differential geometry. In *Proceedings of the Workshop Contemporary Geometry and Related Topics*, pages 301–318. World Scientific Publishing Co., 2004.

[86] E. Malkowsky and V. Veličković. Solutions of some visibility and contour problems in the visualisation of surfaces. *Appl. Sci. (APPS)*, 10:125–140, 2008.

[87] E. Malkowsky and V. Veličković. Topologies of some new sequence spaces, their duals, and the graphical representations of neighborhoods. *Topology Appl.*, 158(12):1369–1380, 2011.

[88] E. Malkowsky and V. Veličković. Some new sequence spaces, their duals and a connection with Wulff's crystals. *MATCH Commun. Math. Comput. Chem.*, 67(3):589–607, 2012.

[89] E. Malkowsky and V. Veličković. The duals of certain matrix domains of factorable triangles and some related visualisations. *Filomat*, 27(5):821–829, 2013.

[90] Z. Marković, J. Dimitrić-Marković, and Ć. Dolićanin. Mechanistic pathways for the reaction of quercetin with hydroperoxy radical. *Theor. Chem. Acc.*, 127(1–2):69–80, 2010.

[91] J. Mikeš. *Differential Geometry of Special Mappings*. Palacký University, Faculty of Science, Olomouc, 2015.

[92] R. S. Millman. *Elements of Differential Geometry*. Pearson, first edition, 1977.

[93] S. Minčić and L. Velimirović. *Generalized Riemannian Spaces and Spaces of Non-Symmetric Affine Connection*. Faculty of Sciences and Mathematics, Niš, 2013.

[94] S. Montiel and A. Ros. *Curves and Surfaces*. American Mathematical Society, 2009.

[95] M. Nieser, C. Schulz, and K. Polthier. Patch layout from feature graphs. *Comput.-Aided Des.*, 42(3):213–220, 2010.

[96] V. Nikolić, Ć. Dolićanin, and D. Dimitrijević. Dynamic model for the stress and strain state analysis of a spur gear transmission. *Strojniski Vestnik – Journal of Mechanical Engineering*, 58(1):56–67, 2012.

[97] V. Nikolić, Ć. Dolićanin, and M. Radojković. Application of finite element analysis of thin steel plate with holes. *Tehnički Vjesnik – Technical Gazette*, 18(1):57–62, 2011.

[98] V. Nikolić, Ć. Dolićanin, M. Radojković, and E. Dolićanin. Stress distribution in an anisotropic plane field weakened by an elliptical hole. *Tehnički Vjesnik – Technical Gazette*, 12(2), 2015.

[99] V. Nikolić-Stanojević, L. Veljović, and Ć. Dolićanin. A new model of the fractional order dynamics of the planetary gears. *Math. Probl. Eng.*, 2013. ID 932150.

[100] J. Nitsche. *Lectures on Differential Geometry, Volume I.* Cambridge University Press, Cambridge, 1989.

[101] B. O'Neill. *Elementary Differential Geometry.* Academic Press, New York, second edition, 2006.

[102] J. Oprea. *Differential Geometry and Its Applications.* American Mathematical Society, third edition, 2019.

[103] P. Osmokrović, N. Arsić, M. Vujisić, K. Stanković, and Ć. Dolićanin. Reliability of three-electrode spark gaps. *Plasma Devices Oper.*, 16(4):235–245, 2008.

[104] P. Osmokrović, G. Ilić, Ć. Dolićanin, K. Stanković, and M. Vujisić. Determination of pulse tolerable voltage in gas-insulated systems. *Jpn. J. Appl. Phys.*, 47(12):8928–8934, 2008.

[105] R. Ossermann. *A Survey of Minimal Surfaces.* Dover Publications, Inc., New York, 1986.

[106] A. N. Pressley. *Elementary Differential Geometry.* Springer, 2010.

[107] S. Radenović, T. Došenović, V. Ozturk, and Ć. Dolićanin. A note on the paper: "Nonlinear integral equations with new admissibility types in b–metric spaces". *Fixed Point Theory Appl.*, 19:2287–2295, 2017.

[108] D. Randjelović, Ć, Dolićanin, and M. Randjelović. Classes of block designs and theirs application in the analysis of agricultural experiment organization. In *Proceedings of the 6th WSEAS International Conference on Applied Computer Science*, pages 596–601, 2007.

[109] D. Randjelović, Ć. Dolićanin, M. Milenković, and J. Stiojković. An application of multiple criteria analysis to assess optimal factor combination in one experiment. In *MABE 09: Proceedings of the 5th WSEAS International Conference on Mathematical Biology and Ecology*, pages 21–26, 2009.

[110] S. Rančić, M. Zlatanović, and N. Velimirović. Cutting patterns of membrane structures. *Filomat*, 29(3):651–660, 2015.

[111] S. Rešić, A. B. Antonevich, and Ć. Dolićanin. Convergence of a trajectory of a vector subspace under the action of a linear map: General case. *Novi Sad J. Math.*, 32(2):149–159, 2077.

[112] V. Rovenski. *Geometry of Curves and Surfaces with MAPLE.* Birkhäuser, Boston, 2000.

[113] G. Scheffers. *Anwendung der Differential– und Inegralrechnung auf Geometrie. Bd. 2, Einführung in die Theorie der Flächen.* W. de Gruyter & Co., Berlin, Leipzig, 1922.

[114] R. Schoen and S. T. Yau. *Lectures on Differential Geometry.* International Press of Boston, 2010.

[115] W. Schöne. *Differentialgeometrie.* BSB. B. G. Teubner, Leipzig, 1978.

[116] T. Sochi. *Introduction to Differential Geometry of Space Curves and Surfaces: Differential Geometry of Curves and Surfaces.* CreateSpace Independent Publishing Platform, 2017.

[117] P. Spalević, S. Panić, Ć. Dolićanin, M. Stefanović, and A. Mosić. SSC diversity receiver over correlated $\alpha - \mu$ fading channels in the presence of Cochannel Interference. *EURASIP Journal on Wireless Communications and Networking,* 2010.

[118] M. Spivak. *Differential Geometry*, volume 3. Publish or Perish, Inc., 1976.

[119] S. Sternberg. *Lectures on Differential Geometry.* American Mathematical Society, second edition, 1999.

[120] S. Sternberg. *Curvature in Mathematics and Physics.* Dover Publication, 2012.

[121] J. J. Stoker. *Differential Geometry.* Wiley–Interscience, New York, 1969.

[122] D. J. Struik. *Lectures on Classical Differential Geometry.* Dover Publications, 1988.

[123] I. A. Taimanov. *Lectures on Differential Geometry.* European Mathematical Society, 2008.

[124] K. Tapp. *Differential Geometry of Curves and Surfaces.* Springer, 2016.

[125] J. A. Thorpe. *Elementary Topics in Differential Geometry.* Springer–Verlag, 1979.

[126] A. A. Tikhonov, Ć. Dolićanin, T. A. Partalin, and I. Arandjelović. A new form of equations for rigid body rotational dynamics. *Tehnički Vjesnik – Technical Gazette,* 21(6), 2014.

[127] V. A. Toponogov. *Differential Geometry of Curves and Surfaces: A Concise Guide.* Birkhäuser, 2006.

[128] L. W. Tu. *Differential Geometry: Connections, Curvature, and Characteristic Classes.* Springer, 2017.

[129] O. Tuğ, V. Rakočević, and E. Malkowsky. Domain of generalised difference operator Δ_i^3 of order 3 in the Hahn sequence space h and matrix ransformations. *Linear and Multilinear Algebra,* page doi: 10.1080/03081087.2021.1992875, 2021.

[130] A. Vasić, P. Osmokrović, M. Vujisić, Ć. Dolićanin, and K. Stanković. Possibilities of improvement of silicon solar cell characteristics by lowering noise. *J. Optoelectron. Adv. Mater.*, 10(10):2800–2804, 2008.

[131] L. Velimirović. *Infinitesimal Bending*. Faculty of Science and Mathematics, Niš, 2009.

[132] L. Velimirović and S. Rančić. Higher order infinitesimal bending of a class of toroids. *European J. Combin.*, 31(4):1136–1147, 2010.

[133] L. Velimirović, S. Rančić, and M. Zlatanović. Rigidity and flexibility analysis of a kind of surfaces of revolution and visualization. *Appl. Math. Comput.*, 217:4612–4619, 2011.

[134] V. Veličković. The basic principles and concepts of a software package for visualisation of mathematics. *BSG Proc. Geom. Balkan Press Bucharest*, 13:192–203, 2006.

[135] V. Veličković. On surfaces of rotation of a given constant Gaussian curvature and their visualisations. In *Proceedings of the Conference Contemporary Geometry and Related Topics, 2006*, pages 523–534, 2006.

[136] V. Veličković. Visualisations of mathematics using line graphics. In *Spring School on Visualisation and Discrete Geometry, Berlin, Germany, 10–13 April, 2006*, 2006.

[137] V. Veličković. Visualization of Enneper's surface by line graphics. *Filomat*, 31(2):387–405, 2017.

[138] V. Veličković. Visualization of lines of curvature on quadratic surfaces. *Facta Univ. Ser. Math. Inform.*, 32(1):11–29, 2017.

[139] V. Veličković and E. Dolićanin. Line graphics for visualization of surfaces and curves on them. *Filomat*, 37(4):1017–1027, 2023.

[140] V. Veličković and E. Dolićanin. Visualization of spheres in the generalized Hahn space. *Filomat*, 37(6), 2023.

[141] V. Veličković and E. Malkowsky. Visualization of neighbourhoods in some FK spaces. *Advancements in Mathematical Sciences (AMS 2015)*, 1676, 2015. :020003.

[142] V. Veličković and E. Malkowsky. Visualization of Wulff's crystals. *Advancements in Mathematical Sciences (AMS 2015)*, 1676, 2015. :020087.

[143] V. Veličković, E. Malkowsky, and E. Dolićanin. Modeling spheres in some paranormed sequence spaces. *Mathematics*, 10(6), 2022. 917. https://doi.org/10.3390/math10060917.

[144] V. Veličković, E. Malkowsky, and F. Özger. Visualization of the spaces $W(u, v; l(p))$ and their duals. In *International Conference on Analysis and Applied Mathematics (ICAAM 2016)*, volume 1759 (1), 2016. :020020.

[145] L. Vereb, P. Osmokrovic, M. Vujisić, Ć. Dolićanin, and K. Stanković. Prospects of constructing 20 kV asynchronous motors. *IEEE Trans. Dielectr. Electr. Insul.*, 16(1):251–256, 2009.

[146] K. L. Wardle. *Differential Geometry*. Dover Publications, 2008.

[147] T. J. Willmore. *An Introduction to Differential Geometry*. Dover Publications, 2012.

[148] L. Woodward and J. Bolton. *A First Course in Differential Geometry (Surfaces in Euclidean Space)*. Cambridge University Press, first edition, 2018.

[149] V. Wünsch. *Differentialgeometrie, Kurven und Flächen*. Vieweg+Teubner Verlag, 1997.

Index

For Product Safety Concerns and Information please contact our EU
representative GPSR@taylorandfrancis.com
Taylor & Francis Verlag GmbH, Kaufingerstraße 24, 80331 München, Germany

www.ingramcontent.com/pod-product-compliance
Ingram Content Group UK Ltd.
Pitfield, Milton Keynes, MK11 3LW, UK
UKHW050926180425
457613UK00003B/31